Iron Uptake and Homeostasis in Microorganisms

Edited by

Pierre Cornelis

Laboratory of Microbial Interactions
Vrije Universiteit Brussel
Belgium

and

Simon C. Andrews

School of Biological Sciences
University of Reading
UK

Caister Academic Press

Caister Academic Press
Norfolk, UK

www.caister.com

British Library Cataloguing-in-Publication Data
A catalogue record for this book is available from the British Library

ISBN: 978-1-904455-65-3

Cover image adapted from Figure 4.1

Printed and bound in Great Britain

Contents

Contributors

Sandra K. Armstrong
Department of Microbiology
University of Minnesota
Minneapolis, MN
USA

armst018@umn.edu

Pierre-Louis Blaiseau
Laboratoire Mitochondries, Métaux et Stress Oxydant
Institut Jacques Monod
CNRS-Université Paris Diderot
Paris
France

blaiseau.pierre-louis@ijm.univ-paris-diderot.fr

Timothy J. Brickman
Department of Microbiology
University of Minnesota
Minneapolis, MN
USA

brick011@umn.edu

Herbert Budzikiewicz
Institut für Organische Chemie
Universität zu Köln
Köln
Germany

aco88@uni-koeln.de

James Butcher
Ottawa Institute of Systems Biology
Department of Biochemistry, Microbiology and Immunology
Faculty of Medicine
University of Ottawa
Ottawa, ON
Canada

butcher1@gmail.com; jbut076@uottawa.ca

Jean-Michel Camadro
Laboratoire Mitochondries, Métaux et Stress Oxydant
Institut Jacques Monod
CNRS-Université Paris Diderot
Paris
France

camadro@ijm.jussieu.fr

Dominique Expert
Laboratoire des Interactions Plantes Pathogènes
UMR 217 INRA/Agroparistech/UPMC
Paris
France

expert@agroparistech.fr

Elena Fabiano
Laboratorio de Bioquímica y Genómica Microbiana
Instituto de Investigaciones Biológicas Clemente Estable
Montevideo
Uruguay

efabiano@iibce.edu.uy

Annika Flint
Ottawa Institute of Systems Biology
Department of Biochemistry, Microbiology and Immunology
Faculty of Medicine
University of Ottawa
Ottawa, ON
Canada

aflin053@uottawa.ca

Thierry Franza
Laboratoire des Interactions Plantes Pathogènes
UMR 217 INRA/Agroparistech/UPMC
Paris
France

franza@agroparistech.fr

Ahmed Gaballa
Department of Microbiology
Cornell University
Ithaca, NY
USA

ag67@cornell.edu

John D. Helmann
Department of Microbiology
Cornell University
Ithaca, NY
USA

jdh9@cornell.edu

Manuel L. Lemos
Department of Microbiology and Parasitology
Institute of Aquaculture
University of Santiago de Compostela
Santiago de Compostela
Spain

manuel.lemos@usc.es

Emmanuel Lesuisse
Laboratoire Mitochondries, Métaux et Stress Oxydant
Institut Jacques Monod
CNRS-Université Paris Diderot
Paris
France

lesuisse.emmanuel@ijm.univ-paris-diderot.fr

Alexandra R. Mey
Section of Molecular Genetics and Microbiology
Institute for Cellular and Molecular Biology
University of Texas
Austin, TX
USA

armey@mail.utexas.edu

Kerstin Nicolaisen
Goethe University
Cluster of Excellence Macromolecular Complexes
Centre of Membrane Proteomics
Department of Biosciences
Molecular Cell Biology of Plants
Frankfurt
Germany

k.nicolaisen@bio.uni-frankfurt.de

Mark R. O'Brian
Department of Biochemistry
State University of New York at Buffalo
Buffalo, NY
USA

mrobrian@buffalo.edu

Carlos R. Osorio
Department of Microbiology and Parasitology
Institute of Aquaculture
University of Santiago de Compostela
Santiago de Compostela
Spain

cr.osorio@usc.es

Shelley M. Payne
Section of Molecular Genetics and Microbiology
Institute for Cellular and Molecular Biology
University of Texas
Austin, TX
USA

payne@mail.utexas.edu

Gleb Pishchany
Vanderbilt University Medical Center
Nashville, TN
USA

Gleb.pishchany@vanderbuilt.edu

Girija Ramakrishnan
Department of Medicine/Division of Infectious Diseases
University of Virginia
Charlottesville, VA
USA

gr6q@virginia.edu

Michelle L. Reniere
Vanderbilt University Medical Center
Nashville, TN
USA

Michelle.l.reniere@vanderbilt.edu

Edson R. Rocha
Department of Microbiology and Immunology
East Carolina University Brody School of Medicine
Greenville, NC
USA

rochae@ecu.edu

Enrico Schleiff
Goethe University
Cluster of Excellence Macromolecular Complexes
Centre of Membrane Proteomics
Department of Biosciences
Molecular Cell Biology of Plants
Frankfurt
Germany

schleiff@bio.uni-frankfurt.de

Alexandra Seguin
Laboratoire Mitochondries, Métaux et Stress Oxydant
Institut Jacques Monod
CNRS-Université Paris Diderot
Paris
France
seguin.alexandra@ijm.univ-paris-diderot.fr

Eric P. Skaar
Vanderbilt University Medical Center
Nashville, TN
USA

eric.skaar@vanderbilt.edu

C. Jeffrey Smith
Department of Microbiology and Immunology
East Carolina University Brody School of Medicine
Greenville, NC
USA

smithcha@ecu.edu

Martin Stahl
Ottawa Institute of Systems Biology
Department of Biochemistry
Microbiology and Immunology
Faculty of Medicine
University of Ottawa
Ottawa, ON
Canada

mstah007@uottawa.ca

Alain Stintzi
Ottawa Institute of Systems Biology
Department of Biochemistry, Microbiology and Immunology
Faculty of Medicine
University of Ottawa
Ottawa, ON
Canada

astintzi@uottawa.ca

Cécile Wandersman
Unité des Membranes Bactériennes
Département de Microbiologie Institut Pasteur
Paris
France

cwander@pasteur.fr

Current Books of Interest

Streptomyces: Molecular Biology and Biotechnology	2011
Alphaherpesviruses: Molecular Virology	2011
Recent Advances in Plant Virology	2011
Vaccine Design: Innovative Approaches and Novel Strategies	2011
Salmonella: From Genome to Function	2011
PCR Troubleshooting and Optimization: The Essential Guide	2011
Insect Virology	2010
Environmental Microbiology: Current Technology and Water Applications	2010
Sensory Mechanisms in Bacteria: Molecular Aspects of Signal Recognition	2010
Bifidobacteria: Genomics and Molecular Aspects	2010
Molecular Phylogeny of Microorganisms	2010
Nanotechnology in Water Treatment Applications	2010
Iron Uptake and Homeostasis in Microorganisms	2010
Caliciviruses: Molecular and Cellular Virology	2010
Epstein-Barr Virus: Latency and Transformation	2010
Anaerobic Parasitic Protozoa: Genomics and Molecular Biology	2010
Lentiviruses and Macrophages: Molecular and Cellular Interactions	2010
Microbial Population Genetics	2010
Borrelia: Molecular and Cellular Biology	2010
Influenza: Molecular Virology	2010
RNA Interference and Viruses: Current Innovations and Future Trends	2010
Retroviruses: Molecular Biology, Genomics and Pathogenesis	2010
Metagenomics: Theory, Methods and Applications	2010
Aspergillus: Molecular Biology and Genomics	2010
Environmental Molecular Microbiology	2010
Neisseria: Molecular Mechanisms of Pathogenesis	2010
Frontiers in Dengue Virus Research	2010
ABC Transporters in Microorganisms	2009
Pili and Flagella: Current Research and Future Trends	2009
Lab-on-a-Chip Technology: Biomolecular Separation and Analysis	2009
Lab-on-a-Chip Technology: Fabrication and Microfluidics	2009
Bacterial Polysaccharides: Current Innovations and Future Trends	2009
Microbial Toxins: Current Research and Future Trends	2009
Acanthamoeba	2009
Bacterial Secreted Proteins: Secretory Mechanisms and Role in Pathogenesis	2009
Lactobacillus Molecular Biology: From Genomics to Probiotics	2009
Mycobacterium: Genomics and Molecular Biology	2009
Real-Time PCR: Current Technology and Applications	2009
Clostridia: Molecular Biology in the Post-genomic Era	2009
Plant Pathogenic Bacteria: Genomics and Molecular Biology	2009
Microbial Production of Biopolymers and Polymer Precursors	2009
Plasmids: Current Research and Future Trends	2008
Pasteurellaceae: Biology, Genomics and Molecular Aspects	2008

Preface

Iron is considered to be a minor element employed, in a variety of forms, by nearly all living organisms. In some cases, it is utilized in large quantities, for instance for the formation of magnetosomes within magnetotactic bacteria or during use of iron as a respiratory donor or acceptor by iron-oxidizing or -reducing bacteria. However, in most cases the role of iron is restricted to its use as a cofactor or prosthetic group assisting the biological activity of many different types of protein. The key metabolic processes that are dependent on iron as a cofactor are numerous; they include respiration, light harvesting, nitrogen fixation, the Krebs cycle, redox stress resistance, amino acid synthesis and oxygen transport. Indeed, it is clear that life in its current form would be impossible in the absence of iron. One of the main reasons for the reliance of life upon this metal is the ability of iron to exist in multiple redox states, in particular the relatively stable ferrous (Fe^{2+}) and ferric (Fe^{3+}) forms. The availability of these stable oxidation states allows iron to engage in redox reactions over a wide range of midpoint potentials, depending on the coordination environment, making it an extremely adaptable mediator of electron exchange processes. Iron is also one of the most common elements within the earth's crust (5% abundance) and thus is considered to have been readily available when life evolved on our early anaerobic planet. However, as oxygen accumulated (the 'great oxidation event') within the atmosphere some 2.4 billion years ago, and as the oceans became less acidic, the iron within primordial oceans was converted from its soluble reduced form to its weakly soluble oxidized ferric form, which precipitated (~1.8 billion years ago) to form the 'banded iron formations' (BIFs) observed today in Precambrian sedimentary rocks around the world. These BIFs provide a geological record marking a transition point away from the ancient anaerobic world towards modern aerobic earth. They also indicate a period over which the bioavailability of iron shifted from abundance to limitation, a condition that extends to the modern day. Thus, it is considered likely that the vast majority of extant organisms face the common problem of securing sufficient iron from their environment – a problem that life on earth has had to cope with for some 2 billion years. This struggle for iron is exemplified by the competition for this metal amongst co-habiting microorganisms which resort to stealing (pirating) each other's iron supplies!

The reliance of microorganisms upon iron can be disadvantageous to them, and to our innate immune system it represents a chink in the microbial armour, offering an opportunity that can be exploited to ward off pathogenic invaders. In order to infect body tissues and cause disease, pathogens must secure all their iron from the host. To fight such infections, the host specifically withdraws available iron through the action of various iron-depleting processes (e.g. the release of lactoferrin and lipocalin-2) – this represents an important strategy in our defence against disease. However, pathogens are frequently able to deploy iron acquisition systems that target host iron sources such as transferrin, lactoferrin and haemoproteins, and thus counteract the iron withdrawal approaches of the host. Inactivation of such host-targeting iron uptake systems often attenuates the pathogenicity of the invading microbe, illustrating the importance of 'the battle for iron' in the infection process.

The role of iron sequestration systems in facilitating microbial infections has been a major driving force in research aimed at unravelling the complexities of microbial iron transport processes. But also, the intricacy of such systems offers a challenge that stimulates the curiosity. One such challenge is to understand how balanced levels of free iron within the cytosol are achieved in a way that avoids toxicity whilst providing sufficient levels for metabolic purposes – this is a requirement that all organisms have to meet. Although the systems involved in achieving this balance can be highly variable amongst different microorganisms, the overall strategy is common. On a coarse level, the homeostatic control of cellular iron is maintained through strict control of the uptake, storage and utilisation of available iron, and is co-ordinated by integrated iron regulatory networks. However, much yet remains to be discovered concerning the fine details of these different iron regulatory processes. As already indicated, perhaps the most difficult task in maintaining iron homeostasis is simply the procurement of sufficient iron from external sources. The importance of this problem is demonstrated by the plethora of distinct iron transporters often found within a single bacterium, each targeting different forms (complex or redox state) of iron or a different environmental condition. Thus, microbes devote considerable cellular resource to securing iron from their surroundings, reflecting how successful acquisition of iron can be crucial in the competition for survival.

The aim of this book is provide the reader with an overview of iron transport processes within a range of microorganisms and to provide an indication of how microbial iron levels are controlled. This aim is promoted through the inclusion of expert reviews on several well-studied examples that illustrate the current state of play concerning our comprehension of how iron is translocated into the bacterial (or fungal) cell and how iron homeostasis is controlled within microbes. The first two chapters (1 and 2) consider the general properties of microbial iron-chelating compounds (known as 'siderophores'), and the mechanisms used by bacteria to acquire haem and utilize it as an iron source. The following 12 chapters (3–14) focus on specific types of microorganism that are of key interest, covering both an array of pathogens for humans, animals and plants (e.g. species of *Bordetella, Shigella, Erwinia, Vibrio, Aeromonas, Francisella, Campylobacter, Staphylococcus* and EHEC) as well as a number of prominent non-pathogens (e.g. the rhizobia, *E. coli* K-12, *Bacteroides* spp., cyanobacteria, *Bacillus* spp. and yeasts). The chapters relay the common themes in microbial iron uptake approaches (e.g. the use of siderophores, TonB-dependent transporters, and ABC transport systems), but also highlight many distinctions (such as use of different types iron regulator and the impact of the presence/absence of a cell wall) in the strategies employed.

We hope that those both within and outside the field will find this book useful, stimulating and interesting. We intend that it will provide a source for reference that will assist relevant researchers and provide an entry point for those initiating their studies within this subject. Finally, it is important that we acknowledge and thank wholeheartedly the many contributors who have provided the 14 excellent chapters from which this book is composed. Without their considerable efforts, this book, and the understanding that it relays, would not have been possible.

Simon C. Andrews and Pierre Cornelis

Siderophores from Bacteria and from Fungi

1

Herbert Budzikiewicz

Abstract
In the early days of life on earth due to the emergence of photosynthesis, the atmosphere became oxidative and so iron prevailed in its trivalent form. The consequence was that most microorganisms live today in a surrounding where the concentration of soluble iron compounds is much too low to satisfy their physiological needs. One of the possibilities to circumvent this problem is the production of so-called siderophores, compounds which can form water soluble complexes with Fe^{3+}. They are released in situations of iron deficiency.

Fe^{3+} possesses six coordination sites arranged in the shape of an octahedron which can accommodate three bidentate ligands. The major ligand types are catecholates, hydroxamates and α-hydroxycarboxylates, in the ideal case connected by adequate spacers which keep them in the correct position. Many siderophores have elaborate structures which can bind to the cell surface of the producing species. In this way pirating by competing microorganisms is rendered more difficult.

Introduction

Iron is important for almost all living organisms being involved in many metabolic processes. The redox potential between its two valency states, Fe^{2+} and Fe^{3+}, lies within the range of physiological processes. However, iron is not readily available for most microorganisms though it is not a rare element, being fourth in abundance in the earth crust. In the soil at pH values around 7 it prevails in the form of ferric oxide hydrates. The concentration of free Fe^{3+} is at best 10^{-17} mol/l while about 10^{-6} mol/l would be needed for most microorganisms. In living organisms iron is usually strongly bound to proteins such as transferrins, lactoferrins, ferritins or haemoglobins. One possibility to circumvent this problem and to increase the supply of soluble iron is the production of Fe^{3+} chelating compounds, so-called siderophores.

Siderophores are secondary metabolites with masses below 2000 Da. Small iron–siderophore complexes can enter the cell via unspecific porins, while larger ones need a specific receptor which recognizes the ferri-siderophore at the cell surface and transfers it into the cell. There iron is released in most cases by reduction to the less strongly bound Fe^{2+} state, and the free siderophore is re-exported ('shuttle mechanism'). In few instances the siderophore is degraded in the cell as it is the case for enterobactin. Alternatively Fe^{3+} is transferred at the cell surface from the ferri-siderophore to a *trans*-membrane transport system ('taxi mechanism'). Siderophores frequently have elaborate structures providing recognition only by the cell surface of the producing species.

Due to its high charge density, small ion radius, and low polarizability Fe^{3+} is a hard Lewis acid and can bind strongly to hard Lewis bases such as oxide ions. It forms octahedral d^5 high spin complexes providing six coordination sites which can accommodate three bidentate ligands. The major ligands types are catecholates, hydroxamates, and α-hydroxycarboxylates. Siderophores combining different ligand types are not uncommon. Three bidentate ligands are often connected by aliphatic segments keeping them in place for complexation. Siderophores containing only two binding sites form either $(Fe^{3+})_2(ligand)_3$ complexes or the remaining two octahedral loci accommodate some external ligand (see, for example, pyochelin). The classification used in this review will be based on the structural type of the backbone carrying the chelating groups, viz. (1) siderophores based on peptides, (2) siderophores based on di- and triamino alkanes, (3) siderophores based on citric acid, and (4) miscellaneous siderophores.

For earlier compilations of siderophore literature see Drechsel and Winkelmann (1997), specifically for fungal siderophores Winkelmann (1990), Renshaw *et al.* (2002), and for biosynthesis pathways Crosa and Walsh (2002) and Challis (2005). This chapter is based on a comprehensive review on microbial siderophores (Budzikiewicz, 2010).

Siderophores based on peptides

The peptide chain carrying the ligand sites usually contains both D (underlined in the structural formulas below) and L amino acids. Frequently its two ends are blocked by the formation of cyclic structures or otherwise. This prevents the degradation by proteolytic enzymes. Non-proteinogenic amino acids are encountered (homoserine, Hse; ornithine, Orn; 2,4-diaminobutyric acid, Dab), lysine and ornithine can be incorporated in the chain by their ε/δ- rather than by their α-amino group. Amino acids may be modified to form ligand sites (3-hydroxy-aspartic, OHAsp; 3-hydroxy-histidine, OHHis; N^5-acyl-N^5-hydroxy-ornithine, acylOHOrn; *N*-hydroxy-*cyclo*-ornithine, i.e. 3-amino-1-hydroxy-piperidone-2, cOHOrn). Dab frequently condenses with the preceding amino acid giving a terahydropyrimidine ring (see, for example, corrugatin) (Filsak *et al.*, 1994).

Pyoverdins and related siderophores

The most thoroughly investigated representatives are the **pyoverdins** produced by the fluorescent members of the genus *Pseudomonas* (for reviews see Budzikiewicz, 2004a; Meyer *et al.*, 2008). Pyoverdins consist of three distinct structural parts, a chromophore (Chr**a**) (Fig. 1.1), a dicarboxylic acid (succinic, malic, glutamic or α-ketoglutaric acid) or its monoamide bound to the amino group of the chromophore, and a peptide chain comprising 6–12 amino acids. The approximately 50 pyoverdins for which structures have been proposed can be divided into three structural types, exemplified by the three pyoverdins of *Pseudomonas aeruginosa* (Meyer *et al.*, 1997): (a) pyoverdins with a C-terminal cOHOrn, e.g. ATCC 27853

Suc-Chr**a**-Ser-formylOHOrn-Orn-Gly-aThr-Ser-cOHOrn

(b) pyoverdins with a C-terminal tri- or tetra-cyclopeptidic substructure (lactame formation between the C-terminal carboxyl group and an in-chain Lys or Orn), e.g. ATCC 27853

Suc-Chr**a**-Ser-Arg-Ser-formylOHOrn-(Lys-formylOHOrn-Thr-Thr)

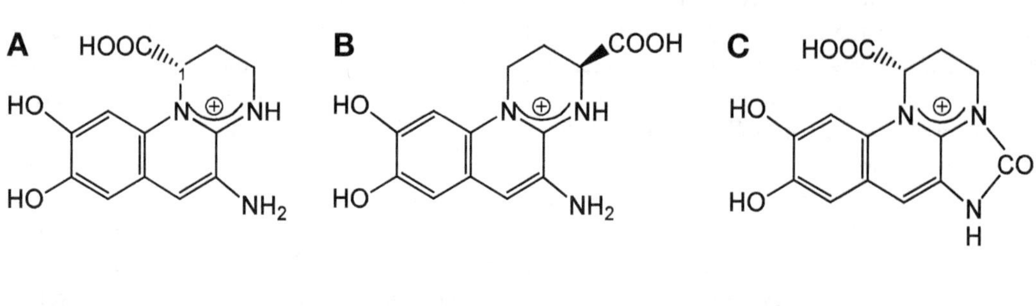

Figure 1.1 Chromophore types: (A) pyoverdin, (B) isopyoverdin, (C) azotobactin.

and (c) pyoverdins with a C-terminal free carboxyl group, e.g. Pa6

Suc-Chr**a**-(Ser-Dab)-formylOHOrn-Gln-Gln-formylOHOrn-Gly

The free carboxyl group is probably the hydrolysis product of a depsipeptidic substructure (ester formation between the C-terminal carboxyl group and an in-chain Ser or Thr); in several cases both the cyclic or the hydrolysed open-chain form were found (e.g. Bultreys *et al.*, 2004). Binding sites for Fe^{3+} are the catecholate part of the chromophore Chr**a** and two units in the peptide chain, hydroxamate (acylOHOrn, cOHOrn) and/or α-hydroxycarboxylate (OHAsp, OHHis) units.

To determine the three-dimensional structure, an X-ray analysis of the Fe^{3+} complex of the pyoverdin B10:

Suc-Chr**a**-εLys-OHAsp-Ala-aThr-Ala-cOHOrn

was performed (Teintze *et al.*, 1981). As an alternative strategy the investigation of the isomorphic Ga^{3+} complexes by NMR analysis was developed (Mohn *et al.*, 1994) for the pyoverdin GM-II:

Suc-Chr**a**-Ala-Lys-Gly-Gly-OHAsp-Gln-Ser-Ala-Ala-Ala-Ala-cOHOrn

and extended to other pyoverdins (H. Budzikiewicz, unpublished). In all cases the metal ion was found to lie at the surface of the complex. This facilitates its uptake and release.

The peptide chains of the pyoverdins are responsible for the recognition of the ferri-siderophore at the cell surface of the producing species. It is usually highly strain specific. Cross-recognition between two strains is only observed when structurally closely related pyoverdins are produced (Georgias *et al.*, 1999; Meyer *et al.*, 1998).

Isopyoverdins contain the siderophore Chr**b** (Fig. 1.1). They have been encountered so far only in isolates from *Pseudomonas putida* strains, e.g. BTP1 (Jacques *et al.*, 1995):

Glu-Chr**b**-Asp-Ala-Asp-acetylOHOrn-Ser-cOHOrn

Also *Azomonas macrozytogenes* produces a siderophore with an isopyoverdin chromophore, **azoverdin**, but with a peptide chain related to those of azotobactins containing homoserine units (Michalke *et al.*, 1996):

Suc-Chr**b**-Hse-(Hse-Dab)-acetylOHOrn-Ser-acetylOHOrn

The three-dimensional structure of the Ga^{3+} complex was determined by NMR techniques as outlined above. Also here the metal ion lies at the surface of the complex (Wasielewski *et al.*, 2002).

From *Azotobacter vinelandii* the structures of **azotobactins** were elucidated. They contain the chromophore Chr**c** (Fig. 1.1) and Hse units as, for example, from the strain ATCC 12837:

Chr**c**-Ser-Ser-Hse-Gly-OHAsp-Hse-Hse-Hse-hydroxybutyrylOHOrn-Hse

(of the three Hse in sequence two are L and one is D) (Schaffner *et al.*, 1996).

Anachelin

Cyanobacteria were probably the first organisms which used photosynthesis initiating the production of O_2. This resulted eventually in the gradual oxidation of Fe^{2+} to Fe^{3+}, thus limiting the supply of soluble iron compounds. To cope with this problem the production of siderophores was initiated. One of them is **anachelin**, a compound whose biosynthesis requires *inter alia* steps from the non-ribosomal peptide and polyketide paths. It exists in an open (anachelin H, Fig. 1.2) and two cyclic forms arising from an interaction of the salicylic acid carbonyl group with one of the neighbouring hydroxyl groups (anachelin 1 and 2) (Beiderbeck *et al.*, 2000; Itou *et al.*, 2001). The absolute stereochemistry of all chiral centres was established (Ito *et al.*, 2004) and confirmed by synthesis (Gademann *et al.*, 2007). Mass spectrometric analysis of the Fe^{3+} complex suggests a 1:1 Fe^{3+} to ligand ratio.

Bacterial hydroxamate siderophores

Exochelins (Sharman *et al.*, 1995a,b) are peptidic siderophores from *Mycobacterium* spp. Exochelin MS from *M. smegmatis* contains β-Ala and three substituted OHOrn units linked by their N^5 atoms, thus forming hydroxamic acids:

N^5-formyl-N^5-OH<u>Orn-</u>β-Ala-N^5-OH<u>Orn</u>-a<u>Thr</u>-N^5-OHOrn

Exochelin MN from *M. neoaurum* contains N^2-methyl-N^5-OHOrn linked by N^5 to β-Ala, Orn linked by its N^2 and cOHOrn, all amino acids being L configured,

OHHis-β-Ala-β-Ala-MeOHOrn-Orn-cOHOrn

Vicibactin from *Rhizobium leguminosarum* is a macrocyclic trilactone consisting of N^2-acetyl-N^5-hydroxy-D-Orn and *R*-3-hydroxybutyric acid (Dilworth *et al.*, 1998),

$[\text{-O-CH(CH}_3\text{)-CH}_2\text{-CO-NOH-(CH}_2\text{)}_3\text{-CH(NHCOCH}_3\text{)-CO-}]_3$

Vicibactin is identical to **neurosporin** produced by the fungus *Neurospora crassa,* for which also X-ray data of the Fe^{3+} complex are available (Eng-Wilmot *et al.*, 1984).

Fungal hydroxamate siderophores

The fungal siderophores to be discussed here can be divided in three groups, the **fusarinines**, the **ferrichromes**, and the **coprogens**, all based on N^5-hydroxy-N^5-acyl-L-Orn. The review by Renshaw *et al.* (2002) should be consulted for details and for a listing of representatives. For a number of siderophores and their Fe^{3+} complexes X-ray and other structural analyses are reported (van der Helm *et al.*, 1989). Relationships between structure of the siderophores and their recognition at the cell surface have been discussed (Huschka *et al.*, 1986).

Fusarinines produced by several fungal genera comprise the acyl unit *Z*-5-hydroxy-3-methyl-pent-2-enoic acid (anhydromevalonic acid). They can be a linear monomer, dimer (**fusarinine A**) or trimer (**fusarinine B**) (the monomer can also be *E*-configured) (Jalal *et al.*, 1986)

$\text{HO-[CO-CHNH}_2\text{-(CH}_2\text{)}_2\text{-NOH-CO-CH=CCH}_3\text{-(CH}_2\text{)}_2\text{-O]}_{1-3}\text{-H}$

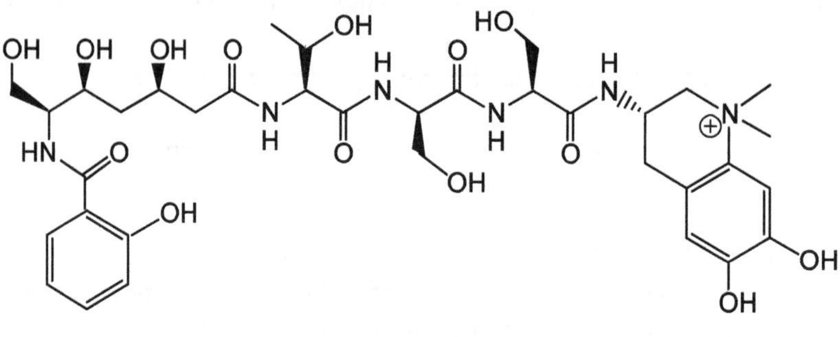

Figure 1.2 Anachelin.

The trimer, by forming an ester bond between the two terminal functions, results in a lactone ring (**fusarinine C** or **fusigen**). Since the ester bonds are rather labile it is not clear whether the open forms are genuine siderophores, precursors of fusigen or just hydrolysis products. The free α-amino groups of the Orn units can be acetylated (Diekmann and Kretzdorn, 1975; Moore and Emery, 1976; Hossein *et al.*, 1980). The members of the **ferrichrome** group are cyclohexapeptides with the general structure

[-N^5-hydroxy-N^5-acyl-L-Orn)$_3$-A-B-Gly-]

where A and B can be Gly, Ala or Ser. The acyl groups vary.

The third group comprises the **coprogen** family. Their characteristic element is a diketopiperazine ring formed by the head-to-head condensation of two N^5-hydroxy-N^5-acyl-L-Orn units. Rhodotorulic acid, isolated from the yeast *Rhodotorula pilimanae,* contains two acetyl groups (Fig. 1.3**a**) (Atkin and Neilands, 1968), dimerum acid (Fig. 1.3**b**) from *Fusarium dimerum* (Diekmann, 1970) and other fungi (Jalal *et al.,* 1986) two *E*-anhydromevalonyl residues. In the coprogens a third variously substituted *E*-fusarinine unit is added by an ester bond. Rhodotorulic and dimerum acid form Fe_2Lig_3 complexes, but also a mixed 1:1:1 complex of Fe^{3+} with dimerum acid and *Z*-fusarinine was observed. Coprogen B yields a 1:1 complex with Fe^{3+} (Carrano and Raymond, 1978; Jalal *et al.*, 1986).

Catecholate siderophores

Ideally suited for Fe^{3+} complexation – exemplified by the extremely high complexing constant of 10^{49} (Loomis and Raymond, 1991) – is **enterobactin** (**enterochelin**) first isolated from *Salmonella typhimurium* (Pollak and Neilands, 1970) and *Escherichia coli* (O'Brien and Gibson, 1970). It is a cyclic trilactone of *N*-2,3-dihydroxybenzoyl-L-serine (DHB-Ser)

[-O-CH_2-CH(NH-DHB)-CO-]$_3$

In the culture medium precursors or degradation products of enterobactin were also found, open-chain compounds comprising two or three constitutional units. Iron release is effected by degradation of enterobactin by an esterase (Abergel *et al.,* 2009). *Salmonella enterica* produces a derivative of enterobactin, **salmochelin** S4, where two of the aromatic rings are β-C-glucosylated in 5-position. Also glycosylated degradation products were found (Bister *et al.*, 2004).

From *Corynebacterium glutamicum* the siderophore **corynebactin** was obtained (Budzikiewicz *et al.,* 1997) differing from enterobactin in being composed of three DHB-Gly-L-Thr units. Later the same siderophore was found to be produced by *Bacillus subtilis* and named **bacillibactin** (May *et al.*, 2001). *B. subtilis* under iron starvation also secretes DHB-Gly (Ito and Neilands, 1958) subsequently named **itoic acid** (Peters and Warren, 1968).

Erwinia chrysanthemi (Persmark *et al.,* 1989) and *Serratia marcescens* (Ehlert *et al.,* 1994) produce DHB-D-Lys-L-Ser named **chrysobactin.** At physiological pH values 2 or 3 chrysobactin residues are associated with Fe^{3+} (Persmark and Neilands, 1992). From *Chryseomonas luteola* in addition to chrysobactin a derivative (**chrysomonin**) was isolated, where C-6 of the DHB unit is substituted with the N-atom of a pyridinium cation (Adolphs *et al.,* 1996).

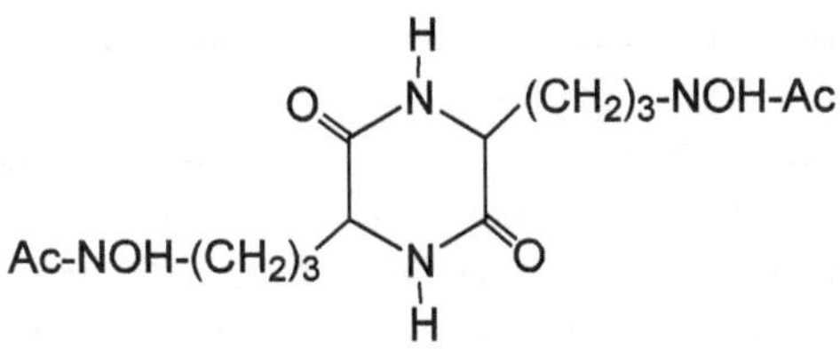

Figure 1.3 (A) Rhodotorulic acid (Ac = $COCH_3$), (B) dimerum acid (Ac = CO-CH=CCH_3-$(CH_2)_2$-OH).

Amonabactins were found to be excreted by *Aeromonas hydrophila* (Telford and Raymond, 1997) and by *Pseudomonas stutzeri* (Zawadzka *et al.*, 2006). They are based on the peptides Lys-Lys-Phe and Lys-Lys-Trp. N^6 of the first Lys residues is derivatized by DHB or by a DHB-Gly-residue, that of the second Lys by a DHB group. At high pH values and excess ligand a $(Fe^{3+})_2Lig_3$ complex is formed, at neutral pH a 1:1 ratio prevails with H_2O molecules satisfying the remaining coordination sites (Telford and Raymond, 1998).

Alterobactin A is a cyclodepsipeptide from *Alteromonas luteoviolacea*:

DHB-NH-$(CH_2)_4$-$CHNH_2$-CHOH-CH_2-CO-Ser-Gly-Arg-β-OHAsp-Gly-β-OHAsp

with an ester bond between the C-terminal β-OHAsp and Ser. It is accompanied by its hydrolysis product alterobactin B without the ester bond (Reid *et al.*, 1993). A synthesis is reported (Deng *et al.*, 1995). Related structures are the **pseudoalterobactins** A and B from *Pseudoalteromonas* sp. (Kanoh *et al.*, 2003), one of the rare examples of bacterial metabolites containing an aromatic sulphonic acid (Budzikiewicz, 2006).

Heterobactin A and B are produced by *Rhodococcus erythropolis* (Carrano *et al.*, 2001). They are based on the sequence Orn-Gly-cOHOrn. The δ-amino group of Orn is substituted by a DHB group. In heterobactin A its α-amino group is free while in B it is substituted probably by a hydroxybenzoxazolyl-carbonyl group.

Rhodobactin was isolated from *Rhodococcus rhodochrous* (Dhungana *et al.*, 2007). A sequence of four Orn derivatized in different ways is linked together. The nitrogen atoms of the N-terminal Orn are substituted by DHB groups, it follows two Orn whose N^5-amino groups are transformed into urea units (NH_2CONH-), and the C-terminus is cOHOrn. The stereochemistry of the Orn units was not determined.

Thermobifida fusca produces three closely related siderophores, the **fuscachelins** (Dimise *et al.*, 2008). Fuscochelin B starts with the sequence DHB-Arg-Gly-Gly-Ser. which is bound to the hydroxylated N^5-amino group of Orn. Its N^2-amino group (the carboxyl group is free) is bound to the C-terminus of the sequence Gly-Gly-Arg-DHB

DHB-Arg-Gly-Gly-Ser-N^5-OHOrn-N^2H-Gly-Gly-Arg-DHB

In fuscachelin C the carboxyl group of Orn forms an amide, in fuscachelin A occurs an ester bond between the carboxyl group of Orn and the hydroxyl group of Ser.

Lipopeptide siderophores

From *Burkholderia cepacia* (formerly *Pseudomonas cepacia*) three siderophores named **ornibactins** were isolated whose structures were determined by degradation and NMR studies (Stephan *et al.*, 1993a,b) as containing 3-hydroxy fatty acid residues and putrescin blocking the C-terminus:

acylOHOrn-OHAsp-Ser-formylOHOrn-NH-$(CH_2)_4$-NH_2

with acyl=R-CHOH-CH_2-CO (R=CH_3, C_3H_7, C_5H_{11}). Ornibactins are the main siderophores of a series of *Burkholderia* strains.

From *Nocardia* strains several closely related compounds (**nocobactins, formobactin, amamistatins**) were isolated (Ratledge and Snow, 1974; Murakami *et al.*, 1996; Kokubo *et al.*, 2000) which contain three typical Fe^{3+} binding sites, 2 hydroxamate units and a hydroxyphenyloxazol structure derived from salicylic acid (see Fig. 1.6). The C-terminus is *N*-hydroxy-*cyclo*-Lys which is bound to a long chain 3-hydroxy fatty acid, whose hydroxy group is esterified by N^6-acyl-N^6-hydroxy-Lys, the α-amino group of which in turn is bound to 2-*o*-hydroxyphenyl-4-methyl-oxazol-3-carboxylic acid. For the amamistatins the stereochemistry of the cyclic Lys was determined as L, of the open one as D, and that of C-3 of the fatty acid as *S*.

Structurally related with the nocobactin family are the **mycobactins** and **carboxymycobactins** from *Mycobacterium* spp. For reviews see (Snow, 1970; Vergne *et al.*, 2000). They have the same basic skeleton as the nocobactins, but the 2,3–double bond of the oxazole ring is saturated. A series of differently substituted representatives have been isolated. The major group carries saturated and unsaturated long-chain fatty acid residues as substituent of the hydroxamic acid unit formed by the ε-amino group of Lys. The stereochemistry of most chiral centres has been determined. For the carboxymycobactins the fatty acid residue is replaced by a dicarboxylic acid.

Corrugatin (Fig. 1.4) is the siderophore of *Pseudomonas corrugata* (Risse *et al.*, 1998). It was also found as secondary siderophore of several pyoverdin producing *Pseudomonas* strains such as *Pseudomonas fluorescens,* occasionally in slightly modified forms such as **ornicorrugatin** where one Dab is replaced by Orn (Matthijs *et al.*, 2008) or with OHHis instead of OHAsp as the C-terminus (S. Matthijs, unpublished).

A group of amphiphilic siderophores was isolated from marine bacteria, the **marinobactins** from *Marinobacter* sp., **aquachelins** from *Halomonas aquamarina* (Martinez *et al.*, 2000; Martinez and Butler, 2007), the **amphibactins** from *Vibrio* sp. (Martinez *et al.*, 2003), and the **loihichelins** from *Halomonas* spp. (Homann *et al.*, 2009b). They all comprise series of related molecules differing in the nature of the saturated or unsaturated fatty acid linked to the N-terminus of peptide chains. The siderophores show a strong affinity to lipid membranes (Xu *et al.*, 2002). The Fe^{3+} complexes of aquachelins and marinobactins suffer degradation under sun light irradiation. For the Fe^{3+}–aquachelin complexes the formation of Fe^{2+}, of hydrophobic and of hydrophylic cleavage products was observed (Barbeau *et al.*, 2001). There is evidence that this type of photolytic degradation is common for siderophores containing α-hydroxycarboxyl ligands (Barbeau *et al.*, 2003).

Siderophores based on diamino- and triaminoalkane skeletons

Rhizobactin

Rhizobactin is the siderophore of *Rhizobium meliloti* (Smith *et al.*, 1985). It contains L-malic acid and two α-amino acid units as probable binding sites for Fe^{3+}. The stereochemistry of the amino acids is not known.

$$HOOC\text{-}CH(CH_3)\text{-}NH\text{-}(CH_2)_2\text{-}NH\text{-}CH(COOH)\text{-}(CH_2)_4\text{-}NH\text{-}CO\text{-}CH_2\text{-}CHOH\text{-}COOH$$

Catecholate siderophores

For a review see Budzikiewicz (2004b).

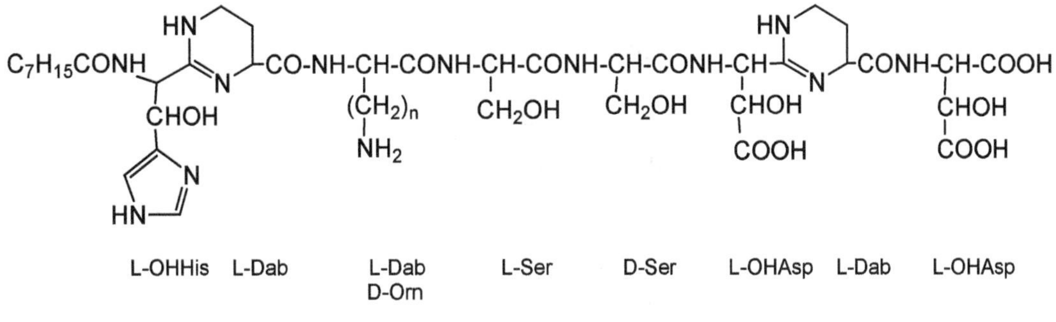

Figure 1.4 Corrugatin (*n*=2, L-Dab) and ornicorrugatin (*n*=3, D-Orn).

The tricatecholate siderophore **protochelin** (Fig. 1.5) was obtained from a methanol-degrading bacterium (Taraz *et al.*, 1990). Subsequently it was also found to be produced by *Azotobacter vinelandii* (Cornish and Page, 1995) together with its constituents 2,3-dihydroxybenzoic acid, **azotochelin** (bis-DHB Lys) (Corbin and Bulen, 1969) and **aminochelin** (mono-DHB cadaverine) (Page and von Tigerstrom, 1988). **Cepaciachelin** from *B. cepacia* (Barelmann *et al.*, 1996) lacks the third DHB residue of protochelin. The amino acid in all these compounds is L-Lys. **Myxochelin A** from *Angiococcus disciformis* can be considered as a reduction product of azotochelin (lysinol instead of Lys) (Kunze *et al.*, 1989). The stereochemistry of lysinol was determined as *S* (Miyanaga *et al.*, 2006).

When DHB is bound to Ser or Thr cyclization may occur by reaction of the hydroxyl with the carbonyl group resulting in an oxazolin ring (see Fig. 1.6). It has been discussed whether the oxazolin nitrogen atom can act as a ligand (Robinson *et al.*, 1984). This would explain why DHB can be replaced by salicyclic acid in some cases.

To this group of siderophores belong **photobactin** from *Photorhabdus luminescens* (Ciche *et al.*, 2003), **agrobactin** from *Agrobacterium tumefaciens* (Fig. 1.6**a**) (Ong *et al.*, 1979) whose structure was confirmed by X-ray analysis (Eng-Wilmot and van der Helm, 1980) and synthesis of the hydrolized form **agrobactin A** (Peterson *et al.*, 1980), and **parabactin** (Fig. 1.6**b**) from *Paracoccus denitrificans* (Peterson and Neilands, 1979). Two syntheses are reported for parabactin (Bergeron and Kline, 1982; Nagao *et al.*, 1984). **Fluvibactin** (Fig. 1.6**c**) from *Vibrio fluvialis* (Yamamoto *et al.*, 1993) differs from agrobactin by replacement of spermidine by norspermidine. **Vibriobactin** from *Vibrio vulnificus* contains two oxazoline substituents formed from DHB and Thr (Griffiths *et al.*, 1984). In **vulnibactin** (Okujo *et al.*, 1994a) two DHB groups are replaced by salicylic acid units. Syntheses of agrobactin, fluvibactin and vibriobactin are published (Bergeron *et al.*, 1985, 2001; Sakakura *et al.*, 2008). L-*allo*-Thr is incorporated in all five siderophores.

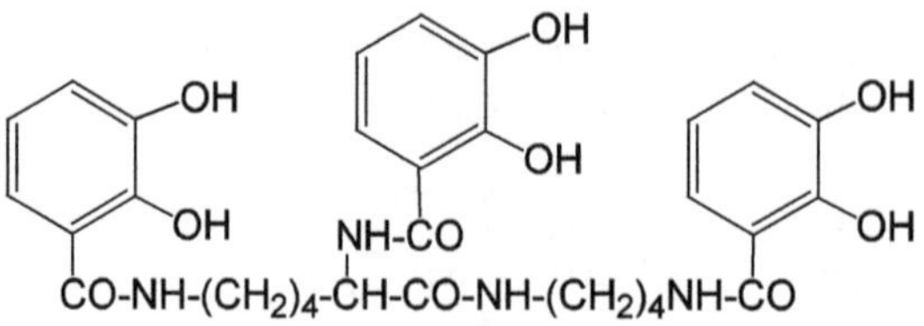

Figure 1.5 Protochelin.

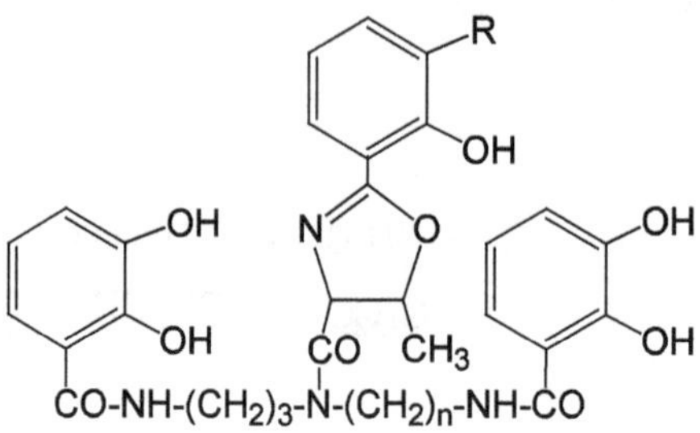

A n = 4, R = OH
B n = 4, R = H
C n = 3, R = OH

Figure 1.6 Agrobactin (A), parabactin (B), and fluvibactin (C).

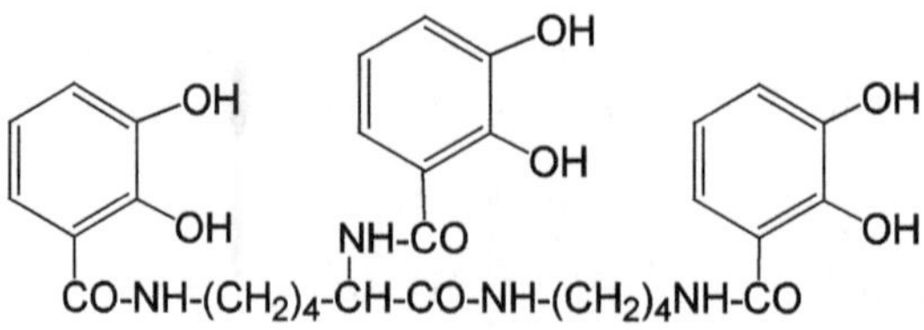

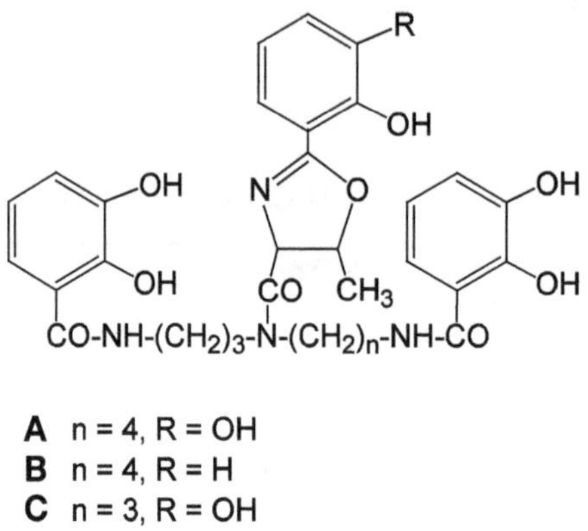

Hydroxamic acid siderophores

Bisucaberin from *Alteromonas haloplanktis* is a cyclic dimer of succinyl-(N-hydroxycadaverin) (Takahashi *et al.*, 1987),

$$[\text{-CO-CH}_2\text{-CH}_2\text{-CO-NH-CH}_2\text{-CH}_2\text{-CH}_2\text{-CH}_2\text{-CH}_2\text{-NOH-}]_2$$

In **putrebactin** from *Shewanella putrefaciens* (Ledyard and Butler, 1997) cadaverin is replaced by putrescin (R = H), while **alcaligin** from *Alcaligenes denitrificans* (Nishio *et al.*, 1988) and from *Bordetella* spp. (Moore *et al.*, 1995) is a cyclic dimer of succinyl-N^1, 3-dihydroxyputrescin (R = OH)

$$[\text{-CO-CH}_2\text{-CH}_2\text{-CO-NOH-CH}_2\text{-CH}_2\text{-CHR-CH}_2\text{-NH-}]_2$$

A group of related siderophores are the **desferriferrioxamines** or **proferrioxamines**. Originally they were obtained from actinomycetes, mainly *Nocardia* and *Streptomyces* (Keller-Schierlein *et al.*, 1964) and later found to be produced also by *Erwinia amylovora* (e.g. Feistner *et al.*, 1993; Kachadourian *et al.*, 1996) and *Pseudomonas stutzeri* (Zawadzka *et al.*, 2006; Mulet *et al.*, 2008). They consist of three mono-*N*-hydroxy-1,4-diaminobutane (putrescine) or mono-*N*-hydroxy-1,5-diaminopentane (cadaverine) units connected by succinic acid units. The hydroxylated terminus carries an acetyl or a succinyl residue,

$$\text{H-[-HN-(CH}_2)_{4-5}\text{-NOH-CO-(CH}_2)_2\text{-CO-]}_3\text{-OH}$$

In the latter case the free carboxyl group and the free N-terminus may form a macrolactam.

Siderophores based on citric acid

For a review see Budzikiewicz (2005).

Siderophores with two hydroxamic acid units

In siderophores of this series 1,3-diaminopropane, 1,5-diaminopentane or Lys with its α-amino group are connected to the outer two carboxyl groups of citric acid. These spacers in turn are acylated by acetyl or fatty acid residues and derivatized by an N-hydroxyl group thus forming hydroxamic acids. Representatives are **schizokinen** (first isolated from *Bacillus megaterium*, Byers *et al.*, 1967; Mullis *et al.*, 1971), **arthrobactin** (from *Arthrobacter* spp., Linke *et al.*, 1972; Lee and Miller, 1983), **acinetoferrin** (from *Acinetobacter haemolyticus*, Okujo *et al.*, 1994b; Wang and Phanstiel, 1998), **aerobactin** (from *Aerobacter aerogenes*, Gibson and Magrath, 1969; Maurer and Miller, 1982), **nannochelin C** (from the myxobacterium *Nannocystis excedens*, Kunze *et al.*, 1992; Bergeron and Phanstiel, 1992), the **ochrobactins** (from *Ochrobactum* sp., Martin *et al.*, 2006), **rhizobactin 1021** (from *Rhizobium meliloti*, Persmark *et al.*, 1993) and the **synechobactins** (from *Synechococcus*, Ito and Butler, 2005).

Siderophores with 2-oxoglutaric acid units

N-alkylated 2-oxoglutaric acid derivatives cyclize at neutral pH values by interaction of the α-carbonyl with the amino group bound to the δ-carboxyl group giving 2-hydroxy-5-oxoproline structures (cf. Fig. 1.7). In this way α-hydroxycarboxylic acid groupings are formed which can act as ligand sites for Fe^{3+}. **Vibrioferrin** was isolated from *Vibrio parahaemolyticus* (Yamamoto *et al.*, 1994). **Staphyloferrin B** is produced by *Staphylococcus hyicus* (Drechsel *et al.*, 1993) and by *Ralstonia eutropha* (Münzinger *et al.*, 1999a). Mass spectral investigations show a 1:1 Fe^{3+}-to-ligand ratio and NMR studies of the Ga^{3+} complex confirm the participation of the two α-hydroxy- and of the α-amino acid functions. **Achromobactin** (Fig. 1.7) is produced by *Erwinia chrysanthemi* (Münzinger *et al.*, 2000).

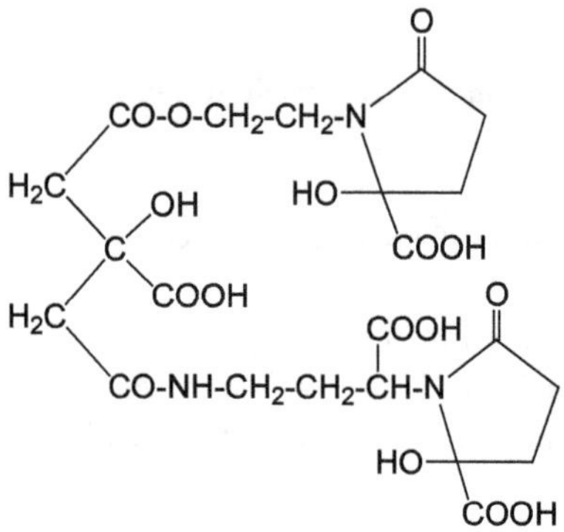

Figure 1.7 Achromobactin.

Siderophores with two catecholate units

In **petrobactin** spermidine residues are bound to citric acid substituted with 3,4-dihydroxybenzoyl (Bergeron *et al.*, 2003) units. One or both of the substituents can also carry a sulphonic acid group in 2-position of the aromatic ring (Homann *et al.*, 2009a; cf. Budzikiewicz, 2006). Petrobactin was originally obtained from *Marinobacter hydrocarbonoclasticus* (Barbeau *et al.*, 2002) and subsequently from *Bacillus anthracis* (Koppisch *et al.*, 2005). It is probably identical with the not fully characterized **anthrachelin** (Garner *et al.*, 2004).

Siderophores with two citric acid units

S,S-(*enantio*)-**rhizoferrin** was obtained from *Ralstonia pickettii* (Münzinger *et al.*, 1999b). It is the optical antipode of the fungal *R,R*-rhizoferrin (Thieken and Winkelmann, 1992; Drechsel *et al.* 1992). Two citric acid residues are connected by putrescine. Despite that fact that it has only two α-hydroxy acid binding sites (central part of the two citric acid units) it forms a 1:1 Fe^{3+}-to-ligand complex. **Staphyloferrin A** is a second siderophore from *Staphylococcus*. D-Orn links the two citric acid parts. Staphyloferrin A forms a 1:1 Fe^{3+}-to-ligand complex (Meiwes *et al.*, 1990).

Miscellaneous siderophores

Pyochelin and related siderophores

For a review see Sansinenea and Ortiz (2009).

This group comprises condensation products of salicyclic acid with Cys giving a thiazoline ring. The structure of **pyochelin** (for a detailed bibliography see Budzikiewicz, 2004a), a secondary siderophore of *Pseudomonas aeruginosa* and of *Burkholderia cepacia* was established (Cox *et al.*,1981) as 2-(2-*o*-hydroxyphenyl-2-thiazolin-4-yl)–3-methylthiazolidine-4-carboxylic acid (Fig. 1.8). It consists of a mixture of two easily interconvertible stereoisomers (pyochelin I and II) differing in the configuration of C-2″. The relative and absolute stereochemistry (4′*R*,2″*R*,4″*R*) of pyochelin I was established by an X-ray analysis of its Fe^{3+} complex (Schlegel *et al.*, 2006). Fe^{3+} is associated with the phenolate and the

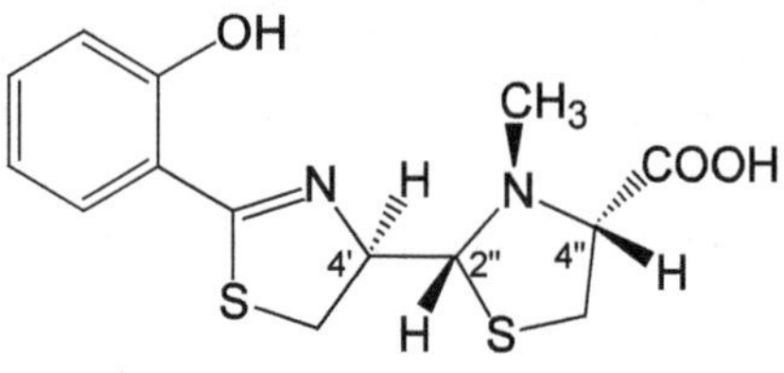

Figure 1.8 Pyochelin I.

carboxylate oxygen ions and with the two nitrogen atoms. Two of these units are bridged by an acetate ion and a water molecule thus satisfying the remaining two ligand loci of Fe^{3+}. Also an X-ray analysis of ferri-pyochelin bound to its outer membrane receptor has been performed (Cobessi *et al.*, 2005). Pyochelin II has the configuration (4′*R*,2″*S*,4″*R*). It does not complex Fe^{3+} (Hayen and Volmer, 2006). Several syntheses resulting in mixtures of stereoisomers have been developed (e.g. Ankenbauer *et al.*, 1988; Zamri and Abdallah, 2000). *Pseudomonas fluorescens* CHA0 produces *enantio*-pyochelin (Youard *et al.*, 2008). The two optical antipodes are not accepted reciprocally by the two species.

Yersiniabactin was obtained from *Yersinia* spp., but it is produced also by *Pseudomonas syringae* and *Escherichia coli* (Bultreys *et al.*, 2006; Jones *et al.*, 2007). Its structure was elucidated independently by two groups (Drechsel *et al.*, 1995; Chambers *et al*, 1996). Yersiniabactin contains an additional thiazoline ring. **Anguibactin** from *Vibrio anguillarum* (Jalal *et al.*, 1989) is derived from DHB condensed with Cys. **Desferri-ferrithiocin** from *Streptomyces antibioticus* (Naegeli and Zähner, 1980) is based on 3-hydroxypyridine-2-carboxylic acid.

Siderophores with thiocarboxylic acid substituents

For a review see Budzikiewicz (2003).

Pyridine-2,6-dimonothiocarboxylic acid was obtained from *Pseudomonas putida* (Ockels *et al.*, 1978) and later from *Pseudomonas stutzeri* (Lee *et al.*, 1999). It forms a brown Fe^{3+} and a blue Fe^{2+} complex (both $FeLig_2$) (Hildebr *et al.*, 1984), which may be accompanied by complexes carrying two additional cyanide ions (Hildebr *et al.*, 1985). There is evidence that a sulphenic acid residue (-CO-SOH) is the biosynthetic link between -COOH and -COSH (Hildebr *et al.*, 1986).

From iron-deficient cultures of *Pseudomonas fluorescens*, 8-hydroxy-4-methoxy-monothioquinaldic acid (thioquinolobactin) together with the corresponding quinaldic acid (quinolobactin), could be isolated (Neuenhaus *et al.*, 1980). Quinolobactin can act as an alternative siderophore of *P. fluorescens* (Mossialos *et al.*, 2000), although it is the hydrolysis product of the thioacid (Matthijs *et al.*, 2007). Its synthesis and complex formation as $(Fe^{3+})Lig_2$ was described (du Moulinet d'Hardemare *et al.*, 2004).

References

Abergel, R.J., Zawadzka, A.M., Hoette, T.M., and Raymond, K.M. (2009). Enzymatic hydrolysis of trilactone siderophores: where chiral recognition occurs in enterobactin and bacillibactin iron transport. J. Am. Chem. Soc. *131*, 12682–12692.

Adolphs, M., Taraz, K., and Budzikiewicz, H. (1996). Catecholate siderophores from *Chryseomonas luteola*. Z. Naturforsch. *51c*, 281–285.

Ankenbauer, R.G., Toyokuni, T., Staley, A., Rinehard, K.L. Jr., and Cox, C.D. (1988). Synthesis and biological activity of pyochelin, a siderophore of *Pseudomonas aeruginosa*. J. Bacteriol. *170*, 5344–5351.

Atkin, C.L., and Neilands, J.B. (1968). Rhodotorulic acid, a diketopiperazine dihydroxamic acid with growth-factor activity. I. Isolation and characterization. Biochemistry 7, 3734–3739.

Barbeau, K., Rue, E.L., Bruland, K.W., and Butler, A. (2001). Photochemical cycling of iron in the surface ocean mediated by microbial iron(III)-binding ligands. Nature *413*, 409–413.

Barbeau, K., Rue, E.L., Trick, C.G., Bruland, K.W., and Butler, A. (2003). Photochemical reactivity of siderophores produced by marine heterotrophic bacteria and cyanobacteria based on characteristic Fe(III) binding groups. Limnol. Oceanogr. *48*, 1069–1078.

Barbeau, K. Zhang, G. Live, D. H., and Butler, A. (2002). Petrobactin, a photoreactive siderophore produced by the oil-degrading marine bacterium *Marinobacter hydrocarbonoclaticus*. J. Am. Chem. Soc. *124*, 378–379.

Barelmann, I., Meyer, J.-M., Taraz, K., and Budzikiewicz, H. (1996). Cepaciachelin, a new catecholate siderophore from *Burkholderia* (*Pseudomonas*) *cepacia*. Z. Naturforsch. *51c*, 627–630.

Beiderbeck, H., Taraz, K., Budzikiewicz, H., and Walsby, A.E. (2000). Anachelin, the siderophore of the cyanobacterium *Anabena cylindrica* CCAP 1403/2A. Z. Naturforsch. *55c*, 681–687.

Bergeron, R.J., Garlich, J.R., and McManis, J.S. (1985). Total synthesis of vibriobactin. Tetrahedron *41*, 507–510.

Bergeron, R.J., Huang, G., Smith, R.E., Bharti, N., McManis J.S., and Butler, A. (2003). Total synthesis and structure revision of petrobactin. Tetrahedron *59*, 2007–2014.

Bergeron, R.J., and Kline, S.J. (1982). Short synthesis of parabactin. J. Am. Chem. Soc. *104*, 4489–4492.

Bergeron, R.J., and Phanstiel, O. IV. (1992). The total synthesis of nannochelin, a novel cinnamoyl hydroxamate-containing siderophore. J. Org. Chem. *57*, 7140–7143.

Bergeron, R.J., Xin, M.G., Weimar, W.R., Smith, R.E., and Wiegand, J. (2001). Significance of asymmetric sites in choosing siderophores as deferration agents. J. Med. Chem. *44*, 2469–2478.

Bister, B., Bischoff, D., Nicholson, G.J., Valdebenito, M., Schneider, K., Winkelmann, G., Hantke, K., and Süssmuth, R.D. (2004). The structure of salmochelins: *C*-glucosylated enterobactins of *Salmonella enterica*. BioMetals *17*, 471–481.

Budzikiewicz, H. (2003). Heteroaromatic monothiocarboxylic acids from *Pseudomonas* spp. Biodegradation *14*, 65–72.

Budzikiewicz, H. (2004a). Siderophores of the Pseudomonadaceae *sensu stricto* (fluorescent and non-fluorescent *Pseudomonas* spp.). Progr. Chem. Org. Nat. Prod. *87*, 81–237.

Buzikiewicz, H. (2004b). Bacterial catecholate siderophores. Mini-Rev. Org. Chem. *1*, 163–168.

Budzikiewicz, H. (2005). Bacterial citrate siderophores. Mini-Rev. Org. Chem. *2*, 119–124.

Budzikiewicz, H. (2006). Bacterial aromatic sulfonates – a Bucherer reaction in nature? Min-Rev. Org. Chem. *3*, 93–97.

Budzikiewicz, H. (2010). Bacterial siderophores. Progr. Chem. Org. Nat. Prod. *92*, 1–95.

Budzikiewicz, H., Bössenkamp, A., Taraz, K., Pandey, A., and Meyer, J.-M. (1997). Corynebactin, a cyclic catecholate siderophore from *Corynebacterium glutamicum* ATCC 14067 (*Brevibacterium* sp. DSM 20411). Z. Naturforsch. *52c*, 551–554.

Bultreys, A., Gheysen, I., and de Hoffmann, E. (2006). Yersiniabactin production by *Pseudomonas syringae* and *Escherichia coli*, and description of a second yersiniabactin locus evolutionary group. Appl. Environ. Microbiol. *72*, 3814–3825.

Bultreys A., Gheysen, I., Wathelet, B., Schäfer, M., and Budzikiewicz, H. (2004). The pyoverdins of *Pseudomonas syringae* and *Pseudomonas cichorii*. Z. Naturforsch. *59c*, 613–618.

Byers, B.R., Powell, M.V., and Lankford, C.E. (1967). Iron-chelating hydroxamic acid (schizokinen) active in initiation of cell division in *Bacillus megaterium*. J. Bacteriol. *93*, 286–294.

Carrano, C.J., Jordan, M., Drechsl, H., Schmid, D.G., and Winkelmann, G. (2001). Heterobactins: a new class of siderophores from *Rhodococcus erythropolis* IGTS8 containing both hydroxamate and catecholate donor groups. BioMetals *14*, 119–125.

Carrano, C.J., and Raymond, K.N. (1978). Coordination chemistry of microbial iron transport. 10. Characterization of the complexes of rhodotorulic acid, a dihydroxamate siderophore. J. Am. Chem. Soc. *100*, 5371–5374.

Challis, G.L. (2005). A widely distributed bacterial pathway for siderophore biosynthesis independant of nonribosomal peptide synthetases. ChemBioChem. *6*, 601–611.

Chambers, C.E., McIntyre, D.D., Mouck, M., and Sokol, P.A. (1996). Physical and structural characterization of yersiniophore, a siderophore produced by clinical isolates of *Yersinia enterocolitica*. BioMetals *9*, 157–167.

Ciche, T.A., Blackburn, M., Carney, J.R., and Ensign, J.B. (2003). Photobactin: a catechol siderophore produced by *Photorhabdus luminescens*, an entomopathogen mutually associated with *Heterorhabditis bacteriophora* NC1 nematodes. Appl. Environ. Microbiol. *69*, 4706–4713.

Cobessi, D., Celia, H., and Pattus, F. (2005). Crystal structure at high resolution of ferric-pyochelin and its membrane receptor FptA from *Pseudomonas aeruginosa*. J. Mol. Biol. *352*, 893–904.

Corbin, J.L., and Bulen, W.A. (1969). The isolation and identification of 2,3-dihydroxybenzoic acid and 2-*N*,6-*N*-di(2,3-dihydroxybenzoyl)-L-lysine formed by iron-deficient *Azozobacter vinelandii*. Biochemistry *8*, 757–762.

Cornish, A.S., and Page, W.J. (1995). Production of the triacetocholate siderophore protochelin by *Azotobacter vinelandii*. BioMetals *8*, 332–338.

Cox, C.D., Rinehart, K.L. Jr., Moore, M.L., and Cook, C.J. Jr. (1981). Pyochelin: novel structure of an iron-chelating growth promotor for *Pseudomonas aeruginosa*. Proc. Natl. Acad. Sci. U.S.A. *78*, 4256–4260.

Crosa, J.H., and Walsh, C.T. (2002). Genetics and assembly line enzymology of siderophore biosynthesis in bacteria. Microbiol. Mol. Biol. Rev. *66*, 223–249.

Deng, J., Hamada, Y., and Shioiri, T. (1995). Total synthesis of alterobactin A, a super siderophore from an open-ocean bacterium. J. Am. Chem. Soc. *117*, 7824–7825.

Dhungana, S., Michalczyk, R., Boukhalfa, H., Lack, J.G., Koppisch, A.T., Fairlee, J.M., Johnson, M.T., Ruggiero, C.E., John, S.G., Cox, M.M., Browder, C.C., Forsythe, J.H., Vanderberg, L.A., Neu, M.P., and Hersman, L.E. (2007). Purification and characterization of rhodobactin: a mixed ligand siderophore from *Rhodococcus rhodochrous* strain OFS. Biometals *20*, 853–867.

Diekmann, H. (1970). Stoffwechselprodukte von Mikroorganismen. 81. Mitteilung. Vorkommen und Strukturen von Coprogen B und Dimerumsäure. Arch. Microbiol. *73*, 65–76.

Diekmann, H., and Krezdorn, E. (1975). Stoffwechselprodukte von Mikroorganismen. 150. Mitteilung. Ferricrocin, Triacetylfusigen und andere Sideramine aus Pilzen der Gattung *Aspergillus*, Gruppe *Fumigatus*. Arch. Microbiol. *106*, 191–194.

Dilworth, M.J., Carson, K.C., Giles, R.G.F., Byrne, L.T., and Glenn, A.R. (1998). *Rhizobium leguminosarum* bv. *viciae* produces a novel cyclic trihydroxamate siderophore, vicibactin. Microbiology *144*, 781–791.

Dimise, E.J., Widboom, P.F., and Bruner, S.D. (2008). Structure elucidation and biosynthesis of fuscachelins, peptide siderophores from the moderate thermophile *Thermobifida fusca*. Proc. Natl. Acad. Sci. U.S.A. *105*, 15311–15316.

Drechsel, H., Freund, S., Nicholson, G., Haag, H., Jung, O., Zähner, H., and Jung, G. (1993). Purification and chemical characterization of staphyloferrin B, a hydrophilic siderophore from staphylococci. BioMetals *6*, 185–192.

Drechsel, H., Jung, G., and Winkelmann, G. (1992). Stereochemical characterization of rhizoferrin and identification of its dehydration product. BioMetals *5*, 141–148.

Drechsel, H., Stephan, H., Lotz, R., Haag, H., Zähner, H., Hantke, K., and Jung, G. (1995). Structure elucidation of yersiniabactin, a siderophore from highly virulent *Yersinia* strains. Liebigs Ann. Chem. 1727–1733.

Drechsel, H., and Winkelmann, G. (1997). Iron chelation and siderophores. In *Transition Metals in Microbial Metabolism*, G. Winkelmann and C.J. Carrano, eds. (Amsterdam, NL: Harwood Academic Publishers), pp. 1–50.

du Moulinet d'Hardemare, A., Serratrice, G., and Pierre, J.L. (2004). Synthesis and iron-binding properties of quinolobactin, a siderophore from a pyoverdine-deficient *Pseudomonas fluorescens*. BioMetals *17*, 691–697.

Ehlert, G., Taraz, K., and Budzikiewicz, H. (1994). Serratiochelin, a new catecholate siderophore from *Serratia marcescens*. Z. Naturforsch. *49c*, 11–17.

Eng-Wilmot, D.L., Rahman, A., Mendenhall, J.V., Grayson, S.L., and van der Helm, D. (1984). Molecular structure of ferric neurosporin, a minor siderophore-like compound containing N^{δ}-hydroxy-D-ornithine. J. Am. Chem. Soc. *106*, 1285–1290.

Eng-Wilmot, D.L., and van der Helm, D. (1980). Molecular and crystal structure of the linear tricatechol siderophore, agrobactin. J. Am. Chem. Soc. *102*, 7719–7725.

Feistner, G.J., Stahl, D.C., and Gabrik, A.H. (1993). Proferrioxamine siderophores of *Erwinia amylovora*. A capillary liquid chromatographic/electrospray tandem mass spectrometric study. Org. Mass Spectrom. *28*, 163–175.

Filsak, G., Taraz, K., and Budzikiewicz, H. (1994). Untersuchungen zur Struktur und Derivatisierung von Kondensationsprodukten der 2,4-Diaminobuttersäure mit anderen Aminosäuren. Z. Naturforsch. *49c*, 18–25.

Gademann, K., Bethuel, Y., Locher, H.H., and Hubschwerlen, C. (2007). Biomimetic total synthesis and antimicrobial evaluation of anachelin H. J. Org. Chem. *72*, 8361–8370.

Garner, B.L., Arceneaux, J.E.L., and Byers, R.B. (2004). Temperature control of a 3,4-dihydroxybenzoate (protocatechuate)-based siderophore in *Bacillus anthracis*. Curr. Microbiol. *49*, 89–94.

Georgias, H., Taraz, K., Budzikiewicz, H., Geoffroy, V., and Meyer, J.-M. (1999). The structure of the pyoverdin from *Pseudomonas fluorescens* 1.3. Structural and biological relationship of pyoverdins from different strains. Z. Naturforsch. *54c*, 301–308.

Gibson, F., and Magrath, D.I. (1969). The isolation and characterization of a hydroxamic acid (aerobactin) formed by *Aerobacter aerogenes* 62-I. Biochim. Biophys. Acta *192*, 175–184.

Griffiths, G.L., Sigel, S.P., Pyne, S.M., and Neilands, J.B. (1984). Vibriobactin, a siderophore from *Vibrio cholerae*. J. Biol. Chem. *10*, 383–385.

Hayen, H., and Volmer, D.A. (2006). Different iron-chelating properties of pyochelin diastereomers revealed by LC/MS. Anal. Bioanal. Chem. *385*, 606–611.

Hildebrand, U., Lex, J., Taraz, K., Winkler, S., Ockels, W., and Budzikiewicz, H. (1984). Untersuchungen zum Redox-System Bis(pyridin-2,6-dicarbothioato)-ferrat(II)/ferrat(III). Z. Naturforsch. *39b*, 1607–1613.

Hildebrand, U., Taraz, K., and Budzikiewicz, H. (1986). 6-(Hydroxythio)carbonylpyridin-2-carbonsäure und Pyridin-2-carbonsäure-6-monothiocarbonsäure als biosynthetische Zwischenstufen bei der Bildung von Pyridin-2,6-di(monothiocarbonsäure) aus Pyridin-2,6-dicarbonsäure. Z. Naturforsch. *41c*, 691–694.

Hildebrand, U., Taraz, K., Budzikiewicz, H., Korth, H., and Pulverer, G. (1985). Dicyano-bis(pyridin-2,6-dicarbothioato)-ferrat (II)/ferrat (III), ein weiteres eisenhaltiges Redoxsystem aus der Kulturlösung eines Pseudomonas-Stammes. Z. Naturforsch. *40c*, 201–207.

Homann, V.V., Edwards, K.J., Webb, E.A., and Butler, A. (2009a). Siderophores of *Marinobacter aquaeolei*: petrobactin and its sulfonated derivatives. Biometals *22*, 565–571.

Homann, V.V., Sandy, M., Tincu, J.A., Templeton, A.S., Tebo, B.M., and Butler, A. (2009b). Loihichelins A-F, a suite of amphiphilic siderophores produced by the marine bacterium *Halomonas* LOB-5. J. Nat. Prod. *72*, 884–488.

Hossein, M.B., Eng-Wilmot, D.E., Loghry, R.A., and van der Helm., D. (1980). Circular dichroism, crystal structure, and absolute configuration of the siderophore ferric *N,N',N"*-triacetylfusarinine, $FeC_{39}H_{57}N_6O_{15}$. J. Am. Chem. Soc. *102*, 5766–5773.

Huschka, H.-G., Jalal, M.A.F., van der Helm, D., and Winkelmann, G. (1986). Molecular recognition of siderophores in fungi: role of iron-surrounding *N*-acyl residues and the peptide backbone during membrane transport in *Neurospora crassa*. J. Bacteriol. *167*, 1020–1024.

Ito, T., and Neilands, J.B. (1958). Products of 'low-iron fermentation' with *Bacillus subtilis*: isolation, characterization and synthesis of 2,3-dihydroxybenzoylglycine. J. Am. Chem. Soc. *80*, 4645–4647.

Ito, Y., and Butler, A. (2005). Structure of synechobactins, new siderophores of the marine cyanobacterium *Synechococcus* sp. PCC 7002. Limnol. Oceanogr. *50*, 1918–1923.

Ito, Y., Ishida, K., Okada, S., and Murakami, M. (2004). The absolute stereochemistry of anachelins, siderophores from the cyanobacterium *Anabena cylindrica*. Tetrahedron. *60*, 9075–9080.

Itou, Y., Okada, S., and Murakami, M. (2001). Two structural isomeric siderophores from the freshwater cyanobacterium *Anabena cylindrica* (NIES-19). Tetrahedron 57, 9093–9099.

Jacques, P., Ongena, M., Gwose, I., Seinsche, D., Schröder, H., Delphosse, P., Thonart, P., Taraz, K., and Budzikiewicz, H. (1995). Structure and characterization of isopyoverdin from *Pseudomonas putida* BTP1 and its relation to the biogenetic pathway leading to pyoverdins. Z. Naturforsch. *50c*, 622–629.

Jalal, M.A.F., Hossain, M.B., van der Helm, D., Sanders-Loehr, J., Actis, L.A., and Crosa, J.H. (1989). Structure of anguibactin, a unique plasmid-related siderophore from fish-pathogen *Vibrio anguillarum*. J. Am. Chem. Soc. *111*, 292–296.

Jalal, M.A.F., Love, S.K., and van der Helm, D. (1986). Siderophore mediated iron(III) uptake in *Gliocladium virens*. 1. Properties of *cis*-fusarinine, *trans*-fusarinine, dimerum acid, and their ferric complexes. J. Inorg. Biochem. *28*, 417–430.

Jones, A.M., Lindow, S.E., and Wildermuth, M.C. (2007). Salicylic acid, yersiniabactin, and pyoverdin production by the model phytopathogen *Pseudomonas syringae* pv. tomato DC3000: synthesis, regulation, and impact on tomato and *Arabidopsis* host plants. J. Bacteriol. *189*, 6773–6786.

Kachadourian, R., Dellagi, A., Laurent, J., Bricard, L., Kunesch, G., and Expert, D. (1996). Desferrioxamine-dependent iron transport in *Erwinia amylovora* CFBP1430: cloning of the gene encodimg the ferrioxamine receptor FoxR. BioMetals *9*, 143–150.

Kanoh, K., Kamino, K., Leleo, G., Adachi, K., and Shizuri, Y. (2003). Pseudoalterobactin A and B, new siderophores excreted by marine bacterium *Pseudoalteromonas* sp. KP20-4. J. Antibiotics *56*, 871–875.

Keller-Schierlein, W., Prelog, V., and Zähner, H. (1964). Siderochrome (Natürliche Eisen(III)-trihydroxamat-Komplexe). Progr. Chem. Org. Nat. Prod. *22*, 279–322.

Kokubo, S., Suenaga, K., Shinohara, C., Tsuji, T., and Uemura, D. (2000). Structures of amamistatins A and B, novel growth inhibitors of human tumor cell lines fron *Nocardia asteroides*. Tetrahedron *56*, 6435–6440.

Koppisch, A.T., Browder, C.C., Moe, A.L., Shelley, J.T., Kinkel, B.A., Hersman, L.E., Iyer, S., and Ruggiero, C.E. (2005). Petrobactin is the primary siderophore synthesized by *Bacillus anthracis* str. *Sterne* under conditions of iron starvation. BioMetals *18*, 577–585.

Kunze, B., Bedorf, N., Kohl, W., Höfle, G., and Reichenbach, H. (1989). Myxochelin A, a new iron-chelating compound from *Angiococcus disciformis* (Myxobacterales). Production, isolation, physicochemical and biological properties. J. Antibiotics. *42*, 14–17.

Kunze, B., Trowitzsch-Kienast, W., Höfle, G., and Reichenbach, H. (1992). Nannochelins A, B and C, new iron chelating compounds from *Nannocystis excedens* (mycobacteria). Production, isolation, physico-chemical and biological properties. J. Antibiotics *45*, 147–150.

Ledyard, K.M., and Butler, A. (1997). Structure of putrebactin, a new dihydroxamate siderophore produced by *Shewanella putrefaciens*. J. Biol. Inorg. Chem. *2*, 93–97.

Lee, B.H., and Miller, M.J. (1983). Natural ferric ionophores: total synthesis of schizokinen, schizokinen A, and arthrobactin. J. Org. Chem. *48*, 24–31.

Lee, C.H., Lewis, T.A., Paszczynski, A., and Crawford, R.L. (1999). Identification of an extracellular catalyst of carbon tetrachloride dehalogenation from Pseudomonas stutzeri strain KC as pyridine-2,6-bis(thiocarboxylate). Biochem. Biophys. Res. Commun. *261*, 562–566.

Linke, W.-D., Crueger, A., and Diekmann, H. (1972). Stoffwechselprodukte von Mitroorganismen. 106. Mitteilung. Zur Konstitution des Terregens-Faktors. Arch. Mikrobiol. *85*, 44–50.

Loomis, L.D., and Raymond, K.N. (1991). Solution equilibria of enterobactin and metal–enterobactin complexes. Inorg. Chem. *30*, 906–911.

Martin, J.D., Ito, Y., Homann, V.V., Haygood, M.G., and Butler, A. (2006). Structure and membrane affinity of new amphiphilic siderophores produced by *Ochrobactrum* sp. SP18. J. Biol. Inorg. Chem. *11*, 633–641.

Martinez, J.S., and Butler, A. (2007). Marine amphiphilic siderophores: marinobactin structure, uptake, and microbial partitioning. J. Inorg. Biochem. *101*, 1692–1698.

Martinez, J.S., Carter-Franklin, J.N., Mann, E.L., Martin, J.D., Haygood, M.G., and Butler, A. (2003). Structure and membrane affinity of a suite of amphiphilic siderophores produced by a marine bacterium. Proc. Natl. Acad. Sci. U.S.A. *100*, 3754–3759.

Martinez, J.S., Zhang, G.P., Holt, P.D., Jung, H.-T., Carrano, C.J., Haygood, M.G., and Butler, A. (2000). Self-assembling amphiphilic siderophores from marine bacteria. Science *287*, 1245–1247.

Matthijs, S., Budzikiewicz, H., Schäfer, M., Wathelet, B., and Cornelis, P. (2008). Ornicorrugatin, a new siderophore from *Pseudomonas fluorescens* AF76. Z. Naturforsch. *63c*, 8–12.

Matthijs, S., Tehrani, K.A., Laus, G., Jackson, R.W., Cooper, R.M., and Cornelis, P. (2007). Thioquinolobactin, a Pseudomonas siderophore with antifungal and anti-Pythium activity. Environ. Microbiol. *9*, 425–434.

Maurer, P.J., and Miller, M.J. (1982). Microbial iron chelators. Total synthesis of aerobactin and its constituent amino acid, N^6-acetyl-N^6-hydroxylysine. J. Am. Chem. Soc. *104*, 3096–3101.

May, J.J., Wendrich, T.M., and Mahariel, M.A. (2001). The *dhb* operon of *Bacillus subtilis* encodes the biosynthetic template for the catecholic siderophore 2,3-dihydroxybenzoate-glycine-threonine trimeric ester bacillibactin. J. Biol. Chem. *276*, 7209–7217.

Meiwes, J., Fiedler, H.-P., Haag, H., Zähner, H., Konetschny-Rapp, S., and Jung, G. (1990). Isolation and characterization of staphyloferrin A, a compound with siderophore activity from *Staphylococcus hyicus* DSM 20459. FEMS Microbiol. Lett. *67*, 201–206.

Meyer, J.-M., Gruffaz, C., Raharinosy, V., Bezverbnaya, I., Schäfer, M., and Budzikiewicz, H. (2008). Siderotyping of fluorescent *Pseudomonas*: molecular mass determination by mass spectrometry as a powerful pyoverdine siderotyping method. Biometals *21*, 259–271.

Meyer, J.-M., Stintzi, A., Coulanges, V., Shivaji, S., Voss, J.A., Taraz, K., and Budzikiewicz, H. (1998). Siderotyping of fluorescent pseudomonads: characterization of pyoverdines of *Pseudomonas fluorescens* and *Pseudomonas putida* strains from Antartica. Microbiology *144*, 3119–3126.

Meyer, J.-M., Stintzi, A., de Vos, D., Cornelis, P., Tappe, R., Taraz, K., and Budzikiewicz, H. (1997). Use of siderophores to type pseudomonads: the three *Pseudomonas aeruginosa* pyoverdine systems. Microbiology *143*, 35–43.

Michalke, R., Taraz, K., and Budzikiewicz, H. (1996). Azoverdin – an isopyoverdin. Z. Naturforsch. *51c*, 772–780.

Miyanaga, S., Obata, T., Onaka, H., Fujita, T., Saito, N., Sakurai, H., Saiki, I., Furumai, T., and Igarashi, Y. (2006). Absolute configuration and antitumor activity of myxochelin A produced by *Nonomuraea pusilla* TP-A0861. J. Antibiotics *59*, 698–703.

Mohn, G., Koehl, P., Budzikiewicz, H., and Lefèvre, J.-F. (1994). Solution structure of pyoverdin GM-II. Biochemistry *33*, 2843–2851.

Moore, C.H., Foster, L.-A., Gerbig, D.G. Jr., Dyer, D.W., and Gibson, B.W. (1995). Identification of alcaligin as the siderophore produced by *Bordetella pertussis* and *B. bronchiseptica*. J. Bacteriol. *177*, 1116–1118.

Moore, R.E., and Emery, T. (1976). N^{α}-Acetylfusarinines: isolation, characterization, and properties. Biochemistry *15*, 2719–2723.

Mossialos, D., Meyer, J.M., Budzikiewicz, H., Wolf, U., Koedam, N., Baysse, C., Anjaiah, V., and Cornelis, P. (2000). Quinolobactin, a new siderophore of *Pseudomonas fluorescens* ATCC *17400*, the production of which is repressed by the cognate pyoverdine. Appl. Environ. Microbiol. *66*, 487–492.

Mulet, M., Gomila, M., Gruffaz, C., Meyer, J.-M., Palleroni, N.J., Lalucat, J., and García-Valdéz, E. (2008). Phylogenetic analysis and siderotyping as useful tools in the taxonomy of *Pseudomonas stutzeri*: description of a novel genovar. Int. J. System. Evol. Microbiol. *58*, 2309–2315.

Mullis, K.B., Pollak, J.R., and Neilands, J.B. (1971). Structure of schizokinen, an iron transport compound from *Bacillus megaterium*. Biochemistry *10* 4894–4898.

Münzinger, M., Budzikiewicz, H., Expert, D., Enard, C., and Meyer, J.-M. (2000). Achromobactin, a new citrate siderophore of *Erwinia chrysanthemi*. Z. Naturforsch. *55c*, 328–332.

Münzinger, M., Taraz, K., and Budzikiewicz, H. (1999a). Staphyloferrin B, a citrate siderophore from *Ralstonia eutropha*. Z. Naturforsch. *54c*, 867–875.

Münzinger, M., Taraz, K., and Budzikiewicz, H., Drechsel, H., Heymann, P., Winkelmann, G., and Meyer, J.-M. (1999b). S,-S-rhizoferrin (*enantio*-rhizoferrin) a siderophore of *Ralstonia* (*Pseudomonas*) *pickettii* DSM 6297 – the optical antipode of R,R-rhizoferrin isolated from fungi. BioMetals *12*, 189–193.

Murakami, Y., Kato, S., Nakajima, M., Matsuoka, M., Kawai, H., Shin-Ya, K., and Seto, H. (1996). Formobactin, a novel free radical scavenging and neural cell protecting substance from *Nocardia* sp. J. Antibiotics. *39*, 839–845.

Naegeli, H.U., and Zähner, H. (1980). Stoffwechselprodukte von Mikroorganismen. 193. Mitteilung. Ferrithiozin. Helv. Chim. Acta. *63*, 1400–1406.

Nagao, Y., Miyasaka, T., Hagiwara, Y., and Fujita, E. (1984). Total synthesis of parabactin, a spermidine siderophore. J. Chem. Soc. Perkin Trans. I 183–187.

Neuenhaus, W., Budzikiewicz, H., Korth, H., and Pulverer, G. (1980). 8-Hydroxy-4-methoxymonothiochinaldinsäure – eine weitere Thiosäure aus Pseudomonas. Z. Naturforsch. *35b*, 1569–1571.

Nishio, T., Tanaka, N., Hiratake, J., Katsube, Y., Ishida, Y., and Oda, J. (1988). Isolation and structure of the novel dihydroxamate siderophore alcaligin. J. Am. Chem. Soc. *110*, 8733–8734.

O'Brien, I.G., and Gibson, F. (1970). The structure of enterochelin and related 2,3–dihydroxy-*N*-benzoylserine conjugates from *Escherichia coli*. Biochem. Biophys. Acta *215*, 393–402.

Ockels, W., Römer, A., Budzikiewicz, H., Korth, H., and Pulverer, G. (1978). An Fe(II) complex of pyridine-2,6-di(monothiocarboxylic acid) – a novel bacterial metabolic product. Tetrahedron Lett. 3341–3342.

Okujo, N., Saito, M., Yamamoto, S., Yoshida, T., Miyoshi, S., and Shinoda, S. (1994a). Structure of vulnibactin, a new polyamine-containing siderophore from *Vibrio vulnificus*. BioMetals 7, 109–116.

Okujo, N., Sakakibara, Y., Yoshida, T., and Yamamoto, S. (1994b). Structure of acinetoferrin, a new citrate-based dihydroxamate siderophore from *Acinetobacter haemolyticus*. BioMetals 7, 170–176.

Ong, S.A., Peterson, T., and Neilands, J.B. (1979). Agrobactin, a siderophore from *Agrobacterium tumefaciens*. J. Biol. Chem. *25*, 1860–1865.

Page, W.J., and von Tigerstrom, M. (1988). Aminochelin, a catecholamine siderophore produced by *Azotobacter vinelandii*. J. Gen. Microbiol. *134*, 453–460.

Persmark, M., Expert, D., and Neilands, J.B. (1989). Isolation, characterization, and synthesis of chrysobactin, a compound with siderophore activity from *Erwinia chrysanthemi*. J. Biol. Chem. *264*, 3187–3193.

Persmark, M., and Neilands, J.B. (1992). Iron(III) complexes of chrysobactin, the siderophore of *Erwinia chrysanthemi*. BioMetals *5*, 29–36.

Persmark, M., Pittman, P., Buyer, J.S., Schwyn, B., Gill, P.R. Jr., and Neilands, J.B. (1993). Isolation and structure of rhizobactin *1021*, a siderophore from the alfalfa symbiont *Rhizobium meliloti*. J. Am. Chem. Soc. *115*, 3950–3956.

Peters, W.J., and Warren, R.A.J. (1968). Itoic acid synthesis in *Bacillus subtilis*. J. Bacteriol. *95*, 360–366.

Peterson, T., Falk, K.-E., Leong, S.A., Klein, M.P., and Neilands, J.P. (1980). Structure and behavior of spermidine siderophores. J. Am. Chem. Soc. *102*, 7715–7718.

Peterson, T., and Neilands, J.B. (1979). Revised structure of a catecholamide siderophore from *Paracoccus denitrificans*. Tetrahedron Lett. 4805–4808.

Pollak, J.R., and Neilands, J.B. (1970). Enterobactin, an iron transport compound from *Salmonalla typhimurium*. Biochem. Biophys. Res. Commun. *38*, 989–992.

Ratledge, C., and Snow, G.A. (1974). Isolation and structure of nocobactin NA, a lipid-soluble iron-binding compound from *Nocardia asteroides*. Biochem. J. *139*, 407–413.

Reid, R.T., Live, D.H., Faulkner, D.J., and Butler, A. (1993). A siderophore from a marine bacterium with an exceptional ferric ion affinity constant. Nature *366*, 455–458.

Renshaw, J.C., Robson, G.D., Trinci, A.P.J., Wiebe, M.G., Livens, F.R., Collison, D., and Taylor, R.J. (2002). Fungal siderophores: structures, functions and application. Mycol. Res. *106*, 1123–1142.

Risse, D., Beiderbeck, H., Taraz, K., Budzikiewicz, H., and Gustine, D. (1998). Corrugatin, a lipopeptide siderophore from *Pseudomonas corrugata*. Z. Naturforsch. *53c*, 295–304.

Robinson J.P., Wawrousek, E.F., McArdle, J.V., Coyle, G., and Adler, I. (1984). X-ray photoelectron and electron spin resonance spectra of iron(III) parabactin. Inorg. Chim. Acta *92*, L19–L21.

Sakakura, A., Umemura, S., and Ishihara, K. (2008). Convergent total syntheses of fluvibactin and vibriobactin using molybdenum(VI) oxide-catalyzed dehydrative cyclization as a key step. Chem. Commun. 3561–3563.

Sansinenea, E., and Ortiz, A. (2009). Bacterial siderophores containing a thiazoline ring. Mini-Rev. Org. Chem. *6*, 120–127.

Schaffner, E.M., Hartmann, R., Taraz, K., and Budzikiewicz, H. (1996) Structure elucidation of azotobactin *87*, isolated from *Azotobacter vinelandii* ATCC 12837. Z. Naturforsch. *51c*, 139–150.

Schlegel, K., Lex, J., Taraz, K., and Budzikiewicz, H. (2006). The X-ray structure of the pyochelin Fe^{3+} complex. Z. Naturforsch. *61c*, 263–266.

Sharman, G.J., Williams, D.H., Ewing, D.F., and Ratledge, C. (1995a). Isolation, purification and structures of exochelin MS, the extracellular siderophore from *Mycobacterium smegmatis*. Biochem. J. *305*, 187–196.

Sharman, G.J., Williams, D.H., Ewing, D.F., and Ratledge, C. (1995b). Determination of the structure of exochelin MN, the extracellular siderophore from *Mycobacterium neoaurum*. Chem. Biol. *2*, 553–561.

Smith, M.J., Shoolery, J.N., Schwyn, B., Holden, I., and Neilands, J.B. (1985). Rhizobactin, a structurally novel siderophore from *Rhizobium meliloti*. J. Am. Chem. Soc. *197*, 1739–1743.

Snow, G.A. (1970). Mycobactins: iron-chelating growth factors from mycobacteria. Bacteriol. Rev. *34*, 99–125.

Stephan H., Freund, S., Beck, W., Jung, G., Meyer, J.-M., and Winkelmann, G. (1993a). Ornibactins – a new family of siderophores from *Pseudomonas*. BioMetals *6*, 93–100.

Stephan H., Freund, S., Meyer, J.-M., Winkelmann, G., and Jung, G. (1993b). Structure elucidation of the gallium–ornibactin complex by 2D-NMR spectroscopy. Liebigs Ann. Chem. 43–48.

Takahashi, A., Nakamura, H., Kameyama, T., Kurasawa, S., Naganawa, H., Okami, Y., Takeuchi, T., and Umezawa, H., and Iitaka, Y. (1987). Bisucaberin, a new siderophore, sensitizing tumor cells to macrophage-mediated cytolysis. II. Physico-chemical properties and structure determination. J. Antibiotics *40*, 1671–1676.

Taraz, K., Ehlert, G., Geisen, K., Budzikiewicz, H., Korth, H., and Pulverer, G. (1990). Protochelin, ein Catecholat-Siderophor aus einem Bakterium (DMS Nr. 5746). Z. Naturforsch. *45b*, 1327–1332.

Teintze, M., Hossain, M.B., Barnes, C.L., Leong, J., and van der Helm, D. (1981). Structure of ferric pseudobactin, a siderophore from a plant growth promoting *Pseudomonas*. Biochemistry *20*, 6446–6457.

Telford, J.R., and Raymond, K.N. (1997). Amonabactin: a family of novel siderophores from a pathogenic bacterium. J. Biol. Inorg. Chem. *2*, 750–761.

Telford, J.R., and Raymond, K.N. (1998). Coordination chemistry of the amonabactins, bis(catecholate) siderophores from *Aeromonas hydrophila*. Inorg. Chem. *37*, 4578–4583.

Thieken, A., and Winkelmann, G. (1992). Rhizoferrin: a complexon type siderophore of the Mucorales and Entomophthorales (Zygomycetes). FEMS Microbiol. Lett. *94*, 37–42.

van der Helm, D., Jalal, M.A.F., and Hossain, M.B. (1989). The crystal structures and configurations of siderophores. In Iron Transport in Microbes, Plants and Animals, G. Winkelmann, D. van der Helm, and J.B. Neilands, eds. (VCH: Weinheim), pp. 135–165.

Vergne, A.F., Walz, A.J., and Miller, M.J. (2000). Iron chelators from mycobacteria (1954–1999) and potential therapeutic applications. Nat. Prod. Rep. *17*, 99–116.

Wang, Q.X., and Phanstiel, O. IV. (1998). Total synthesis of acinetoferrin. J. Org. Chem. *63*, 1491–1495.

Wasielewski, E., Adkinson, R.A., Abdallah, M.A., and Kieffer, B. (2002). The three-dimensional-structure of the gallium complex of azoverdin, a siderophore of *Azomonas macrozytogenes* ATCC 12334, determined by NMR using residual dipolar coupling constants. Biochemistry *41*, 12488–12497.

Winkelmann, G. (1990). Structural and stereochemical aspects of iron tranport in fungi. Biotech. Adv. *8*, 207–231.

Xu, G., Martinez, J.S., Groves, J.T., and Butler, A. (2002). Membrane affinity of the amphiphilic marinobactin siderophores. J. Am. Chem. Soc. *124*, 13408–13415.

Yamamoto, S., Okujo, N., Fujita, Y., Saito, M., Yoshida, T., and Shinoda, S. (1993). Structures of two polyamine-containing catecholate siderophores from *Vibrio fluvialis*. J. Biochem. *113*, 538–544.

Yamamoto, S., Okujo, N., Yoshida, T., Matsuura, S., and Shinoda, S. (1994). Structure and iron transport activity of vibrioferrin, a new siderophore of *Vibrio parahaemolyticus*. J. Biochem. *115*, 868–874.

Youard, Z.A., Mislin, G.L.A., Majcherczyk, P.A., Schalk. I.J., and Reimmann, C. (2007). *Pseudomonas fluorescens* CHA0 produces enantio-pyochelin, the optical antipode of the *Pseudomonas aeruginosa* siderophore pyochelin. J. Biol. Chem. *282*, 35546–35553.

Zamri, A., and Abdallah, M.A. (2000). An improved stereocontrolled synthesis of pyochelin, siderophore of *Pseudomonas aeruginosa* and *Burkholderia cepacia*. Tetrahedron *56*, 249–256.

Zawadzka, A.M., Vandecasteele, F.P.J., Crawford, R.L., and Paszczynski, J. (2006). Identification of siderophores from *Pseudomonas stutzeri*. Can. J. Microbiol. *52*, 1164–1176.

Haem Uptake and Iron Extraction by Bacteria

2

Cécile Wandersman

Abstract

Haem is ubiquitous, abundant and necessary for energy metabolism. Most bacteria have a haem biosynthesis pathway, but nevertheless, since haem is a major source of iron (an essential metal), microbes take up exogenous haem to retrieve iron. To grab haem, microbes extract it from host haemoproteins. This is achieved by two non-exclusive distinct pathways. One pathway involves proteins secreted by bacteria (haemophores) that scavenge haem from host haemoproteins. The second pathway involves microbial cell surface receptors that catch haemoproteins circulating in the vicinity of the cell surface. Both pathways lead to haem docking to cell surface receptors. In Gram-negative bacteria, docked haem is transported through the outer membrane by an energy-dependent process. In Gram-positive bacteria, docked haem is transferred to membrane-anchored haem-binding lipoproteins. In all thus far described systems, haem is actively transported through the plasma membrane by an ATP hydrolysis-powered ABC transporter. Haem is either degraded into biliverdin, CO and iron by haem oxygenases, or iron is retrieved from haem, keeping the tetrapyrrole ring intact by recently identified enzymes. As excess haem is toxic, haem uptake, efflux and degradation are usually highly regulated. In most cases, intracytoplasmic haem or iron released during haem degradation are cofactors along with transcriptional regulators. In several cases, haem uptake and efflux are regulated by extracellular haem.

Introduction

Haem is ubiquitous, abundant and necessary for energy metabolism. Most cells (including microorganisms) have a complete haem biosynthetic pathway. Nevertheless, since haem is a major source of iron (an essential metal), microbes take up exogenous haem to retrieve iron. In addition, some bacteria, like *Enterococcus faecalis, Lactococcus lactis* and *Haemophilus influenzae,* which are unable to synthesize the tetrapyrrol ring, use exogenous haem as an environmental detector to activate the respiratory chain, leading to faster growth and more rapid spread of infection. Bacterial requirements for haem as an iron source and those for haem as a respiratory enzyme cofactor are very different: a micromolar concentration is required as an iron source, whereas a nanomolar concentration is sufficient for respiratory chain functioning.

Owing to its high toxicity (it is a hydrophobic molecule which associates with membranes and thus promotes non-enzymatic redox reactions), haem is not free in living hosts but is bound to diverse haemoproteins. Thus, to grab haem, microbes must extract it from host haemoproteins. This is achieved by two non-exclusive distinct pathways. One pathway involves proteins secreted by bacteria (haemophores) that scavenge haem from both extracellular and intracellular host haemoproteins, such as haemopexin and haemoglobin (released after host cell lysis), respectively. The loaded haemophores transfer their haem to specific surface receptors. The second pathway involves microbial cell surface receptors that catch haemoproteins circulating in the vicinity of the cell surface. Both pathways lead to haem docking to cell surface receptors. In Gram-negative bacteria, docked haem is transported through the outer membrane by an energy-dependent process. In Gram-positive bacteria, docked haem is transferred to membrane-anchored haem binding lipoproteins. In all thus far described systems, haem is actively transported through the plasma membrane by an ATP hydrolysis-powered ABC transporter. In the cytosol, iron is retrieved (see Fig. 2.1 for a general outline of haem uptake). An excess of haem uptake which might occur, for example, during bacteraemia is toxic, and haem uptake, efflux and degradation are usually highly regulated. In most cases, intracytoplasmic haem or iron released during haem degradation are cofactors along with transcriptional regulators. In several cases, haem uptake and efflux are regulated by extracellular haem. This review successively describes the following:

1. haem stores;
2. haem scavenging in the extracellular medium;
3. haem uptake through the outer membrane;
4. haem uptake through the cell wall;
5. haem uptake through the plasma membrane;
6. haem intracellular trafficking and degradation;
7. haem efflux;
8. haem-mediated regulation;
9. involvement of haem metabolism in bacterial virulence.

Haem stores

Haem

Haem, the iron protoporphyrin IX molecule, is the prosthetic group of many enzymes. Owing to its high toxicity, haem is rarely found free. However, in laboratory conditions, haem is an iron source for many bacterial species.

Haemoglobin

Haemoglobin is located in red blood cells, where it functions as the oxygen transporter. It is a tetramer with two α- and two β-chains. Each subunit binds a haem molecule. Haem iron is penta-coordinated to four nitrogens in the porphyrin ring and to the imidazole of one histidine residue. The sixth coordination is either free in methaemoglobin (the oxidized form of haemoglobin, FeIII) or is bound to oxygen in oxyhaemoglobin or to carbon dioxide in carboxyhaemoglobin (the reduced forms of haemoglobin, FeII). With a blood concentration of about 350 g/l (packed in erythrocytes) in humans, it is the major iron source for bacteria.

Many bacterial pathogenic and non-pathogenic species have haem–haemoglobin uptake systems. Some receptors, like *Yersinia enterocolitica* HemR, recognize haem and haemoglobin (Bracken *et al.*, 1999), whereas others, such as *Neisseria meningitidis* HmbR and the *Haemophilus ducreyi* HgbA, appear to be more specific for haemoglobin (Richardson and Stojiljkovic, 1999) (Nepluev *et al.*, 2009). The distinction between haem and haemoglobin utilization is rendered even more complicated by the rapid oxidation of haemoglobin to methaemoglobin as soon as it is released from red blood cells by haemolysis. Methaemoglobin rapidly dissociates into αβ dimers, which bind haem with lower affinity.

Haptoglobin–haemoglobin

Haptoglobin is a serum tetrameric glycoprotein. In serum, haptoglobin binds the haemoglobin released by haemolysis with an affinity so high that the complex does not dissociate, but is degraded after clearance in the liver. The haptoglobin–haemoglobin complex is an iron source for several bacteria such as *N. meningitidis* and *Porphyromonas gingivalis*. Haptoglobin–haemoglobin, haemoglobin and apo-haptoglobin bind to bacterial cell surface receptors (Rohde and Dyer, 2004).

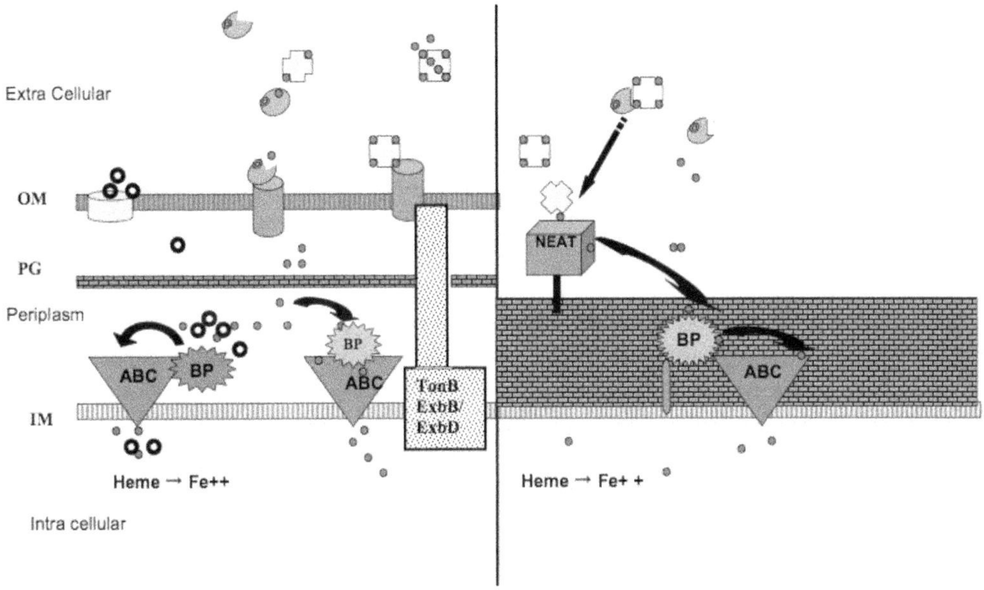

Figure 2.1 Schematic view of haem transport in Gram-negative (left) and -positive (right) bacteria. IM: inner membrane; PG: peptidoglycan layer; OM: outer membrane; ABC: ATPase belonging to the ABC protein super family; BP: binding protein.

Haemopexin

Protection against haem released into plasma is provided by haemopexin, which binds haem with very high affinity. Haemopexin is a 60-kDa glycoprotein with two similarly folded domains which tether the ligand by a bis-histidyl ligation. The haem–haemopexin complex is transported to the liver, where apo-haemopexin is recycled after intracellular discharge of haem. Several bacteria have been reported to bind and/or use haem-haemopexin. However, no direct specific utilization has been demonstrated. As described below, haem is captured from haemopexin by a specific *H. influenzae* haemophore (Cope *et al.*, 2001).

Albumin

Albumin, the most abundant plasma protein (60% of total serum protein), is another serum haem carrier. It is a good haem source for bacteria under laboratory conditions. However, no albumin-specific receptors have been described. As the affinity of albumin for haem might be in the same range as that of bacterial surface receptors for haem, it is likely that haem bound to albumin is recognized by haem–haemoglobin receptors and passively transferred from albumin to the receptors.

Other haemoproteins

In plants and animals, there are many other haemoproteins in which haem is not covalently linked to the apoproteins. They include myoglobin, leghaemoglobin, catalase, and type *b* cytochrome, and most are satisfactory haem sources for bacteria in laboratory media. However, bacteria do not have specific receptors for any of these proteins. Thus, it is likely that haem is the molecule recognized. It is unlikely that haem transfer occurs by equilibrium displacement to a haem bacterial receptor with a higher affinity, because some of these proteins, myoglobin, for example, bind haem with very high affinity. Thus, haem may be stripped off at the cell surface by an as yet uncharacterized mechanism. Alternatively, haem could be released after haemoprotein degradation by surface proteases such as *P. gingivalis* gingipains (Olczak *et al.*, 2005). Many bacterial extracellular proteases are induced in iron-restricted media and might be involved in somewhat specific haemoprotein degradation.

The lipocalin α1-microglobulin

Lipocalin 2, an antimicrobial protein, was shown to bind enterochelin, a siderophore produced by most enterobacteria and hence to deprive bacteria of iron. Another serum protein, α-microglobulin, binds free extracellular haem released after erythrocyte lysis and could interfere with haem acquisition, mediating a host haem sequestration defence (Larsson *et al.*, 2004).

Haem scavenging in the extracellular medium

Haem scavenging in extracellular medium relies on compounds synthesized and released by bacteria into the extracellular medium to scavenge haem from haemoproteins. They function like siderophores (see Chapter 1) and are thus referred to as haemophores. However, whereas known bacterial siderophores are low molecular weight compounds, haemophores are proteins. Moreover, entire siderophores loaded with iron are taken up, whereas haem is stripped of haemophores at the cell surface and only haem is taken up. Haemophores were first identified in Gram-negative bacteria, but more recently, they have been found in a Gram-positive species. They belong to three protein families without sequence similarity, and the members of each family reach the extracellular medium by distinct pathways and obtain haem from haemoproteins using different mechanisms.

HasA-type haemophores

HasA haemophores were first identified in *Serratia marcescens*, but they are highly conserved in several Gram-negative species including *Yersinia pestis, Yersinia enterocolitica, Erwinia carotovora, Pseudomonas aeruginosa* and *Pseudomonas fluorescens* (see Cescau *et al.*, 2007, for review).

HasA proteins are secreted by type I secretion machines which are made up of three proteins: an ABC inner membrane protein providing energy for transport, a membrane fusion protein which bridges the two membranes and an outer membrane protein belonging to the TolC family (Wandersman and Delepelaire, 1990). In addition to being a component of type I secretion apparati, TolC proteins are involved in drug efflux (Zgurskaya and Nikaido, 1999). They form a small β-barrel in the outer membrane and have a long channel domain plunging into the periplasm and contacting the membrane fusion proteins (such as the *E. coli* AcrA protein) and the inner membrane drug efflux components such as the *E. coli* AcrB protein (Koronakis *et al.*, 2000). HasA proteins have a C-terminal secretion signal like most type I secretion substrates (Gray *et al.*, 1989) (see Fig. 2.2 for HasA cycle). The *S. marcescens* HasA protein mechanism of haem scavenging has been studied in detail.

Holo-HasA (loaded with haem) is a globular monomer of 19 kDa with two opposite faces, one made up of four α-helices and the other with seven β-strands (Arnoux *et al.*, 1999). Haem is tightly bound ($K_d = 18$ pM), with a stoichiometry of 1, by two axial iron ligands His32 and Tyr75 located on two loops that separate the two faces (Létoffé *et al.*, 2001). Apo-HasA has the same overall fold, with only a wide displacement of the loop carrying the H32 ligand (Wolff *et al.*, 2008) (Fig. 2.3). The *P. aeruginosa* holo-HasA shows a similar fold (Alontaga *et al.*, 2009). By virtue of its very high affinity for haem, HasA retrieves haem from haemoglobin, haemopexin, leghaemoglobin and myoglobin without forming a stable complex with these haemoproteins (Létoffé *et al.*, 2001). A transient complex between the *P. aeruginosa* haemophore (HasAp) and haemoglobin has been found by NMR spectroscopy (Alontaga *et al.*, 2009). Loaded HasA delivers its haem to a specific outer membrane receptor, HasR (Ghigo *et al.*, 1997) (see below).

HxuA-type haemophores

HxuA haemophore was characterized only in *H. influenzae*, a Gram-negative human pathogen. HxuA is a 100-kDa protein secreted by the two-partner secretion machine (TPS). Due to its N-terminal signal

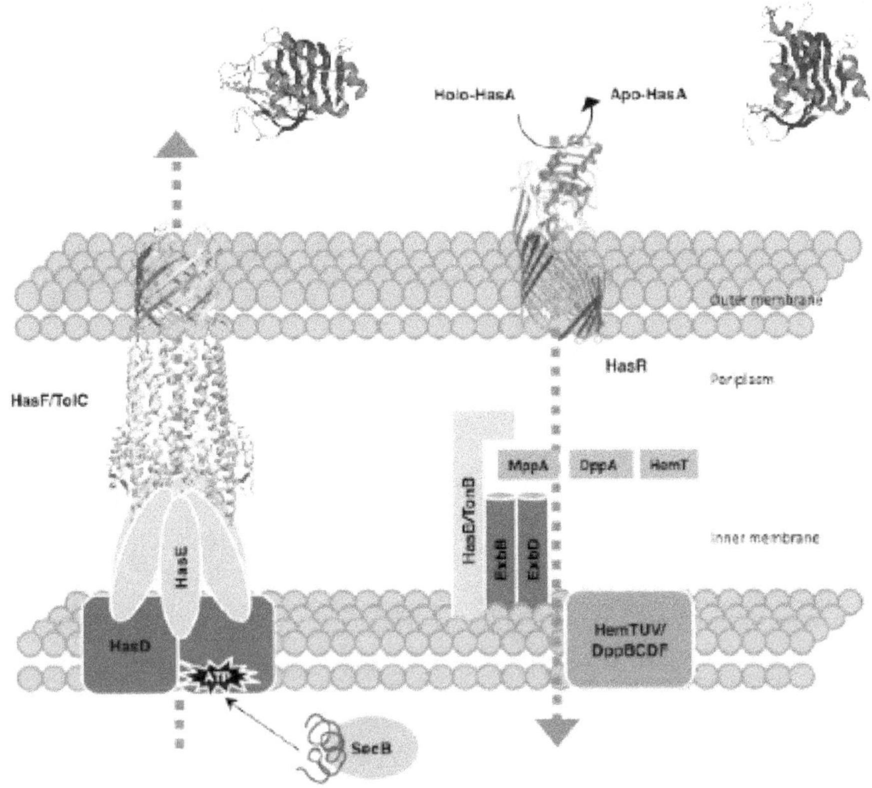

Figure 2.2 *Serratia marcescens* haemophore cycle.

peptide, it reaches the periplasm where it is recognized by HxuB, an outer membrane protein required for HxuA translocation through this membrane. HxuB belongs to the Omp85 protein family involved in outer membrane protein maturation (Dautin and Bernstein, 2007). Release into the extracellular medium does not require a proteolytic step, as is the case for several proteins secreted by the TPS pathway. HxuA is involved only in haem–haemopexin utilization (Cope *et al.*, 1998). But how haem is extracted from haemopexin and how it is transferred to its cognate outer membrane receptor, HxuC, remain a mystery.

NEAT-type haemophores

The IsdX1 and IsdX2 haemophores were characterized only in *Bacillus anthracis*, the mammalian pathogen responsible for anthrax. The proteins are encoded by adjacent genes located in the *isd* iron-regulated operon involved in haem uptake. The *isd* operon also encodes cell wall and membrane-anchored proteins. IsdX1 and IsdX2 have signal peptides, but no motifs that anchor proteins to the cell wall such as LPXTG. And in fact, both proteins are found in the extracellular medium. Each protein contains a 150 amino acid residues long domain named NEAT (NEAr iron transporter) (Grigg *et al.*, 2007). IsdX1 is a 15-kDa protein with one NEAT domain, whereas IsdX2 has five NEAT domains with a molecular weight of 99 kDa. NEAT domains are found in many proteins involved in haem acquisition in Gram-positive bacteria. Up to now, they had not been identified in Gram-negative species. Despite a low level of sequence similarity, the 3D structures of several staphylococcal NEAT proteins reveal that they have a common fold referred to as an immunoglobulin-like fold (see below). IsdX1 binds haem with an affinity much lower than haemoglobin's affinity for haem. Yet IsdX1 is able to remove haem from haemoglobin. Thus, haem retrieval from haemoglobin is not the result of equilibrium displacement of haem towards

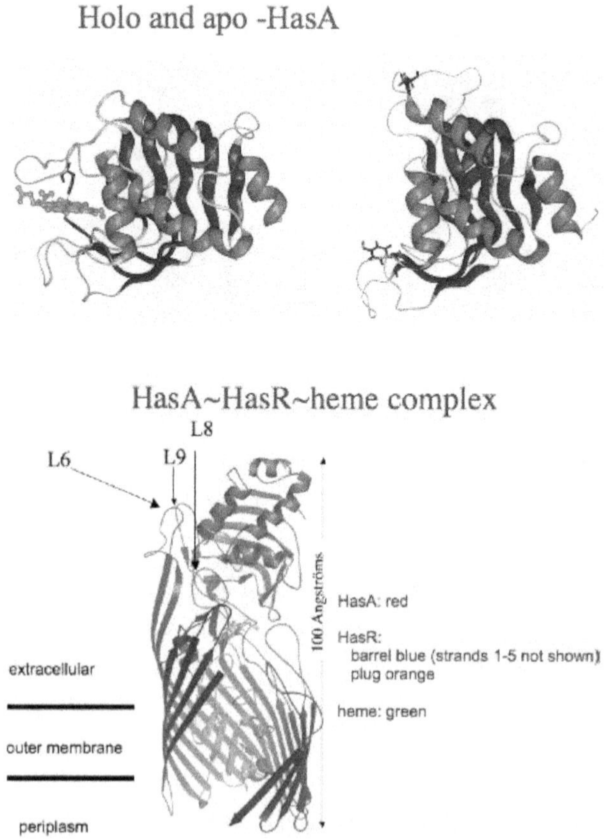

Figure 2.3 3D structure of *the S. marcescens* holo-haemophore and the apo-haemophore (top: left and right, respectively), and of the haemophore–receptor complex (bottom).

the higher affinity protein. Moreover, IsdX1 directly binds haemoglobin and dissociates from it upon haem transfer (Maresso *et al.*, 2008). Thus, the mechanism by which IsdX1 retrieves haem depends on protein–protein interactions and is different from the HasA haem retrieval mechanism.

Other extracellular proteins involved in haem acquisition

Secreted proteases able to degrade haemoglobin and bind haem have been described in several bacteria including *Escherichia coli* EPEC EspC serine protease (Drago-Serrano *et al.*, 2006) and *P. gingivalis* gingipains (Olczak *et al.*, 2005).

Haem uptake through the outer membrane

Haem binding to outer membrane receptors

Gram-negative bacteria have outer membrane haem receptors that actively transport haem through this membrane. All thus far identified haem outer membrane receptors for haem, haemoproteins and haemophores also bind free haem and only haem is internalized, while apoproteins remain outside. Affinities for haem have been determined for several receptors. They are in the micromolar range. HasR, the *S. marcescens* receptor for haem and haemophore, binds haem with a stoichiometry of 1 and a K_d of 0.2

μM (Izadi-Pruneyre *et al.*, 2006). Two conserved histidine residues in the *N. meningitidis* haem receptor HemR and in the *S. marcescens* HasR receptor are required for haem binding and transport (Bracken *et al.*, 1999) (Izadi-Pruneyre *et al.*, 2006). The haem receptors have a low level of sequence similarity amongst themselves and with siderophores and vitamin B12 receptors (see Chapter 1). Nevertheless, structural modelling has predicted that they have a fold similar to that of the few siderophore and vitamin B12 receptors whose 3D structures have already been solved (Barjon *et al.*, 2007). This fold consists of a transmembrane β-barrel with 22 antiparallel β-strands and an N-terminal plug domain inserted in and closing the β-barrel pore. One of the histidine residues essential for haem binding is predicted to be in the external portion of the plug, whereas the other might be positioned in a surface loop. Recently, one haem receptor 3D structure was determined, the *S. marcescens* haem/haemophore receptor HasR in complex with its cognate haemophore (Krieg *et al.*, 2009). It has an overall fold similar to that of siderophore receptors, with a plug and barrel organization. In the HasA-HasR haem complex, haem is bound to HasR by the two conserved histidines, H189 and H603. H189 is located on the extracellular face of the N-terminal plug, whereas H603 is located on one extracellular loop of the β-barrel. The two H residues are close enough to simultaneously ligate haem (see Fig. 2.3). In the *E. coli* UPEC uropathogenic strain, 3D modelling of the haem receptor Hma was used to perform site-directed mutagenesis and showed that a tyrosine residue replaces the plug His residue and is required for haem uptake. However, the external loop β-barrel residue has not yet been identified (Hagan and Mobley., 2009).

Haemoprotein binding to outer membrane receptors

Since free haem is not readily available in the host, but is tightly bound to proteins and often buried inside them, it is likely that some haem receptors might also recognize the carrier proteins. Indeed, the *N. meningitidis* HmbR (Richardson and Stojiljkovic, 1999) and the *H. ducreyi* HgbA (Nepluev *et al.*, 2009) proteins enable better assimilation of haemoglobin-bound haem than free haem. On the basis of HgbA homologies with other siderophore receptors, a three-dimensional model of HgbA was constructed and each of the putative surface-exposed loops was mutagenized by deletion. Only deletions in loop 5 and 7 abolished haemoglobin binding (Nepluev *et al.*, 2009). A similar study on HmbR also indicated a crucial role for loop 7 in haemoglobin binding (Perkins-Balding *et al.*, 2003). However, mechanisms enabling haem transfer from the haemoprotein to the bacterial receptor have not yet been characterized.

N. meningitidis expresses another type of haemoglobin receptor, consisting of two outer membrane proteins: HpuA and HpuB. The HpuB sequence predicts that it is a typical outer membrane receptor with a plug inside a barrel fold. HpuA is an outer membrane lipoprotein crucial for haptoglobin and haemoglobin-haptoglobin binding to the cell surface (Rohde and Dyer, 2004). A similar two-component receptor is present in *P. gingivalis*. The structure of the outer membrane lipoprotein HmuY has recently been elucidated. It shows a new protein fold consisting only of β-strands. Each monomer fixes haem by two histidines which are axial iron ligands (Wójtowicz *et al.*, 2009). The native protein is made up of homotetramers. However, neither the precise role of each protein of these two component receptors nor the mechanism of haem transfer between them (if it does take place) has been clarified.

Haemophore binding to outer membrane receptors

Haemophore binding to its cognate outer membrane receptor was characterized in the *S. marcescens* system. Binding of HasA to HasR is independent of the haem-loaded status of HasA. Both apo-HasA and holo-HasA bind to the receptor with a K_d of 5 nM at common sites on HasR (Létoffé *et al.*, 2001). The interactions involve two HasA β-strands which independently allow efficient HasA binding to HasR. Only proteins with alterations in the two binding sites are no longer able to interact with HasR (Létoffé *et al.*, 2003). The 3D structure of the holo-HasA–HasR complex confirmed the genetically identified HasA binding sites and revealed the two HasR large extracellular loops each of which contacts one of the HasA binding sites (Fig. 2.3) (Krieg *et al.*, 2009).

Haem transfer from holo-HasA to HasR was studied *in vitro* by measuring UV-visible absorption spectroscopy and resonance Raman spectroscopy. Results showed that the proteins loaded with haem have distinct spectroscopic properties resulting from different haem environments in the proteins. When holo-HasA was mixed with HasR, spectroscopic properties were shifted from the HasA to the HasR haem environment in ten minutes. These results demonstrate that haem transfer from its high-affinity site on HasA to its lower affinity site in HasR takes place against a 10^5 affinity gradient *in vitro* and thus is not an energy-driven process. Transfer required the two His haem binding residues of HasR (Izadi-Pruneyre *et al.*, 2006). The 3D structure of the holo-HasA–HasR complex confirmed these results, showing that haem is bound to HasR in the complex via the two His iron ligands. Moreover, it indicated that one HasR Ile residue located in an extracellular loop might contact haem during HasA docking to HasR. This clash could disrupt HasA–Tyr75 haem coordination and thus trigger haem transfer from HasA to HasR. The inability of the HasR Ile 671→Gly variant to acquire haem from HasA *in vitro* and *in vivo* confirms this scenario (Krieg *et al.*, 2009). After haem transfer from HasA to HasR, apo-HasA remains bound to HasR. Only denaturing conditions allow release of HasA from the complex *in vitro*. *In vivo*, apo-HasA release from the receptor is an energy-driven process (Létoffé *et al.*, 2004).

Energy supply for haem uptake through the outer membrane

Haem outer membrane receptors belong to the family of receptors having a common fold characterized by a plug inside a β-barrel organization. Substrate uptake requires an electrochemical proton gradient across the inner membrane and a three-inner-membrane protein complex (ExbB, ExbD, TonB) which couples the proton motive force with receptor conformational changes, enabling substrate transport. Due to such TonB dependence, this family of receptors was named TBDT (TonB-dependent transporter) (Braun, 2006). ExbB and ExbD are inner membrane integral proteins homologous to MotA and MotB proteins, respectively (Cascales *et al.*, 2001). MotA and MotB proteins are part of a proton channel which is required for proton motive force-dependent flagellar rotation (Terashima *et al.*, 2008). By extension, ExbB and ExbD are considered to be involved in proton channel formation, although up to now, experimental data supporting this model are missing. The TonB protein is about 200 amino acid residues long with three domains. An N-terminal membrane anchor region is responsible for the interaction with both ExbB and ExbD, a proline-rich central region allowing TonB periplasmic expansion and a C-terminus region which contacts the TBDT (Ahmer *et al.*, 1995). While each bacterial species has many TBDT (sometimes more than fifty) specific to nutrients encountered in their diverse environments (Schauer *et al.*, 2008), there are only a few copies of the TonB protein per species, with numbers ranging between one and nine unique copies of different TonB within the same organism. (Mey and Payne, 2001; Paquelin *et al.*, 2001; Chu *et al.*, 2007). Genetic and biochemical data have unambiguously established that the TonB C-terminal region interacts with a short conserved region located close to the receptor termini and named the TonB box. The co-crystal structures of the *E. coli* TonB carboxy terminus bound either to FhuA or to BtuB show that the TonB boxes of each receptor (which are disordered on the crystals of the receptor alone) interact with the third TonB β-strand, forming an antiparallel inter-protein β-sheet. This β-sheet exchange results in new interactions between TonB and the receptor plugs (Pawelek *et al.*, 2006; Shultis *et al.*, 2006). How the substrate is taken up through the β-barrel pore is not yet understood. The strong ionic interaction between TonB and the TonB box favours a model which pulls the ligand bound to the TonB–plug complex through the β-barrel, rather than a model in which TonB shuttles from the outer membrane to the inner membrane.

In the case of haemophore-dependent haem uptake, energy transduced by the TonB complex is also necessary to release empty haemophores from the receptor once haem has been captured. Empty haemophore release requires a higher TonB concentration and is more sensitive to protonophore addition, indicating that haemophore release is more energy-consuming than haem uptake (Létoffé *et al.*, 2004). It also suggests that, besides the structural changes in the plug, the TonB complex induces changes in surface-exposed receptor regions responsible for receptor/haemophore interactions.

Unlike *E. coli*, which has a single set of *tonB–exbB–exbD* genes, *S. marcescens* has two functional TonB proteins, TonB and HasB (Paquelin *et al.*, 2001). One is like the *E. coli* TonB, a generalist protein functioning with many TBDTs including HasR and other haem receptors such as *S. marcescens* HemR, whereas the second (HasB) is only able to energize HasR, the haemophore receptor (Benevides-Matos *et al.*, 2008). Nevertheless, HasB binds to FhuA receptor with an affinity in the μM range, comparable to that of TonB for TBDTs (Lefèvre *et al.*, 2008). Thus, failure to transduce energy to these receptors is not the result of the absence of an interaction. In addition, HasB binds to HasR with very high affinity (in the 10 nM range), about fifty times higher than TonB affinity for TBDT receptors. HasB, but not TonB, binds to a HasR mutant protein deleted for five amino acids from the TonB box consensus sequence box, indicating that HasB might form contacts with other HasR plug amino acids. The very high HasB affinity for HasR suggests that the two proteins do not dissociate and raises the question of how TonB recycles each time a new haem molecule binds to the receptor.

Haem uptake through the cell wall

The envelope of Gram-positive bacteria is composed of the peptidoglycan layer with associated proteins, carbohydrates and teichoic acid molecules within some species, crystalline protein S-layers and capsules. *Staphylococcus aureus* has four haem-binding proteins anchored to the cell wall: IsdA, IsdB, IsdC and IsdH (HarA) (Mazmanian *et al.*, 2003). IsdA, IsdB and IsdC are encoded by the *isd* operon, whereas the IsdH structural gene is unlinked to the *isd* operon. IsdA and IsdB have an LPXTG motif and are linked to the peptidoglycan by sortase A (see also Chapter 13). IsdC has an NPQTN motif and is attached to the murein by sortase B, an enzyme specific to IsdC protein anchoring and encoded by the *isd* operon (Mazmanian *et al.*, 2002). IsdA binds haem, IsdB binds haem and haemoglobin and IsdH binds haem and the haemoglobin–haptoglobin complex. IsdC binds haem and is not exposed on the cell surface, but rather is embedded in the cell wall. All four proteins have conserved 125 amino acid residue NEAT domains in different copy numbers. IsdA and IsdC have one NEAT domain, IsdB has two and IsdH has three (Maresso and Schneewind, 2006). The determined 3D structures of IsdA,(Grigg *et al.*, 2007), IsdH (Watanabe *et al.*, 2008) and IsdC (Sharp *et al.*, 2007) NEAT domains revealed an immunoglobulin-like fold composed of eight β-strands forming the two sheets of a β-sandwich. All four NEAT domains have overall similar haem binding sites. The haem is positioned on one β-strand and is covered by loops in IsdA and IsdH. In IsdC, haem is located between two β-sheets. In IsdH, the hydroxyl group of Tyr642 coordinates the iron in the haem (Fig. 2.4). In IsdA, iron is coordinated by the phenolate of Tyr166, and in IsdC, the iron axial ligand could be Trp77.

How do proteins with similar NEAT domains, all binding haem but which are present in diverse copy number, carry on haem extraction from different haemoproteins?

The observation that proteins that bind only haem have one NEAT domain (IsdA and IsdC), whereas those that bind haem and haemoproteins have several NEAT domains (IsdB and IsdH), has led to an interesting hypothesis. The simultaneous binding of a dimeric or tetrameric haemoprotein to two NEAT domains present on one Isd protein might increase haemoprotein unfolding, decreasing the affinity for haem and leading to its capture by the Isd protein (Pilpa *et al.*, 2009). Haem bound to one Isd protein is then transferred between cell-anchored Isd proteins to reach the most deeply buried IsdC protein. Haem transfer between IsdA and IsdC proteins was studied *in vitro*. Holo-IsdA forms a complex with apo-IsdC that weakens haem–IsdA interactions and permits haem transfer to IsdC (Liu *et al.*, 2008). This energy-independent haem transfer is reminiscent of haem transfer from HasA to HasR.

This fascinating model of haem scavenging from haemoproteins and transfer across the cell wall by modular NEAT domains nonetheless requires more experimental data for confirmation. (See Fig. 2.4 for a schematic view of haem transfer via modular NEAT domains.)

Haem transport through the cytoplasmic membrane

Once in the periplasm in Gram-negative bacteria or in the inner face of the cell wall in Gram-positive bacteria, haem is transported through the cytoplasmic membrane by haem permease, constituted by a haem binding protein that catches haem and transfers it to an ABC transporter made up of an integral membrane protein and a membrane-associated ATPase belonging to the ABC protein family.

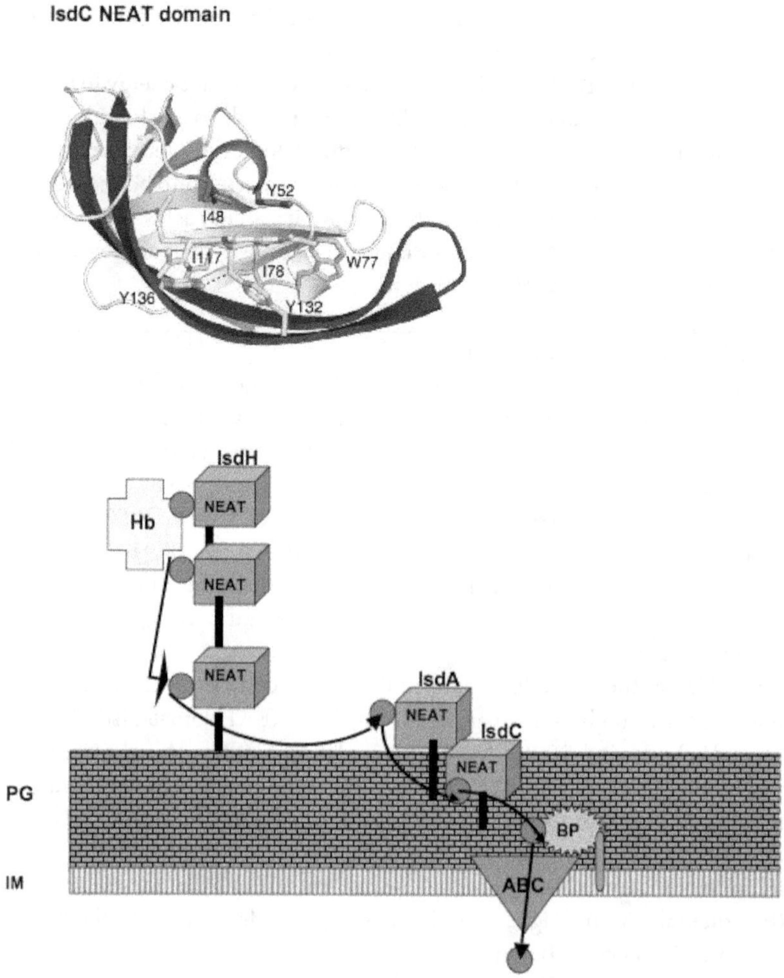

Figure 2.4 The IsdC NEAT protein domain (A) and NEAT-dependent haem transport in Gram-positive bacteria (B).

Transport through the inner membrane in Gram-negative bacteria

Permeases are usually less specific than outer membrane receptors. In the case of siderophore transport, one permease gathers several siderophores that have independently crossed the outer membrane. However, haem permeases are in most cases specific to haem. They are similar to the first haem permease identified in *Yersinia enterocolitica*, constituted by the periplasmic binding protein (HemT), HemU, the integral membrane protein, and HemV, the ATPase (Stojiljkovic and Hantke, 1994). All three proteins are, in most cases, encoded by genes clustered in an operon together with the haem outer membrane receptor gene and up-regulated in iron-restricted conditions. The haem acquisition operon is, in many cases, localized at a pathogenicity island (Wyckoff *et al.*, 1998). Whereas some crystal structures of complete ABC transporters have been recently determined (Dawson *et al.*, 2007), most data concern the soluble periplasmic binding proteins (PBPs). PBPs have a similar fold with two N- and C-terminal lobes separated by a ligand binding cleft. Based on the size and structure of each domain and of the cleft, the PBPs have been divided into three classes. Class I, whose prototype is MalE, the maltose binding protein, has six β-strands in each domain; class II, whose prototype is DppA, the dipeptide binding protein, has five β-strands in each domain. Both classes have large clefts and the two lobes close upon ligand binding (see Fig. 2.5). The third class is constituted by smaller PBPs with an interdomain containing only one helix whose prototype is the siderophore binding protein FhuD. In class III PBPs, ligand binding does not induce substantial structural change. The structures of the haem-unloaded and -loaded *P. aeruginosa* PhuT and *S. dysenteria* ShuT binding proteins show that haem binding periplasmic proteins belong to class III PBPs and undergo a discrete change upon haem binding. Haem is pentacoordinated with a Tyr as the iron axial ligand (Fig. 2.5). Nevertheless, other PBPs which do not belong to class III have been shown to bind haem. The *H. influenzae* HbpA protein sharing 50% identity with DppA binds haem and is involved in haem acquisition (Hanson *et al.*, 1992) (Morton *et al.*, 2005). An alternative haem permease was found in *E. coli* K-12. This laboratory strain, which lacks a haem outer membrane receptor and a specific haem permease, is made competent for exogenous haem iron utilization by the expression of a

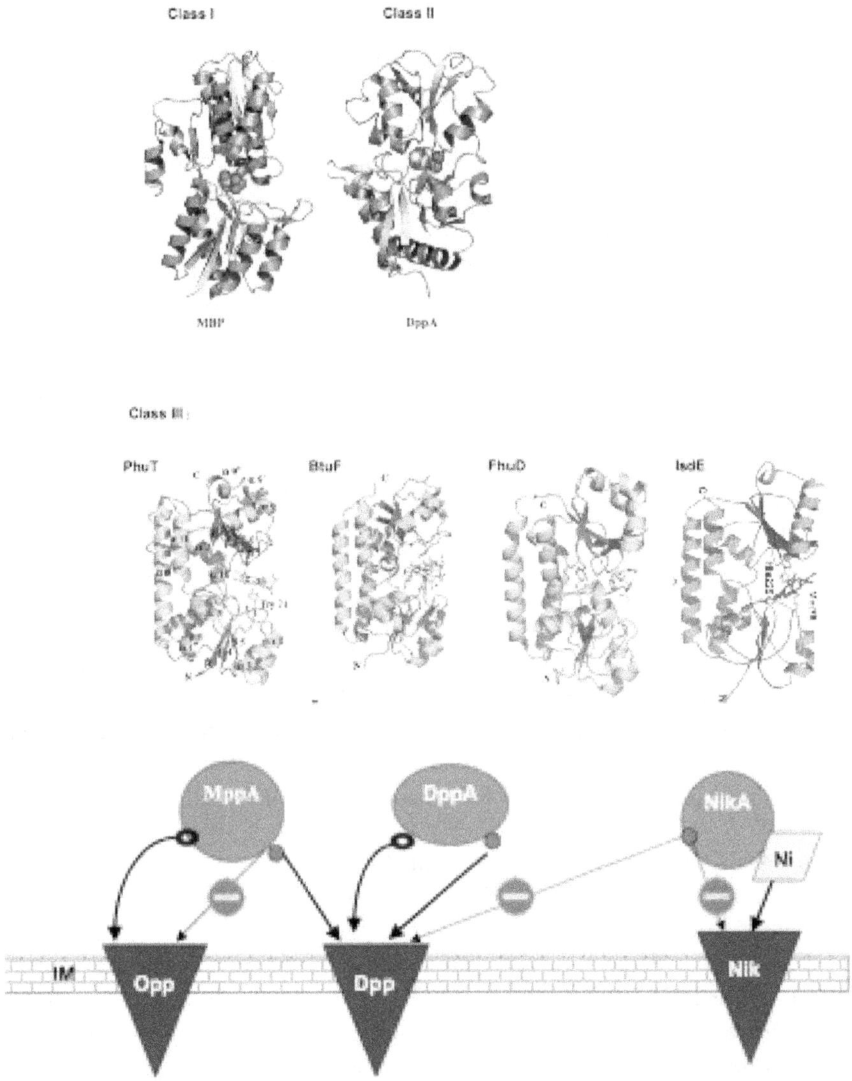

Figure 2.5 Proteins involved in haem transport through the cytoplasmic membrane. (A) Periplasmic binding proteins and haem binding proteins. (B) Specificity of periplasmic haem binding proteins.

foreign outer membrane receptor. This utilization requires an alternative haem permease constituted by the dipeptide ABC transporter consisting of DppB and DppC membrane protein subunits and DppD and DppF ATPase subunits. The DppBCDF transporter functions for haem transport with one of two optional periplasmic binding proteins, either MppA, the L-alanyl-D-glutamyl-meso-diaminopimelate binding protein, or DppA, the dipeptide binding protein (Chen *et al.* 2001). Haem utilization is inhibited *in vivo* by a mixture of di- and tripeptides. When only DppA is functional, dipeptides inhibit haem acquisition, and when only MppA is functional, haem uptake is inhibited by tripeptides. Thus, in *E. coli* K-12, haem and peptides share a common transporter. Both purified binding proteins DppA and MppA bind haem *in vitro*. Haem binding is competed for by dipeptides in the case of DppA and by tripeptides in the case of MppA, suggesting that peptides and haem may share common binding sites on DppA and MppA binding proteins (Létoffé *et al.*, 2006). NikA is another example of a class II PBP that binds haem. NikA is an *E. coli* periplasmic nickel binding protein with a large ligand cleft sharing high sequence and structure similarities with DppA. It is induced under anaerobic conditions via the FNR regulatory protein. It binds nickel, haem and protoporphyrin IX. Binding of nickel does not inhibit haem binding, indicating that the binding sites are distinct (Shepherd *et al.*, 2007). NikA is involved in nickel uptake via the NikBCDE ABC transporter and in haem delivery to the anaerobic nickel-dependent hydrogenases.

In most cases, specificity of the ABC transporter is usually determined by the PBP and not by the ligand they present to the transporter. However, results on NikA and DppA/MppA haem permeases question this rule. Tripeptide-loaded MppA delivers its tripeptides to the OppBCDF transporter (Park *et al.*, 1998), whereas haem-loaded MppA delivers its haem to the DppBCDF transporter. NikA delivers haem to hydrogenase and nickel to the NikBCDE transporter. Thus, in addition to the PBPs, the haem molecule might be involved in recognition of membrane ABC transporter components (see Fig. 2.5). Another puzzling result concerning haem PBP specificity was obtained in *Sinorhizobium meliloti*, in which two PBPs, HemT loaded with haem and FhuP loaded with ferrichrome, delivered their ligand to a common ABC transporter HemU, HemV (Cuív *et al.*, 2008). Finally, the *S. aureus* HstABC transporter could be involved in staphyloferrin and haem uptake (Beasley *et al.*, 2009). Further work is required to understand the respective roles of haem/PBPs in ligand delivery.

Transport through the plasma membrane in Gram-positive bacteria

Gram-positive haem permeases have an overall architecture similar to Gram-negative haem permeases with haem binding protein HemT, membrane transporter HemU and membrane-associated ABC protein HemV. HemT is a lipoprotein anchored by fatty acids in the plasma membrane. It belongs to class III PBPs with bilobed architecture (Fig. 2.5). The two lobes are separated by a small interdomain constituted of one long α-helix. Haem binding does not induce a major conformational change. Haem is hexacoordinated with His and Met as iron axial ligands. The haem permease components are usually encoded by one iron-regulated operon and are clustered with genes encoding cell-wall-anchored NEAT proteins. The lipoprotein does not have NEAT domains (Pluym *et al.*, 2007). Haem permeases belonging to the dipeptide permease family have not been identified in Gram-positive species.

Cytoplasmic haem trafficking and metabolism

Haem oxygenases

Haem oxygenases (HO) are found in most organisms, including bacteria, plants and animals. They all catalyse oxidation of haem into biliverdin, CO and free iron in the presence of O_2 and an electron donor. The overall catalytic reaction requires three O_2 molecules and seven electrons. Mamalian HOs are 35 kDa membrane-anchored proteins, whereas bacterial HOs are smaller and cytoplasmic. HO orthologues are present in both Gram-positive and Gram-negative species such as *Corynebacterium diphtheriae*, *P. aeruginosa* and *N. meningitidis* (see Frankenberg-Dinkel, 2004, for review). Sequence similarities between prokaryotic and eukaryotic HOs are low, but the various HO structures show that they have similar overall structures mainly consisting of α-helices, with apo- and holo-structures being similar. The enzymes interact with NADPH-cytochrome P450 reductase to transfer electrons required for haem degradation. Most HOs oxidize the α-meso-carbon of the haem, resulting in formation of biliberdin IXa. Other HOs such as *P. aeruginosa* PigA cleave the tetrapyrrol ring at the β- and δ-meso-carbon position. Other haem-degrading enzymes without homology with the HO enzymes, but sharing weak similarity between themselves, are present in few Gram-positive and Gram-negative bacteria such as *S. aureus*, *B. anthracis* and *Bradyrhizobium leguminosarum* (Skaar *et al.*, 2006) (Puri and O'Brian, 2006). This class of enzymes also degrades haem into biliverdin, CO and iron.

Bacterial haem oxygenases are involved in several biological functions. They provide iron from haem and degrade haem into biliverdin which, in turn, is reduced to bilirubin, a component with potent anti-oxidant properties. CO is an important regulatory molecule in mammals. Yet it is not known whether it has any function in bacterial physiology. Some haem oxygenases are involved in phytochrome biosynthesis. Phytochromes interact with membrane sensor kinase transducing light signals.

Haem iron-retrieving proteins

Two *E. coli* paralogs, YfeX and EfeB, without any previously known physiological functions and belonging to the DyP family of haem peroxidases, promote iron extraction from haem, preserving the tetrapyrrol ring intact. This novel enzymatic reaction corresponds to deferrochelation of haem. Both proteins bind haem and protoporphyrin IX with very high affinity (in the nanomolar range). YfeX and EfeB are the sole proteins able to provide iron from exogenous haem sources to *E. coli*. YfeX is located in the cytoplasm. EfeB is periplasmic and enables iron extraction from haem in the periplasm and iron uptake in the absence of any haem permease. Iron is transported by a Fe^{3+} permease homologous to the yeast high-affinity iron permease Frtp1 (Grosse *et al.*, 200; Cao *et al.*, 2007). YfeX and EfeB are widespread and highly conserved in bacteria. Their physiological function might be to retrieve iron from haem without production of CO, an antibacterial gas. In the case of EfeB orthologues, iron might be extracted at the cell surface without haem internalization (Létoffé *et al.*, 2009).

Haem chaperones

A class of highly similar haem binding proteins (named ChuS, ShuS, PhuS or HemS) are found in many Gram-negative bacteria. They are encoded by genes clustering with haem uptake genes. They were shown to protect against haem toxicity resulting from high haem uptake (Stojiljkovic and Hantke, 1994). They are 35- to 37-kDa molecular weight proteins with a unique fold consisting of two similar halves that most likely originated from gene duplication. They bind one haem per molecule with micromolar affinity in the cleft between the two domains. A histidine residue belonging to the C-terminal domain is the iron axial ligand (ChuS H193, HemS 196 and PhuS H209) (Suits *et al.*, 2006), (Schneider *et al.*, 2006), (Block *et al.*, 2007).

Purified *E. coli* O157:H7 ChuS protein has haem oxygenase activity in the presence of reductants leading to CO formation (Suits *et al.*, 2005). On the other hand, the *P. aeruginosa* PhuS purified protein has no haem oxygenase activity and releases iron only in the presence of peroxide. The haem-PhuS protein forms a complex with *P. aeruginosa* Pa-HO haem oxygenase and increases its haem oxygenase activity *in vitro* (Bhakta and Wilks, 2006). Such different properties for highly conserved proteins is surprising, and further work is required to understand the functioning of these proteins.

Haem toxicity

Haem toxicity resulting from lipid and DNA damage was observed in several species. Although the role of haem oxygenase in protection against haem toxicity has not been tested, two haem resistance mechanisms have been characterized. In Gram-negative species, haem toxicity is highly increased in strains lacking the HemS type chaperones (Stojiljkovic and Hantke, 1994). HemS proteins are found only in species able to take up exogenous haem. In two Gram-positive species, *S. aureus* and *B. anthracis*, haem resistance involves an ABC efflux pump consisting of HtrA, the ABC protein and HtrB, the membrane component. The *htrAB* operon is regulated by a two-component system, HssRS, which senses haem (see

Chapter 13). Haem binds to the PAS domain of HssS kinase, which phosphorylates and thus activates the positive regulator HssR (Torres *et al.*, 2007). It is currently not clear as to whether efflux pumps exist in Gram-negative species and haem chaperones in Gram-positive species.

Haem uptake and efflux regulation

Negative regulation by DxtR/Fur metalloproteins

Haem uptake and haem degradation genes, like most genes involved in iron acquisition, are negatively regulated by metalloproteins such as Fur and DxtR using iron II as a corepressor, preventing iron overload. In addition to iron homeostasis, Fur and DxtR control virulence genes such as extracellular toxins. DxtR and Fur proteins do not share sequence similarity, but they do share substantial structural similarities (Lee and Helmann, 2007). Both proteins are homodimers regardless of the presence or absence of iron. They have an N-terminal DNA binding domain and a C-terminal metal binding domain responsible for dimerization. The Fur and DxtR DNA binding sites have palindromic AT rich 19 bp conserved sequences.

Fur is a histidine-rich protein and binds haem with significant affinity (Smith *et al.*, 1996). It is not known whether this binding has any relevant physiological function.

Haem-positive regulation by two-component systems

When iron is abundant, expression of haem uptake and degradation genes are repressed by Fur or DxtR iron-loaded repressors. When iron is limiting, transcription of haem uptake and metabolism genes is in several cases activated by haem. In *C. diphteriae*, this activation is mediated by a two-component system constituted by ChrS, a membrane-associated sensor with histidine kinase activity, and a response regulator, ChrA. Haem binding activates ChrS autophosphorylation and phosphoryl transfer to ChrA (Schmitt, 1999). Phosphorylated ChrA activates haem oxygenase gene transcription and represses haem biosynthetic genes (Bibb *et al.*, 2007). In Gram-positive bacteria, the histidine kinase sensor protein is surface-exposed and thus enables detection of extracellular haem.

Regulation by extracytoplasmic sigma factors

In Gram-negative bacteria, in which haem is not in direct contact with TCS sensors, regulation by haem is achieved via a mechanism involving haem outer membrane receptors. Haem binding to its cognate outer membrane receptor releases an extracytoplasmic sigma factor (ECF) from anti-sigma inhibition and thereby leads to transcription of target operons. This type of regulation has been most extensively described in the *E. coli* ferric dicitrate uptake system, but is also found in other siderophore and haem uptake systems (Braun *et al.*, 2003). In all thus so far described systems, the receptors which promote signal transduction belong to the TBDT family. In addition, they have an N-terminal extension located in the periplasm that is required for signalling but not for transport.

In *Bordetella* species (se Chapter 4), ECF signalling is mediated by haem binding to its receptor (Vanderpool and Armstrong, 2003.). In the *S. marcescens* haemophore-dependent haem acquisition system, holo-HasA, but not haem, is the signalling molecule (Rossi *et al.*, 2003). The inducing stimulus can be provided by two independent molecules together: haem and HasA in the apo form. Residues 51–55 of HasA, which contribute to one of the two HasA binding sites, are necessary for induction. These experiments also showed that induction could occur without empty haemophore ejection, whereas haem uptake requires this step (Cwerman *et al.*, 2006).

Unlike known iron starvation ECF anti-sigma factors, transcription of the anti-sigma gene *hasS* is directly regulated by sigma protein HasI; consequently, it is also subject to the holo-haemophore signaling cascade (Biville *et al.*, 2004). A regulatory network similar to this has been found for the heat shock ECF factor of Xylella *fastidiosa* (da Silva Neto *et al.*, 2007). Such repression of the sigma factor by its anti-sigma factor leads to a controlled level of target operon expression, whereas sigma autoregulation, which is usually found for other ECFs such as heat shock and stress sigma factors, is a means of amplifying the response to a given stimulus. Other regulatory activities involving haem are described in Chapter 3.

Haem metabolism and bacterial virulence

Haem is the most abundant intracellular iron source, in particular for pathogens able to invade vertebrates. Indeed, several reports have demonstrated the importance of haem uptake in virulence. For example, haem uptake was shown to be essential for virulence of uropathogenic *E. coli* and *S. aureus* strains in a murine model (Hagan and Mobley, 2009; Torres *et al.*, 2007). Moreover, in group B streptococci, haem auxotroph pathogens, haem uptake enables the shift from fermentative growth to respiration, and thus infection dissemination (Yamamoto *et al.*, 2005). A more puzzling example is that of *H. ducreyi*, which has three TonB-dependent haem receptors. Two of them are not necessary for virulence, whereas the haemoglobin receptor HgbA was shown to be necessary for virulence in humans (Leduc *et al.*, 2008). In addition, in a swine model of chancroid infection, the HgbA protein was able to generate protective antibodies and might thus be a potential candidate for a vaccine (Afonina *et al.*, 2006). It is likely that the two other haem receptors are not expressed or are not functional *in vivo*.

Nevertheless, since most bacterial pathogens have several surface-exposed functional haem receptors, inactivation of one such receptor generally does not result in loss of haem utilization. In addition, despite their somewhat redundant functions, haem surface receptors share low-level sequence similarities; thus, inhibitors common to several haem receptors might be difficult to design. On the other hand, haem ABC transporters enabling haem uptake through the inner membrane and haem-degrading enzymes releasing iron are better conserved, but are less readily accessible to inhibitors.

Concluding remarks

Haem is present and synthesized in almost all living organisms. The biosynthetic pathway is largely conserved, with only one major discrepancy for the first step of D-aminolaevulinic acid biosynthesis. This reaction is achieved either by the Shemin pathway, i.e. condensation of succinyl-CoA and glycine, or by the C5 pathway starting with the glutamyl-tRNA. The Shemin pathway is present in most non-photosynthetic eukaryotes, whereas the C5 pathway is used by most bacteria, archaea and plants.

The haem degradation pathway by haem-oxygenase into biliverdine and iron is also highly conserved from bacteria to man. However, some bacteria have enzymes belonging to the Dye peroxidase family that retrieve iron from haem without breaking the tetrapyrrol ring. Haem ligation to apocytochromes is achieved by three distinct systems present in all life kingdoms: the first, present in most Gram-negative bacteria and plant mitochondria; the second in Gram-positive bacteria, chloroplasts and archaea; and a third system found in mitochondria of most non-photosynthetic eukaryotes.

On the other hand, haem transport across membranes (import and export) is not achieved by conserved mechanisms. In eukaryotes, haem is uptaken by Hrg proteins which are conserved in vertebrates, but have no homologies with bacterial haem uptake systems.

Haem export systems across the plasma membrane for ligation to cytochromes are poorly characterized in both eukaryotes and prokaryotes.

Haem uptake in bacteria is mediated by a wide range of bacterial proteins that have different sequences, folds and affinity for haem. There is no sequence consensus for the haem binding pocket. Nevertheless, some common features emerge:

1. Haemophores are present in Gram-positive and Gram-negative species. They help to scavenge haem from host haemoproteins. Some of them achieve this capture through their higher affinity for haem, such as *S. marcescens* HasA. For others, haem capture requires a protein–protein interaction, as is the case for IsdX, the newly discovered *B. anthracis* haemophore.
2. In several documented cases, haem transfer between proteins does not require energy and takes place against an affinity gradient. This is the case for haem transfer from haemophore HasA to its

cell surface receptor, and between peptidoglycan-anchored cell surface NEAT proteins. In both examples, transfer requires protein–protein interactions.

3 Haem transport through the inner membrane is an energy-dependent process and requires ABC transporters and haem binding proteins which belong to two classes: either small haem-specific binding proteins or large binding proteins which also accommodate ligands other than haem.

Acknowledgements

I thank Dr Nadia Izadi (Institut Pasteur) and Dr Muriel Masi (Institut Pasteur)from the National Center for Scientific Research, Mixed Unit Research 8619, the University of Paris XT, for helpful advice.

References

Afonina, G., Leduc, I., Nepluev, I., Jeter, C., Routh, P., Almond, G., Orndorff, P., Hobbs, M., and Elkins, C. (2006). Immunization with the *Haemophilus ducreyi* hemoglobin receptor HgbA protects against infection in the swine model of chancroid. Infect. Immun. *74*, 2224–2232.

Ahmer, B.M., Thomas, M.G., Larsen, R.A., and Postle, K. (1995). Characterization of the *exbBD* operon of *Escherichia coli* and the role of ExbB and ExbD in TonB function and stability. J. Bacteriol. *177*, 4742–4747.

Alontaga, A., Rodriguez, J., Schönbrunn, E., Becker, A., Funke, T., Yukl, E., Hayashi, T., Stobaugh, J., Moënne-Loccoz, P., and Rivera, M. (2009). Structural characterization of the hemophore HasAp from *Pseudomonas aeruginosa*: NMR spectroscopy reveals protein–protein interactions between Holo-HasAp and hemoglobin. Biochemistry. *48*, 96–109.

Arnoux, P., Haser, R., Izadi, N., Lecroisey, A., Delepierre, M., Wandersman, C., and Czjzek, M. (1999). The crystal structure of HasA, a hemophore secreted by *Serratia marcescens*. Nature Struct. Biol. *6*, 516–520.

Barjon, C., Wecker, K., Izadi-Pruneyre, N., and Delepelaire, P. (2007). Mutagenesis and molecular modeling reveal three key extracellular loops of the membrane receptor HasR that are involved in hemophore HasA binding. J. Bacteriol. *189*, 5379–5382.

Beasley, F., Vinés, E., Grigg, J., Zheng, Q., Liu, S., Lajoie, G., Murphy, M., and Heinrichs, D. (2009). Characterization of staphyloferrin A biosynthetic and transport mutants in *Staphylococcus aureus*. Mol. Microbiol. *72*, 947–963.

Benevides-Matos, N., Wandersman, C., and Biville, F. (2008). HasB, the *Serratia marcescens* TonB paralog, is specific to HasR. J. Bacteriol. *190*, 21–27.

Bhakta, M., and Wilks, A. (2006). The mechanism of heme transfer from the cytoplasmic heme binding protein PhuS to the delta-regioselective heme oxygenase of *Pseudomonas aeruginosa*. Biochemistry. *45*, 11642–11649.

Bibb, L., Kunkle, C., and Schmit, T.M. (2007). The ChrA-ChrS and HrrA-HrrS signal transduction systems are required for activation of the *hmuO* promoter and repression of the *hemA* promoter in *Corynebacterium diphtheriae*. Infect. Immun. *75*, 2421–2431.

Biville, F., Cwerman, H., Létoffé, S., Rossi, M., Drouet, V., Ghigo, J., and Wandersman, C. (2004). Hemophore-mediated signaling in *Serratia marcescens:* a new mode of regulation for an Extra Cytoplasmic Function (ECF) sigma factor involved in heme-acquisition. Mol. Microbiol. *53*, 1267–1277.

Block, D., Lukat-Rodgers, G., Rodgers, K., Wilks, A., Bhakta, M., and Lansky, I. (2007). Identification of two heme-binding sites in the cytoplasmic heme-trafficking protein PhuS from *Pseudomonas aeruginosa* and their relevance to function. Biochemistry. *46*, 14391–14402.

Bracken, C.S., Baer, M.T., Abdur-Rashid, A., Helms, W., and Stojiljkovic, I. (1999). Use of heme–protein complexes by the *Yersinia enterocolitica* HemR receptor: histidine residues are essential for receptor function. J. Bacteriol. *181*, 6063–6072.

Braun, V. (2006). Energy transfer between biological membranes. ACS Chem. Biol. *1*, 352–354.

Braun, V., Mahren, S., and Ogierman, M. (2003). Regulation of the FecI-type ECF sigma factor by transmembrane signalling. Curr. Opin. Microbiol.6, 173–180.

Cao, J., Woodhall, M., Alvarez, J., Cartron, M., and Andrews, S. (2007). EfeUOB (YcdNOB) is a tripartite, acid-induced and CpxAR-regulated, low-pH Fe^{2+} transporter that is cryptic in *Escherichia coli* K-12 but functional in *E. coli* O157:H7. Mol. Microbiol. *65*, 857–875.

Cascales, E., Lloubes, R., and Sturgis, J. (2001). The TolQ-TolR proteins energize TolA and share homologies with the flagellar motor proteins MotA-MotB. Mol. Microbiol. *42*, 795–807.

Cescau, S., Cwerman, H., Létoffé, S., Delepelaire, P., Wandersman, C., and Biville, F. (2007). Heme acquisition by hemophores. Biometals *20*, 603–613.

Chen, J., Sharma, S., Quiocho, F.A., and Davidson, A.L. (2001). Trapping the transition state of an ATP-binding cassette transporter: evidence for a concerted mechanism of maltose transport. Proc. Natl. Acad. Sci. U.S.A. *98*, 1525–1530.

Chu, B., Peacock, R., and Vogel, H. (2007). Bioinformatic analysis of the TonB protein family. Biometals *20*, 467–483.

Cope, L., Love, R., Guinn, S., Gilep, A., Usanov, S., Estabrook, R., Hrkal, Z., and Hansen, E. (2001). Involvement of HxuC outer membrane protein in utilization of hemoglobin by *Haemophilus influenzae*. Infect. Immun. *69*, 2353–2363.

Cope, L., Thomas, S., Hrkal, Z., and Hansen, E. (1998). Binding of heme–hemopexin complexes by soluble HxuA protein allows utilization of this complexed heme by *Haemophilus influenzae*. Infect. Immun. *66*, 4511–4516.

Cuív, P., Keogh, D., Clarke, P., and O'Connell, M. (2008). The *hmuUV* genes of *Sinorhizobium meliloti* 2011 encode the permease and ATPase components of an ABC transport system for the utilization of both haem and the hydroxamate siderophores, ferrichrome and ferrioxamine B. Mol. Microbiol. *70*, 1261–1273.

Cwerman, H., Wandersman, C., and Biville, F. (2006). Heme and a five-amino-acid hemophore region form the bipartite stimulus triggering the Has signaling cascade. J. Bacteriol. *188*, 3357–3364.

da Silva Neto, J., Koide, T., Gomes, S., and Marques, M. (2007). The single extracytoplasmic-function sigma factor of *Xylella fastidiosa* is involved in the heat shock response and presents an unusual regulatory mechanism. J. Bacteriol. *189*, 551–560.

Dautin, N., and Bernstein, H. (2007). Protein secretion in gram-negative bacteria via the autotransporter pathway. Annu. Rev. Microbiol. *61*, 89–112.

Dawson, R., Hollenstein, K., and Locher, K. (2007). Uptake or extrusion: crystal structures of full ABC transporters suggest a common mechanism. Mol. Microbiol. *65*, 250–257.

Drago-Serrano, M., Parra, S., and Manjarrez-Hernández, H. (2006). EspC, an autotransporter protein secreted by enteropathogenic *Escherichia coli* (EPEC), displays protease activity on human hemoglobin. FEMS Microbiol. Lett. *265*, 35–40.

Frankenberg-Dinkel, N. (2004). Bacterial heme oxygenases. Antioxid. Redox Signal. *5*, 825–834.

Ghigo, J.M., Létoffé, S., and Wandersman, C. (1997). A new type of hemophore-dependent heme acquisition system of *Serratia marcescens* reconstituted in *Escherichia coli*. J. Bacteriol. *179*, 3572–3579.

Gray, L., Baker, K., Kenny, B., Mackman, N., Haigh, R., and Holland, I.B. (1989). A novel C-terminal signal sequence targets *Escherichia coli* haemolysine directly to the medium. J. Cell. Sci. Suppl. *11*, 45–57.

Grigg, J., Vermeiren, C., Heinrichs, D., and Murphy, M. (2007). Haem recognition by a *Staphylococcus aureus* NEAT domain. Mol. Microbiol. *63*, 139–149.

Grosse, C., Scherer, J., Koch, D., Otto, M., Taudte, N., and Grass, G. (2006). A new ferrous iron-uptake transporter, EfeU (YcdN), from *Escherichia coli*. Mol. Microbiol. *62*, 120–131.

Hagan, E., and Mobley, H.L. (2009). Haem acquisition is facilitated by a novel receptor Hma and required by uropathogenic *Escherichia coli* for kidney infection. Mol. Microbiol. *71*, 79–91.

Hanson, M., Slaughter, C., and Hansen, E. (1992). The *hbpA* gene of *Haemophilus influenzae* type b encodes a heme-binding lipoprotein conserved among heme-dependent *Haemophilus* species. Infect. Immun. *60*, 2257–2266.

Izadi-Pruneyre, N., Huche, F., Lukat-Rodgers, G., Lecroisey, A., Gilli, R., Rodgers, K., Wandersman, C., and Delepelaire, P. (2006). The heme transfer from the soluble HasA hemophore to its membrane bound receptor HasR is driven by protein–protein interaction from a high to a lower affinity binding site. J. Biol. Chem. *281*, 25541–25550.

Koronakis, V., Sharff, A., Koronakis, E., Luisi, B., and Hughes, C. (2000). Crystal structure of the bacterial membrane protein TolC central to multidrug efflux and protein export. Nature *405*, 914–919.

Krieg, S., Huché, F., Diederichs, K., Izadi-Pruneyre, N., Lecroisey, A., Wandersman, C., Delepelaire, P., and Welte, W. (2009). Heme uptake across the outer membrane as revealed by crystal structures of the receptor–hemophore complex. Proc. Natl. Acad. Sci. U.S.A. *106*, 1045–1050.

Larsson, J., Allhorn, M., and Kerström, B. (2004). The lipocalin alpha(1)-microglobulin binds heme in different species. Arch. Biochem. Biophys. *432*, 196–204.

Leduc, I., Banks, K., Fortney, K., Patterson, K., Billings, S., Katz, B., Spinola, S., and Elkins, C. (2008). Evaluation of the repertoire of the TonB-dependent receptors of *Haemophilus ducreyi* for their role in virulence in humans. J. Infect. Dis. *197*, 1103–1109.

Lee, J., and Helmann, J. (2007). Functional specialization within the Fur family of metalloregulators. Biometals *20*, 485–499.

Lefèvre, J., Delepelaire, P., Delepierre, M., and Izadi-Pruneyre, N. (2008). Modulation by substrates of the interaction between the HasR outer membrane receptor and its specific TonB-like protein, HasB. J. Mol. Biol. *378*, 840–851.

Létoffé S., Heuck, G., Delepelaire, P., Lange, N., and Wandersman, C. (2009). Bacteria capture iron from heme by keeping tetrapyrrol skeleton intact. Proc. Natl. Acad. Sci. U.S.A. *106*, 11719–11724.

Létoffé, S., Debarbieux, L., Izadi, N., Delepelaire, N., and Wandersman, C. (2003). Ligand delivery by heme carrier proteins: the binding of *Serratia marcescens* hemophore to its outer membrane receptor is mediated by two distinct peptide regions. Mol. Microbiol. *50*, 77–88.

Létoffé, S., Delepelaire, P., and Wandersman, C. (2004). Free and hemophore-bound heme acquisitions through the outer membrane receptor HasR have different requirements for the TonB–ExbB-ExbD complex. J. Bacteriol. *186*, 4067–4074.

Létoffé, S., Delepelaire, P., and Wandersman, C. (2006). The housekeeping dipeptide permease is the *Escherichia coli* heme transporter and functions with two optional peptide binding proteins. Proc. Natl. Acad. Sci. U.S.A. *103*, 12891–12896.

Létoffé, S., Deniau, C., Wolff, N., Dassa, E., Delepelaire, P., Lecroisey, A., and Wandersman, C. (2001). Haemophore-mediated bacterial haem transport: evidence for a common or overlapping site for haem-free and haem-loaded haemophore on its specific outer membrane receptor. Mol. Microbiol. *41*, 439–450.

Liu, M., Tanaka, W., Zhu, H., Xie, G., Dooley, D., and Lei, B. (2008). Direct hemin transfer from IsdA to IsdC in the iron-regulated surface determinant (Isd) heme acquisition system of *Staphylococcus aureus*. J. Biol. Chem. *283*, 6668–6676.

Maresso, A., Garufi, G., and Schneewind, O. (2008). *Bacillus anthracis* secretes proteins that mediate heme acquisition from hemoglobin. PLoS Pathog. *4*, e 1000132.

Maresso, A., and Schneewind, O. (2006). Iron acquisition and transport in *Staphylococcus aureus*. Biometals *19*, 193–203.

Mazmanian, S., Skaar, E., Gaspar, A., Humayun, M., Gornicki, P., Jelenska, J., Joachmiak, A., Missiakas, D., and Schneewind, O. (2003). Passage of heme-iron across the envelope of *Staphylococcus aureus*. Science. *299*, 906–909.

Mazmanian, S., Ton-That, H., Su, K., and Schneewind, O. (2002). An iron-regulated sortase anchors a class of surface protein during *Staphylococcus aureus* pathogenesis. Proc. Natl. Acad. Sci. U.S.A. *99*, 2293–2298.

Mey, A., and Payne, S. (2001). Haem utilization in *Vibrio cholerae* involves multiple TonB-dependent haem receptors. Mol. Microbiol. *42*, 835–849.

Morton, D., Madore, L., Smith, A., Vanwagoner, T., Seale, T., Whitby, P., and Stull, T. (2005). The heme-binding lipoprotein (HbpA) of *Haemophilus influenzae*: role in heme utilization. FEMS Microbiol. Lett. *253*, 193–199.

Nepluev, I., Afonina, G., Fusco, W., Leduc, I., Olsen, B., Temple, B., and Elkins, C. (2009). An immunogenic, surface exposed domain of *Haemophilus ducreyi* outer membrane protein HgbA is involved in hemoglobin binding. Infect. Immun. *77*, 3065–3074.

Olczak, T., Simpson, W., Liu, X., and Genco, C. (2005). Iron and heme utilization in *Porphyromonas gingivalis*. FEMS Microbiol. Rev. *29*, 119–144.

Paquelin, A., Ghigo, J.M., Bertin, S., and Wandersman, C. (2001). Characterization of HasB, a *Serratia marcescens* TonB-like protein specifically involved in the haemophore-dependent haem acquisition system. Mol. Microbiol. *42*, 995–1005.

Park, J., Raychaudhuri, D., Li, H., Normark, S., and Mengin-Lecreulx, D. (1998). MppA, a periplasmic binding protein essential for import of the bacterial cell wall peptide L-alanyl-gamma-D-glutamyl-meso-diaminopimelate. J. Bacteriol. *180*, 1215–1223.

Pawelek, P., Croteau, N., Ng-Thow-Hing, C., Khursigara, C., Moiseeva, N., Allaire, M., and Coulton, J. (2006). Structure of TonB in complex with FhuA, *E. coli* outer membrane receptor. Science. *312*, 1399–1402.

Perkins-Balding, D., Baer, M.T., and Stojiljkovic, I. (2003). Identification of functionally important regions of a haemoglobin receptor from *Neisseria meningitidis*. Microbiology *149*, 3423–3435.

Pilpa, R., Robson, S., Villareal, V., Wong, M., Phillips, M., and Clubb, R.T. (2009). Functionally distinct NEAT (NEAr Transporter) domains within the *Staphylococcus aureus* IsdH/HarA protein extract heme from methemoglobin. J. Biol. Chem. *284*, 1166–1176.

Pluym, M., Vermeiren, C., Mack, J., Heinrichs, D., and Stillman, M.J. (2007). Heme binding properties of *Staphylococcus aureus* IsdE. Biochemistry. *46*, 12777–12787.

Puri, S., and O'Brian, M. (2006). The *hmuQ* and *hmuD* genes from *Bradyrhizobium japonicum* encode heme-degrading enzymes. J. Bacteriol. *188*, 6476–6482..

Richardson, A.R., and Stojiljkovic, I. (1999). HmbR, a hemoglobin-binding outer membrane protein of *Neisseria meningitidis*, undergoes phase variation. J. Bacteriol. *181*, 2067–2074.

Rohde, K., and Dyer, D. (2004). Analysis of haptoglobin and hemoglobin–haptoglobin interactions with the *Neisseria meningitidis* TonB-dependent receptor HpuAB by flow cytometry. Infect. Immun. *72*, 2494–2506.

Rossi, M., Paquelin, A., Ghigo, J., and Wandersman, C. (2003). Haemophore-mediated signal transduction across the bacterial cell envelope in *Serratia marcescens*: the inducer and the transported substrate are different molecule. Mol. Microbiol. *48*, 1467–1480.

Schauer, K., Rodionov, D., and de Reuse, H. (2008). New substrates for TonB-dependent transport: do we only see the 'tip of the iceberg'? Trends Biochem. Sci. *33*, 330–338.

Schmitt, M.P. (1999). Identification of a two-component signal transduction system from *Corynebacterium diphtheriae* that activates gene expression in response to the presence of heme and hemoglobin, J. Bacteriol. *181*, 5330–5340.

Schneider, S., Sharp, K., Barker, P., and Paoli, M. (2006). An induced fit conformational change underlies the binding mechanism of the heme transport proteobacteria-protein HemS. J. Biol. Chem. *281*, 32606–32610.

Sharp, K., Schneider, S., Cockayne, A., and Paoli, M. (2007). Crystal structure of the heme–IsdC complex, the central conduit of the Isd iron/heme uptake system in *Staphylococcus aureus*. J. Biol. Chem. *282*, 10625–10631.

Shepherd, M., Heath, M., and Poole, R. (2007). NikA binds heme: a new role for an *Escherichia coli* periplasmic nickel-binding protein. Biochemistry. *46*, 5030–5037.

Shultis, D., Purdy, M., Banchs, C., and Wiener, M. (2006). Outer membrane active transport: structure of the BtuB:TonB complex. Science. *312*, 1396–1399.

Skaar, E., Gaspar, A., and Schneewind, O. (2006). *Bacillus anthracis* IsdG, a heme-degrading monooxygenase. J. Bacteriol. *188*, 1071–1080..

Smith, A., Hooper, N., Shipulina, N., and Morgan, W. (1996). Heme binding by a bacterial repressor protein, the gene product of the ferric uptake regulation (*fur*) gene of *Escherichia coli*. J. Protein. Chem. *15*, 575–583.

Stojiljkovic, I., and Hantke, K. (1994). Transport of haemin across the cytoplasmic membrane through a haemin-specific periplasmic binding-protein-dependent transport system in *Yersinia enterocolitica*. Mol. Microbiol. *13*, 719–732.

Suits, M., Jaffe, N., and Jia, Z. (2006). Structure of the *Escherichia coli* O157:H7 heme oxygenase ChuS in complex with heme and enzymatic inactivation by mutation of the heme coordinating residue His-193. J. Biol. Chem. *281*, 36776–36782.

Suits, M., Pal, G., Nakatsu, K., Matte, A., Cygler, M., and Jia, Z. (2005). Identification of an *Escherichia coli* O157:H7 heme oxygenase with tandem functional repeats. Proc. Natl. Acad. Sci. U.S.A. *102*, 16955–16960..

Terashima, H., Kojima, S., and Homma, M. (2008). Flagellar motility in bacteria structure and function of flagellar motor. Int. Rev. Cell Mol. Biol. *270*, 39–85.

Torres, V., Stauff, D., Pishchany, G., Bezbradica, J., Gordy, L., Iturregui, J., Anderson, K., Dunman, P., Joyce, S., and Skaar, E. (2007). A *Staphylococcus aureus* regulatory system that responds to host heme and modulates virulence. Cell Host Microbe. *1*, 109–119.

Vanderpool, C.K., and Armstrong, S.K. (2003). Heme-responsive transcriptional activation of *Bordetella bhu* genes. J. Bacteriol. *185*, 909–917.

Wandersman, C., and Delepelaire, P. (1990). TolC, an *Escherichia coli* outer membrane protein required for hemolysin secretion. Proc. Natl. Acad. Sci. U.S.A. *87*, 4776–4780.

Watanabe, M., Tanaka, Y., Suenaga, A., Kuroda, M., Yao, M., Watanabe, N., Arisaka, F., Ohta, T., Tanaka, I., Tsumoto, K., *et al.* (2008). Structural basis for multimeric heme complexation through a specific protein–heme interaction: the case of the third neat domain of IsdH from *Staphylococcus aureus*. J. Biol. Chem. *283*, 28649–28659.

Wójtowicz, H., Guevara, T., Tallant, C., Olczak, M., Sroka, A., Potempa, J., Solà, M., Olczak, T., and Gomis-Rüth, F. (2009). Unique structure and stability of HmuY, a novel heme-binding protein of *Porphyromonas gingivalis*. PLoS Pathog. *5*, e 1000419.

Wolff, N., Izadi-Pruneyre, N., Couprie, J., Habeck, M., Linge, J., Rieping, W., Wandersman, C., Nilges, M., Delepierre, M., and Lecroisey, A. (2008). Comparative analysis of structural and dynamic properties of the loaded and unloaded hemophore HasA: functional implications. J. Mol. Biol. *376*, 517–525.

Wyckoff, E., Duncan, D., Torres, A., Mills, M., Maase, K., and Payne, S. (1998). Structure of the *Shigella dysenteriae* haem transport locus and its phylogenetic distribution in enteric bacteria. Mol. Microbiol. *28*, 1139–1152.

Yamamoto, Y., Poyart, C., Trieu-Cuot, P., Lamberet, G., Gruss, A., and Gaudu, P. (2005). Respiration metabolism of group B *Streptococcus* is activated by environmental haem and quinone and contributes to virulence. Mol. Microbiol. *56*, 525–534.

Zgurskaya, H.I., and Nikaido, H. (1999). Bypassing the periplasm: reconstitution of the AcrAB multidrug efflux pump of *Escherichia coli*. Proc. Natl. Acad. Sci. U.S.A. *96*, 7190–7195.

Mechanisms and Regulation of Iron Homeostasis in the Rhizobia

3

Mark R. O'Brian and Elena Fabiano

Abstract

The rhizobia live as free-living soil bacteria or in symbiosis with leguminous plants. The success of these organisms in each milieu involves the ability to sense the environment to assess the availability of nutrients, and to optimize cellular systems for their acquisition. Iron in the rhizosphere is mostly inaccessible due to low solubility, and microorganisms must compete for this limited nutrient. Rhizobia belong to the α-Proteobacteria, a diverse taxonomic group that includes numerous species that form close or intracellular associations with eukaryotic hosts in a symbiotic or pathogenic context. Thus, in addition to their agricultural and economic importance, rhizobia are model organisms that have given new insights into related, but less tractable animal pathogens. In particular, genetic control of iron homeostasis in the rhizobia and other α-Proteobacteria has moved away from the Fur paradigm to an iron sensing mechanism responding to the metal indirectly. Moreover, utilization of haem as an iron source is not unique to animal pathogens, but is an acquisition strategy employed by the rhizobia with some interesting novel features. This chapter will review recent advances in our understanding of iron metabolism in the rhizobia.

Introduction

General considerations

The rhizobia are a diverse group of Gram-negative soil bacteria that can form a symbiosis with leguminous plants. In general, a rhizobial species recognizes one or a few legume hosts; the molecular basis for the specificity between the bacterium and host, and the events leading to the symbiotic state are understood in considerable detail (Downie and Walker, 1999; Gibson *et al.*, 2008; Oke and Long, 1999; Oldroyd and Downie, 2008; Spaink, 2000). Rhizobia elicit the formation of a symbiotic organ called a root nodule comprised of differentiated plant and bacterial cells. In this context, rhizobia are endosymbionts within the plant cells of the nodule.

Differentiated rhizobia within nodules are termed bacteroids, and acquire the ability to convert atmospheric nitrogen to ammonia by a process called nitrogen fixation. Nitrogen fixed by the endosymbiont is exported out of the cell in the form of ammonium, and subsequently assimilated by the plant host to fulfil its nutritional nitrogen requirement. In return, the plant provides the bacteroids with a carbon source ultimately derived from photosynthesis. The rhizobia–legume symbioses account for nearly half of the global biological nitrogen fixation (Gruber and Galloway, 2008), and is therefore an important agricultural and environmental process. Moreover, the rhizobia are phylogenetically related to several plant and animal pathogens, and serve as a model to understand the molecular basis of bacterial–eukaryote interactions.

Rhizobia include the genera *Rhizobium*, *Bradyrhizobium*, *Sinorhizobium*, *Mesorhizobium*, *Azorhizobium* and *Allorhizobium*. These genera belong to the α-Proteobacterial subdivision of the purple bacteria, an extremely diverse group that includes pathogens, symbionts, photosynthetic organisms, bacteria that degrade environmental pollutants, and the abundant marine organism *Pelagibacter ubique* (Ettema and Andersson, 2009). The bacterial ancestor of mitochondria belongs to this group as well. More recently, numerous β-Proteobacterial species have been also identified that form symbiosis with legumes (Chen *et al.*, 2003; Moulin *et al.*, 2001). Although these bacteria are also referred to as rhizobia, they are phylogenetically distinct from the α-Proteobacterial species, and are not considered further here.

In recent years, studies on rhizobial iron metabolism have focused to a large extent on three species: *Rhizobium leguminosarum*, *Sinorhizobium meliloti* and *Bradyrhizobium japonicum*. *R. leguminosarum* and *S. meliloti* belong to the family Rhizobiaceae, whereas *Bradyrhizobium* belongs to the *Bradyrhizobiaceae*. These two families diverged approximately 500 million years ago (Turner and Young, 2000), and thus it is not unexpected that important differences between them have been observed with respect to iron metabolism.

The problem of iron acquisition

As free-living organisms, rhizobia must compete with other soil microbiota for nutrients and other resources, including iron. Iron is an abundant element, but it is primarily in the oxidized form, which has an extremely low solubility in aqueous environments at neutral pH. As described below, bacteria have developed numerous strategies to acquire iron from the environment, including highjacking other microbial iron acquisition systems. At present, much more is known about iron acquisition and regulation in free-living rhizobia than in symbiosis.

Bacteroids within a functional nodule are surrounded by the peribacteroid membrane, a plant structure which separates the bacteria from the host within the plant cell. The peribacteroid membrane encompasses one or many bacteroids, depending on the type of nodule formed by the plant, and this structure is called a symbiosome. This means that the supply of iron and other nutrients to the bacteria is ultimately controlled by the plant. Whereas bacteria in the rhizosphere probably experience iron stress due to the low solubility of ferric iron, it cannot be assumed that the symbiotic prokaryote is iron-limited despite the large demand. At present, the mode of transport of iron into bacteroids is not known, nor is it known whether a high affinity transporter is necessary. Evidence for an iron pool in the peribacteroid space of soybean nodules could suggest that the bacteroids perceive a high iron environment (LeVier *et al.*, 1996), but none of these issues have been resolved.

Roles of iron in symbiosis

Iron is a major constituent of two crucial proteins for the N_2-fixing symbiosis process, bacterial nitrogenase and plant leghaemoglobin. Moreover, other iron-containing proteins contribute to the N_2 fixation process are induced symbiotically, such as rhizobial hydrogenase and cytochromes that allow respiration in the low-oxygen milieu of the nodule (Baginsky *et al.*, 2002; Hennecke, 1992; Sangwan and O'Brian, 1992).

The nitrogenase complex contains about 34 atoms of iron. It comprises the iron protein (the nucleotide binding and electron-donating element) and the iron–molybdenum protein which contains the N_2-reducing site. The iron protein is a homodimer of two subunits that coordinate one [4Fe–4S] cluster. The iron–molybdenum protein is a tetramer of two polypeptides ($\alpha_2\beta_2$) in which the α subunit contains a [Mo–7Fe–9S-homocitrate] cluster and the β subunit a [8Fe–7S] cluster (Georgiadis *et al.*, 1992; Rees and Howard, 2000). Nitrogenase constitutes about 30% of the soluble proteins (11% of total proteins) in the bacteroid (Verma and Nadler, 1984).

In N_2-fixing nodules, leghaemoglobin can represent about 20% of total nodule iron (1–3 mg of leghaemoglobin per gram of fresh weight depending on plant species) (Dilworth, 1980). As previously mentioned, once in the plant cytosol, differentiated bacteroids are enveloped in a peribacteroid membrane and immersed in a leghaemoglobin-rich cytosol. It is interesting to note that peribacteroid membrane is one of the first structures to be degraded during the nodule senescence (Herrada *et al.*, 1993). In this scenario, it is possible that that iron requirement is greater for nodulated plants than for host plants alone. In fact, it has been reported that iron deficiency negatively affects the rhizobia-legume symbiosis. Application of the metal to iron-stressed plants increases nodule number and plant mass. Furthermore, *S. meliloti* iron-starved cells are less competitive for nodulation, and iron deficiency prevents nodule initiation and limits nodule development (Battistoni *et al.*, 2002b; Expert and Gill, 1992; O'Hara *et al.*, 1988; Tang *et al.*, 1990).

Iron transport systems

Siderophore-mediated iron transport

A common strategy used by many bacteria to obtain iron from their environment when it is scarce involves the production and secretion of ferric chelating compounds termed siderophores (Greek for iron carriers) (Lankford, 1973; Neilands, 1973, 1981; Ratledge and Dover, 2000). Siderophores are low molecular weight compounds (around 300–1500 Da) that bind iron specifically and with high affinity (see also Chapter 1). The secreted siderophore, bound to iron, is then taken up by the cell through a ferric siderophore transport system. In Gram-negative bacteria, translocation of ferric–siderophore complexes across the outer membrane requires energized transport via outer membrane receptors. The energy is provided by the proton motive force, and it is transduced to outer membrane receptors by the TonB–ExbB–ExbD complex (Postle and Kadner, 2003).

Although siderophores differ widely in their overall structure, variation in metal-binding groups is more limited. According to the iron-chelating group, siderophores are classified as catecholates, hydroxamates, α-hydroxycarboxylates or mixed siderophores. Although these are the most common groups, other metal binding groups have been identified, for example oxazoline, thiazoline, hydroxypyridine and β-hydroxyacids. Siderophores can be also characterized according their biosynthetic mechanism as non-ribosomal peptide synthetase (NRPS)-dependent or NRPS-independent (Challis, 2005; Donadio *et al.*, 2007; Raymond and Dertz, 2004).

Three rhizobial endogenous siderophores have been chemically characterized to date (Fig. 3.1):

1 vicibactins, trihydroxamate NRPS-dependent siderophores synthesized by *R. leguminosarum*;
2 rhizobactin, an α-hydroxycarboxylate siderophore produced by *S. meliloti* strain DM;
3 rhizobactin 1021, a citrate-derivative dihydoxamate and NRPS-independent siderophore synthesized by *S. meliloti* strain 1021 (Carter *et al.*, 2002; Dilworth *et al.*, 1998; Persmark *et al.*, 1993; Smith *et al.*, 1985).

Siderophore biosynthesis has been studied in *R. leguminosarum* 8401(pRL1JI) and *S. meliloti* 1021. In both cases biosynthetic genes are located on plasmids and they are clustered close to the genes encoding their respective outer membrane receptors (Fig. 3.2). In addition, some *R. leguminosarum* species produce trihydroxymates, while some sinorhizobia produce dihydroxamates or α-hydroxycarboxylates although chemical structures for them are unavailable (Carson *et al.*, 2000). Catechol-like siderophore production has been reported in *R. leguminosarum* IARI 102 and *Rhizobium sp* (cowpea) RA-1 (Modi *et al.*, 1985; Patel *et al.*, 1988). Unfortunately, none of these structures has been resolved. Anthranilate, a precursor of tryptophan, has been reported to promote the uptake of ferric iron into iron-starved cells in *R. leguminosarum* (Rioux *et al.*, 1986). Anthranilate was synthesized and transported in cultures grown with iron, and therefore is not a specific response to iron deficiency. Anthranilate synthesis mutants of *S. meliloti* form defective nodules that cannot be accounted for by a tryptophan auxotrophy since downstream mutants in the tryptophan pathway form effective nodules (Barsomian *et al.*, 1992). Interestingly, the *B. japonicum* 110 genome is a single chromosome with no plasmids, and no endogenous siderophore has been identified in that species. Nevertheless, *B. japonicum* does have genes encoding siderophore receptors, as discussed below.

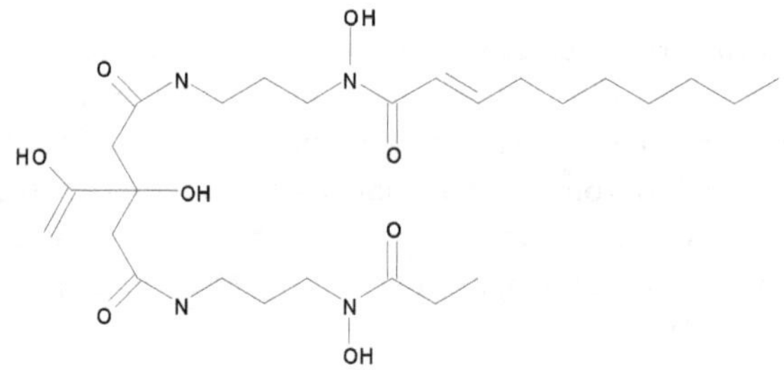

Rhizobactin 1021 (siderophore produced by *S. meliloti* 1021)

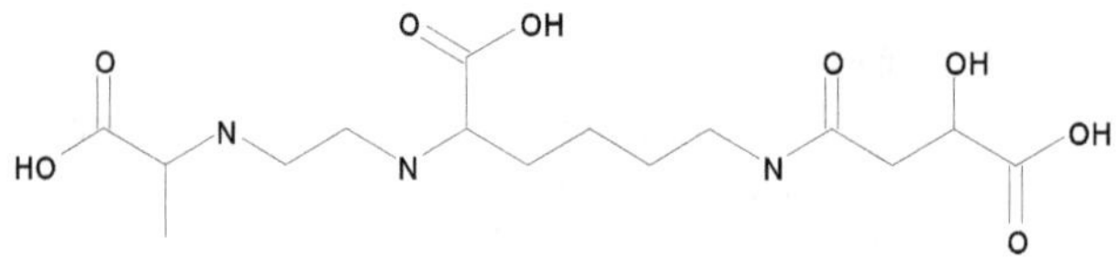

Rhizobactin (siderophore produced by *S. meliloti* DM4)

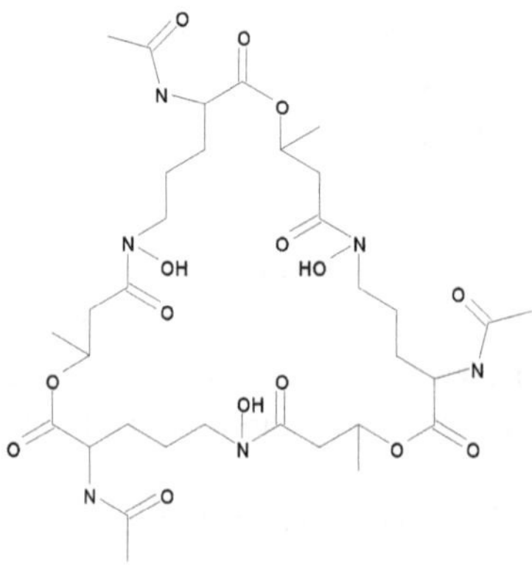

Vicibactin (siderophore produced by *R. leguminosarum* WSM710)

Figure 3.1 Chemical structures of three characterized siderophores produced by rhizobia.

Vicibactin

Vicibactin, the siderophore made by *R. leguminosarum,* is a cyclic trihydroxamate containing three residues each of D-3-hydroxybutanoic acid and *N*2-acetyl-*N*5-hydroxy-D-ornithine arranged in alternate ester and peptide bonds (Dilworth *et al.,* 1998) (Fig. 3.1). Vicibactin was identified from three different *R. leguminosarum* strains, WSM709, WSM710 and WU235. The sequenced *R. leguminosarum* 3841 strain also contains genes presumably involved in vicibactin biosynthesis located on plasmid pRL12 and close to the putative vicibactin receptor gene (Fig. 3.2).

Vicibactin biosynthesis involves a NRPS gene homologue, *vbsS* (Carter *et al.,* 2002). NRPSs are assembly lines of specialized domains that form peptide bonds independently of ribosomes (Donadio *et al.,* 2007; Finking and Marahiel, 2004). NRPSs use amino or hydroxy-acids as building blocks, catalysing the formation of amide or ester bonds, respectively. Each NRPS module consists of three basic domains: an adenylation domain (A) that activates the selected amino acid as an acyl adenylate, a peptide carrier (PCP) or T domain responsible for the formation of the thioester bond and, a condensation domain (C) which catalyses peptide bond formation between two activated amino acids elongating the peptide and translocating it to the PCP domain. These three domains form a module, and NRPSs are consecutive

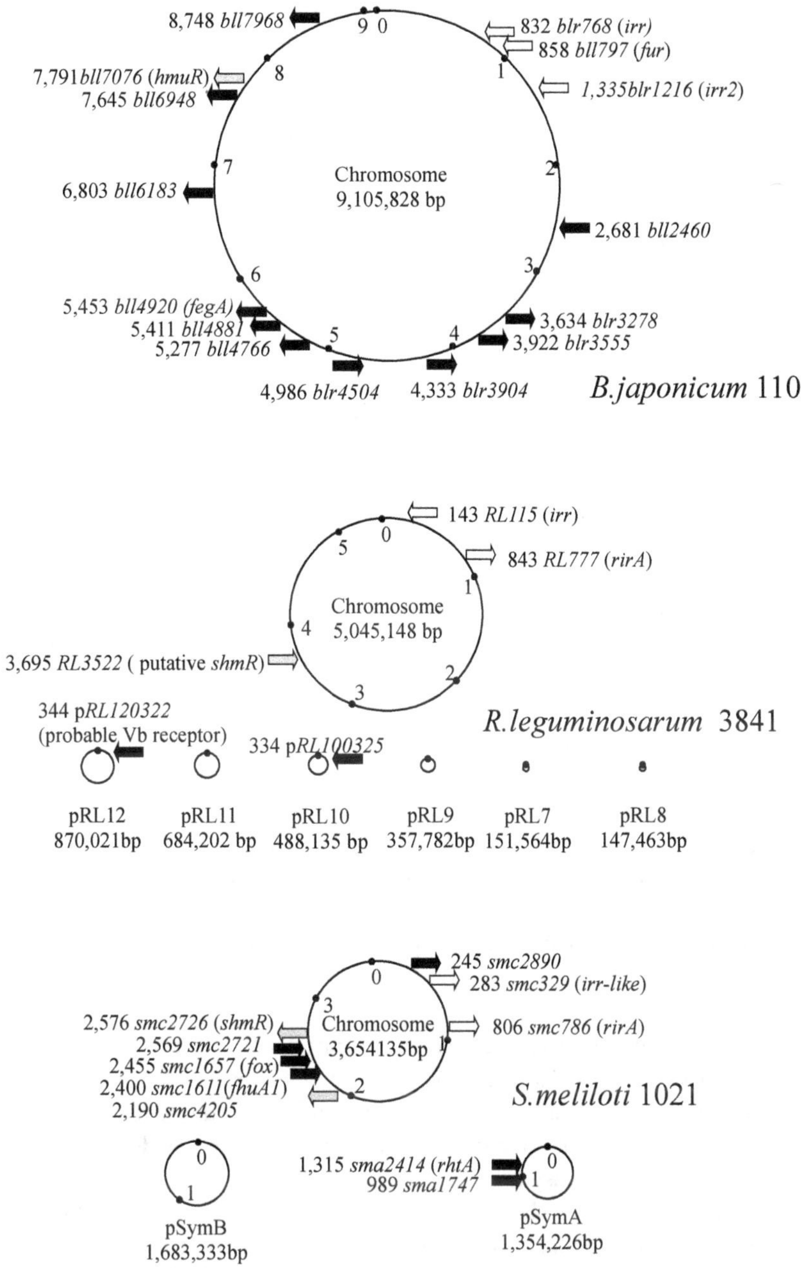

Figure 3.2 Genetic maps of genes coding putative TonB-dependent receptors and iron regulators in *B. japonicum* 110, *R. leguminosarum* 3841 and *S. meliloti* 1021. Genes coding haem receptors are shown as grey arrows, other putative TonB-dependent receptor genes as black arrows and iron regulator genes as empty arrows. The directions of transcription are indicated by the arrows. The distances in Mbp from the origins of replication are indicated as dots. The approximate locations of the genes are shown in kbp before the name of the genes. Sizes of chromosomes are indicated into the circles, and sizes and names of plasmids at the bottom of each plasmid.

tandems of these modules with each module being responsible for the addition of a new amino acid. Modules are collinear with the amino acids they link. A final module containing a thioesterase (TE) domain, instead of the C domain, completes the peptide formation. Additional activities may be catalysed by specialized domains such as cyclization (Cy domain), epimerization (E domain), methylation (M domain) or oxidation or reduction (Donadio *et al.*, 2007; Finking and Marahiel, 2004).

Biosynthesis of vicibactin involves a cluster of eight genes arranged in four operons, *vbsGSO*, *vbsADL*, *vbsC* and *vbsP* (Carter *et al.*, 2002; Dilworth *et al.*, 1998). A biosynthetic pathway has been proposed based on some mutant phenotypes and *in silico* studies. According to the model, the first step in biosynthesis is the hydroxylation of L-ornithine to give L-N^5-hydroxyornithine probably catalysed by VbsO. L-N^5-hydroxyornithine is then the substrate for VbsS, an enzyme similar to NRPS. VbsS activates the substrate by adenylation and transfer it to the PCP domain. The location and sequence of *vbsP* suggests that VbsP is responsible for the addition of phosphopantotheinate to the PCP domain, although this gene is not essential probably due to the presence of another phosphopantetheinyl transferase. The addition of butyryl coenzyme A to hydroxyornithine attached to VbsS may be then catalysed by VbsA. The resulting N^5-hydroxy-N^5-(D-3-hydroxybutyryl)-ornithine could be subsequently epimerized by VbsL. After the acetylation catalysed by VbsC, the N^2-acetyl-N^5-hydroxy-N^5-(D-3- hydroxybutyryl)-ornithine bound as a thioester to the PCP domain of VbsS may be trimerized and cyclized by the TE domain of VbsS. The *vbsG* gene is essential for vicibactin biosynthesis although its function is still unknown. Based on sequence similarity, it has been suggested that the *vbsD* gene could be involved in siderophore export, although it is not essential in *R. leguminosarum* as *vbsD* mutants accumulate normal amounts of siderophores (Yeoman *et al.*, 2000). Vicibactin secretion is a topic that requires further investigation.

In *R. leguminosarum* 8401(pRL1JI) a gene, *fhuA1*, similar to the ferrichrome receptor gene from *E. coli*, is found adjacent to the *vbs* genes (Yeoman *et al.*, 2000). In *E. coli*, the ferrichrome uptake system is encoded by the *fhuACDB* operon. FhuA is the outer membrane receptor while FhuBCD is the ABC transport system involved in ferrichrome uptake: FhuB is a permease, FhuC an ATP-binding cassette protein and FhuD the periplasmic protein (Burkhardt and Braun, 1987; Coulton *et al.*, 1987; Koster and Braun, 1990). Upstream and transcribed in the same orientation as *fhuA1*, is located a gene, *fhuF*, whose predicted protein is similar to the *E. coli* putative ferric reductase. It has been suggested that it may be involved in iron release from the ferric–vicibactin complex. The *fhuCDB* operon is predicted to encode a periplasmic vicibactin-binding protein (FhuD), an inner membrane transporter (FhuB), and an ATPase (FhuC) involved in vicibactin transport through the inner membrane (Stevens *et al.*, 1999). Vicibactin internalization requires a functional TonB protein (Wexler *et al.*, 2001).

Rhizobactin 1021

The siderophore produced by *S. meliloti* 1021 was named rhizobactin 1021 to differentiate it from rhizobactin, an amino polycarboxylic acid siderophore produced by *S. meliloti* DM4 (Persmark *et al.*, 1993; Smith *et al.*, 1985) (Fig. 3.1). Rhizobactin was the first siderophore produced by *S. meliloti* to be structurally characterized, but production of rhizobactin 1021 seems to be more widespread among sinorhizobial strains. Rhizobactin 1021 is an asymmetrical citrate-derivative hydroxamate carrying two different acyl groups, acetyl and 2-decenoyl. Other citrate-hydroxamate siderophores are: schizokinen produced by some *Bacillus* species and cyanobacteria, aerobactin from enteric bacteria, arthrobactin from *Arthrobacter pascens*, acinetoferrin from *Acinetobacter haemolyticus* and nannochelin A from *Nannocystis exedens* (Hu and Boyer, 1995; Kunze *et al.*, 1992; Neilands and Leong, 1986; Okujo *et al.*, 1994; Wilhelm and Trick, 1994; Williams, 1979). Citrate-hydroxamate siderophores represent two type of ferric ligands in one molecule since the ferric iron is coordinated with two hydroxamates, and the 2-hydroxycarboxylate group from the citric acid.

In *S. meliloti* 1021, genes required for rhizobactin 1021 synthesis are located in the pSymA plasmid (symbiotic plasmid A) which also contains genes for nodulation (*nod* genes) and nitrogen fixation (*nif* genes). Rhizobactin biosynthesis involves at least a cluster of six genes arranged in one operon, *rhbABCDEF* (Lynch *et al.*, 2001). A plausible biosynthetic pathway was proposed on the basis of gene disruption, predicted function of gene products and, the chemical structure of rhizobactin 1021. It was suggested that the first step involves the synthesis of 1,3-diaminopropane. This compound is produced from L-glutamic acid and L-aspartic-B-semialdehyde by a two steps reaction, catalysed first by RhbA to produce L-2,4-diamino butyric acid and 2-ketoglutaric acid and then by RhbB, responsible for decarboxylation of L-2,4-diaminobutyric acid. The 1,3-diaminopropane is then hydroxylated by RhbE to produce N^4-hydroxy-1-aminopropane which is in turn acetylated by RhbD to form N^4-acetyl-N^4-hydroxy-1-aminopropane. The coupling of two molecules of N^4-acetyl-N^4-hydroxy-1-aminopropane to citrate may occur in a similar way as aerobactin production in *E. coli*. By analogy to aerobactin synthesis, it is proposed that this reaction consists in two steps catalysed by a synthetase complex of two subunits encoded by RhbC and RhbF.

One step of the rhizobactin 1021 synthesis pathway requires an N-acylase for addition of the lipid moiety. RhbG is similar to some acyl-CoA dependent acyl transferases and therefore it seems likely to be responsible for acylation with 2-decenoyl-CoA. Lynch *et al.* (2001) propose that acylation could be the last step of the pathway; nonetheless, Challis (2005) proposes a modified biosynthetic pathway, where N^4-hydroxy-1-aminopropane could be acetylated with acetyl-CoA through RhbD or alternatively with 2-decenoyl-CoA through RhbG. Disruption of the *rhbG* gene did not affect siderophore production or uptake as determined by chrome azurol S (CAS) or bioassays, respectively. However it could not be discounted that the non-lipidic derivative is still used as a siderophore (Lynch *et al.*, 2001). In summary, the role of RhbG in rhizobactin 1021 biosynthesis and how the lipid moiety is incorporated, await further investigations. The secretion system used by rhizobactin 1021 is also an open question.

The outer membrane receptor for rhizobactin 1021 is encoded by the *rhtA* gene, localized upstream of the siderophore biosynthetic genes but transcribed in the opposite orientation. The RhtA amino acid sequence reveals a TonB box and is homologous with IutA, the aerobactin receptor of *E. coli* (Lynch *et al.*, 2001).

Upstream of the *rhbABCDEF* genes and in the same operon is localized the *rhtX* gene. Interestingly, disruption of *rhtX* gene abolishes rhizobactin 1021 utilization. Moreover, RhtX (a member of the major facilitator superfamily of solute transporters) alone could substitute for FhuCDB in rhizobactin 1021 transport in *E. coli*. On these bases, Cuiv *et al.*(2004) propose RhtX as a permease that belongs to a novel family of siderophore transport systems not related to the ABC family of transporters (Fig. 3.3).

Xenosiderophore transport

Soil is an important niche for many diverse genera of microbes, and rhizobia must compete with these organisms for nutritional iron in the rhizosphere. One strategy employed by soil microbes for iron acquisition is to take up ferric-siderophores synthesized and secreted by other organisms. These are referred to as xenosiderophores (or exogenous siderophores). In the upper layer of humic soil, streptomycetes and fungi are predominant microorganisms, and they produce the trihydroxamate siderophores desferrioxamines and desferrichrome, respectively (Winkelmann, 2007). Ferrichrome and ferrioxamines can be used as iron sources by many fungi and by some Gram-positive and Gram-negative bacteria as is the case for some rhizobia strains. The use of xenosidephores was found to be a variable trait in rhizobia. It has been reported that ferrichrome and rhodotorulic acid (a trihydroxamate siderophore produced by some yeasts) stimulate the growth of *B. japonicum* U110 and 61A152 strains (Plessner *et al.*, 1993). Some, but not all, sinorhizobia strains tested are able to use ferrichrome, ferric rhodotorulate and ferrioxamine B as iron sources (Reigh and O'Connell, 1993; Smith and Neilands, 1984). One out of eight *R. leguminosarum* strains tested of can utilize ferrioxamine B, whereas the remaining strains could not (Skorupska *et al.*, 1989). Only 3 out of 17 *Mesorhizobium* strains tested used ferrichrome as a sole iron source (Carlton *et al.*, 2007). To explain the variation in the ability of using ferrichrome as iron source in *Mesorhizobium* strains, Carlton *et al.* (2007) suggest that ferrichrome transport systems evolved through cycles of gene acquisition and deletion, with the positive selection pressure of an iron-poor or siderophore-rich environment.

There are several reports of xenosiderophores use by rhizobia, but only a few are well characterized. Inspection of the genomes of *B. japonicum* 110, *R. leguminosarum* 3841 and *S. meliloti* 1021 reveals the presence of numerous putative TonB-dependent receptors (Fig. 3.2). In addition to endogenous

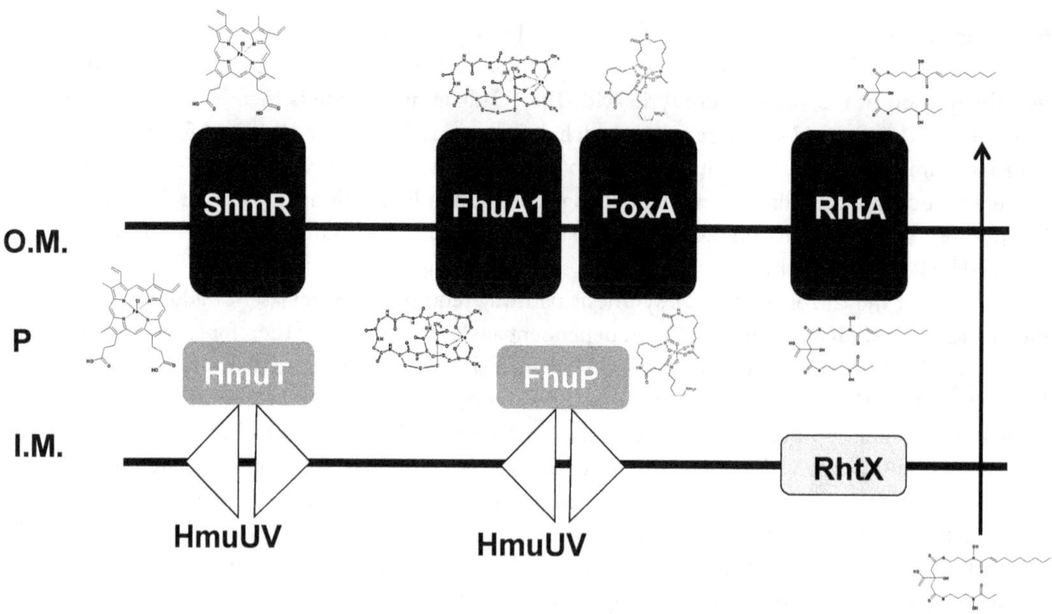

Figure 3.3 Model of iron acquisition systems expressed by *S. meliloti*. Haem transport system involves ShmR, the outer membrane haem receptor; HmuT, a periplasmic binding protein; HmuU, a membrane permease, and HmuV, an ATPase. Ferrichrome and ferrioxamine B transport systems consist of FhuA1, the ferrichrome outer membrane receptor and FoxA, the ferrioxamine B outer membrane receptor; FhuP, a periplasmic binding protein for both siderophores and the HmuUV transporter. The ferric rhizobactin1021 transport system requires RhtA, the siderophore outer membrane receptor and RhtX, a permease that belong to a novel family of siderophore transporters. Abbreviations: O.M., outer membrane; P, perisplasm; I.M., inner membrane.

siderophore receptors and putative haem receptors, the number of TonB-dependent receptors differs in these three rhizobia strains; there is only one in *R. leguminosarum* 3841, five in *S. meliloti* 1021 and 11 in *B. japonicum* 110. As only a few of them have been characterized, it cannot be assumed that they are all involved in iron transport. In fact, two *B. japonicum* TonB-dependent receptor genes, *bll4766* and *bll6848*, have been suggested to be involved in cobalamin and nickel transport, respectively (Schauer *et al.*, 2008). Based on growth assays, iron uptake experiments and *in silico* analysis, no obvious receptors for catechol-like siderophore have been identified in the rhizobia, while ferrichrome seems to be the xenosiderophore more widely used.

Five putative TonB-dependent outer membrane receptor genes in *B. japonicum* strains 110 and LO are strongly induced by iron limitation (Small *et al.*, 2009; Yang *et al.*, 2006b), and thus are the best candidates to encode receptors for iron siderophores. One of them, *bll4920*, is highly similar to the ferrichrome receptor *fegA* gene identified previously for *B. japonicum* strain 61A152 (Benson *et al.*, 2005; LeVier *et al.*, 1996), and is essential for ferrichrome utilization (Small *et al.*, 2009). None of the five putative outer membrane receptor genes in *B. japonicum* are proximal to other ABC transporter genes necessary for uptake into the cytoplasm as is found in *E. coli*, and thus no ferric iron transport system has been completely characterized in that species. The *fegA* gene in strain 61A152 is found in an operon with *fegB* located downstream of it (Benson *et al.*, 2005), but the *fegB* gene is not found in strains 110 or LO (Small *et al.*, 2009). The function of *fegB* is unknown.

O'Connell and colleagues have revealed a novel route of ferrichrome and ferrioxamine B transport in *S. meliloti* (Cuiv *et al.*, 2008). In *S. meliloti* 2011, a strain closely related to the sequenced *S. meliloti* 1021 strain, ferrichrome and ferrioxamine B outer membrane receptors are encoded by homologues of *smc01611* and *smc01657*. These genes were termed *fhuA1* and *foxA*, respectively, by analogy to enteric bacteria. A third *fhuA* homologue, *sma1747*, is also found (Fig. 3.2), although it is not involved in ferrichrome or ferrioxamine B uptake, and its role remains unknown. A predicted periplasmic protein encoded by the *fhuP* gene localized downstream the *foxA* gene was shown to be involved in both ferrichrome and ferrioxamine B uptake. Intriguingly, *hmuU* and *hmuV* mutants were impaired in utilization of these xenosiderophores as iron sources. The ABC system *hmuTUV* has been shown to be necessary for haemin and haemoglobin utilization in rhizobia (Fig. 3.3) (Cuiv *et al.*, 2008). The ability of one permease complex, *hmuUV*, to transport two structurally unrelated compounds, hydroxamate siderophores and haem, is unusual. From this, it follows that more than one periplasmic binding protein should be able to interact with a single inner membrane permease, which allows the delivery of both siderophore and haem to the cytoplasm. Alternatively, iron may be released from iron compounds in the periplasm, and then transported by a common permease system. It has been proposed that in *Pseudomonas aeruginosa*, iron may be removed from some siderophores in the periplasm and that an iron-specific rather than a siderophore-specific transporter delivers the metal to the cytoplasm (Poole, 2004). Whereas a ferric reductase activity may be sufficient to remove iron complexed to a siderophore, reduction of iron from haem does not release the metal, and thus a periplasmic haem-degrading enzyme would be necessary (see below). From this HmuUV probably transports two different substrates.

Haem and haem proteins as iron source

Haem is an iron-containing porphyrin compound essential for many cellular processes including oxygen transport, energy generation, cellular redox reactions, and metabolic regulation (O'Brian and Thony-Meyer, 2002). It can also be a source of iron to bacteria that have access to it. Bacterial pathogens of animals can use haemoglobin and other haem proteins as iron sources (Genco and Dixon, 2001; Wandersman and Delepelaire, 2004; Wandersman and Stojiljkovic, 2000). The concentration of free haem is usually very low, as it is bound tightly to haemoproteins or sequestered by serum albumin or haemopexin. Similar to siderophore transport, the use of haem or haem compounds relies on TonB-dependent outer membrane receptors able to transport haem, but not the apoprotein, into the periplasm and on ABC transporters to internalize it into the cytosol. Haem transport systems are maximally expressed when iron is scarce. Haem receptor proteins can be classified into two families. The first family comprises receptors that directly interact with free haem or haemoproteins. The second family includes receptors that depend on extracellular haem binding proteins called haemophores (Wandersman and Delepelaire, 2004). Haemophores are proteins produced and secreted by the bacterium that sequester haem from diverse sources and then binds to an outer membrane haemophore receptor, followed by transport into the periplasm. Once in the periplasm, haem is then transported into the cytoplasm by an ABC transporter. The Hem system of *Yersinia enterocolitica* (HemTUV) was amongst the first ABC transporters involved in haem uptake to be characterized, and is considered as a prototype for haem transport. Its homologue in *Y. pestis* is called HmuTUV. According to the model of Hmu transport, once the haem moiety is translocated into the periplasm it is bound by a periplasmic binding protein HmuT which in turn presents it to the inner membrane permease-ATP hydrolase complex of HmuU/HmuV. Intact haem is then delivered to the bacterial cytoplasm. Interestingly, mutations in haem-specific ABC transport systems do not always result in abrogation of haem utilization by the bacterium. Recently, an alternative inner membrane system involved in haem uptake was discovered by Wandersman and co-workers (Letoffe *et al.*, 2006, 2008), who found that the dipeptide periplasmic binding protein DppA or the L-alanyl-γ-D-glutamyl-meso diaminopimelate periplasmic binding protein MppA is implicated in haem transport together with the dipeptide permease ABC, DppBCDF complex.

For many years, transport and degradation of haem as a mechanism to acquire iron was thought to be exclusive to pathogens. However, Noya *et al.* (1997) discovered that rhizobia and other non-pathogenic bacteria can use haem or haemoglobin as iron sources. Subsequently, genes involved in haem transport were identified in *B. japonicum*, *R. leguminosarum* and *S. meliloti* (Amarelle *et al.*, 2008; Battistoni *et al.*, 2002a; Nienaber *et al.*, 2001; Wexler *et al.*, 2001). Notably, while the use of xenosiderophores was found to be variable, haem transport systems are a common trait of rhizobia strains studied so far. Moreover, rhizobia haem transport systems are present in the chromosome and not in plasmids (Fig. 3.2). As mentioned before, leghaemoglobin is the most abundant protein in the cytosol of symbiotic root nodule. Although rhizobia are not in direct contact with leghaemoglobin within the nodule cytosol, they may have access to it during nodule senescence. The possibility of leghaemoglobin as an iron source under physiological conditions has not been addressed experimentally.

The *B. japonicum bll07076* gene and the *S. meliloti smc02726* gene coding for haem receptor proteins HmuR and ShmR respectively, are required for utilization of haem, haemoglobin and leghaemoglobin. Both predicted proteins are similar to outer membrane TonB-dependent receptors, and display the typical FRAP/NPNL motif of haem receptors, although the highly conserved histidine residue in the FRAP/NPNL region is substituted by asparagine in ShmR. Despite the similarity of these proteins with other haem receptors, the identity of these two rhizobial proteins to each other is only 23%. In *R. leguminosarum* a predicted protein coded by *rll3522* and showing 65% identity with ShmR was identified by *in silico* studies, even thought its function in haem-uptake has not yet been described. Analysis of the *S. meliloti* 1021 genome also reveals the presence of a second putative haem receptor protein: Smc04205. The predicted sequence of Smc04205 indicates that this protein is homologous to HasR-like outer membrane receptors which recognize haem-bound haemophores. Moreover, the genetic region around *smc04205* is syntenous with the *hasR* regions of numerous other bacteria (Cescau *et al.*, 2007). Nonetheless, no haemophore like protein has yet been detected in rhizobia and no *hasR* promoter activity could be found.

In *B. japonicum, hmuR* is proximal and divergently transcribed from *hmuTUV* genes. In *S. meliloti* and in *R. leguminosarum, hmuTUV* genes are distal from *shmR*. By heterologous expression in *E. coli*, Cuiv *et al.* (2008) confirmed that *S. meliloti hmuTUV* genes are implicated in haem transport in *E. coli*. Intriguingly, *hmuTUV* mutants in *B. japonicum, R. leguminosarum* and *S. meliloti* presented a reduced but not abolished use of haem and haemoglobin as iron sources indicating that there is more than one haem transport system in those organisms (Nienaber *et al.*, 2001; Wexler *et al.*, 2001).

ExbBD–TonB

The energy for the transport of haem or siderophore-mediated iron transport across the outer membrane is provided by the TonB–ExbB–ExbD complex, where ExbBD somehow use the proton motive force to induce conformational changes in TonB. This energy coupling protein in turn interacts directly with the outer membrane TonB dependent receptors (Postle and Larsen, 2007). The precise mechanism involved is not completely understood. Moreover, it has been recently found that TonB is needed for transport of substrates in addition to iron (Cornelis *et al.*, 2009; Schauer *et al.*, 2008). In some organisms there are many TonB-dependent receptors and only one TonB protein, while in other bacteria there are a few TonB dependent receptors and more than one TonB protein (Schauer *et al.*, 2008).

In *B. japonicum* 110, there are two clusters encoding TonB–ExbB–ExbD like proteins, *bll7071–73* and *blr3906–08*. It is worth noting that *bll7071–bll7072* mutants are impaired in utilization of haem as an iron source but can still use ferrichrome or ferric citrate as iron sources suggesting different function for both TonB systems (Nienaber *et al.*, 2001). A different situation is found in *R. leguminosarum* where the disruption of the *tonB* homologue impedes the use of vicibactin as well as haem as iron sources (Wexler *et al.*, 2001). In *S. meliloti* 1021 there is only one gene encoding a TonB-like protein and its function has not yet been reported.

Haem degradation for iron release

Once exogenous haem is taken up by bacterial cells, the tetrapyrrole ring is cleaved by haem oxygenase in order to release the iron. The pioneering work on bacterial haem oxygenases was carried out with the proteins from *Corynebacterium diphtheriae* (Schmitt, 1997; Wilks and Schmitt, 1998) and *Neisseria meningitidis* (Ratliff *et al.*, 2001; Schuller *et al.*, 2001; Zhu *et al.*, 2000a; Zhu *et al.*, 2000b). These bacterial haem oxygenases have limited sequence similarity to each other and to eukaryotic haem oxygenases, but have similar structures overall (Hirotsu *et al.*, 2004; Schuller *et al.*, 1999; Schuller *et al.*, 2001). More recently, a structurally unrelated haem-degrading oxygenase was described in *Staphylococcus aureus*

(Skaar *et al.*, 2004; Wu *et al.*, 2005), and is also found in a limited number of Gram-positive bacteria, including *Bacillus anthracis* (Skaar *et al.*, 2006). Both families of haem degradation enzymes use the same substrates and produce biliverdin, iron, and CO (see also Chapter 13).

Because numerous rhizobia can use haem as an iron source (Noya *et al.*, 1997), they should be able to cleave haem and release iron. The photosynthetic bacterium *Bradyrhizobium* sp. strain ORS578 cleaves haem for the synthesis of the biliverdin chromophore, and contains a gene encoding a classical HmuO protein adjacent to a phytochrome gene (Giraud *et al.*, 2002; Giraud *et al.*, 2000). It is not known whether this haem oxygenase is involved in iron utilization. *Rhodopseudomonas palustris, Agrobacterium tumefaciens* and *R. leguminosarum* have a similar gene arrangement in their genomes. *B. japonicum* does not have a *hmuO* gene homologue, but it does have two genes, *hmuD* and *hmuQ*, encoding weak homologues to the *S. aureus* IsdG (Puri and O'Brian, 2006). *hmuQ* is within the gene cluster encoding the haem transport system identified by Nienaber *et al.* (2001). HmuQ binds haem with high affinity and catalyses the cleavage of haem to biliverdin (Puri and O'Brian, 2006). Homologues of *hmuQ* and *hmuD* were identified in many bacterial genera, and the recombinant homologue from *Brucella melitensis* has haem degradation activity as well. Thus, *hmuQ* and *hmuD* encode haem oxygenases, and the IsdG family of haem-degrading monooxygenases is not restricted to Gram-positive pathogenic bacteria.

Ferric dicitrate

Citric acid can complex ferric ion and it can be used for iron transport into bacterial cells, although the affinity of iron for citrate is not as high as it is for siderophores. Some bacteria can use citrate as a xeno-siderophore (e.g. *E. coli, Mycobacterium smegmatis, Neisseria meningitidis* and *Pseudomonas aeruginosa*) and, occasionally, it can be used as an endogenous siderophore. For instance *B. japonicum* strains 61A152 and 110 can use citrate as siderophore but only *B. japonicum* 61A152 strain is able to secrete it (Guerinot *et al.*, 1990). In *E. coli* and *P. aeruginosa* iron uptake from ferric-dicitrate is inducible by environmental citrate and this effect requires an extracytoplasmic function (ECF) sigma factor. The ferric dicitrate uptake system requires the outer membrane receptor FecA and the periplasmic binding protein and ABC transporter system FecBCDE (Enz *et al.*, 2000).

Although some rhizobia strains can use ferric citrate as an iron source, *B. japonicum* 110, *R. leguminosarum* 3841 and *S. meliloti* 1021 genomes have no homologues of genes encoding a FecABCDE citrate transport system. In *Streptomyces coelicolor*, exogenous citrate serves not only as an iron chelator, but also as an energy source, and thus ferric citrate is taken up for both purposes (Lensbouer *et al.*, 2008). In this case, ferric citrate is taken up by a member of the CitMHS family of secondary transporters, a family found in numerous Gram-positive bacteria. No obvious homologues are found in sequenced rhizobial genomes.

Role of iron transport in symbiosis

Symbiotic bacteroids have a large iron demand, but little is known about how iron gets into cells or the form of the metal that is transported. Mutations in genes encoding iron or haem transport systems described thus far do not have symbiotic phenotypes, indicating that such activities are not required for symbiotic function, or else they are redundant (Amarelle *et al.*, 2008; Cuiv *et al.*, 2008; Fabiano *et al.*, 1995; Gill and Neilands, 1989; Lynch *et al.*, 2001; Nienaber *et al.*, 2001; Puri and O'Brian, 2006; Small *et al.*, 2009; Wexler *et al.*, 2001). One exception is the *fegA* gene of *B. japonicum* strain 61A152, which is required for symbiosis (Benson *et al.*, 2005). It was noted that accessibility to ferrichrome within a nodule is unlikely, so another function was suggested for this gene (Nienaber *et al.*, 2001; Wexler *et al.*, 2001; Benson *et al.*, 2005). However, the corresponding gene in the USDA 122 derivative LO, *bll4920*, is not required for symbiosis, nor are the other four iron-responsive putative ferric siderophore receptor genes (Small *et al.*, 2009). Similarly, *S. meliloti* mutants defective in ferrichrome and ferrioxamine B uptake have no symbiotic phenotypes on alfalfa nodules (Cuiv *et al.*, 2008).

The abundance of leghaemoglobin in the infected plant cells of nodules makes it tempting to speculate that it is a potential iron source for bacteroids. Bacteroids are surrounded by a peribacteroid membrane to form symbiosomes, and thus the endosymbiont is physically separated from the plant globin. Haem transport across the peribacteroid membrane has not been demonstrated, and the haem transport systems employed by free living rhizobia are not necessary for symbiosis (Nienaber *et al.*, 2001; Wexler *et al.*, 2001). However, leghaemoglobin haem may be accessible to the bacteria during nodule senescence.

Isolated soybean symbiosomes can take up both ferric chelates and ferrous iron (LeVier *et al.*, 1996; Moreau *et al.*, 1995). Ferric chelate substrates are probably reduced to the ferrous form by a peribacteroid ferric reductase activity (LeVier *et al.*, 1996), and ferrous iron can be taken up by bacteroids (Moreau *et al.*, 1998). Soybeans express the divalent ion transporter GmDMT1 on the bacteroid membrane within nodules that may account for direct ferrous iron uptake by symbiososmes (Kaiser *et al.*, 2003). Thus, it is possible that ferrous iron is the form taken up by the endosymbiont.

Ferrous iron is transported by FeoB in *E. coli* and other bacteria, and homologous systems have been identified in other organisms (Hantke, 2003). A putative FeoB protein encoded by *blr6523* is present in the genome of *B. japonicum* 110, but its function remains unknown. There are no homologues of the Feo system encoded on the *S. meliloti* 1021 or *R. leguminosarum* 3481 genomes. The existence of a novel ferrous iron uptake system or other divalent metal transporter involved in ferrous iron uptake requires further investigation. As noted earlier, symbiotic bacteroids may receive an adequate supply of iron from the plant host, and may not normally be iron limited. In that case, a high affinity bacterial iron transport system would appear to be unnecessary.

Regulation of iron homeostasis in the Rhizobia

Studies on the control of bacterial iron homeostasis have focused largely on Fur (ferric uptake regulator), a regulatory protein that responds to cellular iron. Perhaps the biggest surprise in elucidating iron metabolism in the rhizobia is that genetic regulation differs greatly from the bacterial paradigm established by *E. coli* and other model systems. Although the reductionist approach has been enormously successful, this aspect of rhizobial biology reminds us of the diversity of bacterial biology, and the need for restraint in extrapolating data from a few model systems to an entire kingdom of life. Moreover, there is substantial diversity within the rhizobia. We now know that rhizobial Fur homologues are manganese-responsiveness regulators, at least in some species, and that the job of iron perception and regulation is carried out by two other regulators, Irr and RirA. Furthermore, some species have only Irr, whereas others have both Irr and RirA.

The Fur protein

Brief overview of bacterial Fur

In *E. coli* and many other bacteria, Fur represses genes involved in high-affinity iron transport under iron-replete conditions, and which are derepressed when the metal is scarce. In addition, Fur is involved in numerous other facets of iron metabolism, and also in processes not obviously linked to iron, such as acid shock response (Hall and Foster, 1996), synthetic pathways (Stojiljkovic *et al.*, 1994), and the production of toxins and other virulence factors (Litwin and Calderwood, 1993). Fur is the founding member of a family of regulators which also includes Zur (Gaballa and Helmann, 1998; Patzer and Hantke, 1998), PerR (Bsat *et al.*, 1998) and Irr (Hamza *et al.*, 1998; Qi *et al.*, 1999). These proteins differ in function and have different DNA binding sites, but are all involved in metal-dependent control of gene expression.

Fur homologues are found in many bacterial genomes. Structural analysis of Fur and its DNA binding properties have been most extensively studied in *E. coli*, *P. aeruginosa* and *B. subtilis*, whereas analyses of *fur* mutants and the identification of genes under Fur control have also been studied in those bacteria and in several other organisms as well. The working model for Fur function posits that, when bound by ferrous (Fe^{2+}) iron, Fur binds its target DNA within the promoter of the regulated gene to repress transcription. However, when iron is limiting in the cell, Fur protein is unbound by iron and no longer binds DNA with high affinity, hence gene expression is derepressed. Although generally regarded as a repressor, an apparent activator activity of Fur from *E. coli* and *P. aeruginosa* is actually an indirect effect of its repression of a small RNA molecule that negatively regulates genes in an iron-dependent manner (Massé and Gottesman, 2002; Wilderman *et al.*, 2004). However, Fur from *Neisseria meningitidis* has bona fide activator function as well as repressor activity (Delany *et al.*, 2004). Interestingly, the metal-free form of Fur from *Helicobacter pylori* appears to have a bona fide anti-repressor activity to control its own expression (Delany *et al.*, 2001; 2003).

Rhizobial Fur proteins are manganese-responsive regulators

The *sitABCD* operon of *S. meliloti* was identified in a screen for mutants that could not grow in the presence of metal chelator, and was shown to be necessary for manganese acquisition (Platero *et al.*, 2003). These genes encode an ABC system belonging to the metal transport family found in many bacteria (Claverys, 2001). A *fur* gene homologue lies adjacent to the *sitABCD* genes, which led those authors to speculate that Fur may control *sitABCD*. Subsequently, three reports came out showing that this is indeed the case, but the regulatory metal is Mn^{2+}, not Fe^{2+} (Chao *et al.*, 2004; Diaz-Mireles *et al.*, 2004; Platero *et al.*, 2004). Manganese-responsiveness of a *sitA* reporter fusion is lost and constitutively high in a *fur* mutant of *R. leguminosarum* or *S. meliloti*, and Fur binds to the *sitA* promoter *in vitro* (Chao *et al.*, 2004; Diaz-Mireles *et al.*, 2004; Platero *et al.*, 2007; Platero *et al.*, 2004). Furthermore, the *sitABCD* operon is not regulated by iron. A microarray analysis of a *S. meliloti fur* mutant identified 23 genes that are normally down-regulated by Fur (Chao *et al.*, 2004). Other than *sitABCD*, it is not known whether these genes are regulated by manganese, or whether they have manganese-related functions.

Although *B. japonicum* Fur has been described in terms of iron control (see below), it is now known that this protein mediates manganese control of the *mntH* gene encoding the NRAMP1 family of Mn^{2+} transporter MntH (Hohle and O'Brian M, 2009). Fe^{2+} is neither a substrate for MntH nor a regulator of the *mntH* gene in *B. japonicum*.

It is now clear from these findings, and other studies described below, that Fur is not the primary iron-responsive transcriptional regulator in the rhizobia, and that this function is taken over by Irr, RirA or both. Thus the rhizobia, and apparently the α-proteobacteria as a whole, control iron homeostasis by a mechanism quite different than the paradigm established in *E. coli* and other model organisms. The Fur homologue has been renamed Mur in *R. leguminosarum* and *S. meliloti* to denote its prominent role in manganese control. In this review, we refer to it as Fur/Mur since some classical roles of Fur have been preserved, at least in *B. japonicum*, and the complete repertoire of functions for this protein have not been sufficiently elucidated in the rhizobia to rule out certain functions for it.

Characterization of the Fur/Mur-binding cis*-acting regulatory element*

Central to the model of Fur function in *E. coli* and other model organisms is the so-called Fur box, a DNA binding element for Fur that contains similarity to a 19-bp, AT-rich palindromic consensus sequence (Fig. 3.4). Sequence similarity to a Fur box consensus within promoter regions of genes is taken as *ab initio* evidence for regulation by Fur. The binding site of Fur has been interpreted as two 9-bp inverted repeats, as three shorter hexameric repeats in a head-to-head-to-tail orientation and as two 7–1–7 inverted repeat motifs (Baichoo and Helmann, 2002; Escolar *et al.*, 1998)

B. japonicum Fur was originally identified based on its ability to complement an *E. coli fur* gene mutant, thus its ability to bind to a Fur box consensus is not surprising (Friedman and O'Brian, 2003; Hamza *et al.*, 1999). Similarly, the *E. coli* Fur-regulated *bfd* gene is regulated by *R. leguminosarum* Fur/Mur in an iron-dependent manner in *E. coli* (Diaz-Mireles *et al.*, 2004). However, these organisms do not contain an *E. coli*-like Fur box sequence. Characterization of the promoters of the Fur-regulated genes *irr* and *mntH* in *B. japonicum* identified the binding site as three imperfect direct repeat hexamers that are all required for normal occupancy by Fur. This site is dissimilar to the *E. coli* Fur box consensus (Friedman and O'Brian, 2003; Hohle and O'Brian, 2009). Although *B. japonicum* Fur binds both this element and Fur box DNA with high affinity, *E. coli* Fur does not bind to the *irr* Fur binding site. Fur

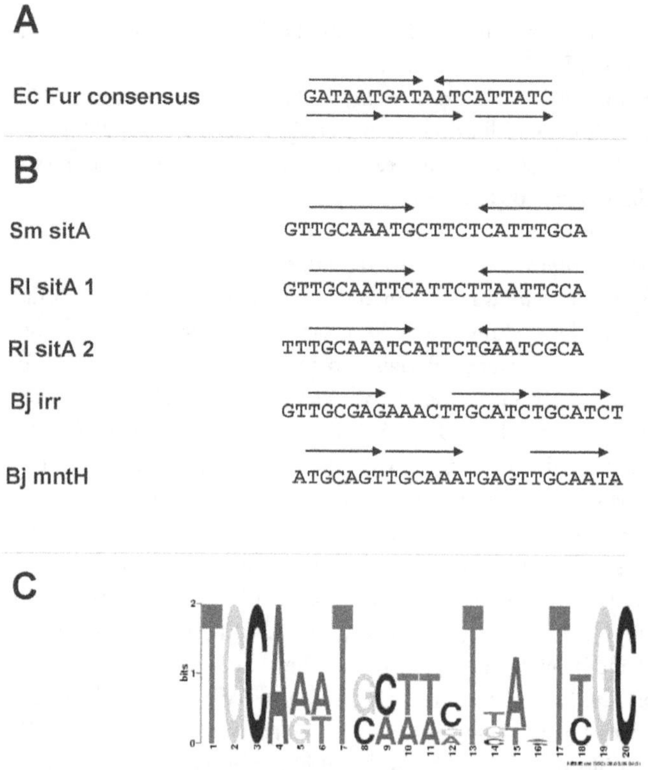

Figure 3.4 Fur/Mur binding sites in rhizobial promoters. (A) The *E. coli* Fur box consensus sequence. The arrows show interpretation of the sequence as either a palindromic sequence or as three hexameric repeats. (B) Binding sites of experimentally-known target promoters for Fur/Mur in *S. meliloti* (Sm), *R. leguminosarum* (Rl) and *B. japonicum* (Bj). The genes are *sitA*, *irr* and *mntH*. The *R. leguminosarum sitA* promoter has two Fur/Mur binding sites. (C) Graphical representation of Fur/Mur binding site using the known elements in part B.

binds to the promoter as one or two dimers, and this binding is sufficient to inhibit transcription *in vitro* in a metal-dependent manner (Friedman and O'Brian, 2003).

The *R. leguminosarum* and *S. meliloti sitA* promoters have two and one Fur/Mur binding sites, respectively, with similar sequences (Diaz-Mireles *et al.*, 2004; Diaz-Mireles *et al.*, 2005; Platero *et al.*, 2007). These *cis*-acting elements are described as palindromic, and although they are somewhat similar to the Fur consensus sequence, Fur/Mur binds the Fur box with lower affinity (Bellini and Hemmings, 2006; Platero *et al.*, 2007; Wexler *et al.*, 2003). *R. leguminosarum* Mur/Fur binds each *sitA* binding site as one or two dimers (Bellini and Hemmings, 2006), but *S. meliloti* Fur/Mur occupies its cognate element as a single dimer (Platero *et al.*, 2007). The basis for these differences is unknown, but the *S. meliloti* study used much less protein in the *in vitro* binding analysis than the *R. leguminosarum* Fur/Mur work. Mutation of a single hexamer within the *B. japonicum irr* or *mntH* Fur binding site can result in occupancy of only one dimer (Friedman and O'Brian, 2003; Hohle and O'Brian, 2009), suggesting that differences in the DNA targets may account for differential occupancy, as has been described for *E. coli* Fur (Escolar *et al.*, 1999). A *B. japonicum* Fur mutant was identified that can only occupy the *irr* promoter as a single dimer (Friedman and O'Brian, 2004), but the residues that were altered are conserved in the *S. meliloti* and *R. leguminosarum* proteins. Despite differences between the *B. japonicum* Fur/Mur binding sites and those of *S. meliloti* and *R. leguminosarum*, computational analysis of known and putative Fur/Mur binding sites indicates a core of common residues in all of the *cis*-acting elements (Rodionov *et al.*, 2006). A comparison of the five known Fur/Mur binding sites yields a similar consensus (Fig. 3.4).

Metal-binding properties of Fur/Mur

Fur/Mur regulators function by binding the cognate regulatory metal directly, which confers DNA-binding activity on the protein (Bellini and Hemmings, 2006; Friedman and O'Brian, 2004; Platero *et al.*, 2007). A regulatory Fe^{2+}-binding site (site 1) and a structural Zn^{2+}-binding site (site 2) implicated from the recent crystal structure of Fur from *P. aeruginosa* are comprised of amino acids highly conserved in many Fur proteins, including those in the rhizobia. *B. japonicum* Fur mutants containing substitutions in site 1 or site 2 bound DNA with high affinity and repressed transcription *in vitro* in a Fe^{2+}-dependent manner. Interestingly, only a single dimer of site 2 mutant occupied the *irr* promoter, whereas the wild type and site 1 mutant displayed one or two dimer occupancy. Both mutants were able to repress transcription from the *irr* promoter *in vitro*. Furthermore, both DNA binding and transcriptional repression were strictly metal dependent. It appears that the putative functions for metal-binding sites deduced from the structure of *P. aeruginosa* Fur cannot be extrapolated to other bacterial Fur proteins as a whole.

Are rhizobial Fur proteins responsive to both Mn^{2+} and Fe^{2+} in vivo*?*

It is clear that Fur is not the major iron responsive regulator in the rhizobia, but rather this function is carried out by Irr and/or RirA, as will be discussed below. Nevertheless, *B. japonicum* Fur controls the *irr* gene directly, and many genes are aberrantly regulated in a *B. japonicum fur* mutant (Hamza *et al.*, 1999; Yang *et al.*, 2006c). In addition, Fur controls iron-dependent expression of the *fegA* gene encoding the ferrichrome receptor in *B. japonicum* strain 61A152 (Benson *et al.*, 2004), but not in strain 110 or 122 (Small *et al.*, 2009). Although the question of whether Fur can be both a Mn^{2+} and Fe^{2+} regulator is most pressing in *B. japonicum* where evidence for control by both metals is available, it is worth considering in the rhizobia as a whole.

Rhizobial Fur proteins bind both Mn^{2+} and Fe^{2+} *in vitro*, and other divalent metal ions as well (Bellini and Hemmings, 2006; Friedman and O'Brian, 2004; Platero *et al.*, 2007), and therefore metal discrimination does not appear to be the basis of regulatory specificity. *B. japonicum* Fur/Mur was isolated based on its ability to complement an *E. coli fur* mutant (Hamza *et al.*, 1999), and *R. leguminosarum* Fur/Mur can also partially complement a *fur* strain as well (Diaz-Mireles *et al.*, 2004). Microarray analysis shows that many iron-regulated proteins are affected in a *B. japonicum fur* mutant (Yang *et al.*, 2006c). In the classical role of Fur as an iron-dependent repressor, the microarray studies should identify up-regulated genes in the mutant compared to the parent strain grown in iron-replete media, which was found. However, most of these iron regulated genes were only weakly affected in the *fur* strain, suggesting indirect control or the participation of an additional regulatory factor. Only in low iron media did the whole genome screen find iron-regulated genes that are strongly controlled by Fur. In *S. meliloti*, the iron-regulated rhizobactin synthesis genes are down-regulated in a *fur/mur* mutant (Chao *et al.*, 2004). Further experiments that distinguish direct from indirect control of Fur/Mur on gene expression and the molecular basis of the integration of iron and manganese metabolism should provide answers to these questions.

The Irr protein

The *irr* gene was initially identified in *B. japonicum* in a screen for loss of control of haem biosynthesis by iron (Hamza *et al.*, 1998), and it has been most extensively characterized in that organism. Haem is the end product of a biosynthetic pathway, culminating with the insertion of iron into a protoporphyrin ring to produce protohaem. Irr coordinates the haem biosynthetic pathway with iron availability to prevent the accumulation of toxic porphyrin precursors under iron limitation (Hamza *et al.*, 1998). Loss of function of the *irr* gene is sufficient to uncouple the pathway from iron-dependent control, as discerned by the accumulation of protoporphyrin. This accumulation is due to derepression of *hemB* and probably *hemA* (Hamza *et al.*, 1998; Yang *et al.*, 2006b). Similarly, an *irr* mutant of *R. leguminosarum* has a fluorescent colony phenotype and is deregulated for the *hemA* gene (Todd *et al.*, 2006; Wexler *et al.*, 2003), and a *Brucella abortus irr* mutant accumulates protoporphyrin (Martinez *et al.*, 2005).

The Irr protein belongs to the Fur family of metalloregulators that includes Fur, PerR, Zur, Nur and Mur (Ahn *et al.*, 2006; Bsat *et al.*, 1998; Chao *et al.*, 2004; Diaz-Mireles *et al.*, 2004; Gaballa and

Helmann, 1998; Hamza *et al.*, 1998; Patzer and Hantke, 1998; Platero *et al.*, 2004). However, Irr behaves differently than these and other regulatory proteins in fundamentally different ways, and allows novel control of iron metabolism.

The cellular Irr level is controlled primarily by haem-dependent protein degradation

The Irr protein accumulates in cells under iron limitation, with very low levels in iron-replete cells. Thus, Irr is distinct from other Fur family proteins because it functions in the absence of the regulatory metal, whereas the other members require direct metal-binding for activity.

The control of iron on *irr* expression is primarily post-translational; Irr is a conditionally stable protein that degrades in cells exposed to iron (Qi *et al.*, 1999). *B. japonicum* Irr contains a haem-regulatory motif (HRM) near the N-terminus that binds haem and is necessary for rapid degradation. Accordingly, Irr is stabilized in a haem-deficient background or by mutagenesis of Cys-29 within the HRM. Since the discovery of haem-dependent degradation of Irr, numerous other eukaryotic proteins have been identified that degrade in response to haem by binding to HRM motifs (Hu *et al.*, 2008; Ishikawa *et al.*, 2005; Jeong *et al.*, 2004; Yang *et al.*, 2008; Zenke-Kawasaki *et al.*, 2007).

A fundamental problem with haem as a signalling molecule is that it is reactive and lipophilic. Haem can catalyse the formation of reactive oxygen species, and binds non-specifically to lipids, proteins and other macromolecules. Thus, a regulatory free haem pool is unlikely. The discovery of new and novel roles for haem as a regulatory molecule in eukaryotes and prokaryotes begs for reconciliation between these functions and the cytotoxicity of haem. This problem has been partially resolved for the Irr protein from *B. japonicum*. Ferrochelatase catalyses the insertion of iron into protoporphyrin to form haem in the final step of the haem biosynthetic pathway. Irr interacts directly with ferrochelatase and responds to iron via the status of haem and protoporphyrin localized at the site of haem synthesis (Qi and O'Brian, 2002) (Fig. 3.5). Competition of the wild type ferrochelatase with a catalytically inactive one inhibits iron-dependent degradation of Irr even though the cell is not haem-defective. This means that Irr does not respond to a free haem pool, but rather to haem locally where it is synthesized. The dissociation binding constant (K_d) of haem for Irr is about 1 nM, which is less than one free haem molecule per cell. Irr may represent the simplest type of haem signalling mechanism because there is no obvious need for a factor to chaperone haem from the site of synthesis to its target.

The interaction of Irr with ferrochelatase is affected by the immediate haem precursor protoporphyrin (Fig. 3.5). The porphyrin-bound enzyme does not bind to Irr, which is the state of ferrochelatase when iron is limiting, and allows Irr to be active and affect the genes under its control. Thus, Irr is affected by haem and by its substrates so that haem synthesis does not exceed iron availability. In the presence of iron, ferrochelatase inactivates Irr, followed by Irr degradation to derepress the pathway. Irr is present but inactive in cells that express a catalytically inactive ferrochelatase, but active in a *hemH* deletion strain (Qi and O'Brian, 2002). It is possible that inactivation of Irr allows loss of function that is faster than its degradation. Indeed, the *hemB* mRNA is elevated by iron more rapidly than Irr degrades (Chauhan *et al.*, 1997; Qi *et al.*, 1999).

Regulated degradation of B. japonicum *Irr requires both redox states of haem*

B. japonicum Irr fused to glutathione S transferase (GST) confers iron-dependent instability on GST, but a GST fusion containing only the N-terminal 36 amino acids of Irr, which includes the HRM, is stable (Yang *et al.*, 2005). This means that the HRM is necessary but not sufficient for rapid degradation of Irr. *In vitro* and *in vivo* studies identified an instability domain that includes three consecutive histidines at positions 117–119, with His-117 and His-119 being invariant residues in Irr proteins. This domain is part of a haem-binding region distinct from the HRM. Whereas the HRM binds specifically to ferric (Fe^{3+}) haem, the histidine-rich domain binds ferrous (Fe^{2+}) haem (Yang *et al.*, 2005). An Irr mutant in which the three histidines are replaced by alanines is stable *in vivo* under iron replete conditions (Yang *et al.*, 2005). Irr decay follows first order kinetics (Qi *et al.*, 1999), indicating a single mechanism for degradation. Hence the two haems probably participate in a single degradation process rather than

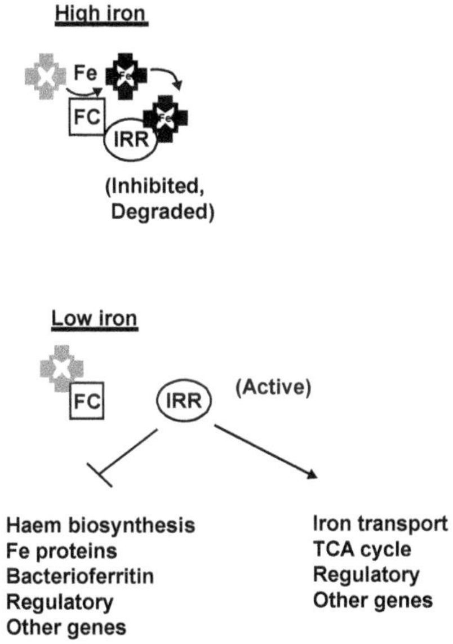

Figure 3.5 Model of Irr function. Irr senses the iron status through haem biosynthesis. In the presence of iron, Irr interacts with ferrochelatase (FC), which synthesizes haem. Haem binds to Irr, which triggers Irr degradation. Under iron deprivation FC is bound only to protoporphyrin (light porphyrin), which promotes its dissociation with Irr. Irr is then active to regulate genes under its control. Irr is both a positive and negative regulator.

independent processes that occur at different rates. These findings implicate a role for the redox activity of haem in Irr degradation, and further evidence suggests that this activity leads to protein oxidation (Yang *et al.*, 2006a). *B. japonicum* Irr degrades in response to cellular oxidative stress by a mechanism that involves haem and iron (Yang *et al.*, 2006a). Furthermore, Irr degradation is strictly O_2-dependent *in vivo* (Yang *et al.*, 2006a). Irr oxidation was demonstrated *in vitro*, requiring haem, O_2 and a reductant. An Irr truncation that does not bind ferrous haem *in vitro* does not degrade *in vivo*. Thus, it was proposed that reactive oxygen species participate in Irr degradation not only as part of an oxidative stress response (see below), but also in normal degradation in response to iron. Protein oxidation can result in hydrolysis of peptide bonds (Berlett and Stadtman, 1997) and thus, in principle, oxidation of Irr could be sufficient for degradation. However, *in vivo* degradation of Irr is rapid whereas carbonylation *in vitro* is slow. It is probable that oxidized Irr is recognized by cellular proteases as a damaged protein that is subsequently degraded. A candidate protease has not been described thus far. Degradation of both IRP2 and Bach1 require the ubiquitin ligase HOIL-1, which interacts with the haem bound form of the respective protein (Ishikawa *et al.*, 2005; Zenke-Kawasaki *et al.*, 2007). The ubiquitin-tagged protein is then degraded. Arginyl transferase is tagged by N-end rule ubiquitin ligases in yeast and mouse for haem-dependent degradation (Hu *et al.*, 2008).

Redox-dependent ligand switching, although not associated with protein degradation, occurs with the transcriptional regulator CooA from *Rhodospirillum rubrum* (Roberts *et al.*, 2004) and the redox sensor EcDos from *E. coli* (Kurokawa *et al.*, 2004). IRP2 binds ferric haem at the HRM and ferrous haem at a conserved histidine residue (Ishikawa *et al.*, 2005; Jeong *et al.*, 2004), and protein oxidation is haem-dependent (Yamanaka *et al.*, 2003). Thus, there are similarities between IRP2 and Irr.

Examination of the Irr homologues reveals that only homologues within the *Bradyrhizobiaceae* have the Cys-Pro sequence and an HRM-like domain, whereas His-117 and His-119 of *B. japonicum* Irr are completely conserved in all of the homologues. This raises the question of whether Irr degradation as described for *B. japonicum* occurs in other rhizobia. A *B. japonicum* Irr derivative lacking an HRM degrades, but the rate is much slower than is found for the wild type protein (Yang *et al.*, 2005). In addition, *B. japonicum* has a lower affinity ferric haem binding site (Qi and O'Brian, 2002) that could possibly serve a similar function as the HRM, albeit less efficiently. By analogy, Irr homologues lacking an HRM may have a compensatory mechanism that allows turnover.

Oxidative stress promotes degradation of the Irr protein

Bacteria have multiple defence strategies against oxidative stress, including the direct detoxification of ROS by catalase, peroxidases and superoxide dismutase. Oxidative stress responses require the activation of regulatory proteins and the induction of genes under their control. In many bacteria, the transcriptional regulator OxyR (Christman *et al.*, 1989; Tao *et al.*, 1989) senses hydrogen peroxide (Zheng *et al.*, 1998) and induces numerous genes whose products are involved in peroxide defence (Altuvia *et al.*, 1994; Tartaglia *et al.*, 1989), redox balance (Prieto-Alamo *et al.*, 2000; Ritz *et al.*, 2000) and other factors (Altuvia *et al.*, 1997; Zheng *et al.*, 1999). In *B. subtilis*, PerR is the major peroxide regulator and represses a large PerR regulon (Herbig and Helmann, 2001). The OhrR family of antioxidant regulators is responsible for organic hydroperoxide resistance (Mongkolsuk *et al.*, 1998). *B. japonicum* contains an OxyR homologue, but it may function differently in that organism than in other systems (Panek and O'Brian, 2004). Evidence points to Irr as an oxidative stress response regulator. Irr degrades in response to H_2O_2 produced endogenously in a catalase-deficient (*katG*) strain, or to H_2O_2 applied exogenously to culture media (Yang *et al.*, 2006a). A *Brucella abortus irr* mutant displays elevated catalase activity, and resistance to killing by H_2O_2 (Martinez *et al.*, 2006). The Irr deficiency causes derepression of *hemB* in *B. japonicum* and elevated haem in *B. abortus*. Catalases and peroxidases are haem proteins that detoxify H_2O_2 and peroxides, respectively, and elevated *hemB* may contribute to the synthesis of those enzymes. Other examples of elevated expression of haem biosynthesis genes are noted. The *Bacillus subtilis* PerR protein mediates the induction of the *hemAXCDBL* operon encoding enzymes for the early steps of haem synthesis (Chen *et al.*, 1995; Mongkolsuk and Helmann, 2002). In *E. coli* and *Salmonella*, the *hemH* gene encoding the haem biosynthetic enzyme ferrochelatase is induced in response to H_2O_2 in an OxyR-dependent manner (Elgrably-Weiss *et al.*, 2002; Zheng *et al.*, 2001). However, it has not been established in *B. japonicum* or any other bacterium that synthesis of catalase or peroxidase substantially increases the overall haem demand in the cell, and thus the physiological relevance of elevated haem synthesis genes is uncertain.

Irr is a global regulator of iron homeostasis

Although Irr was initially described in the context of haem biosynthesis, it is now clear that Irr is a global regulator of iron homeostasis and metabolism (Rudolph *et al.*, 2006; Todd *et al.*, 2006; Yang *et al.*, 2006b). Irr was initially discovered as a positive effector of ferric iron transport as well as a negative regulator of haem biosynthesis (Hamza *et al.*, 1998), and similar roles have been described for *B. abortus* Irr (Martinez *et al.*, 2005, 2006). Microarray analysis of a *B. japonicum irr* mutant shows that Irr affects the expression of many iron-regulated genes in a positive and negative manner. In addition, several iron-regulated genes in *R. leguminosarum* are derepressed in an *irr* strain (Todd *et al.*, 2006). However, iron transport is controlled by RirA in *R. leguminosarum* and *S. meliloti*, and thus the extent of the Irr regulon may be more limited in those species. A *cis*-acting DNA element called an iron control element (ICE) was found in the promoters of the divergently transcribed genes *hmuR* and *hmuT* from *B. japonicum*, and shown to be necessary for activation of those genes under iron limitation (Nienaber *et al.*, 2001). This element was found to bind Irr in a yeast one hybrid screen (Rudolph *et al.*, 2006), and *hmuR* and *hmuT* promoter activity is attenuated in an *irr* mutant (Nienaber *et al.*, 2001). Furthermore, bioinformatic analyses identified ICE-like motifs upstream of many open reading frames in *B. japonicum* and other α-Proteobacteria (Rodionov *et al.*, 2006; Rudolph *et al.*, 2006). Evidence for control by Irr in the absence of an ICE motif was described in the α-Proteobacteria *Brucella abortus* (Martinez *et al.*, 2006) and *Bartonella quintana* (Battisti *et al.*, 2007; Parrow *et al.*, 2009).

Although Irr activity has been characterized extensively in the context of its negative control of *hemB*, its promoter does not contain an ICE motif, and direct control by Irr remains unknown for that gene.

However, *in vitro* Irr binds to ICE motifs in the promoters of *blr7895*, a gene of unknown function, and the bacterioferritin gene. These genes that are down-regulated under iron limitation (Rudolph *et al.*, 2006), consistent with a repressor role for Irr. Moreover, these genes are derepressed in an *irr* mutant, and Irr occupies the promoters of those genes *in vivo* (Sangwan *et al.*, 2008). In addition, Irr represses transcription *in vitro* (Sangwan *et al.*, 2008). Thus, Irr is a transcriptional repressor. Irr activity requires the presence of divalent metal *in vitro* for high affinity DNA binding, although the role of the metal in Irr activity is not known (Sangwan *et al.*, 2008).

Direct positive control of gene expression by Irr has been demonstrated in *B. japonicum* for the *hmuR* and *hmuT* genes as described above (Rudolph *et al.*, 2006) and also for ferric siderophore receptor genes (Small *et al.*, 2009). This work has been hampered by the fact that Irr does not bind well to most ICE motifs *in vitro*. Nevertheless, direct binding was established by various techniques, and Irr is very likely to be an activator (Rudolph *et al.*, 2006; Small *et al.*, 2009; Yang *et al.*, 2006b).

Iron homeostasis is controlled by the status of haem via Irr

Irr interacts directly with the haem biosynthesis enzyme ferrochelatase, resulting in degradation under iron replete conditions, or accumulation of active protein under iron limitation (Qi and O'Brian, 2002). Thus, the discovery that Irr is a global regulator of iron-regulated genes indicates that iron homeostasis is controlled by the status of haem. Indeed, a haem-deficient strain of *B. japonicum* cannot maintain normal iron homeostasis. Control of Irr-regulated genes is aberrant in a haem-defective *B. japonicum* mutant, resulting in iron replete cells behaving as if they are iron-limited (Yang *et al.*, 2006b). The haem mutant has abnormally high cellular iron content, probably because iron transport genes are constitutively activated due to persistence of Irr in that strain. Accordingly, under iron limitation an *irr* mutant behaves as if it were iron replete even though cellular iron levels are lower than found in the wild type (Yang *et al.*, 2006b).

Most bacteria studied to date sense and respond to iron directly to regulate gene expression. That is, iron binds directly to a regulatory protein to modulate its activity. Iron binding to Fur confers DNA-binding activity on the protein, as also occurs for the DtxR regulator from *Corynebacterium diphtheriae* and its homologue, the IdeR protein from *Mycobacterium tuberculosis* (Escolar *et al.*, 1999; Pohl *et al.*, 1999a,b). However, *B. japonicum*, and perhaps other α-Proteobacteria, do not sense iron directly, but rather sense and respond to an iron-dependent process, namely the biosynthesis of haem. Is there an advantage to this type of control? Approximately one-half of the total iron in iron-limited *B. japonicum* cells is found in haem (unpublished observations). Since haem biosynthesis places such a high energy demand on the cell, this synthesis may serve as a sensitive indicator of the overall iron status. Also, many iron-dependent processes such as electron transport, tricarboxylic acid cycle and detoxification are associated with aerobic metabolism, which also requires haem. Therefore, it may allow a better coordination of cellular events.

Irr is prevalent in the α-Proteobacteria

Irr is prevalent in the Alpha subdivision of the proteobacterial phylum. Amongst the sequenced genomes, it is ubiquitious in the order *Rhizobiales* and *Rhodobacteriales*, and found in some *Rhodospirilles* as well (reviewed in Rodionov *et al.*, 2006). It is also present in the marine bacterium *Pelagibacter ubique*, which is in the order *Rickettsiales*, but is not present in its obligate intracellular relatives *Rickettsia*, *Wolbachia* and *Ehrlichia*. Interestingly, an Irr homologue is also found in *Acidothiobacillus ferrooxidans*, a γ-Proteobacterium that lives in acidic environments and is exposed to iron predominantly in the ferrous form. Microarray analysis shows that the vast majority of *B. japonicum* genes that are strongly regulated by iron are under the control of Irr. Thus, Irr is the major iron regulator in that bacterium and probably in other *Bradyrhizobiaceae*, with Fur having a much less prominent role. However, many rhizobia contain RirA in addition to Irr (Chao *et al.*, 2005; Todd *et al.*, 2005; Todd *et al.*, 2002; Viguier *et al.*, 2005; Yeoman *et al.*, 2004). Numerous cellular systems controlled by Irr in *B. japonicum* are regulated by RirA in *S. meliloti* and *R. leguminosarum* (Chao *et al.*, 2005; Todd *et al.*, 2002; Viguier *et al.*, 2005). Notably, haem and ferric iron transport genes are controlled by RirA in those two species, but by Irr in *B. japonicum*. These systems are negatively controlled by RirA in the presence of iron in species, whereas Irr appears to positively control them under low iron conditions. *B. abortus* contains both Irr and RirA, but siderophore genes in that organism are controlled by Irr (Martinez *et al.*, 2006), therefore it may not be possible to assume a priori the role of RirA in organisms that contain it. Nevertheless, Irr and RirA appear to usurp the role of Fur to varying degrees in organisms that contain these novel regulators.

The RirA protein

The realization that Fur is not a global regulator of iron transport in *Rhizobium leguminosarum* led Johnston and colleagues to search for mutants that showed deregulation of genes that are transcriptionally controlled by iron, leading to the discovery of RirA (Todd *et al.*, 2002). They found that numerous iron-regulated genes are constitutively high in a *rirA* mutant, indicating negative control by RirA. Similarly, RirA controls expression of the rhizobactin siderophore synthesis operon in *S. meliloti* (Viguier *et al.*, 2005), and microarray analysis reveals a large RirA regulon that includes numerous genes involved in iron transport, energy metabolism and exopolysaccharide production (Chao *et al.*, 2005). Thus, systems transcriptionally repressed by Fur in *E. coli* are negatively controlled by RirA in *S. meliloti* and *R. leguminosarum*. Although most genes down-regulated by iron limitation in the microarray study are independent of RirA in *S. meliloti* (Chao *et al.*, 2005), an analysis of the *R. leguminosarum* proteome reveals 17 proteins that are diminished in an *rirA* strain (Todd *et al.*, 2005).

A *S. meliloti rirA* mutant has a growth phenotype in iron replete media that is partially restored under iron limitation (Chao *et al.*, 2005). The growth deficiency of the mutant in the presence of iron may be caused by the accumulation of iron to toxic levels as a result of elevated iron transport activity. Consistent with this, the *rirA* strain is more sensitive to H_2O_2 in the presence of iron, presumably due to the generation of reactive oxygen species via the Fenton reaction (Chao *et al.*, 2005). Similarly, a *rirA* mutant of *Agrobacterium tumefaciens* constitutively expresses iron uptake genes, and is more sensitive than the wild type to the iron-activated antibiotic streptonigrin and to hydrogen peroxide (Ngok-Ngam *et al.*, 2009).

As described above, the rhizobia can use haem as an iron source, which requires both its transport into cells, and cleavage of the macrocycle to release iron. A *rirA* mutant is unable to use haem as an iron source (Chao *et al.*, 2005). The *hemPSTU* operon is constitutively expressed in the mutant, indicating that lack of growth is due to haem toxicity rather than iron deficiency (Chao *et al.*, 2005). This conclusion is supported by the inability to rescue the growth phenotype of the *rirA* mutant in the presence of haem by addition of inorganic iron.

RirA has not yet been studied *in vitro*, and thus its exact mechanism of function has not been elucidated. RirA belongs to the Rrf2 family of putative transcriptional regulators, a family that is not well characterized as a whole. The best described member is IscR, a transcriptional regulator that represses iron–sulphur cluster assembly gene operons, and activates the *suf* operon (Giel *et al.*, 2006; Schwartz *et al.*, 2001; Yeo *et al.*, 2006). IscR is an iron–sulphur protein, and the three cysteines are conserved in RirA (Todd *et al.*, 2002). IscR is an active protein both in the absence and presence of the iron–sulphur cluster. Repression of the *iscRSUA* operon requires the iron–sulphur centre, but the demetallated protein activates *sufABCDFE*. The lability of the iron–sulphur cluster links its assembly to the control of O_2-regulated genes (Giel *et al.*, 2006). Perhaps more relevant to the current discussion is that IscR responds to the iron and sulphur status (Gyaneshwar *et al.*, 2005; Outten *et al.*, 2004). Johnston and colleagues speculate that RirA may respond to iron through the status of an iron–sulphur cluster (Todd *et al.*, 2006). The generality of IscR function to Rrf2 family members as a whole is not yet known. It will be important to determine whether RirA is an iron–sulphur protein, and whether the status of that moiety is sensitive to the cellular iron status.

By examining the promoters of several *R. leguminosarum* genes deregulated in an *rirA* mutant, a consensus RirA-dependent *cis*-acting regulatory element was identified (Yeoman *et al.*, 2004). Deletion or mutation of these elements results in loss of iron responsiveness. Furthermore, a bioinformatic search identified RirA-responsive elements upstream of numerous genes in *R. leguminosarum* and *S. meliloti* known to be regulated by RirA (Rodionov *et al.*, 2006). Several iron-regulated *R. leguminosarum* genes are constitutive when introduced into *Paracoccus denitrificans* on a plasmid, but iron-responsiveness is restored when the *rirA* gene is also introduced (Yeoman *et al.*, 2004). Collectively, it is very likely that RirA binds directly to these promoters to repress gene expression in the presence of iron.

RirA is prevalent in the *Rhizobiaceae,* which includes *Brucella, Bartonella* and *Agrobacterium,* as well as *Sinorhizobium* and *Rhizobium,* but homologues are not confined to this taxonomic group. Proteins with high similarity (at least 60% similarity) to *R. leguminosarum* RirA are present in *Nitrobacter* (Bradyrhizobiaceae), *Labrenzia* (Rhodobacterales), and in species within the more distant γ- and β-Proteobacteria, and Firmicutes. The prevalence of RirA within the Rhizobiaceae might suggest that it is the primary iron responsive regulator within this group as a whole, as is found for *R. leguminosarum* and *S. meliloti.* However, studies in *Brucella abortus* and *Bartonella quintana* do not confirm this idea according to their authors. As described above, synthesis of the *B. abortus* siderophores brucebactin and 2,3-dihydroxybenzoic acid is controlled by Irr (Martinez *et al.*, 2005, 2006). However, a putative RirA binding site upstream of the *dhbCEBAD* operon encoding 2,3 dihydroxybenzoic acid synthesis proteins was identified in a bioinformatic study (Rodionov *et al.*, 2006). Thus, a role for RirA in control of *B. abortus* siderophore gene expression is possible. In *B. quintana,* genes encoding haem-binding proteins (Battisti *et al.*, 2007) and a haem utilization locus (Parrow *et al.*, 2009) are regulated by Irr rather than by RirA. In the latter study, control of gene expression was assessed in wild type cells that overexpressed *fur, irr* or *rirA in trans.*

Acknowledgements

Research from the authors' laboratories was supported by a grant from PEDECIBA-Uruguay to E.F., NIH grant GM067966 to M.R.O'B and an NIH Fogarty International Research Collaboration Award R03TW007353 to M.R.O'B with E.F. as the foreign collaborator.

References

Ahn, B.E., Cha, J., Lee, E.J., Han, A.R., Thompson, C.J., and Roe, J.H. (2006). Nur, a nickel-responsive regulator of the Fur family, regulates superoxide dismutases and nickel transport in *Streptomyces coelicolor.* Mol. Microbiol. *59,* 1848–1858.

Altuvia, S., Almiron, M., Huisman, G., Kolter, R., and Storz, G. (1994). The *dps* promoter is activated by OxyR during growth and by IHF and sigma S in stationary phase. Mol. Microbiol. *13,* 265–272.

Altuvia, S., Weinstein-Fischer, D., Zhang, A., Postow, L., and Storz, G. (1997). A small, stable RNA induced by oxidative stress: role as a pleiotropic regulator and antimutator. Cell. *90,* 43–53.

Amarelle, V., O'Brian, M.R., and Fabiano, E. (2008). ShmR is essential for utilization of heme as a nutritional iron source in *Sinorhizobium meliloti.* Appl. Environ. Microbiol. *74,* 6473–6475.

Baginsky, C., Brito, B., Imperial, J., Palacios, J.M., and Ruiz-Argueso, T. (2002). Diversity and evolution of hydrogenase systems in rhizobia. Appl. Environ. Microbiol. *68,* 4915–4924.

Baichoo, N., and Helmann, J.D. (2002). Recognition of DNA by Fur: a reinterpretation of the Fur box consensus sequence. J. Bacteriol. *184,* 5826–5832.

Barsomian, G.D., Urzainqui, A., Lohman, K., and Walker, G.C. (1992). *Rhizobium meliloti* mutants unable to synthesize anthranilate display a novel symbiotic phenotype. J. Bacteriol. *174,* 4416–4426.

Battisti, J.M., Smitherman, L.S., Sappington, K.N., Parrow, N.L., Raghavan, R., and Minnick, M.F. (2007). Transcriptional regulation of the heme binding protein gene family of *Bartonella quintana* is accomplished by a novel promoter element and iron response regulator. Infect. Immun. *75,* 4373–4385.

Battistoni, F., Platero, R., Duran, R., Cervenansky, C., Battistoni, J., Arias, A., and Fabiano, E. (2002a). Identification of an iron-regulated, hemin-binding outer membrane protein in *Sinorhizobium meliloti.* Appl. Environ. Microbiol. *68,* 5877–5881.

Battistoni, F., Platero, R., Noya, F., Arias, A., and Fabiano, E. (2002b). Intracellular Fe content influences nodulation competitiveness of *Sinorhizobium meliloti* strains as inocula of alfalfa. Soil Biol. Biochem. *34,* 593–597.

Bellini, P., and Hemmings, A.M. (2006). *In vitro* characterization of a bacterial manganese uptake regulator of the Fur superfamily. Biochemistry. *45,* 2686–2698.

Benson, H.P., Boncompagni, E., and Guerinot, M.L. (2005). An iron uptake operon required for proper nodule development in the *Bradyrhizobium japonicum*-soybean symbiosis. Mol. Plant Microbe Interact. *18,* 950–959.

Benson, H.P., LeVier, K., and Guerinot, M.L. (2004). A dominant-negative *fur* mutation in *Bradyrhizobium japonicum.* J. Bacteriol. *186,* 1409–1414.

Berlett, B.S., and Stadtman, E.R. (1997). Protein oxidation in aging, disease and oxidative stress. J. Biol. Chem. *272,* 20313–20316.

Bsat, N., Herbig, A., Casillas-Martinez, L., Setlow, P., and Helmann, J.D. (1998). *Bacillus subtilis* contains multiple Fur homologs: identification of the iron uptake (Fur) and peroxide regulon (PerR) repressors. Mol. Microbiol. *29*, 189–198.

Burkhardt, R., and Braun, V. (1987). Nucleotide sequence of the *fhuC* and *fhuD* genes involved in iron (III) hydroxamate transport: domains in FhuC homologous to ATP-binding proteins. Mol. Gen. Genet. *209*, 49–55.

Carlton, T.M., Sullivan, J.T., Stuart, G.S., Hutt, K., Lamont, I.L., and Ronson, C.W. (2007). Ferrichrome utilization in a mesorhizobial population: microevolution of a three-locus system. Environ. Microbiol. *9*, 2923–2932.

Carson, K.C., Meyer, J.-M., and Dilworth, M.J. (2000). Hydroxamate siderophores of root nodule bacteria. Soil Biol. Biochem. *32*, 11–21.

Carter, R.A., Worsley, P.S., Sawers, G., Challis, G.L., Dilworth, M.J., Carson, K.C., Lawrence, J.A., Wexler, M., Johnston, A.W., and Yeoman, K.H. (2002). The *vbs* genes that direct synthesis of the siderophore vicibactin in *Rhizobium leguminosarum*: their expression in other genera requires ECF sigma factor RpoI. Mol. Microbiol. *44*, 1153–1166.

Cescau, S., Cwerman, H., Letoffe, S., Delepelaire, P., Wandersman, C., and Biville, F. (2007). Heme acquisition by hemophores. Biometals *20*, 603–613.

Challis, G.L. (2005). A widely distributed bacterial pathway for siderophore biosynthesis independent of nonribosomal peptide synthetases. Chem. Biochem. *6*, 601–611.

Chao, T.C., Becker, A., Buhrmester, J., Pühler, A., and Weidner, S. (2004). The *Sinorhizobium meliloti fur* gene regulates, with dependence on Mn(II), transcription of the *sitABCD* operon, encoding a metal-type transporter. J. Bacteriol. *186*, 3609–3620.

Chao, T.C., Buhrmester, J., Hansmeier, N., Pühler, A., and Weidner, S. (2005). Role of the regulatory gene *rirA* in the transcriptional response of *Sinorhizobium meliloti* to iron limitation. Appl. Environ. Microbiol. *71*, 5969–5982.

Chauhan, S., Titus, D.E., and O'Brian, M.R. (1997). Metals control activity and expression of the heme biosynthesis enzyme δ-aminolevulinic acid dehydratase in *Bradyrhizobium japonicum*. J. Bacteriol. *179*, 5516–5520.

Chen, L., Keramati, L., and Helmann, J.D. (1995). Coordinate regulation of *Bacillus subtilis* peroxide stress genes by hydrogen peroxide and metal ions. Proc. Natl. Acad. Sci. U.S.A. *92*, 8190–8194.

Chen, W.M., Moulin, L., Bontemps, C., Vandamme, P., Bena, G., and Boivin-Masson, C. (2003). Legume symbiotic nitrogen fixation by beta-proteobacteria is widespread in nature. J. Bacteriol. *185*, 7266–7272.

Christman, M.F., Storz, G., and Ames, B.N. (1989). OxyR, a positive regulator of hydrogen peroxide-inducible genes in *Escherichia coli* and *Salmonella typhimurium*, is homologous to a family of bacterial regulatory proteins. Proc. Natl. Acad. Sci. U.S.A. *86*, 3484–3488.

Claverys, J.P. (2001). A new family of high-affinity ABC manganese and zinc permeases. Res. Microbiol. *152*, 231–243.

Cornelis, P., Matthijs, S., and Van Oeffelen, L. (2009). Iron uptake regulation in *Pseudomonas aeruginosa*. Biometals *22*, 15–22.

Coulton, J.W., Mason, P., and Allatt, D.D. (1987). *fhuC* and *fhuD* genes for iron (III)-ferrichrome transport into *Escherichia coli* K-12. J. Bacteriol. *169*, 3844–3849.

Cuiv, P.O., Clarke, P., Lynch, D., and O'Connell, M. (2004). Identification of *rhtX* and *fptX*, novel genes encoding proteins that show homology and function in the utilization of the siderophores rhizobactin 1021 by *Sinorhizobium meliloti* and pyochelin by *Pseudomonas aeruginosa*, respectively. J. Bacteriol. *186*, 2996–3005.

Cuiv, P.O., Keogh, D., Clarke, P., and O'Connell, M. (2008). The *hmuUV* genes of *Sinorhizobium meliloti* 2011 encode the permease and ATPase components of an ABC transport system for the utilization of both haem and the hydroxamate siderophores, ferrichrome and ferrioxamine B. Mol. Microbiol. *70*, 1261–1273.

Delany, I., Rappuoli, R., and Scarlato, V. (2004). Fur functions as an activator and as a repressor of putative virulence genes in *Neisseria meningitidis*. Mol. Microbiol. *52*, 1081–1090.

Delany, I., Spohn, G., Rappuoli, R., and Scarlato, V. (2001). The Fur repressor controls transcription of iron-activated and -repressed genes in *Helicobacter pylori*. Mol. Microbiol. *42*, 1297–1309.

Delany, I., Spohn, G., Rappuoli, R., and Scarlato, V. (2003). An anti-repression Fur operator upstream of the promoter is required for iron-mediated transcriptional autoregulation in *Helicobacter pylori*. Mol. Microbiol. *50*, 1329–1338.

Díaz-Mireles, E., Wexler, M., Sawers, G., Bellini, D., Todd, J.D., and Johnston, A.W. (2004). The Fur-like protein Mur of *Rhizobium leguminosarum* is a Mn^{2+}-responsive transcriptional regulator. Microbiology *150*, 1447–1456.

Díaz-Mireles, E., Wexler, M., Todd, J.D., Bellini, D., Johnston, A.W., and Sawers, R.G. (2005). The manganese-responsive repressor Mur of *Rhizobium leguminosarum* is a member of the Fur-superfamily that recognizes an unusual operator sequence. Microbiology *151*, 4071–4078.

Dilworth, M.J. (1980). [74] Leghemoglobins. In Methods in Enzymology (Academic Press), pp. 812–823.

Dilworth, M.J., Carson, K.C., Giles, R.G.F., Byrne, L.T., and Glenn, A.R. (1998). *Rhizobium leguminosarum* bv. viciae produces a novel cyclic trihydroxamate siderophore, vicibactin. Microbiology *144*, 781–791.

Donadio, S., Monciardini, P., and Sosio, M. (2007). Polyketide synthases and nonribosomal peptide synthetases: the emerging view from bacterial genomics. Nat. Prod. Rep. *24*, 1073–1109.

Downie, J.A., and Walker, S.A. (1999). Plant responses to nodulation factors. Curr. Opin. Plant Biol. *2*, 483–489.

Elgrably-Weiss, M., Park, S., Schlosser-Silverman, E., Rosenshine, I., Imlay, J., and Altuvia, S. (2002). A *Salmonella enterica* serovar typhimurium *hemA* mutant is highly susceptible to oxidative DNA damage. J. Bacteriol. *184*, 3774–3784.

Enz, S., Mahren, S., Stroeher, U.H., and Braun, V. (2000). Surface signaling in ferric citrate transport gene induction: interaction of the FecA, FecR, and FecI regulatory proteins. J. Bacteriol. *182*, 637–646.

Escolar, L., Perez-Martin, J., and de Lorenzo, V. (1998). Binding of the fur (ferric uptake regulator) repressor of *Escherichia coli* to arrays of the GATAAT sequence. J. Mol. Biol. *283*, 537–547.

Escolar, L., Perez-Martin, J., and de Lorenzo, V. (1999). Opening the iron box: transcriptional metalloregulation by the Fur protein. J. Bacteriol. *181*, 6223–6229.

Ettema, T.J., and Andersson, S.G. (2009). The α-proteobacteria: the Darwin finches of the bacterial world. Biol. Letters. doi: 10.1098/rsbl.2008.0793

Expert, D., and Gill, P.R.J. (1992). Iron: a modulator in bacterial virulence and symbiotic nitrogen-fixation (Boca Raton, Florida, CRC Press).

Fabiano, E., Gill Jr., P.R., Noya, F., Bagnasco, P., De La Fuente, L., and Arias, A. (1995). Siderophore-mediated iron acquisition mutants in *Rhizobium meliloti* 242 and its effect on the nodulation kinetic of alfalfa nodules. Symbiosis. *19*, 197–211.

Finking, R., and Marahiel, M.A. (2004). Biosynthesis of nonribosomal peptides1. Annu. Rev. Microbiol. *58*, 453–488.

Friedman, Y.E., and O'Brian, M.R. (2003). A novel DNA-binding site for the ferric uptake regulator (Fur) protein from *Bradyrhizobium japonicum*. J. Biol. Chem. *278*, 38395–38401.

Friedman, Y.E., and O'Brian, M.R. (2004). The ferric uptake regulator (Fur) protein from *Bradyrhizobium japonicum* is an iron-responsive transcriptional repressor *in vitro*. J. Biol. Chem. *279*, 32100–32105.

Gaballa, A., and Helmann, J.D. (1998). Identification of a zinc-specific metalloregulatory protein, Zur, controlling zinc transport operons in *Bacillus subtilis*. J. Bacteriol. *180*, 5815–5821.

Genco, C.A., and Dixon, D.W. (2001). Emerging strategies in microbial haem capture. Mol. Microbiol. *39*, 1–11.

Georgiadis, M.M., Komiya, H., Chakrabarti, P., Woo, D., Kornuc, J.J., and Rees, D.C. (1992). Crystallographic structure of the nitrogenase iron protein from *Azotobacter vinelandii*. Science. *257*, 1653–1659.

Gibson, K.E., Kobayashi, H., and Walker, G.C. (2008). Molecular determinants of a symbiotic chronic infection. Annu. Rev. Genet. *42*, 413–441.

Giel, J.L., Rodionov, D., Liu, M., Blattner, F.R., and Kiley, P.J. (2006). IscR-dependent gene expression links iron–sulphur cluster assembly to the control of O_2-regulated genes in *Escherichia coli*. Mol. Microbiol. *60*, 1058–1075.

Gill, P.R., Jr., and Neilands, J.B. (1989). Cloning a genomic region required for a high-affinity iron-uptake system in *Rhizobium meliloti* 1021. Mol. Microbiol. 3, 1183–1189.

Giraud, E., Fardoux, J., Fourrier, N., Hannibal, L., Genty, B., Bouyer, P., Dreyfus, B., and Vermeglio, A. (2002). Bacteriophytochrome controls photosystem synthesis in anoxygenic bacteria. Nature *417*, 202–205.

Giraud, E., Hannibal, L., Fardoux, J., Vermeglio, A., and Dreyfus, B. (2000). Effect of *Bradyrhizobium* photosynthesis on stem nodulation of *Aeschynomene sensitiva*. Proc. Natl. Acad. Sci. U.S.A. 97, 14795–14800.

Gruber, N., and Galloway, J.N. (2008). An Earth-system perspective of the global nitrogen cycle. Nature *451*, 293–296.

Guérinot, M.L., Meidl, E.J., and Plessner, O. (1990). Citrate as a siderophore in *Bradyrhizobium japonicum*. J. Bacteriol. *172*, 3298–3303.

Gyaneshwar, P., Paliy, O., McAuliffe, J., Popham, D.L., Jordan, M.I., and Kustu, S. (2005). Sulfur and nitrogen limitation in *Escherichia coli* K-12: specific homeostatic responses. J. Bacteriol. *187*, 1074–1090.

Hall, H.K., and Foster, J.W. (1996). The role of *fur* in the acid tolerance response of *Salmonella typhimurium* is physiologically and genetically separable from its role in iron acquisition. J. Bacteriol. *178*, 5683–5691.

Hamza, I., Chauhan, S., Hassett, R., and O'Brian, M.R. (1998). The bacterial Irr protein is required for coordination of heme biosynthesis with iron availability. J. Biol. Chem. *273*, 21669–21674.

Hamza, I., Hassett, R., and O'Brian, M.R. (1999). Identification of a functional *fur* gene in *Bradyrhizobium japonicum*. J. Bacteriol. *181*, 5843–5846.

Hantke, K. (2003). Is the bacterial ferrous iron transporter FeoB a living fossil? Trends Microbiol. *11*, 192–195.

Hennecke, H. (1992). The role of respiration in symbiotic nitrogen fixation (Dordrecht, The Netherlands, Kluwer Academic Publishers).

Herbig, A.F., and Helmann, J.D. (2001). Roles of metal ions and hydrogen peroxide in modulating the interaction of the *Bacillus subtilis* PerR peroxide regulon repressor with operator DNA. Mol. Microbiol. *41*, 849–859.

Herrada, G., Puppo, A., Moreau, S., Day, D.A., and Rigaud, J. (1993). How is leghemoglobin involved in peribacteroid membrane degradation during nodule senescence? FEBS Lett. *326*, 33–38.

Hirotsu, S., Chu, G.C., Unno, M., Lee, D.S., Yoshida, T., Park, S.Y., Shiro, Y., and Ikeda-Saito, M. (2004). The crystal structures of the ferric and ferrous forms of the heme complex of HmuO, a heme oxygenase of *Corynebacterium diphtheriae*. J. Biol. Chem. *279*, 11937–11947.

Hohle, T.H., and O'Brian M.R. (2009). The *mntH* gene encodes the major Mn^{2+} transporter in *Bradyrhizobium japonicum* and is regulated by manganese via the Fur protein. Mol. Microbiol. 72, 399–409.

Hu, R.G., Wang, H., Xia, Z., and Varshavsky, A. (2008). The N-end rule pathway is a sensor of heme. Proc. Natl. Acad. Sci. U.S.A. *105*, 76–81.

Hu, X., and Boyer, G.L. (1995). Isolation and characterization of the siderophore *N*-deoxyschizokinen from *Bacillus megaterium* ATCC 19213. Biometals 8, 357, 364.

Ishikawa, H., Kato, M., Hori, H., Ishimori, K., Kirisako, T., Tokunaga, F., and Iwai, K. (2005). Involvement of heme regulatory motif in heme-mediated ubiquitination and degradation of IRP2. Mol. Cell. *19*, 171–181.

Jeong, J., Rouault, T.A., and Levine, R.L. (2004). Identification of a heme-sensing domain in iron regulatory protein 2. J. Biol. Chem. *279*, 45450–45454.

Kaiser, B.N., Moreau, S., Castelli, J., Thomson, R., Lambert, A., Bogliolo, S., Puppo, A., and Day, D.A. (2003). The soybean NRAMP homologue, GmDMT1, is a symbiotic divalent metal transporter capable of ferrous iron transport. Plant J. *35*, 295–304.

Koster, W., and Braun, V. (1990). Iron (III) hydroxamate transport into *Escherichia coli*. Substrate binding to the periplasmic FhuD protein. J. Biol. Chem. *265*, 21407–21410.

Kunze, B., Trowitzsch-Kienast, W., Hofle, G., and Reichenbach, H. (1992). Nannochelins A, B and C, new iron-chelating compounds from *Nannocystis exedens* (myxobacteria). Production, isolation, physico-chemical and biological properties. J. Antibiot. (Tokyo) *45*, 147–150.

Kurokawa, H., Lee, D.S., Watanabe, M., Sagami, I., Mikami, B., Raman, C.S., and Shimizu, T. (2004). A redox-controlled molecular switch revealed by the crystal structure of a bacterial heme PAS sensor. J. Biol. Chem. *279*, 20186–20193.

Lankford, C.E. (1973). Bacterial assimilation of iron. Crit. Rev. Microbiol. *2*, 273–331.

Lensbouer, J.J., Patel, A., Sirianni, J.P., and Doyle, R.P. (2008). Functional characterization and metal ion specificity of the metal–citrate complex transporter from *Streptomyces coelicolor*. J. Bacteriol. *190*, 5616–5623.

Létoffé, S., Delepelaire, P., and Wandersman, C. (2006). The housekeeping dipeptide permease is the *Escherichia coli* heme transporter and functions with two optional peptide binding proteins. Proc. Natl. Acad. Sci. U.S.A. *103*, 12891–12896.

Letoffe, S., Delepelaire, P., and Wandersman, C. (2008). Functional differences between heme permeases: *Serratia marcescens* HemTUV permease exhibits a narrower substrate specificity (restricted to heme) than the *Escherichia coli* DppABCDF peptide-heme permease. J. Bacteriol. *190*, 1866–1870.

LeVier, K., Day, D.A., and Guerinot, M.L. (1996). Iron uptake by symbiosomes from soybean root nodules. Plant Physiol. *111*, 893–900.

Litwin, C.M., and Calderwood, S.B. (1993). Role of iron in regulation of virulence genes. Clin. Microbiol. Rev. *6*, 137–149.

Lynch, D., O'Brien, J., Welch, T., Clarke, P., Cuiv, P.O., Crosa, J.H., and O'Connell, M. (2001). Genetic organization of the region encoding regulation, biosynthesis, and transport of rhizobactin 1*021*, a siderophore produced by *Sinorhizobium meliloti*. J. Bacteriol. *183*, 2576–2585.

Martinez, M., Ugalde, R.A., and Almiron, M. (2005). Dimeric *Brucella abortus* Irr protein controls its own expression and binds haem. Microbiology *151*, 3427–3433.

Martinez, M., Ugalde, R.A., and Almiron, M. (2006). Irr regulates brucebactin and 2,3-dihydroxybenzoic acid biosynthesis, and is implicated in the oxidative stress resistance and intracellular survival of *Brucella abortus*. Microbiology *152*, 2591–2598.

Massé, E., and Gottesman, S. (2002). A small RNA regulates the expression of genes involved in iron metabolism in *Escherichia coli*. Proc. Natl. Acad. Sci. U.S.A. *99*, 4620–4625.

Modi, M., Shah, K.S., and Modi, V.V. (1985). Isolation and characterisation of catechol-like siderophore from cowpea *Rhizobium* RA-1. Arch. Microbiol. *141*, 156 – 158.

Mongkolsuk, S., and Helmann, J.D. (2002). Regulation of inducible peroxide stress responses. Mol. Microbiol. *45*, 9–15.

Mongkolsuk, S., Praituan, W., Loprasert, S., Fuangthong, M., and Chamnongpol, S. (1998). Identification and characterization of a new organic hydroperoxide resistance (*ohr*) gene with a novel pattern of oxidative stress regulation from *Xanthomonas campestris* pv. phaseoli. J. Bacteriol. *180*, 2636–2643.

Moreau, S., Day, D.A., and Puppo, A. (1998). Ferrous iron is transported across the peribacteroid membrane of soybean nodules. Planta. *207*, 83–87.

Moreau, S., Meyer, J.M., and Puppo, A. (1995). Uptake of iron by symbiosomes and bacteroids from soybean nodules. FEBS Lett. *361*, 225–228.

Moulin, L., Munive, A., Dreyfus, B., and Boivin-Masson, C. (2001). Nodulation of legumes by members of the beta-subclass of Proteobacteria. Nature *411*, 948–950.

Neilands, J.B. (1973). Microbial iron transport compounds (siderochromes) (Amsterdam, The Netherlands, Elsevier).

Neilands, J.B. (1981). Microbial iron compounds. Annu. Rev. Biochem. *50*, 715–731.

Neilands, J.B., and Leong, S.A. (1986). Siderophores in relation to plant growth and disease. Annu. Rev. Plant Physiol. *37*, 187–208.

Ngok-Ngam, P., Ruangkiattikul, N., Mahavihakanont, A., Virgem, S.S., Sukchawalit, R., and Mongkolsuk, S. (2009). Roles of *Agrobacterium tumefaciens* RirA in iron regulation, oxidative stress response, and virulence. J. Bacteriol. *191*, 2083–2090.

Nienaber, A., Hennecke, H., and Fischer, H.M. (2001). Discovery of a haem uptake system in the soil bacterium *Bradyrhizobium japonicum*. Mol. Microbiol. *41*, 787–800.

Noya, F., Arias, A., and Fabiano, E. (1997). Heme compounds as iron sources for nonpathogenic *Rhizobium* bacteria. J. Bacteriol. *179*, 3076–3078.

O'Brian, M.R., and Thöny-Meyer, L. (2002). Biochemistry, regulation and genomics of haem biosynthesis in prokaryotes. Adv. Microb. Physiol. *46*, 257–318.

O'Hara, G.W., Dilworth, M.J., Boonkerd, J.N., and Parkpian, P. (1988). Iron-deficiency specifically limits nodule development in peanut inoculated with *Bradyrhizobium* sp. New Phytologist. *108*, 51–57.

Oke, V., and Long, S.R. (1999). Bacteroid formation in the *Rhizobium*-legume symbiosis. Curr. Opin. Microbiol. *2*, 641–646.

Okujo, N., Sakakibara, Y., Yoshida, T., and Yamamoto, S. (1994). Structure of acinetoferrin, a new citrate-based dihydroxamate siderophore from *Acinetobacter haemolyticus*. Biometals 7, 170–176.

Oldroyd, G.E., and Downie, J.A. (2008). Coordinating nodule morphogenesis with rhizobial infection in legumes. Annu. Rev. Plant Biol. *59*, 519–546.

Outten, F.W., Djaman, O., and Storz, G. (2004). A *suf* operon requirement for Fe–S cluster assembly during iron starvation in *Escherichia coli*. Mol. Microbiol. *52*, 861–872.

Panek, H.R., and O'Brian, M.R. (2004). KatG is the primary detoxifier of hydrogen peroxide produced by aerobic metabolism in *Bradyrhizobium japonicum*. J. Bacteriol. *186*, 7874–7880.

Parrow, N.L., Abbott, J., Lockwood, A.R., Battisti, J.M., and Minnick, M.F. (2009). Function, regulation, and transcriptional organization of the hemin utilization locus of *Bartonella quintana*. Infect. Immun. *77*, 307–316.

Patel, H.N., Chakraborty, R.N., and Desai, S.B. (1988). Isolation and partial characterization of phenolate siderophore from *Rhizobium leguminosarum* IARI 102. FEMS Microbiol. Lett. *56*, 131–134.

Patzer, S.I., and Hantke, K. (1998). The ZnuABC high-affinity zinc uptake system and its regulator Zur in *Escherichia coli*. Mol. Microbiol. *28*, 1199–1210.

Persmark, M., Pittman, P., Buyer, J.S., Schwyn, B., Gill Jr., P.R., and Neilands, J.B. (1993). Isolation and structure of rhizobactin *1021*, a siderophore from the alfalfa symbiont *Rhizobium meliloti* 1021. J. Am. Chem. Soc. *115*, 3950–3956.

Platero, R., de Lorenzo, V., Garat, B., and Fabiano, E. (2007). *Sinorhizobium meliloti fur*-like (Mur) protein binds a *fur* box-like sequence present in the *mntA* promoter in a manganese-responsive manner. Appl. Environ. Microbiol. *73*, 4832–4838.

Platero, R., Peixoto, L., O'Brian, M.R., and Fabiano, E. (2004). Fur is involved in manganese-dependent regulation of *mntA* (*sitA*) expression in *Sinorhizobium meliloti*. Appl. Environ. Microbiol. *70*, 4349–4355.

Platero, R.A., Jaureguy, M., Battistoni, F.J., and Fabiano, E.R. (2003). Mutations in *sitB* and *sitD* genes affect manganese-growth requirements in *Sinorhizobium meliloti*. FEMS Microbiol. Lett. *218*, 65–70.

Plessner, O., Klapatch, T., and Guerinot, M.L. (1993). Siderophore utilization by *Bradyrhizobium japonicum*. Appl. Environ. Microbiol. *59*, 1688–1690.

Pohl, E., Holmes, R.K., and Hol, W.G. (1999a). Crystal structure of a cobalt-activated diphtheria toxin repressor–DNA complex reveals a metal-binding SH3-like domain. J. Mol. Biol. *292*, 653–667.

Pohl, E., Holmes, R.K., and Hol, W.G. (1999b). Crystal structure of the iron-dependent regulator (IdeR) from *Mycobacterium tuberculosis* shows both metal binding sites fully occupied. J. Mol. Biol. *285*, 1145–1156.

Poole, K. (2004). Iron transport systems in pathogenic bacteria: *Pseudomonas*. (Washington, DC: ASM Press).

Postle, K., and Kadner, R.J. (2003). Touch and go: tying TonB to transport. Mol. Microbiol. *49*, 869–882.

Postle, K., and Larsen, R. (2007). TonB-dependent energy transduction between outer and cytoplasmic membranes. Biometals *20*, 453–465.

Prieto-Alamo, M.J., Jurado, J., Gallardo-Madueno, R., Monje-Casas, F., Holmgren, A., and Pueyo, C. (2000). Transcriptional regulation of glutaredoxin and thioredoxin pathways and related enzymes in response to oxidative stress. J. Biol. Chem. *275*, 13398–13405.

Puri, S., and O'Brian, M.R. (2006). The *hmuQ* and *hmuD* genes from *Bradyrhizobium japonicum* encode heme-degrading enzymes. J. Bacteriol. *188*, 6476–6482.

Qi, Z., Hamza, I., and O'Brian, M.R. (1999). Heme is an effector molecule for iron-dependent degradation of the bacterial iron response regulator (Irr) protein. Proc. Natl. Acad. Sci. U.S.A. *96*, 13056–13061.

Qi, Z., and O'Brian, M.R. (2002). Interaction between the bacterial iron response regulator and ferrochelatase mediates genetic control of heme biosynthesis. Mol. Cell. *9*, 155–162.

Ratledge, C., and Dover, L.G. (2000). Iron metabolism in pathogenic bacteria. Annu. Rev. Microbiol. *54*, 881–941.

Ratliff, M., Zhu, W., Deshmukh, R., Wilks, A., and Stojiljkovic, I. (2001). Homologues of neisserial heme oxygenase in gram-negative bacteria: degradation of heme by the product of the *pigA* gene of *Pseudomonas aeruginosa*. J. Bacteriol. *183*, 6394–6403.

Raymond, K.N., and Dertz, E.A. (2004). Biochemical and physical properties of siderophores (Washington, DC, ASM Press).

Rees, D.C., and Howard, J.B. (2000). Nitrogenase: standing at the crossroads. Curr. Opin. Chem. Biol. *4*, 559–566.

Reigh, G., and O'Connell, M. (1993). Siderophore-mediated iron transport correlates with the presence of specific iron-regulated proteins in the outer membrane of *Rhizobium meliloti*. J. Bacteriol. *175*, 94–102.

Rioux, C.R., Jordan, D.C., and Rattray, J.B. (1986). Iron requirement of *Rhizobium leguminosarum* and secretion of anthranilic acid during growth on an iron-deficient medium. Arch. Biochem. Biophys. *248*, 175–182.

Ritz, D., Patel, H., Doan, B., Zheng, M., Aslund, F., Storz, G., and Beckwith, J. (2000). Thioredoxin 2 is involved in the oxidative stress response in *Escherichia coli*. J. Biol. Chem. *275*, 2505–2512.

Roberts, G.P., Youn, H., and Kerby, R.L. (2004). CO-sensing mechanisms. Microbiol. Mol. Biol. Rev. *68*, 453–473.

Rodionov, D.A., Gelfand, M.S., Todd, J.D., Curson, A.R., and Johnston, A.W. (2006). Computational reconstruction of iron- and manganese-responsive transcriptional networks in alpha-Proteobacteria. PLoS Comput. Biol. *2*, e163.

Rudolph, G., Semini, G., Hauser, F., Lindemann, A., Friberg, M., Hennecke, H., and Fischer, H.M. (2006). The Iron control element, acting in positive and negative control of iron-regulated *Bradyrhizobium japonicum* genes, is a target for the Irr protein. J. Bacteriol. *188*, 733–744.

Sangwan, I., and O'Brian M.R. (1992). Characterization of δ-Aminolevulinic acid formation in soybeanroot nodules. Plant Physiol. *98*, 1074–1079.

Sangwan, I., Small, S.K., and O'Brian, M.R. (2008). The *Bradyrhizobium japonicum* Irr protein is a transcriptional repressor with high-affinity DNA-binding activity. J. Bacteriol. *190*, 5172–5177.

Schauer, K., Rodionov, D.A., and de Reuse, H. (2008). New substrates for TonB-dependent transport: do we only see the 'tip of the iceberg'? Trends Biochem. Sci. *33*, 330–338.

Schmitt, M.P. (1997). Utilization of host iron sources by *Corynebacterium diphtheriae*: identification of a gene whose product is homologous to eukaryotic heme oxygenases and is required for acquisition of iron from heme and hemoglobin. J. Bacteriol. *179*, 838–845.

Schuller, D.J., Wilks, A., Ortiz de Montellano, P.R., and Poulos, T.L. (1999). Crystal structure of human heme oxygenase-1. Nat. Struct. Biol. *6*, 860–867.

Schuller, D.J., Zhu, W., Stojiljkovic, I., Wilks, A., and Poulos, T.L. (2001). Crystal structure of heme oxygenase from the gram-negative pathogen *Neisseria meningitidis* and a comparison with mammalian heme oxygenase-1. Biochemistry. *40*, 11552–11558.

Schwartz, C.J., Giel, J.L., Patschkowski, T., Luther, C., Ruzicka, F.J., Beinert, H., and Kiley, P.J. (2001). IscR, an Fe–S cluster-containing transcription factor, represses expression of *Escherichia coli* genes encoding Fe–S cluster assembly proteins. Proc. Natl. Acad. Sci. U.S.A. *98*, 14895–14900.

Skaar, E.P., Gaspar, A.H., and Schneewind, O. (2004). IsdG and IsdI, heme-degrading enzymes in the cytoplasm of *Staphylococcus aureus*. J. Biol. Chem. *279*, 436–443.

Skaar, E.P., Gaspar, A.H., and Schneewind, O. (2006). *Bacillus anthracis* IsdG, a heme-degrading monooxygenase. J. Bacteriol. *188*, 1071–1080.

Skorupska, A., Deryło, M., and Lorkiewicz, Z. (1989). Siderophore production and utilization by*Rhizobium trifolii* Biometals 2, 45–49.

Small, S.K., Puri, S., Sangwan, I., and O'Brian, M.R. (2009). Positive control of ferric siderophore receptor gene expression by the Irr protein in *Bradyrhizobium japonicum*. J. Bacteriol. *191*, 1361–1368.

Smith, M.J., and Neilands, J.B. (1984). Rhizobactin, a siderophore from *Rhizobium meliloti* J. Plant Nutr. 7, 449 – 458.

Smith, M.J., Shoolery, J.N., Schwyn, B., Holden, I., and Neilands, J.B. (1985). Rhizobactin, a structurally novel siderophore from *Rhizobium meliloti*. J. Am. Chem. Soc. *107*, 1739–1743.

Spaink, H.P. (2000). Root nodulation and infection factors produced by rhizobial bacteria. Annu. Rev. Microbiol. *54*, 257–288.

Stevens, J.B., Carter, R.A., Hussain, H., Carson, K.C., Dilworth, M.J., and Johnston, A.W. (1999). The *fhu* genes of *Rhizobium leguminosarum*, specifying siderophore uptake proteins: FhuDCB are adjacent to a pseudogene version of *fhuA*. Microbiology *145*, 593–601.

Stojiljkovic, I., Baumler, A.J., and Hantke, K. (1994). Fur regulon in gram-negative bacteria. Identification and characterization of new iron-regulated *Escherichia coli* genes by a *fur* titration assay [published erratum appears in J. Mol. Biol. 1994; *240*, 271]. J. Mol. Biol. *236*, 531–545.

Tang, C., Robson, A.D., and Dilworth, M.J. (1990). The role of iron in nodulation and nitrogen fixation in *Lupinus angustifolius* L. New Phytologist. *114*, 173–182.

Tao, K., Makino, K., Yonei, S., Nakata, A., and Shinagawa, H. (1989). Molecular cloning and nucleotide sequencing of *oxyR*, the positive regulatory gene of a regulon for an adaptive response to oxidative stress in *Escherichia coli*: homologies between OxyR protein and a family of bacterial activator proteins. Mol. Gen. Genet. *218*, 371–376.

Tartaglia, L.A., Storz, G., and Ames, B.N. (1989). Identification and molecular analysis of *oxyR*-regulated promoters important for the bacterial adaptation to oxidative stress. J. Mol. Biol. *210*, 709–719.

Todd, J.D., Sawers, G., and Johnston, A.W. (2005). Proteomic analysis reveals the wide-ranging effects of the novel, iron-responsive regulator RirA in *Rhizobium leguminosarum* bv. viciae. Mol. Genet. Genomics. *273*, 197–206.

Todd, J.D., Sawers, G., Rodionov, D.A., and Johnston, A.W. (2006). The *Rhizobium leguminosarum* regulator IrrA affects the transcription of a wide range of genes in response to Fe availability. Mol. Genet. Genomics. *275*, 564–577.

Todd, J.D., Wexler, M., Sawers, G., Yeoman, K.H., Poole, P.S., and Johnston, A.W. (2002). RirA, an iron-responsive regulator in the symbiotic bacterium *Rhizobium leguminosarum*. Microbiology *148*, 4059–4071.

Turner, S.L., and Young, J.P. (2000). The glutamine synthetases of rhizobia: phylogenetics and evolutionary implications. Mol. Biol. Evol. *17*, 309–319.

Verma, D.P.S., and Nadler, K.D. (1984). The *Rhizobium*-legume symbiosis: the host's point of view (New York, Springer-Verlag).

Viguier, C., Cuiv, P.O., Clarke, P., and O'Connell, M. (2005). RirA is the iron response regulator of the rhizobactin 1021 biosynthesis and transport genes in *Sinorhizobium meliloti* 2011. FEMS Microbiol. Lett. *246*, 235–242.

Wandersman, C., and Delepelaire, P. (2004). Bacterial iron sources: from siderophores to hemophores. Annu. Rev. Microbiol. *58*, 611–647.

Wandersman, C., and Stojiljkovic, I. (2000). Bacterial heme sources: the role of heme, hemoprotein receptors and hemophores. Curr. Opin. Microbiol. 3, 215–220.

Wexler, M., Todd, J.D., Kolade, O., Bellini, D., Hemmings, A.M., Sawers, G., and Johnston, A.W. (2003). Fur is not the global regulator of iron uptake genes in *Rhizobium leguminosarum*. Microbiology *149*, 1357–1365.

Wexler, M., Yeoman, K.H., Stevens, J.B., de Luca, N.G., Sawers, G., and Johnston, A.W. (2001). The *Rhizobium leguminosarum tonB* gene is required for the uptake of siderophore and haem as sources of iron. Mol. Microbiol. *41*, 801–816.

Wilderman, P.J., Sowa, N.A., FitzGerald, D.J., FitzGerald, P.C., Gottesman, S., Ochsner, U.A., and Vasil, M.L. (2004). Identification of tandem duplicate regulatory small RNAs in *Pseudomonas aeruginosa* involved in iron homeostasis. Proc. Natl. Acad. Sci. U.S.A. *101*, 9792–9797.

Wilhelm, S.W., and Trick, C.G. (1994). Iron-limited growth of cyanobacteria: Multiple siderophore production is a common response. Limnology and Oceanography. *39*, 1979–1984.

Wilks, A., and Schmitt, M.P. (1998). Expression and characterization of a heme oxygenase (Hmu O) from *Corynebacterium diphtheriae*. Iron acquisition requires oxidative cleavage of the heme macrocycle. J. Biol. Chem. *273*, 837–841.

Williams, P.H. (1979). Novel iron uptake system specified by ColV plasmids: an important component in the virulence of invasive strains of *Escherichia coli*. Infect. Immun. *26*, 925–932.

Winkelmann, G. (2007). Ecology of siderophores with special reference to the fungi. Biometals *20*, 379–392.

Wu, R., Skaar, E.P., Zhang, R., Joachimiak, G., Gornicki, P., Schneewind, O., and Joachimiak, A. (2005). *Staphylococcus aureus* IsdG and IsdI, heme-degrading enzymes with structural similarity to monooxygenases. J. Biol. Chem. *280*, 2840–2846.

Yamanaka, K., Ishikawa, H., Megumi, Y., Tokunaga, F., Kanie, M., Rouault, T.A., Morishima, I., Minato, N., Ishimori, K., and Iwai, K. (2003). Identification of the ubiquitin-protein ligase that recognizes oxidized IRP2. Nat. Cell Biol. *5*, 336–340.

Yang, J., Ishimori, K., and O'Brian, M.R. (2005). Two heme binding sites are involved in the regulated degradation of the bacterial iron response regulator (Irr) protein. J. Biol. Chem. *280*, 7671–7676.

Yang, J., Kim, K.D., Lucas, A., Drahos, K.E., Santos, C.S., Mury, S.P., Capelluto, D.G., and Finkielstein, C.V. (2008). A novel heme-regulatory motif mediates heme-dependent degradation of the circadian factor period 2. Mol. Cell. Biol. *28*, 4697–4711.

Yang, J., Panek, H.R., and O'Brian, M.R. (2006a). Oxidative stress promotes degradation of the Irr protein to regulate haem biosynthesis in *Bradyrhizobium japonicum*. Mol. Microbiol. *60*, 209–218.

Yang, J., Sangwan, I., Lindemann, A., Hauser, F., Hennecke, H., Fischer, H.M., and O'Brian, M.R. (2006b). *Bradyrhizobium japonicum* senses iron through the status of haem to regulate iron homeostasis and metabolism. Mol. Microbiol. *60*, 427–437.

Yang, J., Sangwan, I., and O'Brian, M.R. (2006c). The *Bradyrhizobium japonicum* Fur protein is an iron-responsive regulator *in vivo*. Mol. Genet. Genomics. *276*, 555–564.

Yeo, W.S., Lee, J.H., Lee, K.C., and Roe, J.H. (2006). IscR acts as an activator in response to oxidative stress for the *suf* operon encoding Fe–S assembly proteins. Mol. Microbiol. *61*, 206–218.

Yeoman, K.H., Curson, A.R., Todd, J.D., Sawers, G., and Johnston, A.W. (2004). Evidence that the *Rhizobium* regulatory protein RirA binds to cis-acting iron-responsive operators (IROs) at promoters of some Fe-regulated genes. Microbiology *150*, 4065–4074.

Yeoman, K.H., Wisniewski-Dye, F., Timony, C., Stevens, J.B., deLuca, N.G., Downie, J.A., and Johnston, A.W. (2000). Analysis of the *Rhizobium leguminosarum* siderophore-uptake gene *fhuA*: differential expression in free-living bacteria and nitrogen-fixing bacteroids and distribution of an *fhuA* pseudogene in different strains. Microbiology *146*, 829–837.

Zenke-Kawasaki, Y., Dohi, Y., Katoh, Y., Ikura, T., Ikura, M., Asahara, T., Tokunaga, F., Iwai, K., and Igarashi, K. (2007). Heme induces ubiquitination and degradation of the transcription factor Bach1. Mol. Cell. Biol. *27*, 6962–6971.

Zheng, M., Aslund, F., and Storz, G. (1998). Activation of the OxyR transcription factor by reversible disulfide bond formation. Science. *279*, 1718–1721.

Zheng, M., Doan, B., Schneider, T.D., and Storz, G. (1999). OxyR and SoxRS regulation of *fur*. J. Bacteriol. *181*, 4639–4643.

Zheng, M., Wang, X., Templeton, L.J., Smulski, D.R., LaRossa, R.A., and Storz, G. (2001). DNA microarray-mediated transcriptional profiling of the *Escherichia coli* response to hydrogen peroxide. J. Bacteriol. *183*, 4562–4570.

Zhu, W., Hunt, D.J., Richardson, A.R., and Stojiljkovic, I. (2000a). Use of heme compounds as iron sources by pathogenic neisseriae requires the product of the *hemO* gene. J. Bacteriol. *182*, 439–447.

Zhu, W., Wilks, A., and Stojiljkovic, I. (2000b). Degradation of heme in gram-negative bacteria: the product of the *hemO* gene of *Neisseriae* is a heme oxygenase. J. Bacteriol. *182*, 6783–6790.

Iron Uptake Systems in Pathogenic *Bordetella*

4

Timothy J. Brickman and Sandra K. Armstrong

Abstract

Upon colonization of the mammalian respiratory epithelium by mucosal pathogens of the genus *Bordetella*, the host–pathogen interaction causes inflammatory changes, immune activation, and host cell injury. In this dynamic environment, *Bordetella* cells scavenge the nutritional iron necessary for growth. The three classical *Bordetella* species produce the siderophore alcaligin. In addition, they can utilize xenosiderophores that could be produced by commensals or other microbes that transiently inhabit the nasopharynx. As infection progresses, extravasation of immune cells, erythrocytes and serum to the mucosal surface can occur, exacerbated by the damaging action of *Bordetella* toxins, thus providing iron sources such as transferrin and haem compounds to the microbe. The three characterized *Bordetella* iron systems for utilization of alcaligin, enterobactin and haem are each inducible by the cognate iron source. The ability to sense and respond to the presence of available iron sources allows these pathogens to adapt to temporal changes in iron source availability, and this ability is important for successful *in vivo* growth.

Introduction

Bordetella species and pathogenesis

The genus *Bordetella* includes eight currently recognized species of the *Alcaligenaceae* family of β-proteobacteria. The three so-called classical species, *Bordetella pertussis, Bordetella parapertussis,* and *Bordetella bronchiseptica,* are highly genetically related respiratory pathogens of mammals (reviewed in Mattoo and Cherry, 2005). The strictly human-adapted species *B. pertussis* is the etiological agent of whooping cough or pertussis. *B. parapertussis* strains can be divided into two distinct lineages: those that cause a pertussis-like illness in humans and those that infect ovine hosts. *B. bronchiseptica* causes respiratory diseases such as canine kennel cough and swine atrophic rhinitis, and is also capable of causing chronic asymptomatic infection of a broad range of other mammalian hosts. Comparative genomics studies suggest that *B. bronchiseptica* or a *B. bronchiseptica*-like organism is the progenitor of both *B. pertussis* and *B. parapertussis;* accordingly, the three classical *Bordetella* species have been grouped as the *B. bronchiseptica* taxonomic cluster. The genomes of *B. pertussis* and, to a lesser extent, *B. parapertussis,* are characterized by marked gene loss and rearrangement compared to *B. bronchiseptica* (Parkhill *et al.*, 2003). This genetic streamlining is likely to be a consequence of evolution of *B. pertussis* and *B. parapertussis* toward niche specialization.

B. pertussis, B. parapertussis, and *B. bronchiseptica* colonize the upper respiratory tract of the host, specifically targeting the ciliated cells of the respiratory epithelium. Induction of ciliostasis is an early event in *Bordetella* colonization, which together with epithelial cell damage and mucus accumulation may hinder bacterial clearance and increase host susceptibility to secondary infection (Anderton *et al.*, 2004). Since the isolation of *B. pertussis* in 1906 by Bordet and Gengou (Bordet and Gengou, 1906), pertussis has been widely studied. Pertussis is a vaccine-preventable disease characterized by paroxysmal coughing, which may be followed by sudden involuntary inspiration resulting in an audible whoop. Pertussis-associated lymphocytosis (Lagergren, 1963) and hypoglycaemia (Regan and Tolstoouhov, 1936) are attributed to the activity of released pertussis toxin (Munoz *et al.*, 1981; Cherry and Heininger, 2004). Immunity to *B. pertussis* appears to be complex, with both branches of adaptive immunity playing a role. Passive immunization studies have demonstrated the protective properties of antibodies in *Bordetella* infection (Granstrom *et al.*, 1991; Bruss and Siber, 1999), and experimental infections using athymic mice have shown the importance of cell-mediated immunity (particularly CD4-positive T cells) in combating *B. pertussis* infection (Mills *et al.*, 1993). Since the start of widespread vaccination during the 1940s using whole-cell preparations, and more recently with immunization using acellular pertussis vaccines, the annual incidence of reported pertussis cases in the USA has decreased dramatically (CDC, 2002). *B. pertussis* infection is recognized as a cause of persistent cough in adults, who may also serve as an important environmental reservoir for the disease. Waning immunity from childhood vaccination during adolescence and the potential for disease transmission from the adult population has led to recent development and implementation of adult acellular pertussis vaccines (Ward *et al.*, 2005; CDC, 2006).

Classical virulence determinants

Bacterial cell surface proteins including fimbriae, filamentous haemagglutinin and pertactin are implicated in adherence and host colonization by the species of the *B. bronchiseptica* taxonomic cluster. These bacteria produce several toxins that contribute to their virulence, and their lipopolysaccharides have typical endotoxic properties. Tracheal cytotoxin, a disaccharide tetrapeptide that is released during peptidoglycan synthesis, is responsible for much of the respiratory cytopathology found in infected hosts (Flak and Goldman, 1999). This toxin causes activation of the inducible nitric oxide synthase by interleukin 1, resulting in an increase in host nitric oxide production. Nitric oxide causes intracellular iron depletion and inactivation of iron-containing enzymes important for mitochondrial respiration and DNA synthesis. Subsequently, epithelial cellular tight junctions are compromised, leading to extrusion of the ciliated cells and denudation of the epithelium. The bifunctional adenylate cyclase toxin has both adenylate cyclase and haemolytic activities. Upon internalization by the host cell, the toxin is activated by calmodulin to catalyse the hyperproduction of cyclic AMP, leading to dysregulation of cellular function. The pore-forming haemolytic activity also contributes to the cytoxicity of adenylate cyclase (Hewlett *et al.*, 2006). Dermonecrotic toxin, another toxin causing considerable cytopathology, enzymatically modifies eukaryotic Rho regulatory GTPases, resulting in their constitutive activation and subsequent alterations in cellular morphology and differentiation. All three classical species of *Bordetella* have the genes encoding pertussis toxin; however, only *B. pertussis* produces and secretes pertussis toxin. Pertussis toxin catalyses the transfer of an ADP ribose group to the alpha subunit of host regulatory GTP-binding proteins, thereby inactivating the subunit and causing dysregulation of host cell function. Pertussis toxin, in toxoided form, is an essential protective antigen present in all acellular pertussis vaccines in current use.

In *B. pertussis*, *B. bronchiseptica* and *B. parapertussis*, expression of the genes encoding most of the known virulence factors is regulated by a two-component phosphorelay system consisting of the BvgS transmembrane sensor kinase and the DNA-binding response regulator BvgA (Weiss *et al.*, 1983; Uhl and Miller, 1994). Phosphorylated BvgA activates transcription of certain virulence-associated genes such as those encoding adhesins and protein toxins. Some genes are repressed by the BvgR repressor protein, which itself is encoded by a BvgA-activated gene (Merkel and Stibitz, 1995). Although BvgAS controls the expression of many virulence-associated genes, the natural signals affecting BvgAS activity in the infected host environment are unknown.

In vivo growth and iron acquisition

Upon colonization of the host respiratory epithelium, the ensuing interaction between host and pathogen results in inflammatory changes, activation of immune responses, and host cell injury. In this complex and dynamic environment, *Bordetella* cells scavenge the nutritional iron necessary for growth. The three classical *Bordetella* species produce the siderophore alcaligin. In addition, they can utilize xenosiderophores that could be produced by commensals or other microbes that transiently inhabit the nasopharynx. As *Bordetella* infection progresses, extravasation of immune cells, erythrocytes and serum to the mucosal surface can occur, exacerbated by the damaging action of *Bordetella* toxins on epithelial and other host cells (Persson *et al.*, 1991). Serum and cellular constituents released upon host cell lysis may provide iron sources including transferrin and haem compounds such as haemoglobin and haptoglobin, all of which can be readily assimilated by *Bordetella* species.

The initial environment encountered by these bacteria after transmission to a new host is the respiratory mucosa, where lactoferrin acts to withhold iron from invading microbes. As successful respiratory pathogens, the bordetellae must be able to utilize lactoferrin as an iron source or defeat its iron-sequestering effect. Redhead and colleagues reported that *B. pertussis* tightly bound lactoferrin, ovotransferrin and transferrin, and could use ovotransferrin as a sole iron source (Redhead *et al.*, 1987). Other groups found that growth of *B. pertussis* and *B. bronchiseptica* in the presence of human lactoferrin or transferrin did not require bacterial cell contact with the glycoproteins, suggesting the involvement of a siderophore (most likely alcaligin, the identity of which was unknown at that time) (Gorringe, 1990; Menozzi *et al.*, 1991). Two transferrin- and lactoferrin-binding proteins were isolated using affinity chromatography (Menozzi *et al.*, 1991); however, results of subsequent *Bordetella* genomic sequencing projects revealed no apparent homologues of lactoferrin or transferrin receptor genes. Currently it is thought that acquisition of transferrin- and lactoferrin-bound iron by bordetellae probably requires a siderophore- either alcaligin or a xenosiderophore that may be present in host mucus.

Of the known *Bordetella* iron sources, only the genetic systems for utilization of the native alcaligin siderophore, the enterobactin xenosiderophore, and haem have been characterized. These organisms also have multiple genes encoding predicted TonB-dependent iron receptors of unknown specificity. A key feature shared by the three known iron retrieval systems is that each is inducible by the cognate iron source. The ability to sense and respond to the presence of available iron sources suggested that the ability to discriminate between iron sources might be important for successful *in vivo* growth. Using a mouse respiratory infection model, it was established that the alcaligin, enterobactin, and haem utilization systems each make essential contributions to growth and survival of *B. pertussis* during infection (Brickman *et al.*, 2006, 2007, 2008; Brickman and Armstrong, 2007). Furthermore, it was found that these inducible iron systems are differentially expressed over the course of infection (Brickman *et al.*, 2008; Brickman and Armstrong, 2009), presumably in response to temporal changes in iron source availability.

Alcaligin siderophore utilization

The alcaligin siderophore

Alcaligin is a 20-membered macrocyclic dihydroxamate siderophore [1,8(*S*),11,18(*S*)-tetrahydroxy-1,6,11,16-tetraazacycloeicosane-2,5,12,15-tetrone] (Fig. 4.1) produced by *B. pertussis* and *B. bronchiseptica* (Moore *et al.*, 1995; Brickman *et al.*, 1996). *B. parapertussis* produces a siderophore (Pradel *et al.*, 1998) that is probably alcaligin, but it has not been conclusively identified. Nishio and coworkers first purified the alcaligin siderophore, produced by *Alcaligenes (Achromobacter) denitrificans* subspecies *xylosoxydans* KN3-1 (Nishio *et al.*, 1988; Nishio and Ishida, 1990) isolated from freshwater lake sediment. Later, Bergeron and coworkers successfully synthesized alcaligin (Bergeron *et al.*, 1991). Alcaligin is structurally similar to the macrocyclic siderophores bisucaberin, produced by the marine bacterium *Pseudoalteromonas haloplanktis* (Kameyama *et al.*, 1987; Takahashi *et al.*, 1987), and putrebactin produced by *Shewanella putrefaciens* (*oneidensis*) (Ledyard and Butler, 1997).

Alcaligin binds ferric iron at a 3:2 molar ratio (Fe_2Alc_3) at pH 6.0 (Fig. 4.1), and at a 1:1 molar ratio (FeAlc) at pH 2.0 (Nishio *et al.*, 1988). The Fe_2Alc_3 complex has a remarkably high stability constant that is estimated at 10^{37}/M at pH 6.0 (Nishio *et al.*, 1988). Crystallographic studies of the Fe_2Alc_3 complex

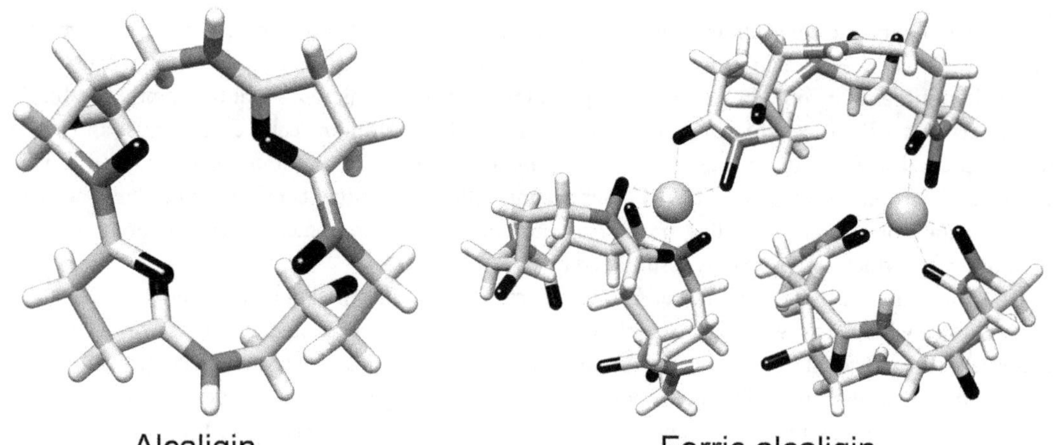

Figure 4.1 Three-dimensional molecular structures of alcaligin and its ferric complex, Fe_2Alc_3. Key: Oxygen, black; nitrogen, dark grey; carbon, light grey; hydrogen, white; iron atoms of the ferric alcaligin complex are depicted as light grey spheres.

determined that the free alcaligin ring is spatially preorganized for ferric iron binding (Hou *et al.*, 1996, 1998). This structural preorganization precludes the requirement for marked conformational changes to bind iron, providing a 100-fold increase in metal complex stability over ferric complexes formed by the more flexible linear dihydroxamate siderophore rhodotorulic acid.

Alcaligin biosynthesis

The alcaligin siderophore biosynthesis, regulation and transport gene cluster spans a contiguous 10.6-kb chromosomal region (Fig. 4.2) in *B. pertussis*, *B. bronchiseptica* and *B. parapertussis* (Parkhill *et al.*, 2003). Of the 'non-classical' *Bordetella* species with sequenced genomes, only the opportunistic pathogen *Bordetella holmesii* has an alcaligin gene cluster (Diavatopoulos *et al.*, 2006); *B. avium* and *B. petrii* do not. In *B. holmesii*, the alcaligin genes reside on a genomic island apparently acquired by horizontal transfer from *B. pertussis*, a process that may have contributed to the emergence of *B. holmesii* and its adaptation to the human host. All the known alcaligin biosynthesis functions are encoded by the *alcABCDER* operon (Table 4.1) (Armstrong and Clements, 1993; Giardina *et al.*, 1995; Kang *et al.*, 1996; Giardina *et al.*, 1997; Kang and Armstrong, 1998; Beaumont *et al.*, 1998; Pradel *et al.*, 1998). Although polypeptide siderophores and most microbial peptide antibiotics are produced by non-ribosomal peptide synthetase (NRPS) multienzymes, certain other siderophores including alcaligin are assembled by an NRPS-independent process from alternating dicarboxylic acid and diamine or amino alcohol units that are joined through amide or ester bonds. The prototypic NRPS-independent siderophore biosynthesis pathway for the production of the siderophore aerobactin uses the IucA and IucC synthetases to join the dicarboxylic acid and diamine units containing the iron-chelating functional groups (de Lorenzo *et al.*,

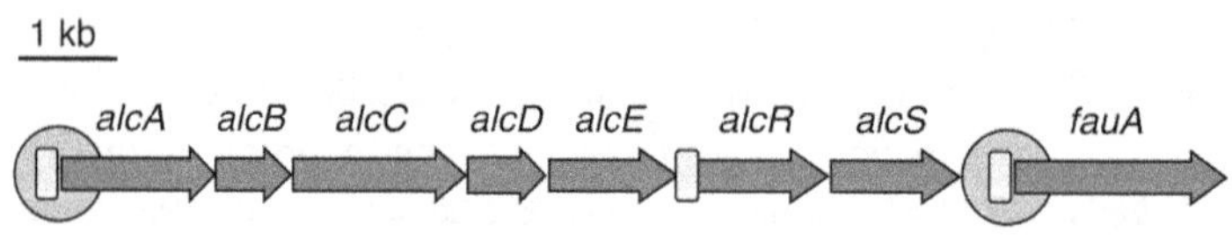

Figure 4.2 The *Bordetella* alcaligin siderophore system gene cluster. A 10.6-kb chromosomal DNA region is depicted. The arrows indicate the transcriptional orientations and spatial limits of genes, and rectangles represent known promoter regions. Known Fur binding sites are located upstream of *alcA*, *alcR* and *fauA*. The circles represent AlcR-responsive control regions upstream of *alcA* and *fauA*. A 1-kb scale bar is shown.

Table 4.1 *Bordetella* alcaligin siderophore system genes and proteins

Gene Symbol	Locus tag[a]	Product
alcA	BP2456	Alcaligin biosynthesis protein
alcB	BP2457	Alcaligin biosynthesis protein
alcC	BP2458	Alcaligin biosynthesis protein
alcD	BP2459	Hypothetical protein predicted to be involved in alcaligin biosynthesis
alcE	BP2460	Putative iron–sulphur protein predicted to be involved in alcaligin biosynthesis
alcR	BP2461	Transcriptional regulator of alcaligin biosynthesis and transport genes
alcS	BP2462	Alcaligin exporter
fauA	BP2463	Ferric alcaligin siderophore outer membrane receptor

[a]Locus tag designation for *Bordetella pertussis* Tohama I orthologue.

1986; de Lorenzo and Neilands, 1986). Other NRPS-independent siderophore biosynthesis pathways have been identified in other bacteria, all of which involve at least one enzyme with similarity to the aerobactin synthetases (Challis, 2005).

The enzymology of alcaligin biosynthesis has not been studied in detail; thus, most of the pathway remains largely hypothetical, based on similarities of Alc proteins with aerobactin synthetases and other NRPS-independent siderophore biosynthesis enzymes (Brickman and Armstrong, 1996; Giardina *et al.*, 1995; Kang *et al.*, 1996; Pradel *et al.*, 1998; Challis, 2005). Production of the essential precursor putrescine is catalysed by the pyridoxal phosphate-dependent ornithine decarboxylase Odc (Brickman and Armstrong, 1996) (**1**) (Fig. 4.3). Putrescine is then thought to be converted to *N*-hydroxyputrescine (**2**) by the putative FAD-dependent monooxygenase AlcA in a reaction similar to that catalysed by IucD in aerobactin biosynthesis. AlcB may catalyse the *N*-acylation of the hydroxylamine group in **2** with succinyl CoA in a manner similar to the IucB-dependent acylation of *N*6-hydroxylysine with acetyl CoA in the aerobactin pathway. Compound **3** could undergo C-hydroxylation by AlcE, which has significant similarity to proteins of the ring-hydroxylating dioxygenase family of iron–sulphur cluster proteins. AlcC, a member of the siderophore synthetase family that includes IucC, could catalyse nucleotide triphosphate-dependent dimerization of compound **4**, followed by macrocyclization to yield alcaligin (**5**). The function of AlcD in alcaligin biosynthesis is uncertain (Table 4.1). AlcD shows very limited amino acid sequence similarity with a putative iron reductase of *Vibrio mimicus*, with ArgH, an argino-succinate lyase of a marine *Synechococcus* species (Palenik *et al.*, 2003), as well as with a gene encoding a hypothetical iron–sulphur cluster protein that maps adjacent to a *Vibrio parahaemolyticus iutA* ferric aerobactin receptor gene homologue.

Alcaligin export

The *alcS* gene (Fig. 4.2) (previously designated as *orfX* or *bcr*) (Beaumont *et al.*, 1998; Pradel *et al.*, 1998) encodes a permease of the major facilitator superfamily class of proton motive force-dependent membrane efflux pumps (Brickman and Armstrong, 2005). Expression of *alcS* is independent of the *alcABCDER* control region and is not iron-repressible (Kang and Armstrong, 1998). AlcS exports monomeric alcaligin from the *Bordetella* cell and is important for the excretion of newly synthesized alcaligin (Brickman and Armstrong, 2005). In laboratory cultures, *alcS* null mutants produce less extra-cellular alcaligin than wild-type bacteria, but have significantly elevated levels of cell-associated alcaligin. This intracellular accumulation of alcaligin is deleterious for growth; as a consequence, *alcS* mutants have a growth defect under iron starvation conditions that can be suppressed by mutations that abrogate

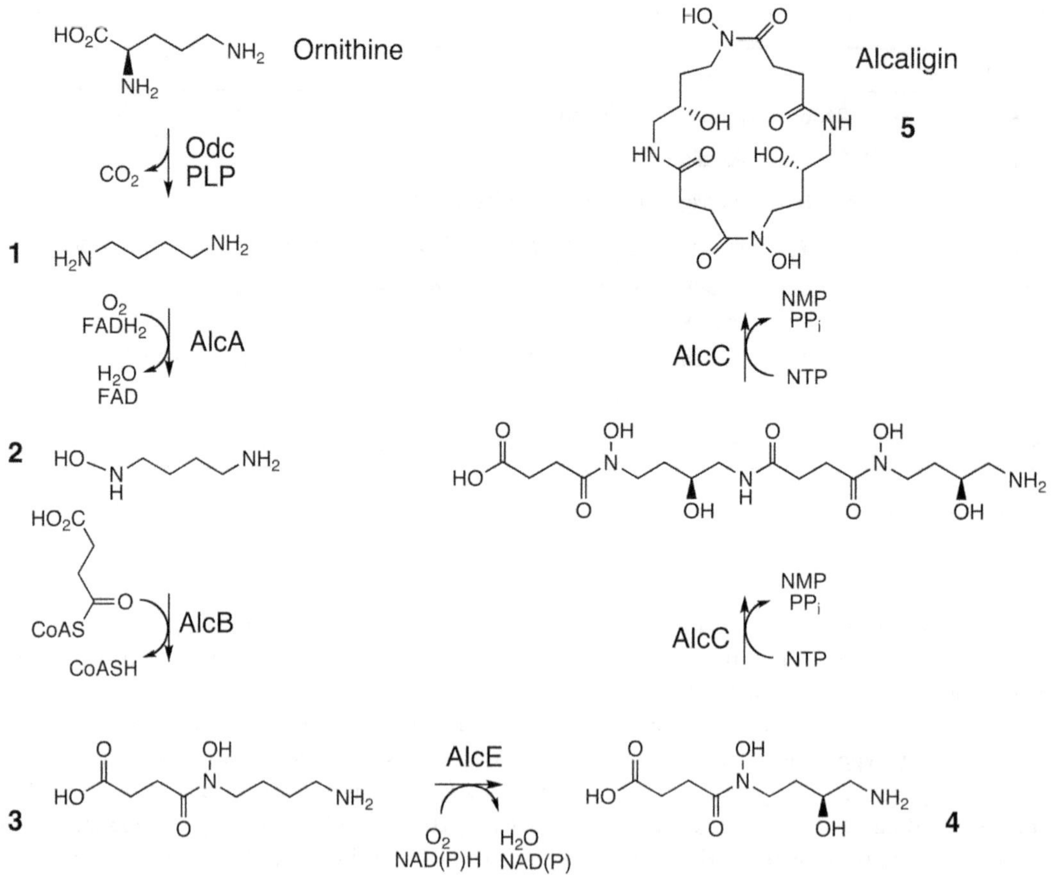

Figure 4.3 Biosynthesis of alcaligin. A proposed pathway for the biosynthesis of alcaligin in *Bordetella* species is shown, based on amino acid sequence similarities between alcaligin biosynthesis proteins and enzymes of other non-ribosomal peptide synthase-independent siderophore biosynthesis pathways. The initial conversion of ornithine to putrescine by the ornithine decarboxylase Odc is the only biochemically characterized step.

alcaligin production. Transcription studies established that the exporter activity of AlcS is required to maintain appropriate intracellular alcaligin levels for normal regulation of alcaligin system genes by the AlcR positive transcriptional regulator. Consequently, AlcS is central to the autogenous control process that regulates alcaligin gene expression.

Ferric alcaligin transport

Ferric alcaligin uptake in *Bordetella* species is dependent on the TonB system (Nicholson and Beall, 1999; Pradel *et al.*, 2000) and only one *tonB* gene has been identified in each of these species (Parkhill *et al.*, 2003). FauA is the 79-kDa TonB-dependent outer membrane receptor protein required for ferric alcaligin uptake (Brickman and Armstrong, 1999). The *fauA* gene (Fig. 4.2) is transcribed from its own Fur-regulated promoter and is dependent on the AlcR regulator and alcaligin inducer for maximal expression. The FauA receptor is believed to specifically recognize the Fe_2Alc_3 complex, since this complex predominates under the conditions thought to exist in the host niche, and preformed Fe_2Alc_3 complexes are rapidly transported by *Bordetella* cells *in vitro* by a FauA-dependent mechanism (Brickman and Armstrong, 1999). The fate of ferric alcaligin after it transits the outer membrane is unknown. Ferric alcaligin-specific transporters or permeases have not yet been identified in *Bordetella* species; but they are not encoded within the alcaligin system gene cluster.

Iron release from ferric alcaligin

The natural process of iron release from ferric alcaligin complexes is not fully understood, but a possible reductive mechanism has been proposed based on work from the Crumbliss laboratory group (Spasojevic *et al.*, 1999). Comparing alcaligin with other hydroxamate siderophores, siderophore denticity and structural preorganization due to cyclization were both found to influence redox potentials, but an increase in ligand denticity was shown to have a greater effect on redox potential than cyclization of the ligand backbone of the same denticity. Kinetic analysis of iron release from alcaligin determined that the structural preorganization of the iron-binding ligands of alcaligin strongly influenced the ligand dissociation paths for the Fe_2L_3 ferric complexes (Fig. 4.1), compared with iron release from the tetradentate siderophore rhodotorulic acid that lacks such structural preorganization (Boukhalfa *et al.*, 2000; Boukhalfa and Crumbliss, 2000).

Regulation of alcaligin system gene expression

The ferric uptake regulator protein, Fur, mediates iron repressibility of *Bordetella* alcaligin siderophore system genes. *B. bronchiseptica fur* mutants were identified by screening for constitutive alcaligin production after selection for manganese resistance (Brickman and Armstrong, 1995). The *B. pertussis fur* gene was cloned by complementation of the *B. bronchiseptica fur* mutant phenotype, restoring iron-repressible alcaligin production, and its identity was confirmed by nucleotide sequencing. At the same time, another group also identified and characterized the *B. pertussis fur* gene (Beall and Sanden, 1995b). The alcaligin siderophore gene cluster (Fig. 4.2) consists of four distinct transcriptional units: *alcABCDER*, *alcR*, *alcS* and *fauA* (Beaumont *et al.*, 1998; Kang and Armstrong, 1998; Pradel *et al.*, 1998; Brickman and Armstrong, 1999, 2005). Transcription of the *alcABCDER* operon initiates at a Fur and iron-repressible promoter upstream of *alcA* (Brickman and Armstrong, 1995; Kang *et al.*, 1996; Kang and Armstrong, 1998). In electrophoretic mobility shift assays using purified *B. pertussis* Fur protein, the *alcA* upstream DNA region was found to have multiple high-affinity Fur-binding sites that display a significant degree of positive cooperativity (Brickman and Armstrong, unpublished). The *alcR* gene is also expressed from a monocistronic transcript originating at a weak promoter immediately upstream of the *alcR* gene. This secondary *alcR* promoter is strongly repressed by Fur and iron, but is independent of AlcR activation (Beaumont *et al.*, 1998); metallated Fur protein specifically binds operator sequences at the *alcR* promoter region *in vitro*. A third Fur-binding region resides upstream of the *fauA* receptor gene (Brickman and Armstrong, 1999). The *fauA* transcript is monocistronic and, as with *alcABCDER* regulation, *fauA* transcripts emanate from a Fur and iron-repressible promoter that is activated by AlcR in response to alcaligin induction. In summary, Fur and iron-mediated negative control of transcription is exerted at three known promoter-operator regions within the alcaligin system gene cluster- upstream of the *alcABCDER* operon, upstream of *alcR* and upstream of *fauA*. The *alcS* alcaligin exporter gene appears to be constitutively expressed (Kang and Armstrong, 1998; Brickman and Armstrong, 2005).

When Fur repression is relieved by iron starvation, *fauA* and *alcABCDER* transcription is activated by the AlcR regulator by a mechanism requiring induction by alcaligin (Brickman *et al.*, 2001). AlcR (Beaumont *et al.*, 1998; Pradel *et al.*, 1998) is a member of the AraC/XylS family of transcriptional regulators (Gallegos *et al.*, 1997), and is most similar to the *Bordetella* BfeR regulator of enterobactin utilization genes (Anderson and Armstrong, 2004), the pyochelin siderophore system regulator PchR of *Pseudomonas aeruginosa* and the yersiniabactin siderophore system regulator YbtA of *Yersinia pestis*. AlcR is a 36-kDa protein that requires activation by alcaligin (Brickman *et al.*, 2001; Brickman and Armstrong, 2002), and has the N-terminal inducer-binding and multimerization domain and C-terminal DNA-binding domain structure typical of most AraC/XylS family regulators. Alcaligin does not induce transcription under iron-replete conditions, indicating that Fur derepression is a prerequisite for induction and activation of transcription (Brickman *et al.*, 2001).

Alcaligin sensing

AlcR-mediated transcriptional activation is exquisitely sensitive to alcaligin. *Bordetella* cells can perceive and transcriptionally respond to nanomolar concentrations of alcaligin under iron starvation conditions (Brickman and Armstrong, 2002). In contrast, at least 1000-fold higher alcaligin concentrations are required for detectable growth stimulation *in vitro* (Brickman *et al.*, 1996; Brickman and Armstrong, 1999). *Bordetella fauA* and *tonB* null mutants that cannot transport ferric alcaligin show no defects in induction by alcaligin (Brickman and Armstrong, unpublished), suggesting that the cytoplasmic AraC-type regulator AlcR is responsive to free alcaligin rather than ferric alcaligin complexes. It is possible that deferri-alcaligin could cross the bacterial outer membrane by passive diffusion through porin channels; however, the mechanism of subsequent inducer uptake across the cytoplasmic membrane is unknown.

Alcaligin-sensing varies among *Bordetella* species. In genetic complementation studies using multicopy *alcR*$^{+}$ plasmids, it was found that overproduction of *B. bronchiseptica* AlcR partially suppressed its inducer requirement for transcriptional activation of *alcA*, whereas *B. pertussis* AlcR retained strict alcaligin inducer dependence when overproduced (Brickman and Armstrong, 2002). This difference in multicopy suppression phenotype was due to a single amino acid difference in the N-terminal inducer-binding and multimerization domain of the AlcR proteins of the two species. Using that information, a mutant AlcR protein was constructed that was fully constitutive (*i.e.* inducer-independent) for activation of transcription at single-copy gene dosage, thus defining a critical AlcR determinant of inducer responsiveness. The interspecies difference in AlcR signal requirements for activation may have biological relevance for signal perception. The fact that modest AlcR overproduction results in inducer independent AlcR function in *B. bronchiseptica* suggests that the alcaligin inducer may normally be required only for initial activation of transcription of the alcaligin system genes. Once activated, elevated levels of AlcR resulting from increased expression of the *alcABCDER* operon might suppress any need for additional inducer. In contrast, *B. pertussis* AlcR may require the persistence of signal from the inducer to maintain AlcR in an active conformation. Consequently, if inducer were no longer perceived by *B. pertussis*, activation of alcaligin system gene transcription would cease and the bacterium's energy and precursors could be appropriately channelled to the utilization of alternative sources of iron. The more stringent genetic regulation of the alcaligin siderophore system by the AlcR protein of *B. pertussis* may contribute to the success of *B. pertussis* as an obligate pathogen.

Enterobactin xenosiderophore utilization

Enterobactin as a xenosiderophore

The hexadentate catechol siderophore enterobactin is an ester-linked cyclic trimer of 2,3-dihydroxybenzoylserine. The ferric enterobactin complex has an extremely high stability constant of 10^{49} at physiological pH (Loomis and Raymond, 1991). Enterobactin is produced by *E. coli* (O'Brien and Gibson, 1970) and several other members of the *Enterobacteriaceae*, as well as by certain *Streptomyces* isolates (Fiedler *et al.*, 2001). Although *Bordetella* cells are incapable of producing enterobactin, they can effectively utilize enterobactin excreted by other bacterial species. The *B. pertussis* TonB-dependent outer membrane receptor for ferric enterobactin, BfeA (Beall and Sanden, 1995a) (Fig. 4.4), shares amino acid sequence similarity with other enterobactin receptors such as *Pseudomonas aeruginosa* PfeA, *Salmonella enterica* IroN, and *E. coli* FepA (Table 4.2). All three classical *Bordetella* species encode BfeA orthologues (Beall and Sanden, 1995a; Parkhill *et al.*, 2003). Several other extraintestinal mucosal colonizers, including *Neisseria gonorrhoeae* (Carson *et al.*, 1999), *Neisseria meningitidis* (Rutz *et al.*, 1991), *Haemophilus parainfluenzae*, and *Haemophilus paraphrophilus* (Williams *et al.*, 1990) are also able to use enterobactin as a xenosiderophore. Xenosiderophore utilization spares the energetic cost of siderophore biosynthesis and secretion. Moreover, microbes with the ability to utilize the strong and ubiquitous siderophore enterobactin are likely to have a significant competitive advantage over those using weaker siderophores when competing for limited iron resources.

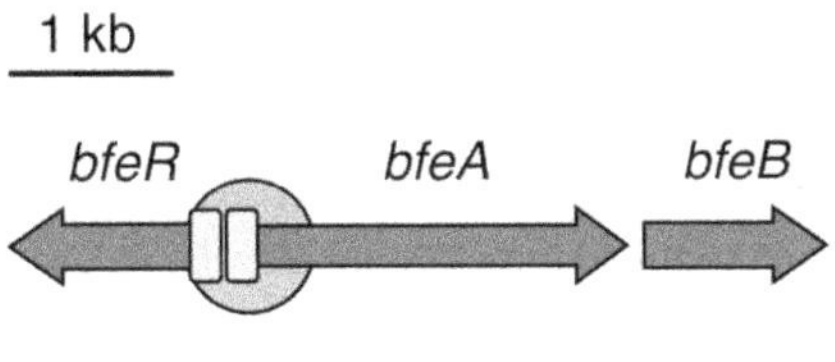

Figure 4.4 The *Bordetella* enterobactin utilization system gene cluster. A 3.4-kb chromosomal DNA region is depicted. The arrows indicate the spatial limits and transcriptional orientations of genes, and the rectangles represent known promoter regions. Fur-binding sites reside in the intergenic region between *bfeR* and *bfeA*. A BfeR-responsive control region indicated by the circle is located upstream of *bfeA*. A 1-kb scale bar is shown.

The *Bordetella* enterobactin utilization system

The *bfeA* gene, encoding the outer membrane receptor for ferric enterobactin, is the promoter-proximal cistron of a putative bicistronic operon that includes the downstream *bfeB* gene (Fig. 4.4). Since only two nucleotides separate the open reading frames, BfeB production is predicted to be translationally coupled to BfeA. BfeB is a member of the α/β superfamily of hydrolases (Table 4.2), and carries the GXSXGG serine active site esterase family signature shared by *P. aeruginosa* PfeE, *S. enterica* IroE, and the Fes protein of *E. coli*. The genes encoding *bfeB*, *pfeA*, *iroE*, and *fes* each reside adjacent to known enterobactin or catechol siderophore transport genes, strongly suggesting a common function in catechol siderophore utilization. IroE and PfeE can hydrolyse ferric salmochelin and enterobactin *in vitro*, and *iroE* mutants are defective in salmochelin utilization (Zhu *et al.*, 2005). Hydrolysis of the ester backbone of enterobactin by the cytoplasmic Fes protein is required for release of the bound iron for assimilation by *E. coli* (Langman *et al.*, 1972; Brickman and McIntosh, 1992). BfeB and IroE are predicted to localize to the periplasm based on the presence of a secretion signal, distinguishing these enzymes from the well-characterized cytoplasmic Fes enzyme, and suggesting that BfeB and IroE may have additional roles in enterobactin transport across the cytoplasmic membrane. The *bfeR* regulatory gene is divergently transcribed from the putative *bfeAB* operon. The BfeR protein is an AraC/XylS family regulator with similarity to the already described *Bordetella* AlcR (Beaumont *et al.*, 1998; Pradel *et al.*, 1998; Brickman *et al.*, 2001), *P. aeruginosa* PchR (Heinrichs and Poole, 1993; Heinrichs and Poole, 1996), *Y. pestis* YbtA (Fetherston *et al.*, 1996), and *Sinorhizobium meliloti* RhrA (Lynch *et al.*, 2001).

Regulation of enterobactin system gene expression

Like the alcaligin utilization system, transcription of the enterobactin utilization genes in *B. pertussis* and *B. bronchiseptica* was found to be Fur and iron-repressible, and iron source substrate-inducible. Analysis of *bfeA–lacZ* transcription and BfeA production established the requirement of the BfeR regulator for enterobactin-responsive *bfeA* expression under iron starvation conditions (Anderson and Armstrong, 2004). Three putative Fur-binding sites in the intergenic control region between *bfeR* and *bfeAB* (Fig.

Table 4.2 *Bordetella* enterobactin/catechol siderophore utilization system genes and proteins

Gene symbol	Locus tag[a]	Product
bfeR	BP2900	Transcriptional regulator of enterobactin and catechol siderophore utilization genes
bfeA	BP2901	Ferric enterobactin, catechol outer membrane siderophore receptor
bfeB	BP2902	Hypothetical protein

[a]Locus tag designation for *Bordetella pertussis* Tohama I orthologue.

4.4) are implicated in iron repression. As with AlcR regulation of alcaligin system genes, high-affinity siderophore transport does not appear to be required for uptake of the enterobactin inducer. Neither the BfeA receptor nor the TonB system is necessary for enterobactin-responsive transcriptional activation of *bfeA* by BfeR, despite the strict requirement for BfeA and the TonB system for ferric enterobactin transport and utilization (Anderson and Armstrong, 2004, 2006).

BfeA mediates utilization of multiple catechol siderophores

Certain structurally similar siderophores can be recognized and transported by a single iron acquisition system, thus obviating the requirement for independent uptake mechanisms. In *S. enterica*, both IroN and FepA receptors can recognize and transport multiple catechol siderophores (Rabsch *et al.*, 1999). Two catechol siderophores with structural similarity to enterobactin (see also Chapter 5 on *E. coli* and *Shigella*), salmochelin (Bister *et al.*, 2004) and corynebactin (Budzikiewicz *et al.*, 1997), were found to stimulate growth of iron-restricted *B. bronchiseptica* (Anderson and Armstrong, 2006). TRENCAM (Rodgers *et al.*, 1987) and MECAM (Venuti *et al.*, 1979; Weitl and Raymond, 1979), two synthetic enterobactin analogues, also stimulated *Bordetella* growth. Growth stimulation by all these compounds was *bfeA* dependent, indicating that BfeA facilitates utilization of several different catechol siderophores (Anderson and Armstrong, 2006). In addition, both salmochelin and MECAM showed inducer activity for BfeR-dependent *bfeA* transcription, although neither catechol was as potent an inducer as enterobactin.

Induction of enterobactin utilization genes by neuroendocrine catecholamines

Host neuroendocrine catecholamines such as epinephrine, norepinephrine, and dopamine compounds have overt structural similarity to enterobactin. Epinephrine and norepinephrine cause elevated expression of a subset of *E. coli* locus of enterocyte effacement (LEE) and flagellar genes (Sperandio *et al.*, 2003), and norepinephrine stimulates production of the FepA ferric enterobactin receptor protein of *E. coli* (Burton *et al.*, 2002). In *B. bronchiseptica*, *bfeA* transcription was induced more than 28-fold by epinephrine, norepinephrine, or dopamine. Induction of the enterobactin utilization genes by norepinephrine was shown to be iron-regulated and dependent on the BfeR regulator, similar to induction of *bfeA* expression by enterobactin. Furthermore, norepinephrine stimulated the growth of alcaligin-deficient *B. bronchiseptica* cells in iron-restricted medium supplemented with serum (Anderson and Armstrong, 2006) or transferrin (Anderson and Armstrong, 2008); however, this growth stimulation, although TonB dependent, does not require the BfeA receptor protein and probably involves one or more unidentified receptor proteins. Norepinephrine is thought to mediate or facilitate the release of transferrin-bound iron in serum, as reported by Freestone and colleagues (Freestone *et al.*, 2000). In addition to their potential role in *Bordetella* iron acquisition *in vivo*, host catecholamines may serve as important host environmental cues for *Bordetella* cells in the context of an infection.

Haem utilization

Taxonomic classification originally placed *B. pertussis* in the genus *Haemophilus* because of similarities in cell and colony morphologies, and because of its apparent growth requirement for blood. It was later established that blood allowed growth of *B. pertussis* on solid agar media by binding inhibitors such as peroxides and fatty acids (Pollock, 1947), and that none of the classical *Bordetella*e were actually haem auxotrophs. Nonetheless, no bloodless solid culture medium surpasses blood-containing media such as the standard Bordet-Gengou agar (Bordet and Gengou, 1906) in supporting optimal growth of virulent-phase *Bordetella* strains. *B. pertussis* has since been shown to use haemin (Agiato and Dyer, 1992; Nicholson and Beall, 1999) and haemoglobin (Nicholson and Beall, 1999) as sole sources of iron, and the utilization of these iron sources by *B. pertussis* and *B. bronchiseptica* is TonB dependent (Nicholson and Beall, 1999; Pradel *et al.*, 2000). Vanderpool and Armstrong are credited with identification and characterization of the haem utilization gene clusters of *B. pertussis* and *B. bronchiseptica* (Vanderpool and Armstrong, 2001).

The Bordetella haem utilization system

The genetic system encoding haem utilization and regulatory functions consists of the *hurIR bhuRSTUV* genes, which are contiguous on the bacterial chromosome (Fig. 4.5) (Vanderpool and Armstrong, 2001). *Bordetella avium*, a more distantly related avian pathogen, has an orthologous gene cluster, *rhuIR bhuRSTUV* (Murphy *et al.*, 2002). The *Bordetella* Bhu proteins (Table 4.3) are similar to haem transport and utilization proteins of other Gram-negative pathogens such as *P. aeruginosa*, *Yersinia pestis*, *Y. enterocolitica*, *Shigella dysenteriae* and *Vibrio cholerae*.

BhuR is the TonB-dependent outer membrane receptor protein required for the utilization of haemin, haemoglobin–haptoglobin, and haem–bovine serum albumin, as well as human haemoglobin and haemoglobins from diverse animal sources, including turkey, rabbit, cow and pig (Vanderpool and Armstrong, 2001). A recent report suggested that haem, spontaneously released from haemoglobin or myoglobin, was the actual ligand for BhuR (Mocny *et al.*, 2007). In addition to having a TonB box C motif, BhuR has the FRAP/NPNL amino acid motif that is conserved within the bacterial haem receptor family. Unique among bacterial haem receptors, BhuR possesses the N-terminal extension domain that is characteristically associated with extracytoplasmic function (ECF) sigma factor-regulated receptors such as the FecA ferric citrate receptor of *E. coli* and the *P. putida* WCS358 ferric pseudobactin receptor PupB. This N-terminal extension is required for the transmembrane signalling process that relays the receptor occupancy signal to the transcriptional machinery, leading to promoter recognition and transcription initiation involving the ECF sigma factor (Kim *et al.*, 1997; Braun and Mahren, 2005).

The BhuT periplasmic haem-binding protein, BhuU cytoplasmic membrane permease, and the ATP-binding protein BhuV are required for haem uptake. BhuS homologues are found in other bacterial haem utilization systems, but little is known about their function. The best characterized example, ShuS of *Shigella dysenteriae*, binds haem and has been hypothesized to sequester internalized haem, thus preventing toxicity associated with its cytoplasmic accumulation (Wyckoff *et al.*, 2005). A *Shigella shuS* mutant was shown to be deficient in haem utilization, and ShuS overproduction appeared to protect the mutant and parental strains from haem toxicity. In *Bordetella* cells, iron is presumably released from the transported haem for nutritional use; however, no predicted haem oxygenase is encoded in the *bhu* gene cluster and none can be predicted from analysis of *Bordetella* genome sequences.

HurI is an ECF sigma factor with high similarity to the *E. coli* ferric citrate uptake regulator FecI (Enz *et al.*, 2000). HurR has a predicted membrane-spanning domain and has the three tryptophan residues conserved among FecR anti-sigma factor homologues. In the *E. coli* ferric citrate system, FecI activity is controlled by direct interaction with the cytoplasmic membrane anti-sigma factor FecR. FecI initiates transcription of ferric citrate transport genes only when ferric citrate binds the FecA outer membrane receptor under conditions of iron limitation. Similarly, the *Bordetella bhu* genes are regulated by HurI and HurR *via* BhuR in response to haem.

Regulation of the haem utilization system genes and haem sensing

Initial studies found that BhuR production during iron starvation was elevated in response to haemin (Vanderpool and Armstrong, 2001), and that this haemin-responsive BhuR production was dependent

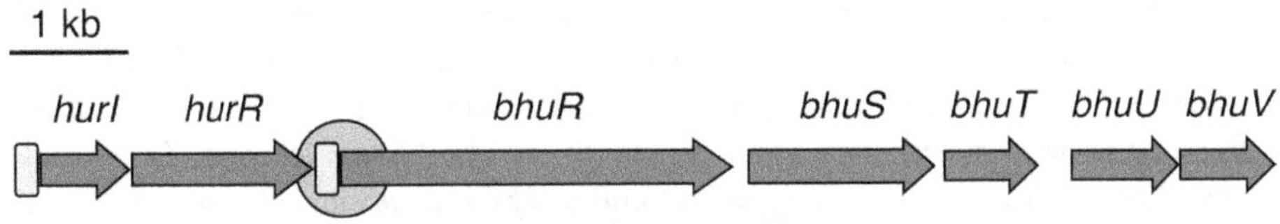

Figure 4.5 The *Bordetella* haem utilization system gene cluster. A 7.9-kb chromosomal DNA region is depicted. The arrows indicate the transcriptional orientations and spatial limits of genes, and the rectangles represent known promoter regions. Fur-binding sites reside upstream of *hurI*, and a HurI ECF sigma factor-responsive promoter region upstream of *bhuR* is represented by the circle. A 1-kb scale bar is shown.

Table 4.3 *Bordetella* haem utilization system genes and proteins

Gene symbol	Locus tag[a]	Product
hurI	BP0349	Haem uptake regulator
hurR	BP0348	Haem uptake transmembrane sensor
bhuR	BP0347	Outer membrane haem receptor
bhuS	BP0346	Putative haemin transport protein
bhuT	BP0345	Putative haemin-binding protein
bhuU	BP0344	Putative haemin permease
bhuV	BP0343	Putative ATP-binding component of haemin transport system

[a]Locus tag designation for *Bordetella pertussis* Tohama I orthologue.

on HurI (Vanderpool and Armstrong, 2003). Consequently, *hurI* mutants are deficient in haem utilization. In genetic complementation analyses, *hurI*$^{+}$ plasmids restored haem-inducible production of BhuR and wild-type haem utilization. Transcriptional analysis using *bhuR-lacZ* gene fusions in *B. bronchiseptica* confirmed that haem-responsive *bhuR* transcription required *hurI*. Overproduction of HurI from high-copy-number *hurI*$^{+}$ plasmids suppressed the haem inducer requirement for *bhuR-lacZ* expression; HurI overproduction is thought to cause in an imbalance in HurI-HurR stoichiometry, thus overcoming the negative influence of HurR. The Connell group has shown that the HurP protein of *B. bronchiseptica*, similar to the RseP protease and encoded outside of the *hurIR bhuRSTUV* gene cluster, is important for haem utilization and haem induction of *bhu* transcription, probably *via* modification of HurR (King-Lyons *et al.*, 2007).

A predicted σ^{70}-like promoter and a Fur-binding site was identified upstream of *hurIR*, and the predicted Fur-binding site was shown to mediate iron-repressible transcription (Vanderpool and Armstrong, 2001) in *E. coli* using an *in vivo* Fur titration assay (Stojiljkovic *et al.*, 1994). In contrast, the *bhuRSTUV* promoter region lacks predicted Fur-binding sequences and has no appreciable Fur-binding activity. Accordingly, transcription from the *hurI* promoter was found to be iron-regulated but independent of HurI and haem induction, whereas maximal activity of the *bhuR* promoter during iron starvation required HurI-dependent haem induction (Vanderpool and Armstrong, 2004). Primer extension and mutational analysis determined that iron-regulated read-through transcription from *hurIR* through the *hurR–bhuR* intergenic region contributes to low-level *bhuR* receptor gene expression in the absence of haem; thus, under iron starvation conditions, low-level transcription of the *hurIR bhuRSTUV* genes initiates at the *hurI* promoter. Exposure to haem activates HurI for initiation of *bhuRSTUV* transcription at the *bhuR* control region, resulting in dramatically increased production of the BhuRSTUV haem utilization machinery. As with the FecA ferric citrate receptor (Harle *et al.*, 1995) and consistent with the hypothesized involvement of BhuR in a signaling cascade activated by receptor occupancy, mutation of *bhuR* was found to prevent both haem uptake and haem-responsive *bhuR* transcription (measured as *bhuR–lacZ* transcription) (Vanderpool and Armstrong, 2003). In this manner, the *bhuRSTUV* genes are maximally expressed only when the cognate haem iron sources are present in the *Bordetella* environment. Similarly, for the related *B. avium* species, the ECF sigma factor RhuI activates transcription of *bhuR* in response to haemin, and this activation requires the BhuR haem receptor (Kirby *et al.*, 2001). The *Bordetella* haem utilization system is an unconventional hybrid system that merges the functionality of the haem uptake systems found in Gram-negative pathogens with regulators of the iron starvation subfamily of ECF regulators.

Xenosiderophores and TonB-dependent receptors

The classical *Bordetella* species can utilize diverse iron sources, and the genomes of *B. pertussis, B. bronchiseptica* and *B. parapertussis* encode a number of putative iron acquisition systems of undetermined specificity. In addition to the three genes encoding the known receptors for ferric alcaligin, ferric enterobactin and haem, there are sixteen additional *Bordetella* genes predicted to encode TonB-dependent receptor proteins (Table 4.4). All 16 putative receptor genes are present in *B. bronchiseptica,* whereas

Table 4.4 *Bordetella* TonB-dependent outer membrane receptors

Receptor	Locus tag[a]	Iron source, homologue, reference[b]
Characterized receptors of known specificity		
FauA	BP2463, BPP3450, BB3900	Ferric alcaligin (Brickman and Armstrong, 1999)
BfeA	BP2901, BPP2495, BB1942	Ferric enterobactin (Beall and Sanden, 1995)
BhuR	BP0347, BPP4185, BB4655	Haem (Vanderpool and Armstrong, 2001)
Putative receptors of unknown specificity		
BfrA	BB4761	*E. coli* Iha catechol receptor ABB17254 (Beall and Hoenes, 1997)
BfrB	BP2016, BPP2396, BB1846	*Y. enterocolitica* ferrichrome receptor FcuA, Q05202 (Beall, 1998)
BfrC	BP3663, BPP0079, BB0078	*Vibrio anguillarum* FatA ferric anguibactin receptor P11461 (Beall, 1998)
BfrD	BP0856, BPP3376, BB3826	*P. aeruginosa* Fiu Pa4514 (Passerini de Rossi *et al.*, 2003)
BfrE	BP0857, BPP3375*[c], BB3825	*P. aeruginosa* Fiu Pa4514
BfrF	BP0736, BPP3688, BB4122	*E. coli* FhuA NP_414692
BfrG	BP2922, BPP1078, BB1294	*E. coli* FhuA NP_414692
BfrH	BP1138, BPP3206, BB3658	*E. coli* FhuA; *Burkholderia cepacia* OrbA, AF013993
BfrI	BP1962, BPP2334, BB1785	*P. aeruginosa* FiuA AF051691; *Salmonella typhimurium* FhuA Y14025
BfrZ	BB4744	COG1629[d] (Pradel and Locht, 2001)
HemC	BP0456, BPP4370*, BB4956	*Haemophilus ducreyi* TdhA haem receptor AAC35765
–	BPP2457, BB1905	COG1629[d]; *Serratia marcescens* HasR haemophore receptor CAE46936
–	BP3101*, BPP0746*, BB0832	COG1629[d]
–	BP3595, BPP3984, BB4457	'BtuB'; TonB-dependent receptor
–	BP3077, BPP0186, BB0189	TonB-dependent receptor
–	BP1760, BPP1991, BB2179	TonB-dependent receptor

[a]Genome sequence database, GeneDB version 2.1 (http://www.genedb.org/), The Wellcome Trust Sanger Institute, Pathogen Sequencing Unit; prefix designations: BP, *B. pertussis*; BPP, *B. parapertussis*; BB, *B. bronchiseptica.*

[b]In parentheses, reference describing the *Bordetella* receptor.

[c]*Indicates pseudogene.

[d]COG1629: outer membrane receptor proteins, mostly Fe transport.

some are absent or are pseudogenes in the more specialized mammalian pathogens *B. pertussis* and *B. parapertussis*. As might be expected, some of the putative TonB-dependent receptors may be involved in the transport of compounds other than iron sources. It is also conceivable that some of the receptors, though once used by an ancestor that lived outside of a mammalian host, are produced but no longer serve a useful function for the host-adapted species that have undergone extensive gene loss during evolution and niche specialization. However, such dispensable genes might be predicted to acquire mutations over time, eventually giving rise to pseudogenes. An alternative view holds that expression of the *Bordetella* receptor genes that appear to remain intact must confer a fitness advantage by contributing to the ability of the organisms to grow in the host or survive transmission.

B. pertussis and *B. bronchiseptica* utilize the xenosiderophores ferrichrome and desferrioxamine B by TonB-dependent mechanisms (Beall and Hoenes, 1997). *B. bronchiseptica* has also been reported to use the siderophores coprogen, schizokinen, ferricrocin, vicibactin, ferrichrysin, ferrirubin, protochelin, aerobactin and several pyoverdins (Pradel and Locht, 2001). Currently, however, enterobactin (Beall and Sanden, 1995a) and related catechols are the only the xenosiderophores for which the receptor and utilization genes are known (Anderson and Armstrong, 2006). The *bfrA* (Beall and Hoenes, 1997), *bfrB*, and *bfrC* (Beall, 1998) receptor genes were identified using a *B. bronchiseptica* Tn*phoA* insertion mutant library in a screen for iron-repressible *phoA* fusion expression. Using feeding assays, the iron source substrates for these putative receptors encoded by these genes could not be identified; mutants retained the ability to use haem, enterobactin, desferrioxamine B and ferrichrome. BfrB was among the iron-regulated outer membrane proteins identified in a study of the iron starvation proteome of *B. pertussis* by Vidakovics and coworkers (Vidakovics *et al.*, 2007). Passerini de Rossi and colleagues identified a 90-kDa outer membrane protein in wild-type *B. pertussis* that was absent in an isogenic *bvgS* sensor kinase mutant (Passerini de Rossi *et al.*, 1999). This protein was later identified as the TonB-dependent receptor protein BfrD, and expression of *bfrD* was activated by the BvgAS signal transduction system involved in virulence gene regulation (Antoine *et al.*, 2000). The substrate specificity was not determined for BfrD (Passerini de Rossi *et al.*, 2003) or for another putative *B. bronchiseptica* TonB-dependent receptor, BfrZ (Pradel and Locht, 2001).

Virulence and iron acquisition

B. pertussis is an obligate human pathogen, however the mouse remains the animal model of choice for respiratory infection studies. For *B. bronchiseptica*, a variety of natural mammalian hosts including mice, rats, rabbits, dogs and swine have been used as experimental animal infection models. Experimental animal infections using iron system mutant strains have established the importance of iron acquisition in *Bordetella* virulence. A limited study of human sera obtained from normal uninfected adults and *B. pertussis* culture-positive patients supplied evidence that *B. pertussis* produces multiple iron-repressible proteins, including the receptors for alcaligin (FauA), enterobactin (BfeA) and haem (BhuR), during natural human infection (Brickman *et al.*, 2008). These results support the hypothesis that *B. pertussis* experiences iron starvation *in vivo*, and implies that the cognate iron source inducers are perceived by *B. pertussis* in the human host environment.

B. bronchiseptica and *B. pertussis tonB* mutants are defective in the uptake of multiple iron sources *in vitro* (Nicholson and Beall, 1999; Pradel *et al.*, 2000). In mouse infections, a *B. pertussis tonB exbB* mutant was unable to efficiently colonize and multiply in the respiratory tract, although clearance by the host occurred at a rate similar to that of the wild-type parental strain (Pradel *et al.*, 2000).

Several infection studies have examined the role of the *Bordetella* alcaligin system in virulence. In one mouse infection study, a *B. pertussis alcR* mutant unable to maximally express the alcaligin system genes showed no significant defect in colonization of the lung (Pradel *et al.*, 1998). It should be noted, however, that *alcR* strains retain some ability to produce and utilize alcaligin (Beaumont *et al.*, 1998). In another study, neonatal swine infected with an alcaligin-deficient *B. bronchiseptica alcA* mutant strain showed marked reduction in colonization of the nasal turbinates, trachea and lung compared with the wild-type parent strain (Register *et al.*, 2001). Furthermore, nasal pathology in swine infected with this siderophore-deficient strain was either mild or absent compared with the pathology commonly associated with atrophic rhinitis. These results indicate that the ability of *Bordetella* to produce and utilize alcaligin is required for maximal virulence in this natural host model. Recent mixed infection competition studies involving co-infection of mice with *B. pertussis* wild-type and *fauA* ferric alcaligin receptor mutant strains determined that the ability to utilize alcaligin was critically important for *in vivo* multiplication and survival of *B. pertussis* in the murine respiratory tract (Brickman and Armstrong, 2007). The attenuating effect of *fauA* inactivation was evident early in the course of the infection, suggesting that the contribution of ferric alcaligin transport to the ecological fitness of *B. pertussis* may be important for adaptation to iron-restricted host conditions that exist at the outset of infection. A somewhat different pattern was found in similar mixed infection experiments aimed at determining the importance of the Bfe enterobactin and catechol siderophore utilization system *in vivo* (Brickman *et al.*, 2008). The *B. pertussis* enterobactin and catechol utilization system mutants showed a marked fitness defect at the initial colonization stage in mice, but the importance of the system appeared to decline after the peak of infection.

The haem utilization system was also shown to be important for *B. pertussis* virulence using *bhuR* haem receptor mutant and wild-type strains in mixed infection competition experiments (Brickman *et al.*, 2006). In contrast to the results of mixed infection studies indicating early involvement of the alcaligin and enterobactin siderophore systems during infection, the haem system mutants showed no growth defect relative to the wild-type strain until after the peak of bacterial multiplication. At this stage, the pathologic effects of *Bordetella* toxic factors on host epithelial cells are thought to liberate haem, making it available to *Bordetella* cells on the mucosal surface. It should be noted that the ability of the *bhuR* mutant strain to produce and utilize its alcaligin siderophore could not compensate for its late-stage haem-related growth defect. This suggests that alcaligin may not effectively scavenge iron in the host environment once extensive tissue damage by *B. pertussis* has occurred, or that the available iron source is not present or accessible to alcaligin. The more distantly-related species *Bordetella avium* does not produce a siderophore, but it has a haem uptake system homologous to that of the three classical *Bordetella* species. In competition infection experiments using mixtures of *B. avium* wild-type and *bhuR* mutant strains, the BhuR haem receptor was found to be important in tracheal colonization of turkey poults (Murphy *et al.*, 2002).

Adaptation to temporal changes in iron source availability

The interaction of *B. pertussis* with its human host characteristically results in inflammation, immune activation, and host tissue injury. These changes in host niche conditions over the course of infection alter the array and abundance of iron sources available to the invading pathogen. The involvement of the iron source substrate-inducible regulators AlcR (Beaumont *et al.*, 1998; Pradel *et al.*, 1998), BfeR (Anderson and Armstrong, 2004), and HurI (Vanderpool and Armstrong, 2001) allow for effective adaptation to changes in iron source availability through differential expression of iron source utilization systems.

The importance of the alcaligin (Brickman and Armstrong, 2007), haem (Brickman *et al.*, 2006) and enterobactin (Brickman *et al.*, 2008) systems for *in vivo* growth of *B. pertussis* was established in mixed infection competition experiments in a mouse respiratory infection model system. Each iron utilization system makes its own distinct and essential contribution to *in vivo* multiplication and survival of *B. pertussis*, albeit at different stages of infection (Fig. 4.6). Although the BfeA enterobactin receptor is important for *B. pertussis* growth in mice early in infection, the relevance of enterobactin to virulence is not clear, since it is unknown whether it is in the host respiratory tract. BhuR was critical to growth during the later stages of infection. The temporal phasing of alcaligin, enterobactin and haem system involvement implies that host iron sources change as infection progresses, and that *B. pertussis* can adapt to those changes by engaging different iron acquisition systems.

Recombinase-based *in vivo* expression technology (RIVET) (Camilli *et al.*, 1994; Veal-Carr and Stibitz, 2005) was used to monitor temporal expression patterns of the three characterized iron

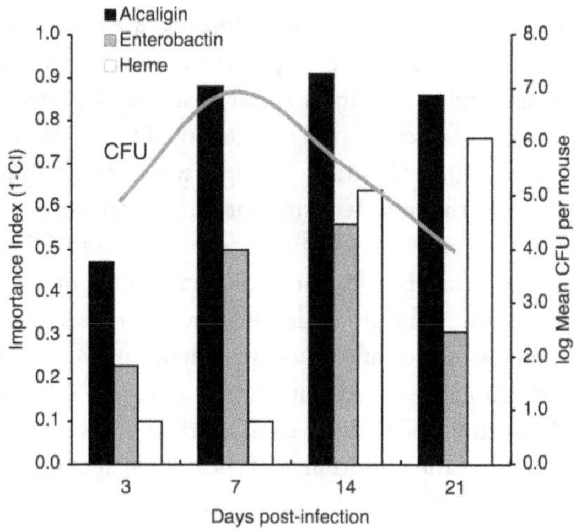

Figure 4.6 Relative importance of *B. pertussis* iron acquisition systems at various stages of infection. The importance index provides a measure of the relative *in vivo* importance or fitness contribution of a particular iron system for multiplication and survival of *B. pertussis* under the existing host niche conditions. Mean competitive index (CI) values (calculated as the mutant/wild-type CFU ratio in the bacterial population recovered from experimentally infected mice, divided by the mutant/wild-type CFU ratio in the infecting dose) from mixed infection competition experiments using wild-type *B. pertussis* with isogenic alcaligin (Brickman and Armstrong, 2007), enterobactin (Brickman *et al.*, 2008), or haem (Brickman *et al.*, 2006) receptor mutant strains were used to calculate importance index values (calculated as 1 – mean CI) at various times post infection. A line plot (CFU) showing CFU levels that are typical of *B. pertussis* mouse respiratory infections is overlaid.

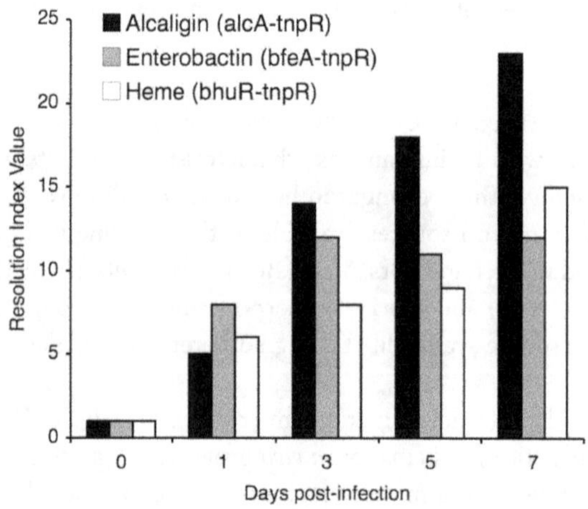

Figure 4.7 Differential expression of *Bordetella* iron systems *in vivo*. Recombinase-based *in vivo* expression technology (RIVET) was used to temporally monitor expression of the alcaligin, enterobactin, and haem utilization systems by *B. pertussis in vivo* in a mouse respiratory infection model (Brickman *et al.*, 2008). In this experimental system, expression of an iron system promoter-*tnpR* recombinase gene fusion results in recombinational excision (resolution) of a selectable antibiotic resistance marker that is flanked by recognition sites for the site-specific TnpR recombinase. The resolution frequency for bacterial populations recovered from infected mice reflects the transcriptional activity of the iron system promoter driving recombinase expression. Resolution index values represent the fold-change in mean resolution frequency for bacterial populations recovered from infected mice relative to the base-line resolution frequency determined for the input population of *B. pertussis*.

systems during mouse respiratory infection (Brickman *et al.*, 2008). The three systems were differentially expressed *in vivo*, showing early induction of the alcaligin and enterobactin siderophore system transcription and delayed induction of the haem system (Fig. 4.7). These expression patterns mirror the results from competition infection experiments that established the importance of siderophores early in infection and the haem system at later stages of infection. These *in vivo* expression studies confirmed that *B. pertussis* expresses these three iron acquisition systems in a host and that the relevant iron sources are present in the host niche, while the competition infection experiments demonstrated that *B. pertussis* requires each of these iron acquisition systems for growth and survival *in vivo*. In sum, these findings evidence a pattern of *Bordetella* iron acquisition in the host that is characterized by early reliance on siderophores and subsequent dependence on haem, based on iron source availability.

Intercellular signalling processes enable a bacterial population to coordinately regulate gene expression in response to environmental signals or cues. *Bordetella* species invoke multiple ultrasensitive iron source-sensing circuits organized in parallel (Fig. 4.8) to acquire iron. In their role as inducers, the alcaligin, haem and enterobactin iron sources function as signalling molecules or 'ferrimones' that activate the cognate iron transport system through the action of the specific substrate-inducible positive regulators, AlcR, BfeR and HurIR. The iron systems are activated independently, and are not subject to interference or signal antagonism. All circuits converge to supply essential iron. When the bacterium's

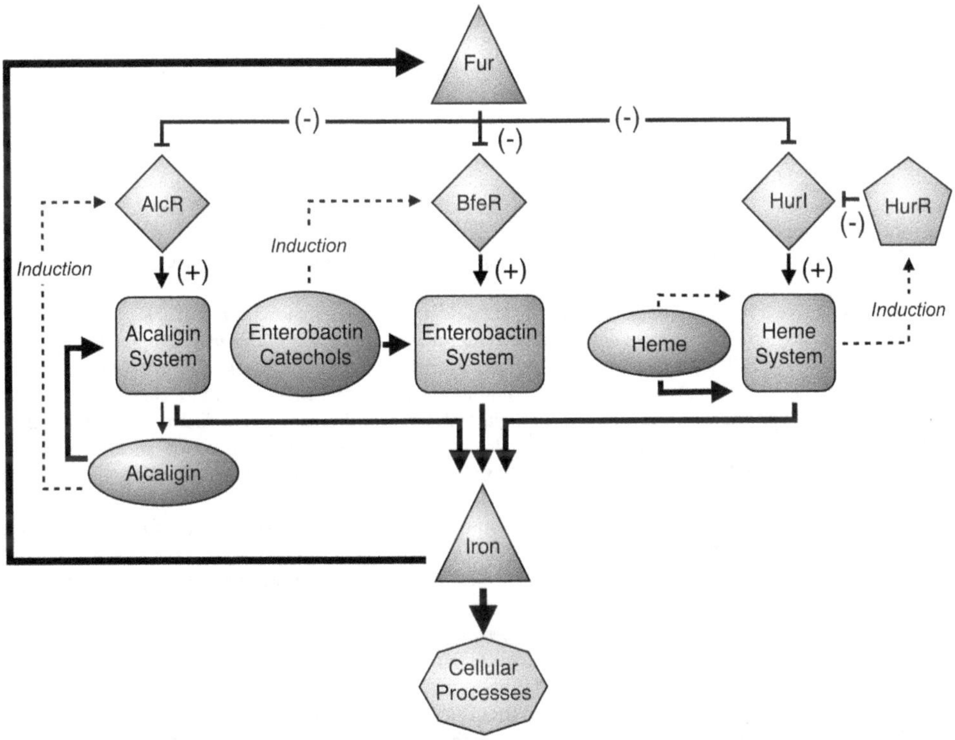

Figure 4.8 Positive and negative regulatory circuits control expression of the alcaligin, enterobactin, and haem utilization systems in *Bordetella* species. The adaptive response of *Bordetella* species to changes in alcaligin, enterobactin, and haem iron source availability involves the Fur repressor and the positive transcriptional regulators AlcR, BfeR and HurI. Expression of the positive regulators is repressible by Fur, depending on availability of its iron corepressor. Positive regulator activity is modulated by the presence of its cognate iron source inducer. The siderophore inducers for the AraC-type regulators AlcR and BfeR are perceived intracellularly, whereas the ECF sigma factor HurI is activated by transduction of the extracellular BhuR haem receptor occupancy signal. The iron utilization systems are not subject to interference or signal antagonism, and all three systems converge to supply nutritional iron. When iron requirements are satisfied, iron activates the DNA-binding activity of Fur, and transport system expression is repressed.

iron requirements are satisfied, the Fur repressor is activated, and iron transport system gene expression is globally repressed. This iron source sensing exploits early and late events in the infection process to assure growth and survival of the pathogen.

References

Agiato, L.A., and Dyer, D.W. (1992). Siderophore production and membrane alterations by *Bordetella pertussis* in response to iron starvation. Infect. Immun. *60*, 117–123.

Anderson, M.T., and Armstrong, S.K. (2004). The BfeR regulator mediates enterobactin-inducible expression of *Bordetella* enterobactin utilization genes. J. Bacteriol. *186*, 7302–7311.

Anderson, M.T., and Armstrong, S.K. (2006). The *Bordetella bfe* system: growth and transcriptional response to siderophores, catechols, and neuroendocrine catecholamines. J. Bacteriol. *188*, 5731–5740.

Anderson, M.T., and Armstrong, S.K. (2008). Norepinephrine mediates acquisition of transferrin iron in *Bordetella bronchiseptica*. J. Bacteriol. *190*, 3940–3947.

Anderton, T.L., Maskell, D.J., and Preston, A. (2004). Ciliostasis is a key early event during colonization of canine tracheal tissue by *Bordetella bronchiseptica*. Microbiology *150*, 2843–2855.

Antoine, R., Alonso, S., Raze, D., Coutte, L., Lesjean, S., Willery, E., Locht, C., and Jacob-Dubuisson, F. (2000). New virulence-activated and virulence-repressed genes identified by systematic gene inactivation and generation of transcriptional fusions in *Bordetella pertussis*. J. Bacteriol. *182*, 5902–5905.

Armstrong, S.K., and Clements, M.O. (1993). Isolation and characterization of *Bordetella bronchiseptica* mutants deficient in siderophore activity. J. Bacteriol. *175*, 1144–1152.

Beall, B. (1998). Two iron-regulated putative ferric siderophore receptor genes in *Bordetella bronchiseptica* and *Bordetella pertussis*. Res. Microbiol. *149*, 189–201.

Beall, B., and Hoenes, T. (1997). An iron-regulated outer-membrane protein specific to *Bordetella bronchiseptica* and homologous to ferric siderophore receptors. Microbiology *143*, 135–145.

Beall, B., and Sanden, G.N. (1995a). A *Bordetella pertussis fepA* homologue required for utilization of exogenous ferric enterobactin. Microbiology *141*, 3193–3205.

Beall, B.W., and Sanden, G.N. (1995b). Cloning and initial characterization of the *Bordetella* pertussis *fur* gene. Curr. Microbiol. *30*, 223–226.

Beaumont, F.C., Kang, H.Y., Brickman, T.J., and Armstrong, S.K. (1998). Identification and characterization of *alcR*, a gene encoding an AraC-like regulator of alcaligin siderophore biosynthesis and transport in *Bordetella pertussis* and *Bordetella bronchiseptica*. J. Bacteriol. *180*, 862–870.

Bergeron, R.J., McManis, J.S., Perumal, P.T., and Algee, S.E. (1991). The total synthesis of alcaligin. J. Org. Chem. *56*, 5560–5563.

Bister, B., Bischoff, D., Nicholson, G.J., Valdebenito, M., Schneider, K., Winkelmann, G., Hantke, K., and Sussmuth, R.D. (2004). The structure of salmochelins: C-glucosylated enterobactins of *Salmonella enterica*. Biometals *17*, 471–481.

Bordet, J., and Gengou, O. (1906). Le microbe de la coqueluche. Ann. Inst. Pasteur (Paris). *20*, 731–741.

Boukhalfa, H., Brickman, T.J., Armstrong, S.K., and Crumbliss, A.L. (2000). Kinetics and mechanism of iron(III) dissociation from the dihydroxamate siderophores alcaligin and rhodotorulic acid. Inorg. Chem. *39*, 5591–5602.

Boukhalfa, H., and Crumbliss, A.L. (2000). Multiple-path dissociation mechanism for mono- and dinuclear Tris(hydroxamato)iron(III) complexes with dihydroxamic acid ligands in aqueous solution. Inorg. Chem. *39*, 4318–4331.

Braun, V., and Mahren, S. (2005). Transmembrane transcriptional control (surface signalling) of the *Escherichia coli* Fec type. FEMS Microbiol. Rev. *29*, 673–684.

Brickman, T.J., Anderson, M.T., and Armstrong, S.K. (2007). *Bordetella* iron transport and virulence. Biometals *20*, 303–322.

Brickman, T.J., and Armstrong, S.K. (1995). *Bordetella pertussis fur* gene restores iron repressibility of siderophore and protein expression to deregulated *Bordetella bronchiseptica* mutants. J. Bacteriol. *177*, 268–270.

Brickman, T.J., and Armstrong, S.K. (1996). The ornithine decarboxylase gene *odc* is required for alcaligin siderophore biosynthesis in *Bordetella* spp.: putrescine is a precursor of alcaligin. J. Bacteriol. *178*, 54–60.

Brickman, T.J., and Armstrong, S.K. (1999). Essential role of the iron-regulated outer membrane receptor FauA in alcaligin siderophore-mediated iron uptake in *Bordetella* species. J. Bacteriol. *181*, 5958–5966.

Brickman, T.J., and Armstrong, S.K. (2002). *Bordetella* interspecies allelic variation in AlcR inducer requirements: identification of a critical determinant of AlcR inducer responsiveness and construction of an *alcR*(Con) mutant allele. J. Bacteriol. *184*, 1530–1539.

Brickman, T.J., and Armstrong, S.K. (2005). *Bordetella* AlcS transporter functions in alcaligin siderophore export and is central to inducer sensing in positive regulation of alcaligin system gene expression. J. Bacteriol. *187*, 3650–3661.

Brickman, T.J., and Armstrong, S.K. (2007). Impact of alcaligin siderophore utilization on *in vivo* growth of *Bordetella pertussis*. Infect. Immun. *75*, 5305–5312.

Brickman, T.J., and Armstrong, S.K. (2009). Temporal signaling and differential expression of *Bordetella* iron transport systems: the role of ferrimones and positive regulators. Biometals *22*, 33–41.

Brickman, T.J., Hanawa, T., Anderson, M.T., Suhadolc, R.J., and Armstrong, S.K. (2008). Differential expression of *Bordetella pertussis* iron transport system genes during infection. Mol. Microbiol. *70*, 3–14.

Brickman, T.J., Hansel, J.G., Miller, M.J., and Armstrong, S.K. (1996). Purification, spectroscopic analysis and biological activity of the macrocyclic dihydroxamate siderophore alcaligin produced by *Bordetella pertussis* and *Bordetella bronchiseptica*. Biometals *9*, 191–203.

Brickman, T.J., Kang, H.Y., and Armstrong, S.K. (2001). Transcriptional activation of *Bordetella* alcaligin siderophore genes requires the AlcR regulator with alcaligin as inducer. J. Bacteriol. *183*, 483–489.

Brickman, T.J., and McIntosh, M.A. (1992). Overexpression and purification of ferric enterobactin esterase from *Escherichia coli*. Demonstration of enzymatic hydrolysis of enterobactin and its iron complex. J. Biol. Chem. *267*, 12350–12355.

Brickman, T.J., Vanderpool, C.K., and Armstrong, S.K. (2006). Heme transport contributes to *in vivo* fitness of *Bordetella pertussis* during primary infection in mice. Infect. Immun. *74*, 1741–1744.

Bruss, J.B., and Siber, G.R. (1999). Protective effects of pertussis immunoglobulin (P-IGIV) in the aerosol challenge model. Clin. Diagn. Lab. Immunol. *6*, 464–470.

Budzikiewicz, H., Bössenkamp, A., Taraz, K., Pandey, A., and Meyer, J.M. (1997). Corynebactin, a cyclic catecholate siderophore from *Corynebacterium glutamicum* ATCC 14067 (*Brevibacterium* sp. DSM 20411). Z. Naturforsch. *52*, 551–554.

Burton, C.L., Chhabra, S.R., Swift, S., Baldwin, T.J., Withers, H., Hill, S.J., and Williams, P. (2002). The growth response of *Escherichia coli* to neurotransmitters and related catecholamine drugs requires a functional enterobactin biosynthesis and uptake system. Infect. Immun. *70*, 5913–5923.

Camilli, A., Beattie, D.T., and Mekalanos, J.J. (1994). Use of genetic recombination as a reporter of gene expression. Proc. Natl. Acad. Sci. U.S.A. *91*, 2634–2638.

Carson, S.D.B., Klebba, P.E., Newton, S.M., and Sparling, P.F. (1999). Ferric enterobactin binding and utilization by *Neisseria gonorrhoeae*. J. Bacteriol. *181*, 2895–2901.

Centers for Disease Control and and Prevention (2002). Pertussis – United States, 1997–2000. MMWR Morb. Mortal Wkly. Rep. *51*, 73–76.

Centers for Disease Control and Prevention (2006). Preventing tetanus, diphtheria, and pertussis among adolescents: use of tetanus toxoid, reduced diphtheria toxoid and acellular pertussis vaccines: recommendations of the Advisory Committee on Immunization Practices (ACIP). MMWR Morb. Mortal Wkly. Rep. *55*, 1–46.

Challis, G.L. (2005). A widely distributed bacterial pathway for siderophore biosynthesis independent of nonribosomal peptide synthetases. Chembiochem. *6*, 601–611.

Cherry, J.D., and Heininger, U. (2004). Pertussis and other *Bordetella* infections. In Textbook of Pediatric Infectious Diseases, 5th ed., R.D. Feigin, J.D. Cherry, G.J. Demmler, and S. Kaplan, eds. (Philadelphia, The W. B. Saunders Co), pp. 1588–1608.

De Lorenzo, V., Bindereif, A., Paw, B.H., and Neilands, J.B. (1986). Aerobactin biosynthesis and transport genes of plasmid ColV-K30 in *Escherichia coli* K-12. J. Bacteriol. *165*, 570–578.

De Lorenzo, V., and Neilands, J.B. (1986). Characterization of *iucA* and *iucC* genes of the aerobactin system of plasmid ColV-K30 in *Escherichia coli*. J. Bacteriol. *167*, 350–355.

Diavatopoulos, D.A., Cummings, C.A., Schouls, L.M., Brinig, M.M., Relman, D.A., and Mooi, F.R. (2005). *Bordetella pertussis*, the causative agent of whooping cough, evolved from a distinct, human-associated lineage of *B. bronchiseptica*. PLoS Pathog. *1*, e45.

Diavatopoulos, D.A., Cummings, C.A., van der Heide, H.G., van Gent, M., Liew, S., Relman, D.A., and Mooi, F.R. (2006). Characterization of a highly conserved island in the otherwise divergent *Bordetella holmesii* and *Bordetella pertussis* genomes. J. Bacteriol. *188*, 8385–8394.

Enz, S., Mahren, S., Stroeher, U.H., and Braun, V. (2000). Surface signaling in ferric citrate transport gene induction: interaction of the FecA, FecR, and FecI regulatory proteins. J. Bacteriol. *182*, 637–646.

Fetherston, J.D., Bearden, S.W., and Perry, R.D. (1996). YbtA, an AraC-type regulator of the *Yersinia pestis* pesticin/yersiniabactin receptor. Mol. Microbiol. *22*, 315–325.

Fiedler, H.P., Krastel, P., Muller, J., Gebhardt, K., and Zeeck, A. (2001). Enterobactin: the characteristic catecholate siderophore of *Enterobacteriaceae* is produced by *Streptomyces* species.(1). FEMS Microbiol. Lett. *196*, 147–151.

Flak, T.A., and Goldman, W.E. (1999). Signalling and cellular specificity of airway nitric oxide production in pertussis. Cell. Microbiol. *1*, 51–60.

Freestone, P.P., Lyte, M., Neal, C.P., Maggs, A.F., Haigh, R.D., and Williams, P.H. (2000). The mammalian neuroendocrine hormone norepinephrine supplies iron for bacterial growth in the presence of transferrin or lactoferrin. J. Bacteriol. *182*, 6091–6098.

Gallegos, M.T., Schleif, R., Bairoch, A., Hofmann, K., and Ramos, J.L. (1997). Arac/XylS family of transcriptional regulators. Microbiol. Mol. Biol. Rev. *61*, 393–410.

Giardina, P.C., Foster, L.-A., Toth, S.I., Roe, B.A., and Dyer, D.W. (1995). Identification of *alcA*, a *Bordetella bronchiseptica* gene necessary for alcaligin production. Gene *167*, 133–136.

Giardina, P.C., Foster, L.A., Toth, S.I., Roe, B.A., and Dyer, D.W. (1997). Analysis of the *alcABC* operon encoding alcaligin biosynthesis enzymes in *Bordetella bronchiseptica*. Gene *194*, 19–24.

Gorringe, A.R., Woods, G., and Robinson, A. (1990). Growth and siderophore production by *Bordetella pertussis* under iron-restricted conditions. FEMS Microbiol. Lett. *54*, 101–105.

Granstrom, M., Olinder-Nielsen, A.M., Holmblad, P., Mark, A., and Hanngren, K. (1991). Specific immunoglobulin for treatment of whooping cough. Lancet. *338*, 1230–1233.

Harle, C., Kim, I., Angerer, A., and Braun, V. (1995). Signal transfer through three compartments: transcription initiation of the *Escherichia coli* ferric citrate transport system from the cell surface. Embo J. *14*, 1430–1438.

Heinrichs, D.E., and Poole, K. (1993). Cloning and sequence analysis of a gene (*pchR*) encoding an AraC family activator of pyochelin and ferripyochelin receptor synthesis in *Pseudomonas aeruginosa*. J. Bacteriol. *175*, 5882–5889.

Heinrichs, D.E., and Poole, K. (1996). PchR, a regulator of ferripyochelin receptor gene (*fptA*) expression in *Pseudomonas aeruginosa*, functions both as an activator and as a repressor. J. Bacteriol. *178*, 2586–2592.

Hewlett, E.L., Donato, G.M., and Gray, M.C. (2006). Macrophage cytotoxicity produced by adenylate cyclase toxin from *Bordetella pertussis*: more than just making cyclic AMP! Mol. Microbiol. *59*, 447–459.

Hou, Z., Raymond, K.N., O'Sullivan, B., Esker, T.W., and Nishio, T. (1998). A preorganized siderophore: thermodynamic and structural characterization of alcaligin and bisucaberin, microbial macrocyclic dihydroxamate chelating agents (1). Inorg. Chem. *37*, 6630–6637.

Hou, Z., Sunderland, C.J., Nishio, T., and Raymond, K.N. (1996). Preorganization of ferric alcaligin, Fe_2L_3. The first structure of a ferric dihydroxamate siderophore. J. Am. Chem. Soc. *118*, 5148–5149.

Kameyama, T., Takahashi, A., Kurasawa, S., Ishizuka, M., Okami, Y., Takeuchi, T., and Umezawa, H. (1987). Bisucaberin, a new siderophore, sensitizing tumor cells to macrophage-mediated cytolysis. I. Taxonomy of the producing organism, isolation and biological properties. J. Antibiot. *40*, 1664–1670.

Kang, H.Y., and Armstrong, S.K. (1998). Transcriptional analysis of the *Bordetella* alcaligin siderophore biosynthesis operon. J. Bacteriol. *180*, 855–861.

Kang, H.Y., Brickman, T.J., Beaumont, F.C., and Armstrong, S.K. (1996). Identification and characterization of iron-regulated *Bordetella pertussis* alcaligin siderophore biosynthesis genes. J. Bacteriol. *178*, 4877–4884.

Kim, I., Stiefel, A., Plantor, S., Angerer, A., and Braun, V. (1997). Transcription induction of the ferric citrate transport genes via the N-terminus of the FecA outer membrane protein, the Ton system and the electrochemical potential of the cytoplasmic membrane. Mol. Microbiol. *23*, 333–344.

King-Lyons, N.D., Smith, K.F., and Connell, T.D. (2007). Expression of *hurP*, a gene encoding a prospective site 2 protease, is essential for heme-dependent induction of *bhuR* in *Bordetella bronchiseptica*. J. Bacteriol. *189*, 6266–6275.

Kirby, A.E., Metzger, D.J., Murphy, E.R., and Connell, T.D. (2001). Heme utilization in *Bordetella avium* is regulated by RhuI, a heme-responsive extracytoplasmic function sigma factor. Infect. Immun. *69*, 6951–6961.

Lagergren, J. (1963). The white blood cell count and the erythrocyte sedimentation rate in pertussis. Acta Paediatr. *52*, 405–409.

Langman, L., Young, I.G., Frost, G.E., Rosenberg, H., and Gibson, F. (1972). Enterochelin system of iron transport in *Escherichia coli*: mutations affecting ferric-enterochelin esterase. J. Bacteriol. *112*, 1142–1149.

Ledyard, K.M., and Butler, A. (1997). Structure of putrebactin, a new dihydroxamate siderophore produced by *Shewanella putrefaciens*. J. Biol. Inorg. Chem. *2*, 93–97.

Loomis, L.D., and Raymond, K.N. (1991). Solution equilibria of enterobactin and metal–enterobactin complexes. Inorg. Chem. *30*, 906–911.

Lynch, D., O'Brien, J., Welch, T., Clarke, P., Cuiv, P.O., Crosa, J.H., and O'Connell, M. (2001). Genetic organization of the region encoding regulation, biosynthesis, and transport of rhizobactin 1021, a siderophore produced by *Sinorhizobium meliloti*. J. Bacteriol. *183*, 2576–2585.

Mattoo, S., and Cherry, J.D. (2005). Molecular pathogenesis, epidemiology, and clinical manifestations of respiratory infections due to *Bordetella pertussis* and other *Bordetella* subspecies. Clin. Microbiol. Rev. *18*, 326–382.

Menozzi, F.D., Gantiez, C., and Locht, C. (1991). Identification and purification of transferrin- and lactoferrin-binding proteins of *Bordetella pertussis* and *Bordetella bronchiseptica*. Infect. Immun. *59*, 3982–3988.

Merkel, T.J., and Stibitz, S. (1995). Identification of a locus required for the regulation of *bvg*-repressed genes in *Bordetella pertussis*. J. Bacteriol. *177*, 2727–2736.

Mills, K.H., Barnard, A., Watkins, J., and Redhead, K. (1993). Cell-mediated immunity to *Bordetella pertussis*: role of Th1 cells in bacterial clearance in a murine respiratory infection model. Infect. Immun. *61*, 399–410.

Mocny, J.C., Olson, J.S., and Connell, T.D. (2007). Passively released heme from hemoglobin and myoglobin is a potential source of nutrient iron for *Bordetella bronchiseptica*. Infect. Immun. *75*, 4857–4866.

Moore, C.H., Foster, L.A., Gerbig, D.G., Jr., Dyer, D.W., and Gibson, B.W. (1995). Identification of alcaligin as the siderophore produced by *Bordetella pertussis* and *B. bronchiseptica*. J. Bacteriol. *177*, 1116–1118.

Munoz, J.J., Arai, H., Bergman, R.K., and Sadowski, P.L. (1981). Biological activities of crystalline pertussigen from *Bordetella pertussis*. Infect. Immun. *33*, 820–826.

Murphy, E.R., Sacco, R.E., Dickenson, A., Metzger, D.J., Hu, Y., Orndorff, P.E., and Connell, T.D. (2002). BhuR, a virulence-associated outer membrane protein of *Bordetella avium*, is required for the acquisition of iron from heme and hemoproteins. Infect. Immun. *70*, 5390–5403.

Nicholson, M.L., and Beall, B. (1999). Disruption of *tonB* in *Bordetella bronchiseptica* and *Bordetella pertussis* prevents utilization of ferric siderophores, hemin and hemoglobin as iron sources. Microbiology *145*, 2453–2461.

Nishio, T., and Ishida, Y. (1990). Production of dihydroxamate siderophore alcaligin by *Alcaligenes xylosoxidans* subsp. *xylosoxidans*. Agr. Biol. Chem. *54*, 1837–1839.

Nishio, T., Tanaka, N., Hiratake, J., Katsube, Y., Ishida, Y., and Oda, J. (1988). Isolation and structure of the novel dihydroxamate siderophore alcaligin. J. Am. Chem. Soc. *110*, 8733–8734.

O'Brien, I.G., and Gibson, F. (1970). The structure of enterochelin and related 2,3-dihydroxy-N-benzoylserine conjugates from *Escherichia coli*. Biochim. Biophys. Acta. *215*, 393–402.

Palenik, B., Brahamsha, B., Larimer, F.W., Land, M., Hauser, L., Chain, P., Lamerdin, J., Regala, W., Allen, E.E., McCarren, J., *et al.* (2003). The genome of a motile marine *Synechococcus*. Nature *424*, 1037–1042.

Parkhill, J., Sebaihia, M., Preston, A., Murphy, L.D., Thomson, N., Harris, D.E., Holden, M.T., Churcher, C.M., Bentley, S.D., Mungall, K.L., *et al.* (2003). Comparative analysis of the genome sequences of *Bordetella pertussis*, *Bordetella parapertussis* and *Bordetella bronchiseptica*. Nat. Genet. *35*, 32–40.

Passerini de Rossi, B.N., Friedman, L.E., Belzoni, C.B., Savino, S., Arico, B., Rappuoli, R., Masignani, V., and Franco, M.A. (2003). Vir90, a virulence-activated gene coding for a *Bordetella pertussis* iron-regulated outer membrane protein. Res. Microbiol. *154*, 443–450.

Passerini de Rossi, B.N., Friedman, L.E., Gonzalez Flecha, F.L., Castello, P.R., Franco, M.A., and Rossi, J.P. (1999). Identification of *Bordetella pertussis* virulence-associated outer membrane proteins. FEMS Microbiol. Lett. *172*, 9–13.

Persson, C.G., Erjefalt, I., Alkner, U., Baumgarten, C., Greiff, L., Gustafsson, B., Luts, A., Pipkorn, U., Sundler, F., Svensson, C., *et al.* (1991). Plasma exudation as a first line respiratory mucosal defence. Clin. Exp. Allergy. *21*, 17–24.

Pollock, M.R. (1947). The growth of *H. pertussis* on media without blood. Br. J. Exp. Pathol. *28*, 295–307.

Pradel, E., Guiso, N., and Locht, C. (1998). Identification of AlcR, an AraC-type regulator of alcaligin siderophore synthesis in *Bordetella bronchiseptica* and *Bordetella pertussis*. J. Bacteriol. *180*, 871–880.

Pradel, E., Guiso, N., Menozzi, F.D., and Locht, C. (2000). *Bordetella pertussis* TonB, a Bvg-independent virulence determinant. Infect. Immun. *68*, 1919–1927.

Pradel, E., and Locht, C. (2001). Expression of the putative siderophore receptor gene *bfrZ* is controlled by the extracytoplasmic-function sigma factor BupI in *Bordetella bronchiseptica*. J. Bacteriol. *183*, 2910–2917.

Rabsch, W., Voigt, W., Reissbrodt, R., Tsolis, R.M., and Baumler, A.J. (1999). *Salmonella typhimurium* IroN and FepA proteins mediate uptake of enterobactin but differ in their specificity for other siderophores. J. Bacteriol. *181*, 3610–3612.

Redhead, K., Hill, T., and Chart, H. (1987). Interaction of lactoferrin and transferrins with the outer membrane of *Bordetella pertussis*. J. Gen. Microbiol. *133*, 891–898.

Regan, J.C., and Tolstoouhov, A. (1936). Relations of acid base equilibrium to the pathogenesis and treatment of whooping cough. NY State J. Med. *36*, 1075–1087.

Register, K.B., Ducey, T.F., Brockmeier, S.L., and Dyer, D.W. (2001). Reduced virulence of a *Bordetella bronchiseptica* siderophore mutant in neonatal swine. Infect. Immun. *69*, 2137–2143.

Rodgers, S.J., Lee, C., Ng, C.Y., and Raymond, K.N. (1987). Ferric ion sequestering agents. 15. Synthesis, solution chemistry, and electrochemistry of a new cationic analogue of enterobactin. Inorg. Chem. *26*, 1622–1625.

Rutz, J.M., Abdullah, T., Singh, S.P., Kalve, V.I., and Klebba, P.E. (1991). Evolution of the ferric enterobactin receptor in gram-negative bacteria. J. Bacteriol. *173*, 5964–5974.

Spasojevic, I., Armstrong, S.K., Brickman, T.J., and Crumbliss, A.L. (1999). Electrochemical behavior of the Fe(III) complexes of the cyclic hydroxamate siderophores alcaligin and desferrioxamine E. Inorg. Chem. *38*, 449–454.

Sperandio, V., Torres, A.G., Jarvis, B., Nataro, J.P., and Kaper, J.B. (2003). Bacteria-host communication: the language of hormones. Proc. Natl. Acad. Sci. U.S.A. *100*, 8951–8956.

Stojiljkovic, I., Baumler, A.J., and Hantke, K. (1994). Fur regulon in gram-negative bacteria. Identification and characterization of new iron-regulated *Escherichia coli* genes by a *fur* titration assay. J. Mol. Biol. *236*, 531–545.

Takahashi, A., Nakamura, H., Kameyama, T., Kurasawa, S., Naganawa, H., Okami, Y., Takeuchi, T., Umezawa, H., and Iitaka, Y. (1987). Bisucaberin, a new siderophore, sensitizing tumor cells to macrophage-mediated cytolysis. II. Physico-chemical properties and structure determination. J. Antibiot. *40*, 1671–1676.

Uhl, M.A., and Miller, J.F. (1994). Autophosphorylation and phosphotransfer in the *Bordetella pertussis* BvgAS signal transduction cascade. Proc. Natl. Acad. Sci. U.S.A. *91*, 1163–1167.

Vanderpool, C.K., and Armstrong, S.K. (2001). The *Bordetella bhu* locus is required for heme iron utilization. J. Bacteriol. *183*, 4278–4287.

Vanderpool, C.K., and Armstrong, S.K. (2003). Heme-responsive transcriptional activation of *Bordetella bhu* genes. J. Bacteriol. *185*, 909–917.

Vanderpool, C.K., and Armstrong, S.K. (2004). Integration of environmental signals controls expression of *Bordetella* heme utilization genes. J. Bacteriol. *186*, 938–948.

Veal-Carr, W.L., and Stibitz, S. (2005). Demonstration of differential virulence gene promoter activation *in vivo* in *Bordetella pertussis* using RIVET. Mol. Microbiol. *55*, 788–798.

Venuti, M.C., Rastetter, W.H., and Neilands, J.B. (1979). 1,3,5-Tris(N,N',N"–2,3-dihydroxybenzoyl)aminomethylbenzene, a synthetic iron chelator related to enterobactin. J. Med. Chem. *22*, 123–124.

Vidakovics, M.L., Paba, J., Lamberti, Y., Ricart, C.A., de Sousa, M.V., and Rodriguez, M.E. (2007). Profiling the *Bordetella pertussis* proteome during iron starvation. J. Proteome. Res. *6*, 2518–2528.

Ward, J.I., Cherry, J.D., Chang, S.J., Partridge, S., Lee, H., Treanor, J., Greenberg, D.P., Keitel, W., Barenkamp, S., Bernstein, D.I., *et al.* (2005). Efficacy of an acellular pertussis vaccine among adolescents and adults. N. Engl. J. Med. *353*, 1555–1563.

Weiss, A.A., Hewlett, E.L., Myers, G.A., and Falkow, S. (1983). Tn5-induced mutations affecting virulence factors of *Bordetella pertussis*. Infect. Immun. *42*, 33–41.

Weitl, F.L., and Raymond, K.N. (1979). Ferric ion sequestering agents. 1. Hexadentate O-bonding N,N',N''-tris(2,3-dihydroxybenzoyl) derivatives of 1,5,9-triazacyclotridecane and 1,3,5-triaminomethylbenzene. J. Am. Chem. Soc. *101*, 2728–2731.

Williams, P., Morton, D.J., Towner, K.J., Stevenson, P., and Griffiths, E. (1990). Utilization of enterobactin and other exogenous iron sources by *Haemophilus influenzae, H. parainfluenzae* and *H. paraphrophilus*. J. Gen. Microbiol. *136*, 2343–2350.

Wyckoff, E.E., Lopreato, G.F., Tipton, K.A., and Payne, S.M. (2005). *Shigella dysenteriae* ShuS promotes utilization of heme as an iron source and protects against heme toxicity. J. Bacteriol. *187*, 5658–5664.

Zhu, M., Valdebenito, M., Winkelmann, G., and Hantke, K. (2005). Functions of the siderophore esterases IroD and IroE in iron-salmochelin utilization. Microbiology *151*, 2363–2372.

Iron uptake in *Shigella* and *Escherichia coli*

5

Shelley M. Payne and Alexandra R. Mey

Abstract

Shigella spp. and pathogenic *E. coli* are characterized by a variety and abundance of iron transport systems. Although members of this group of bacteria are closely related genetically, they differ widely in the iron transport systems they use. This may reflect the different niches occupied by different strains and the nature of the source of iron available in a specific environment. Only the ferrous iron transporter Feo is common to all the commensals and pathogens. All members of this group produce one or more siderophore, but no single siderophore is produced by all. Other iron transport systems include haem transporters and the ferrous iron transporters Sit and Efe. With the exception of the genes for enterobactin and the Feo system, the iron transport genes in the enterics are found within pathogenicity islands or on plasmids, and their presence often increases pathogenicity or colonization of niches within the host.

Introduction

Shigella spp. and *Escherichia coli* are closely related enteric bacteria, whose classification as separate species is largely historical. Members of this group include commensal strains of *E. coli* that are part of the indigenous microbiota of the large intestine as well as a number of human and animal pathogens. They share a common core genome, but acquisitions of plasmids, phages, chromosomal insertions and deletions have enabled different members of the species to occupy distinct niches in the host and to cause very different diseases. Infection by members of the *E. coli/Shigella* group is by the faecal–oral route. The type and severity of disease is determined by the site of infection and the specific set of virulence factors produced. The *Shigella* spp. form a distinct group with respect to pathogenesis and differ from commensal and most pathogenic *E. coli* in their ability to grow within the host cell cytoplasm (Fig. 5.1). *Shigella* spp. cause disease by invading and replicating within epithelial cells in the human colon, provoking the intense inflammatory response and other symptoms associated with bacillary dysentery (Menard *et al.*, 1996). Production of dysentery is associated with the presence of a large plasmid in these strains that encodes a type III secretion system and effector proteins required for invasion of the host cell (Andrews *et al.*, 1991; Sansonetti, 2006). This plasmid is also found in the group of *E. coli* known as the enteroinvasive *E. coli* (EIEC) that, like *Shigella* spp., cause bacillary dysentery (Sansonetti *et al.*, 1982). Most *E. coli* strains, however, do not have the ability to invade and replicate intracellularly, and disease is usually the result of colonization in the intestine and elaboration of exotoxins or other effector proteins. Some strains are able to invade normally sterile compartments of the host such as the urinary tract, kidneys, blood and meninges and cause extraintestinal infections.

One of the notable differences among these enterics is the number and types of iron transport systems they possess (Table 5.1), and these appear to contribute to the ability of some members of the group to colonize particular environments. Some of these pathogens are adapted to growth in multiple environments, e.g. in the lumen of the intestine, in the urinary tract, within host cells and systemically.

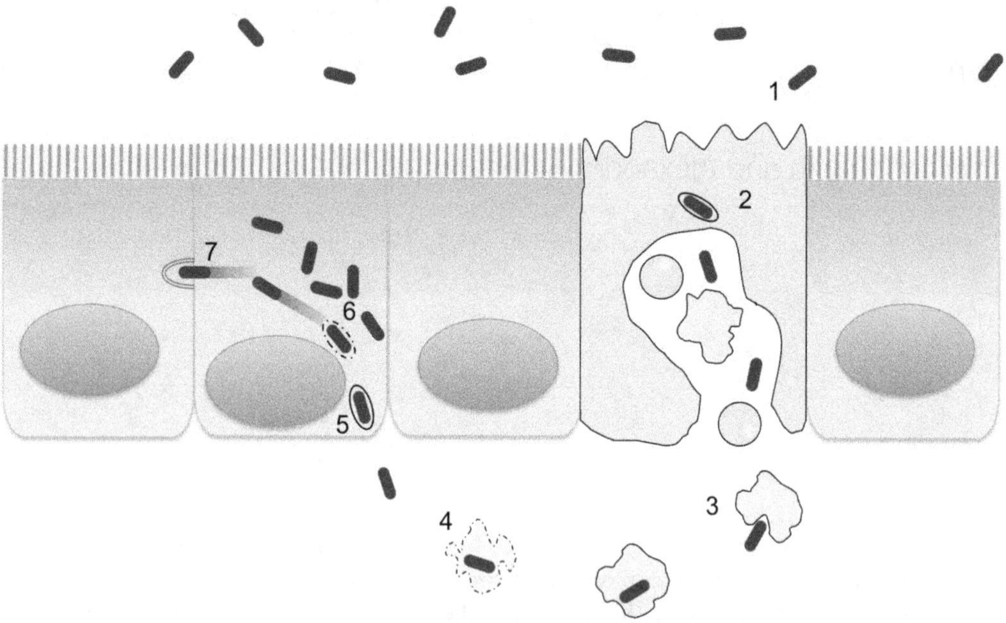

Figure 5.1 *Shigella* in the lumen of the colon (1) cross the colonic epithelial barrier through M cells (2) and are exposed to phagocytic cells in the gut-associated lymphatic tissue in the subepithelial space (3). If ingested by a macrophage, the bacteria cause apoptosis and are released in the vicinity of colonic epithelial cells (4). Shigellae use invasins (Ipas) and a type III secretion system to induce endocytosis and enter the epithelial cells (5). The endocytic vesicle is lysed and the bacteria replicate intracellularly (6). They use actin polymerization to enter adjacent cells (7) and continue replication in an environment protected from the host immune response.

These different environments may present different forms of iron and the amount of available iron may vary considerably in each of the niches that a particular pathogen may inhabit. Thus the relatively large number of iron transport systems probably evolved to enhance survival and competition for iron under a variety of conditions. The following is a review of the types of iron acquisition systems found in representative members of the *E. coli/Shigella* group, and their role in host–pathogen interactions.

Siderophore-mediated iron transport

Shigella and *E. coli*, like other enterics, have high-affinity, siderophore-mediated iron transport systems (Neilands, 1995) (Table 5.2). These bacteria secrete one or more of the low molecular weight, ferric iron chelators and transport the iron–siderophore complex back into the cell via a specific membrane receptor, periplasmic binding protein and cytoplasmic membrane permease. The iron is removed from the siderophore in the cytoplasm by reduction or by degradation of the siderophore. At least four different siderophores are synthesized by members of the *Shigella/E. coli* group. These include enterobactin, salmochelin, aerobactin and yersiniabactin.

Transport of siderophores and other iron complexes across the outer membrane requires the TonB–ExbB–ExbD complex. TonB is a cytoplasmic membrane protein that harvests the energy generated by the cytoplasmic membrane proton gradient and relays it to specific receptors in the outer membrane, allowing active transport of iron complexes into the cell. The ExbB and ExbD accessory proteins are necessary for TonB function (Postle and Larsen, 2007).

Table 5.1 Examples of *E. coli* and *Shigella* colonization sites and iron transport systems

Strain[1]	Site within host	Disease	Iron transport systems[2]
K-12	Intestine	Commensal	Feo, Ent, Fec
UPEC	Urinary tract	Cystitis, nephritis	Feo, Ent, Iro, Iuc, Fec, Sit, Chu, Hma
EHEC	Intestine	Haemorrhagic colitis	Feo, Ent, Fec, Chu, Ybt(Fyu)
NMEC	Systemic	Meningitis, septicaemia	Feo, Ent, Iuc, Ybt[3], Fec[3], Iro[4], Sit[4]
EIEC	Intestine	Dysentery	Feo, Ent, Iuc, Sit[5]
S. dysenteriae type 1	Intracellular in colonic epithelial cells	Dysentery	Feo, Ent, Iro, Fec, Sit, Shu
S. flexneri	Intracellular in colonic epithelial cells	Dysentery	Feo, Iuc, Sit, Fec[6]
S. sonnei	Intracellular in colonic epithelial cells	Dysentery	Feo, Ent, Fec, Sit
S. boydii	Intracellular in colonic epithelial cells	Dysentery	Feo, Ent, Iuc, Sit

[1]K-12, *E. coli* K-12; UPEC, uropathogenic *E. coli*; EHEC, enterohaemorrhagic *E. coli*; NMEC, neonatal meningitis *E. coli*; EIEC, enteroinvasive *E. coli*.

[2]The presence of the indicated intact iron transport systems was determined from analysis of sequenced genomes, or where sequencing and annotation were incomplete, from studies reported in the literature as noted. Xenosiderophore receptors, such as FhuA and FhuE, or systems that primarily transport metals other than iron, MntH and ZupT, are not included in this table.

[3]The systems are present in some, but not all, strains (Yao *et al.*, 2006).

[4]The systems are present in some, but not all, strains (Peigne *et al.*, 2009).

[5]Runyen-Janecky *et al.* (2003).

[6]Found in one strain (Luck *et al.*, 2001).

Enterobactin

Most enterics synthesize and transport enterobactin (Ent), a catechol type siderophore (Earhart, 2004). This siderophore, which has an extremely high affinity for iron, is a cyclic trimer of 2,3-dihydroxybenzoic acid (Fig. 5.2A). The siderophore is secreted into the environment where it binds ferric iron. The iron complex is transported back into the cell via the TonB-dependent outer membrane receptor FepA, the periplasmic binding protein FepB and the cytoplasmic permease composed of FepC, FepD and FepG. Inside the cell the siderophore is broken down by Fes, an esterase, and the iron is released. Ent is a highly efficient iron scavenger and promotes growth of enterics under laboratory conditions. However, its role in iron acquisition in the host is less clear. Early studies to examine the function of Ent *in vivo* were based on testing the virulence of enterobactin mutants of *E. coli* or *Salmonella*. The results were mixed and may have been influenced by the specific mutation and the choice of mouse strains (Yancey *et al.*, 1979; Benjamin *et al.*, 1985). Further, some pathogens, especially *S. flexneri* and *S. boydii*, lack the ability to produce enterobactin due to mutations or deletions of genes within the *ent* cluster (Schmitt and Payne, 1988, 1991), yet these strains are as pathogenic as enterobactin-producing shigellae. Subsequent studies suggest that Ent has limited utility for systemic pathogens, because this siderophore binds to albumin and other serum proteins (Konopka and Neilands, 1984). Further, secreted lipocalin-2 can sequester Ent, preventing the bacteria from using Ent to acquire iron *in vivo* (Fischbach *et al.*, 2006).

Table 5.2 Expression and regulation of iron transport systems in selected *E. coli* and *Shigella* spp.

Iron transport system	Known regulation	K-12[1]	UPEC	EHEC	SD	SF	SS	SB
Enterobactin[2]	Fur	+	+	+	+		+	
Salmochelin	Fur		+		+			
Aerobactin	Fur, ArcA		+			+		+
Yersiniabactin	Fur		+					
Sit	Fur, oxygen		+		+	+	+	+
Feo	Fur, ArcA, Fnr	+	+	+	+	+	+	+
Efe[2]	Fur, CpxAR			+			+[3]	
Fit	Fur, FitR		+	+				
Fec	Fur, FecIR	+			+	+[3]	+	+[3]
Shu/Chu	Fur		+	+	+			
Hma			+	+				

[1]+ indicates the presence of the indicated iron transport system. K-12, *E. coli* K-12; UPEC, uropathogenic *E. coli*; EHEC, enterohaemorrhagic *E. coli*; SD, *S. dysenteriae* type 1; SF, *S. flexneri* 2a; SS, *S. sonnei*; SB, *S. boydii*.

[2]The genes encoding this system are present in most strains, but the system is not functional due to insertions, deletions or other mutations.

[3]The genes for the indicated system are not present in all strains.

Salmochelin

Some pathogens, including uropathogenic *E. coli* (UPEC) and *Shigella dysenteriae* type 1, evade the effects of lipocalin-2 by producing a modified form of enterobactin (Fischbach *et al.*, 2006). This siderophore, first identified in *Salmonella* and named salmochelin, is a C-glucosylated enterobactin (Hantke *et al.*, 2003) (Fig. 5.2B). This glucosylated siderophore is not recognized by lipocalin-2 and can allow the bacteria to overcome the inhibitory effects of lipocalin-2. Salmochelin is transported back into the bacterial cell through the IroN outer membrane receptor (Hantke *et al.*, 2003). The *iro* genes specifying the synthesis and transport of salmochelin map within an island adjacent to the *serX* locus in *S. dysenteriae* type 1 and the UPEC strains (Fig. 5.3A).

Aerobactin

Many pathogenic *E. coli* and *Shigella* spp. produce aerobactin (Fig. 5.2C). This siderophore is a secondary hydroxamate synthesized from citrate and lysine (Gibson and Magrath, 1969). Although the binding affinity of aerobactin for iron is less than that of enterobactin, aerobactin is not sequestered by albumin or lipocalin-2 and may be more stable in the host environment than enterobactin (Konopka and Neilands, 1984; Neilands, 1984). The four genes required for the assembly of this siderophore (*iucABCD*) and the gene encoding the outer membrane receptor (*iutA*) form an operon that maps within pathogenicity islands or on plasmids (Bindereif and Neilands, 1983; Carbonetti and Williams, 1984; Lawlor and Payne, 1984; McDougall and Neilands, 1984). The islands may be found at different locations in different strains (Fig. 5.3A) and are often found adjacent to tRNA genes. The *S. flexneri* 2a aerobactin island is immediately downstream of the *selC* tRNA gene, and the island contains the genes for a phage T4-like integrase, colicin immunity, and numerous insertion sequences and phage-like genes (Moss *et al.*, 1999; Vokes *et al.*, 1999). The *S. boydii* aerobactin island contains a different set of phage and insertion sequences and is inserted downstream of a different tRNA gene, *pheU* (Purdy and Payne, 2001).

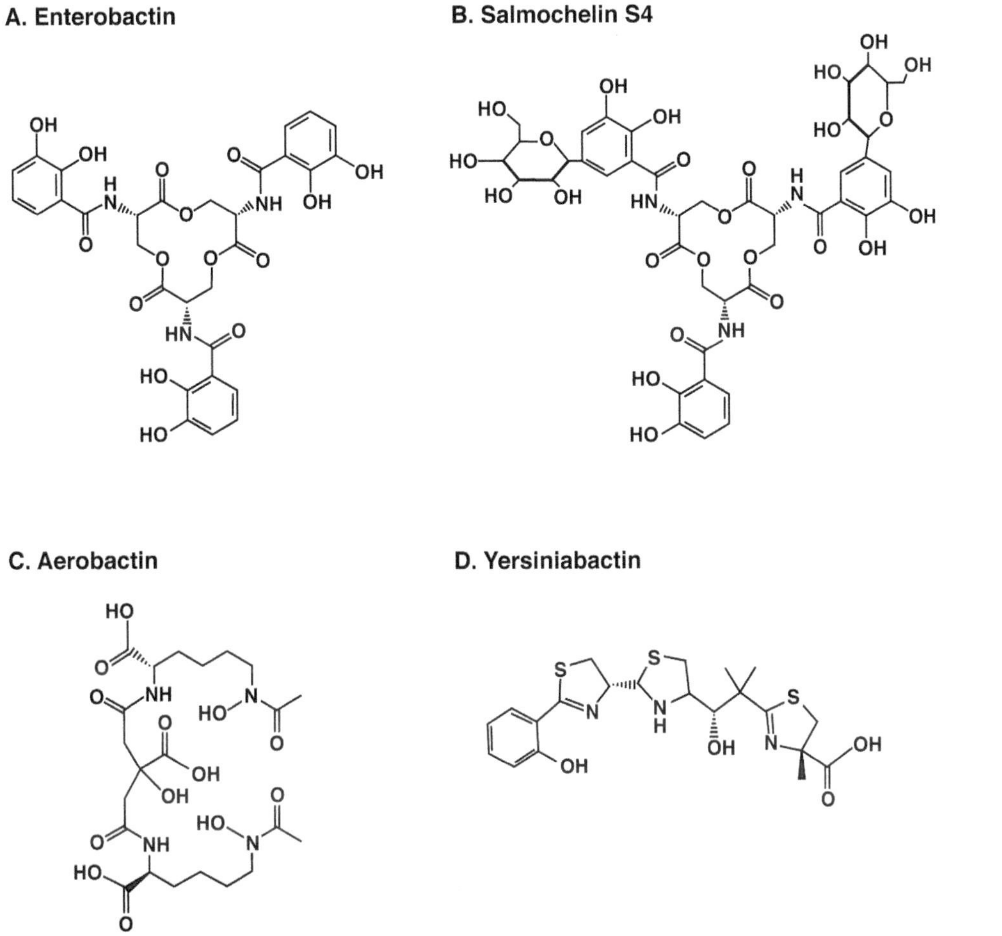

Figure 5.2 Structure of the siderophores commonly produced by *E. coli* and *Shigella*: (A) enterobactin (Earhart, 2004), (B) salmochelin (salmochelin S4 is shown) (Bister *et al.*, 2004), (C) aerobactin (Gibson and Magrath, 1969) and (D) yersiniabactin (Perry *et al.*, 1999).

Yersiniabactin

A third type of siderophore, yersiniabactin, is found in some of the pathogenic *E. coli.* This siderophore is related to pyochelin (produced by *Pseudomonas aeruginosa*) and anguibactin (*Vibrio anguillarum*), and it contains a phenolic group, a thiazolidine ring and two thiazoline rings (Fig. 5.2D). This siderophore was first described in *Yersinia* spp., and its synthesis and transport are encoded within the high pathogenicity island (HPI). Yersiniabactin production is common among extraintestinal *E. coli* (Schubert *et al.*, 2002; Ananias and Yano, 2008) and is found in enteroaggregative *E. coli* and UPEC strains (Schubert *et al.*, 1998; Schubert *et al.*, 1999), but has not been reported in *Shigella* spp. Like the salmochelin and aerobactin genes, the yersiniabactin biosynthesis and transport genes map within an island, which in several isolates has been mapped adjacent to the *asnT* tRNA locus (Schubert *et al.*, 1999) (Fig. 5.3A). In other strains, the location of the genes was not determined but was found to be at a site distinct from the *asnT* locus (Clermont *et al.*, 2001).

Xenosiderophores

Shigella and *E. coli* have the ability to transport siderophores that they do not themselves synthesize, termed xenosiderophores (Hantke, 1983; Koster and Braun, 1990; Braun *et al.*, 2004). For example, FhuA, the receptor for the fungal siderophore ferrichrome is ubiquitous in this group, and other

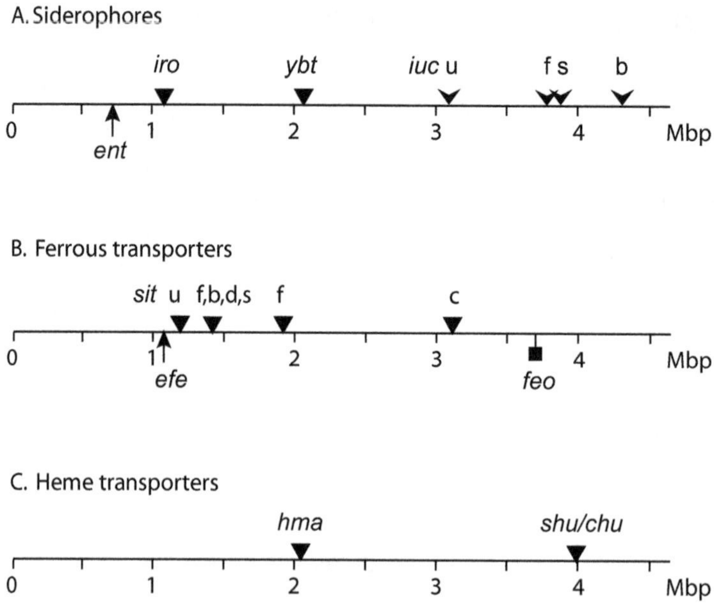

Figure 5.3 Chromosomal locations of the genes encoding the (A) siderophore, (B) ferrous and (C) haem transport systems of *E. coli* and *Shigella*. The locations are shown relative to the *E. coli* K-12 strain MG1655 map and were determined using MAUVE genomic alignments (Darling *et al.*, 2007). The arrows (*ent*, *efe*) and square (*feo*) indicate genes that are present, though not necessarily functional, in all strains. Symbols above the line indicate systems such as *ybt* (*fyu*) (arrowhead in panel A), *iro* (arrowhead in panel A), *iuc* (notched arrowheads in A) and *sit* (arrowhead in B), that are found within genomic islands and, if present, may be found at different locations. Positions of arrowheads indicate the location of sequences adjacent to the insertion site on the K-12 map. Because the genomes vary in size and contain inversions, deletions and insertions relative to each other, the map positions in the *E. coli* K-12 genome do not necessarily correlate with the position on the genome of the pathogen. u, uropathogenic *E. coli*; b, *S. boydii*; d, *S. dysenteriae* type 1; f, *S. flexneri* 2a; s, *S. sonnei*; c, *E. coli* SMS-3-5.

receptors such as FhuE for coprogen transport are common. The periplasmic binding protein and cytoplasmic membrane permease for ferrichrome (FhuB, FhuC and FhuD) are also used by coprogen and aerobactin. Thus, there is specificity of the outer membrane receptors for the hydroxamate siderophores, but the subsequent transport into the cell is mediated by a generic hydroxamate transporter.

Haem transport

Although *E. coli* K-12 lacks the ability to transport haem into the cell, many pathogenic *E. coli* and *Shigella* can use haem as a source of iron. The haem transport systems share features of the siderophore transporters and consist of a TonB-dependent outer membrane receptor, periplasmic binding protein and cytoplasmic membrane permease. One class of receptors, called ShuA in *S. dysenteriae* type 1 (Mills and Payne, 1997) and ChuA in pathogenic *E. coli* strains, (Torres and Payne, 1997; Torres *et al.*, 2001) has been characterized. The gene encoding the receptor, along with seven other ORFs, is found within an approximately 9-kb locus located in the intergenic region between *yhiD* and *yhiF* on the *E. coli* K-12 map (Mills and Payne, 1995; Wyckoff *et al.*, 1998) (Fig. 5.3C).

The location and structure of the island is the same as in *E. coli* O157:H7 and *S. dysenteriae* type 1, and the sequence of the receptor is nearly identical (Wyckoff *et al.*, 1998). A distinct haem transporter, Hma, has been identified in UPEC strain CFT073 (Hagan and Mobley, 2009). This receptor binds haem but lacks the conserved histidine residues typically found in haem receptors and may have a distinct binding mechanism. UPEC strains also have the ChuA haem receptor (Torres *et al.*, 2001), and mutations in both *hma* and *chuA* are required to eliminate haem transport in this uropathogen (Hagan and Mobley,

2009). The Hma receptor is not widespread among the *E. coli/Shigella* group and is found primarily in the UPEC and EHEC strains.

Transport of haem across the inner membrane is facilitated by the Shu/ChuTUV permease, which is encoded within the haem transport locus. In *E. coli* strains that do not possess the haem transport locus, it is possible to confer haem utilization by supplying only an outer membrane receptor for haem, implying that a mechanism other than the ShuTUV permease exists for shunting the haem across the inner membrane (Mills and Payne, 1997). It has now been reported that the dipeptide permease of *E. coli* K-12 can function as a haem-specific inner membrane transporter (Létoffé *et al.*, 2006). The haem is transported intact into the bacterial cell and can serve as both an iron and porphyrin source (Mills and Payne, 1995). The efficient use of haem as an iron source requires the cytoplasmic protein ShuS, which also serves to protect the cell against the toxic effects of high concentrations of haem (Wyckoff *et al.*, 2005). However, ShuS lacks haem oxygenase or other known haem degrading activity (Wilks, 2001). Thus, the fate of haem transported into *S. dysenteriae* or *E. coli* and the mechanism of use of haem iron are as yet unknown.

Ferric citrate

E. coli and some *Shigella* strains can transport ferric iron as the ferric citrate complex. Ferric citrate is transported into the cell via a TonB-dependent outer membrane receptor, FecA, and a dedicated periplasmic binding protein (PBP)-dependent cytoplasmic membrane permease (Braun *et al.*, 2004). Expression of this system is induced by ferric citrate, and regulation involves an extracytoplasmic sigma factor, FecI. Binding of ferric citrate to FecA induces a conformational change in the signalling domain of FecA that is transmitted via the cytoplasmic membrane regulator FecR (an anti-sigma factor) to FecI (an extracytoplasmic sigma factor), inducing expression of the *fec* transport genes (Braun *et al.*, 2005). The Fec system is found in all commensal *E. coli*, but is not uniformly present among the pathogenic *E. coli* strains (Table 5.2). While the *fec* genes are present in at least one representative strain of each *Shigella* spp. (Payne *et al.*, 2006), they do not appear to be widely retained in *Shigella*. Most *S. flexneri* isolates do not use ferric citrate, but in one strain, the *fec* genes have been found within a large island that encodes resistance to a number of antibiotics (Luck *et al.*, 2001).

Fit

The Fit system was originally identified in a screen for genes expressed *in vivo* by an invasive *E. coli* strain during septicaemia in mice (Khan and Isaacson, 2002), and has been shown to be present in several enteroinvasive and uropathogenic strains. However, the Fit system is not found in *E. coli* K-12 (Ouyang and Isaacson, 2006) or in the *Shigella* spp. The Fit system consists of a putative TonB-dependent transporter FitA, and a probable PBP-dependent, ATP-binding cassette (ABC) type transporter system consisting of FitE (PBP), FitCD (permease) and FitB (ATPase). As expected, the *fit* genes are regulated by iron and the Fur repressor as described below (Ouyang and Isaacson, 2006). In addition, the *fit* genes are subject to Fur- and iron-independent repression by FitR, a regulator encoded within the *fit* operon (Ouyang and Isaacson, 2008). The role of the Fit system in iron acquisition is not clear. Mutations in the Fit system do not impair growth in low iron media, and no specific ligand has been identified for this transporter. Expression of the Fit system from a high copy number vector in *E. coli* K-12 was shown to increase the iron content of the cells; however, it is possible that this may be due to sequestration of Fur by the Fur-binding sequences provided by the high copy Fit plasmid, leading to de-repression of the native iron transporters and increased iron uptake. A mutation in *fitB* was associated with a 4-fold decrease in virulence in the mouse septicaemia model (M. A. Khan and R. E. Isaacson, unpublished results cited in Ouyang and Isaacson, 2006).

Ferrous iron transporters

Feo

The ferrous iron transporter Feo is ubiquitous among the enterics and is found in all the *Shigella* and *E. coli* strains examined thus far. This transport system is encoded by three genes, *feoABC*. The best-studied component of this system is FeoB, a cytoplasmic membrane protein that is similar to small regulatory G proteins found in eukaryotes (Marlovits *et al.*, 2002). FeoB is a GTP-binding protein with a G4-motif, NXXD that is responsible for the guanine nucleotide specificity. FeoB and the GTPase activity are essential for Feo-mediated iron uptake (Marlovits *et al.*, 2002), although the mechanism of transport is not known. *feoA* and *feoC* are in an operon with *feoB* and are predicted to encode small, cytoplasmic proteins. However, functions for FeoA and FeoC remain to be established.

Sit

A second ferrous iron transport system, SitABCD, is found in all *Shigella* and some of the *E. coli* strains (Table 5.2). This system is particularly prevalent in the intracellular pathogens. First identified in *Salmonella* (Zhou *et al.*, 1999), Sit can transport both ferrous iron and manganese (Janakiraman and Slauch, 2000; Kehres *et al.*, 2002) and is similar to the Yfe ABC iron transport system of *Yersinia pestis* (Bearden and Perry, 1999). This ABC-type transporter consists of a periplasmic binding protein (SitA), a cytoplasmic membrane permease composed of the putative integral membrane proteins SitC and SitD, and an ATP-binding protein SitB (Zhou *et al.*, 1999). The *S. flexneri* Sit system proteins have homology to the *Salmonella* and *Yersinia* Sit components, and transport assays confirmed that *Shigella* Sit transports ferrous, but not ferric, iron (Fisher *et al.*, 2009).

The *sit* genes map within pathogenicity islands or on plasmids in *E. coli* and *Shigella* (Fig. 5.3B). There is considerable diversity in the location and composition of the islands. Even islands that are found at the same location in the chromosome may differ in size, composition and the orientation of the *sit* genes within the island (Fisher *et al.*, 2009).

Efe

Like the Feo system, the EfeUOB system (formerly known as YcdNOB) transports ferrous iron (Grosse *et al.*, 2006; Cao *et al.*, 2007); however, while the Feo system is most active under anaerobic conditions, the Efe system operates aerobically, at low pH. All three members of the *efeUOB* operon are required for optimal activity of the system (Cao *et al.*, 2007). EfeU is a homologue of the iron permease Ftr1p from yeast, and is a member of the oxidase-dependent iron transporter (OFeT) family. EfeU is an integral cytoplasmic membrane protein that is predicted to have seven transmembrane helices with two REXXE motifs that are important for iron uptake (Grosse *et al.*, 2006). EfeO is predicted to reside in the periplasm, but is not similar to any protein with a known function. EfeB has been shown to be periplasmic and belongs to the haem peroxidase superfamily. It contains a *b*-type haem and requires the twin arginine transport pathway for translocation into the periplasmic space (Sturm *et al.*, 2006). The exact contribution of each Efe component to the iron transport process is not known, but it seems likely that EfeU is itself the iron permease (Grosse *et al.*, 2006).

An intact *efeUOB* operon is found in *E. coli* O157:H7 (Cao *et al.*, 2007) and in *E. coli* strain Nissle 1917 (Grosse *et al.*, 2006), as well as in several *E. coli* B, E, and F strains, and in at least one strain of *S. sonnei* (Ss046). In other *E. coli* and *Shigella* spp., however, the DNA sequences are either absent or contain frameshifts or deletions. In *E. coli* K-12, the *efeUOB* operon is present, but *efeU* is frame-shifted early in the gene (Grosse *et al.*, 2006). When this lesion is repaired, the Efe system is functional and alleviates iron starvation in a strain defective in all other known iron transporters (Cao *et al.*, 2007).

Regulation of iron transport

As in other bacteria, expression of genes encoding the iron transport systems is negatively regulated by Fur in *E. coli* and *Shigella* spp. (Bagg and Neilands, 1987; Schmitt and Payne, 1988; Payne *et al.*, 2006). In conditions where iron is abundant, dimers of iron-bound Fur bind to sequences termed Fur boxes within the promoter (Bagg and Neilands, 1987; deLorenzo *et al.*, 1988). The Fur boxes usually overlap the −10 or −35 region and block transcription. As iron becomes limiting, Fur loses affinity for the DNA and transcription of the iron transport genes is increased. However, derepression of all the iron transport genes is only seen under conditions of extreme iron starvation. The differential expression of the iron transport genes reflects both a difference in the affinities of the Fur boxes for Fur and the presence of additional regulatory sites in the promoters. In addition to controlling expression of iron acquisition genes, Fur regulates a small RNA, RyhB, involved in cellular iron metabolism and homeostasis in *E. coli* (Massé and Gottesman, 2002; Gottesman, 2004). RyhB serves similar functions in *Shigella* spp., but in addition, RyhB is involved in regulating virulence gene expression and acid resistance in *Shigella* (Oglesby *et al.*, 2005; Murphy and Payne, 2007). It is interesting to note that virulence factors may also be directly regulated by iron and Fur. The Shiga toxin of *S. dysenteriae* type 1 and shiga-like toxin, Stx1, have Fur boxes in their promoters and are negatively regulated by iron (Van Heyningen and Gladstone, 1953; Calderwood and Mekalanos, 1987; O'Brien *et al.*, 1992).

Expression of many of the iron transport genes is influenced by the presence of oxygen. In the presence of oxygen, particularly at neutral or higher pH, iron is predominantly ferric, while ferrous iron is more abundant in anaerobic environments. Thus, it is not surprising that ferric iron transport systems are more highly expressed in aerobic conditions while ferrous iron transporters are generally expressed in anaerobic environments.

The control of *E. coli* and *Shigella* gene expression as a function of oxygen tension is mediated by two primary regulators, Fnr and ArcA. The Fnr transcription factor directly senses oxygen through a cubane iron–sulphur cluster. This iron–sulphur cluster, required for the activity of Fnr, is disassembled in the presence of oxygen, causing Fnr to lose affinity for its DNA binding site. ArcA is the response regulator of the ArcAB two component regulatory system. During anaerobiosis, reduced quinones accumulate and activate ArcB as a kinase. ArcB autophosphorylates and then phosphorylates ArcA, which leads to ArcA activation. Both ArcA and Fnr can either induce or repress target genes during anaerobiosis, and their primary function is to activate genes involved in fermentation and anaerobic respiration and repress genes involved in aerobic respiration. Kammler *et al.* (1993) showed that *feo* expression in *E. coli* is regulated by Fnr, which increases its expression anaerobically. *Shigella feo* is also induced by Fnr (Boulette and Payne, 2007). Further, ArcA increases *S. flexneri feo* expression under anaerobic conditions, and an ArcA binding site was identified upstream of the Fur and Fnr binding sites in the *feo* promoter (Boulette and Payne, 2007). ArcA also binds to the *iuc* (aerobactin) promoter, but acts as a repressor for the siderophore genes in anaerobic conditions. Thus, Fnr and ArcA allow *S. flexneri* to switch between ferric and ferrous iron transport systems as a function of the levels of oxygen in the environment.

Unexpectedly, the Sit and Efe systems primarily operate under aerobic conditions, despite the specificity of these systems for ferrous iron (Boulette and Payne, 2007; Cao *et al.*, 2007). Sit expression is increased in an *arcA* mutant but this appears to be an indirect effect, since no binding of ArcA to the *sit* promoter was detected.

Studies of expression of the *efeUOB* genes show that this operon is not only regulated by iron and Fur but also by pH (McHugh *et al.*, 2003; Grosse *et al.*, 2006; Cao *et al.*, 2007). Since acidic pH stabilizes the ferrous form of iron, even in the presence of oxygen, expression of the *efe* genes at low pH allows for ferrous iron utilization aerobically. Regulation of the *efe* genes in response to pH is achieved via the CpxAR two-component sensor regulator, which represses expression of the *efeUOB* operon at alkaline pH (Cao *et al.*, 2007).

Roles of iron transport systems in virulence of *Shigella flexneri* and *E. coli*

There are several ways in which an iron acquisition system can be evaluated for its role in virulence. One way is to assess the prevalence of an iron uptake system among pathogens. If the genes encoding an iron transport system are present in pathogenic strains, but not typically in commensal strains of the same species, it is reasonable to conclude that the system may be involved in some aspect of the pathogenic process. Another method is to determine the expression of an iron transport system in response to the host environment. If the genes are up-regulated in an animal or cell culture model, it may signify that the system is necessary for growth *in vivo*. Although informative, the data obtained from the above methods are correlative but not conclusive. The most direct way to determine the role of a system in virulence is to create a mutation in the gene(s) encoding the system and assess the ability of the mutant to cause disease in an animal model. If the system is one of several that operate in iron uptake in the host, a mutant may not show any disadvantage compared with its parental strain. In this case, it may be necessary to create multiple mutations in order to detect a phenotype. In reality, there are constraints associated with each method. A good animal or cell culture model of disease may not be available, or it may not faithfully represent the disease process in the natural host. Functional redundancy may mask the effect of a given system *in vivo*. The prevalence of certain genes among pathogens may reflect recent common ancestry rather than selective retention of these genes. By combining results from several approaches it is possible, however, to gain insights into the role of individual systems in pathogenesis.

Because of the large number of iron transport systems in most of the enteric pathogens and the incomplete identification of all the systems in most of these bacteria, it has been difficult to assess the role of each system in virulence. Table 5.1 gives an overview of the iron transport systems found in the pathogenic *E. coli*/*Shigella* group, and the site of infection and type of disease associated with each. Some interesting observations can be made by comparing the distribution of individual iron transport systems within this group. The *sit* genes, for example, are found in all the *Shigella* species, as well as in *E. coli* strains capable of invading eukaryotic cells, such as the enteroinvasive and uropathogenic strains, but not in commensal or non-invasive pathogenic *E. coli*. This suggests that the *sit* genes may be important for all members of this group that need to acquire iron within the host cell cytosol. The enteroinvasive *E. coli* and *Shigella* species are not thought to share a recent common ancestor, and the presence of the *sit* genes is therefore probably due to a selective advantage rather than to common ancestry. The *Yersinia* HPI, encoding the yersiniabactin synthesis and transport system (*ybt*/*fyu*), is associated primarily with extraintestinal pathogenic *E. coli*, such as the invasive and uropathogenic strains, and is not commonly found in intestinal pathogenic or commensal *E. coli* (Clermont *et al.*, 2001). It is absent from all species of *Shigella*. This distribution points to a role for the Ybt HPI in extraintestinal infection, and studies show that this island contributes to uropathogenesis (Lloyd *et al.*, 2009) and is critical for the lethality of ExPEC strains in mice (Schubert *et al.*, 2002).

There is not always a direct correlation between the complement of iron transport systems encoded by a pathogen and the site of infection and type of disease caused. *Shigella* species do not all express the same systems; among the *Shigella* species, only *S. dysenteriae* Type 1 carries the haem transport locus and the salmochelin synthesis and transport genes. *S. flexneri* and *S. boydii* produce aerobactin, whereas *S. dysenteriae* and *S. sonnei* secrete enterobactin. All four species, however, infect the colonic epithelium and cause a similar disease in the host. Thus, it becomes necessary to determine not only which genes are present, but which systems are expressed in response to the host environment, and which genes are specifically needed for pathogenesis.

A comparative study of siderophore production by *E. coli* recovered from either the urinary tract or the gut of patients with recurrent urinary tract infection found that yersiniabactin and salmochelin production was high among the urinary tract strains and low in the intestinal strains isolated from the same patient. In contrast, enterobactin production was higher among the intestinal isolates, although the *ent* genes are present in all the uropathogenic strains (Henderson *et al.*, 2009). Since the uropathogens are thought to arise from the gut population, a shift in siderophore production may be an integral part of becoming pathogenic for the urinary tract. Interestingly, aerobactin was not commonly produced by urinary tract strains in any of the patients studied here (Henderson *et al.*, 2009). Although aerobactin transport contributes to urovirulence in mice (Torres *et al.*, 2001), and the aerobactin receptor gene is up-regulated during growth in the murine model of UTI (Snyder *et al.*, 2004), aerobactin may be less important for uropathogenesis in humans than either salmochelin or yersiniabactin.

The role of the salmochelin transport system in virulence is well documented in ExPEC strains. The salmochelin receptor gene *iroN* is up-regulated both in human urine (Walters and Mobley, 2009) and in urine from infected mice (Snyder *et al.*, 2004). IroN promotes invasion of human urothelial cells, which could help the pathogen evade the host immune response (Feldmann *et al.*, 2007). In mice, an *iroN* mutant was outcompeted by the wild type strain in the urine, bladder and kidney of infected mice (Russo *et al.*, 2002), and IroN elicited a protective immune response against kidney infection (Russo *et al.*, 2003). In a rat neonatal meningitis model, IroN was required for bacteraemia (Negre *et al.*, 2004). The *iro* gene cluster is more prevalent among uropathogenic and bacteraemic strains than commensal *E. coli*, further pointing to a role in virulence. As described above, a role for the yersiniabactin locus in virulence has been documented as well.

The ability to acquire haem *in vivo* has been shown to be important for virulence in UPEC strain CFT073 (Torres *et al.*, 2001; Hagan and Mobley, 2009), particularly in colonizing the kidney (Hagan and Mobley, 2009). The kidney is an environment rich in blood, and secretion of haemolysins by UPEC may increase the availability of haemoglobin and haem for the invading pathogen. In contrast, haem utilization by *S. dysenteriae* is not critical for growth in human epithelial cells, but the lack of a suitable mammalian model for shigellosis makes it difficult to determine whether haem is an important iron source *in vivo* for *Shigella*.

The most complete analysis of the role of individual iron transport systems in the host, and their expression in response to specific environmental signals, including in response to the host environment, is available for *S. flexneri* 2a. Most 2a strains have only three iron transport systems: aerobactin, Feo and Sit. Thus it was possible to construct mutants that lacked one, two or all three of the systems and compare them in cultured cell assays and in a mouse lung model of infection (Runyen-Janecky *et al.*, 2003, Fisher *et al.*, 2009). Cell culture assays measure the ability of *Shigella* spp. to invade, multiply intracellularly and spread to adjacent cells, and wild type strains form visible plaques in the epithelial cell monolayers (Oaks *et al.*, 1985). The triple mutant lacking all three iron transport systems was unable to grow *in vitro* in the absence of added siderophore and failed to form plaques in cultured cells (Runyen-Janecky *et al.*, 2003). The single and double mutants generally formed smaller plaques, and plaque formation was influenced by the presence of oxygen in the environment. The *iuc*, *sit* mutant, which has only the Feo iron transport system, formed plaques when the cell cultures were incubated anaerobically but not aerobically, while the *iuc*, *feo* mutant which only has the Sit iron transport system formed plaques aerobically but not anaerobically (Runyen-Janecky *et al.*, 2003; Boulette and Payne, 2007). This indicates that there is a pool of available ferrous iron and sufficient oxygen in the host cell cytoplasm to allow the Sit system to supply iron to the intracellular bacteria.

The genetic analysis is supported by studies of the expression of the iron transport genes in the intracellular environment. A *sit–gfp* fusion was expressed when the bacteria were growing within human epithelial cells but an *iuc–gfp* fusion was not expressed under these conditions (Runyen-Janecky and Payne, 2002). Further, Western blot analysis with antibody against the aerobactin receptor, IutA, confirmed low expression of the receptor in the intracellular environment.

Although there is not a non-primate animal model that mimics dysentery, the mouse lung model allows assessment of invasion and provocation of the host inflammatory response. In this model, the wild type *S. flexneri* was lethal, as was the *iucA* mutant, but the *sitA* mutant was highly attenuated (Fisher *et al.*, 2009). In contrast, an *iuc* mutant was less virulent in the rabbit ligated ileal loop (Nassif *et al.*, 1987). Taken together, these data on expression of the iron transport systems and the effects of mutations in the iron transport systems indicate that Sit is required for efficient iron acquisition in the intracellular environment while the siderophore plays a role in growth in extracellular environments.

Unlike *S. flexneri* 2a, the complete repertoire of iron transport systems has not been defined for other members of this group, and it is not yet possible to test the roles of all their iron transport systems by genetic analysis. However, the data available thus far indicate that no single transport system is sufficient for pathogenesis and that different systems function to supply iron at different stages of disease or in different environmental niches.

References

Ananias, M., and Yano, T. (2008). Serogroups and virulence genotypes of *Escherichia coli* isolated from patients with sepsis. Braz. J. Med. Biol. Res. *41*, 877–883.

Andrews, G.P., Hromockyj, A.E., Coker, C., and Maurelli, A.T. (1991). Two novel virulence loci, *mxiA* and *mxiB*, in *Shigella flexneri* 2a facilitate excretion of invasion plasmid antigens. Infect. Immun. *59*, 1997–2005.

Bagg, A., and Neilands, J.B. (1987). Ferric uptake regulation protein acts as a repressor, employing iron (II) as a cofactor to bind the operator of an iron transport operon in *Escherichia coli*. Biochemistry. *26*, 5471–5477.

Bearden, S.W., and Perry, R.D. (1999). The Yfe system of *Yersinia pestis* transports iron and manganese and is required for full virulence of plague. Mol. Microbiol. *32*, 403–414.

Benjamin, W.H., Jr., Turnbough, C.L., Jr., Posey, B.S., and Briles, D.E. (1985). The ability of *Salmonella typhimurium* to produce the siderophore enterobactin is not a virulence factor in mouse typhoid. Infect. Immun. *50*, 392–397.

Bindereif, A., and Neilands, J.B. (1983). Cloning of the aerobactin-mediated iron assimilation system of plasmid ColV. J. Bacteriol. *153*, 1111–1113.

Bister, B., Bischoff, D., Nicholson, G.J., Valdebenito, M., Schneider, K., Winkelmann, G., Hantke, K., and Sussmuth, R.D. (2004). The structure of salmochelins: C-glucosylated enterobactins of *Salmonella enterica*. Biometals *17*, 471–481.

Boulette, M.L., and Payne, S.M. (2007). Anaerobic regulation of *Shigella flexneri* virulence: ArcA regulates Fur and iron acquisition genes. J. Bacteriol. *189*, 6957–6967.

Braun, V., Braun, M., and Killmann, H. (2004). Ferrichrome- and citrate-mediated iron transport. In Iron Transport in Bacteria, J.H. Crosa, Mey, A.R., and S.M. Payne, ed. (Washington, DC, ASM Press).

Braun, V., Mahren, S., and Sauter, A. (2005). Gene regulation by transmembrane signaling. Biometals *18*, 507–517.

Calderwood, S.B., and Mekalanos, J.J. (1987). Iron regulation of Shiga-like toxin expression in *Escherichia coli* is mediated by the *fur* locus. J. Bacteriol. *169*, 4759–4764.

Cao, J., Woodhall, M.R., Alvarez, J., Cartron, M.L., and Andrews, S.C. (2007). EfeUOB (YcdNOB) is a tripartite, acid-induced and CpxAR-regulated, low-pH Fe^{2+} transporter that is cryptic in *Escherichia coli* K-12 but functional in *E. coli* O157:H7. Mol. Microbiol. *65*, 857–875.

Carbonetti, N.H., and Williams, P.H. (1984). A cluster of five genes specifying the aerobactin iron uptake system of plasmid ColV-K30. Infect. Immun. *46*, 7–12.

Clermont, O., Bonacorsi, S., and Bingen, E. (2001). The *Yersinia* high-pathogenicity island is highly predominant in virulence-associated phylogenetic groups of *Escherichia coli*. FEMS Microbiol. Lett. *196*, 153–157.

Darling, A.E., Treangen, T.J., Messeguer, X., and Perna, N.T. (2007). Analyzing patterns of microbial evolution using the mauve genome alignment system. Methods Mol. Biol. *396*, 135–152.

deLorenzo, V., Giovannini, F., Herrero, M., and Neilands, J.B. (1988). Metal ion regulation of gene expression: Fur repressor–operator interaction at the promoter region of the aerobactin system of pColV-K30. J. Mol. Biol. *203*, 875–884.

Earhart, C.F. (2004). Iron uptake via the enterobactin system. In Iron Transport in Bacteria, J.H. Crosa, Mey, A.R., and S.M. Payne, ed. (Washington, DC, ASM Press).

Feldmann, F., Sorsa, L.J., Hildinger, K., and Schubert, S. (2007). The salmochelin siderophore receptor IroN contributes to invasion of urothelial cells by extraintestinal pathogenic *Escherichia coli in vitro*. Infect. Immun. *75*, 3183–3187.

Fischbach, M.A., Lin, H., Zhou, L., Yu, Y., Abergel, R.J., Liu, D.R., Raymond, K.N., Wanner, B.L., Strong, R.K., Walsh, C.T., *et al.* (2006). The pathogen-associated *iroA* gene cluster mediates bacterial evasion of lipocalin 2. Proc. Natl. Acad. Sci. U.S.A. *103*, 16502–16507.

Fisher, C.F., Davies, N.M., Wyckoff, E.E., Feng, Z., Oaks, E.V., and Payne, S.M. (2009). Genetics and virulence association of the *Shigella flexneri* Sit iron transport system. Infect. Immun. *77*, 1992–1999.

Gibson, F., and Magrath, D.I. (1969). The isolation and characterization of a hydroxamic acid (aerobactin) formed by *Aerobacter aerogenes* 62-I. Biochim. Biophys. Acta. *192*, 175–184.

Gottesman, S. (2004). The small RNA regulators of *Escherichia coli*: roles and mechanisms. Annu. Rev. Microbiol. *58*, 303–328.

Grosse, C., Scherer, J., Koch, D., Otto, M., Taudte, N., and Grass, G. (2006). A new ferrous iron-uptake transporter, EfeU (YcdN), from *Escherichia coli*. Mol. Microbiol. *62*, 120–131.

Hagan, E.C., and Mobley, H.L. (2009). Haem acquisition is facilitated by a novel receptor Hma and required by uropathogenic *Escherichia coli* for kidney infection. Mol. Microbiol. *71*, 79–91.

Hantke, K. (1983). Identification of an iron uptake system specific for coprogen and rhodotorulic acid in *Escherichia coli* K12. Mol. Gen. Genet. *191*, 301–306.

Hantke, K., Nicholson, G., Rabsch, W., and Winkelmann, G. (2003). Salmochelins, siderophores of *Salmonella enterica* and uropathogenic *Escherichia coli* strains, are recognized by the outer membrane receptor IroN. Proc. Natl. Acad. Sci. U.S.A. *100*, 3677–3682.

Henderson, J.P., Crowley, J.R., Pinkner, J.S., Walker, J.N., Tsukayama, P., Stamm, W.E., Hooton, T.M., and Hultgren, S.J. (2009). Quantitative metabolomics reveals an epigenetic blueprint for iron acquisition in uropathogenic *Escherichia coli*. PLoS Pathog. *5*, e1000305.

Janakiraman, A., and Slauch, J.M. (2000). The putative iron transport system SitABCD encoded on SPI1 is required for full virulence of *Salmonella typhimurium*. Mol. Microbiol. *35*, 1146–1155.

Kammler, M., Schon, C., and Hantke, K. (1993). Characterization of the ferrous iron uptake system of *Escherichia coli*. J. Bacteriol. *175*, 6212–6219.

Kehres, D.G., Janakiraman, A., Slauch, J.M., and Maguire, M.E. (2002). SitABCD is the alkaline Mn^{2+} transporter of *Salmonella enterica* serovar Typhimurium. J. Bacteriol. *184*, 3159–3166.

Khan, M.A., and Isaacson, R.E. (2002). Identification of *Escherichia coli* genes that are specifically expressed in a murine model of septicemic infection. Infect. Immun. *70*, 3404–3412.

Konopka, K., and Neilands, J.B. (1984). Effect of serum albumin on siderophore-mediated utilization of transferrin iron. Biochemistry. *23*, 2122–2127.

Koster, W., and Braun, V. (1990). Iron (III) hydroxamate transport into *Escherichia coli*. Substrate binding to the periplasmic FhuD protein. J. Biol. Chem. *265*, 21407–21410.

Lawlor, K.M., and Payne, S.M. (1984). Aerobactin genes in *Shigella* spp. J. Bacteriol. *160*, 266–272.

Létoffé, S., Delepelaire, P., and Wandersman, C. (2006). The housekeeping dipeptide permease is the *Escherichia coli* heme transporter and functions with two optional peptide binding proteins. Proc. Natl. Acad. Sci. U.S.A. *103*, 12891–12896.

Lloyd, A.L., Henderson, T.A., Vigil, P.D., and Mobley, H.L. (2009). Genomic islands of uropathogenic *Escherichia coli* contribute to virulence. J. Bacteriol. *191*, 3469–3481.

Luck, S.N., Turner, S.A., Rajakumar, K., Sakellaris, H., and Adler, B. (2001). Ferric dicitrate transport system (Fec) of *Shigella flexneri* 2a YSH6000 is encoded on a novel pathogenicity island carrying multiple antibiotic resistance genes. Infect. Immun. *69*, 6012–6021.

Marlovits, T.C., Haase, W., Herrmann, C., Aller, S.G., and Unger, V.M. (2002). The membrane protein FeoB contains an intramolecular G protein essential for Fe(II) uptake in bacteria. Proc. Natl. Acad. Sci. U.S.A. *99*, 16243–16248.

Massé, E., and Gottesman, S. (2002). A small RNA regulates the expression of genes involved in iron metabolism in *Escherichia coli*. Proc. Natl. Acad. Sci. U.S.A. *99*, 4620–4625.

McDougall, S., and Neilands, J.B. (1984). Plasmid- and chromosome-coded aerobactin synthesis in enteric bacteria: insertion sequences flank operon in plasmid-mediated systems. J. Bacteriol. *159*, 300–305.

McHugh, J.P., Rodriguez-Quinones, F., Abdul-Tehrani, H., Svistunenko, D.A., Poole, R.K., Cooper, C.E., and Andrews, S.C. (2003). Global iron-dependent gene regulation in *Escherichia coli*. A new mechanism for iron homeostasis. J. Biol. Chem. *278*, 29478–29486.

Menard, R., Dehio, C., and Sansonetti, P.J. (1996). Bacterial entry into epithelial cells: the paradigm of *Shigella*. Trends Microbiol. *4*, 220–226.

Mills, M., and Payne, S.M. (1995). Genetics and regulation of heme iron transport in *Shigella dysenteriae* and detection of an analogous system in *Escherichia coli* O157:H7. J. Bacteriol. *177*, 3004–3009.

Mills, M., and Payne, S.M. (1997). Identification of *shuA*, the gene encoding the heme receptor of *Shigella dysenteriae*, and analysis of invasion and intracellular multiplication of a *shuA* mutant. Infect. Immun. *65*, 5358–5363.

Moss, J.E., Cardozo, T.J., Zychlinsky, A., and Groisman, E.A. (1999). The *selC*-associated SHI-2 pathogenicity island of *Shigella flexneri*. Mol. Microbiol. *33*, 74–83.

Murphy, E.R., and Payne, S.M. (2007). RyhB, an iron-responsive small RNA molecule, regulates *Shigella dysenteriae* virulence. Infect. Immun. *75*, 3470–3477.

Nassif, X., Mazert, M.C., Mounier, J., and Sansonetti, P.J. (1987). Evaluation with an *iuc*::Tn10 mutant of the role of aerobactin production in the virulence of *Shigella flexneri*. Infect. Immun. *55*, 1963–1969.

Negre, V.L., Bonacorsi, S., Schubert, S., Bidet, P., Nassif, X., and Bingen, E. (2004). The siderophore receptor IroN, but not the high-pathogenicity island or the hemin receptor ChuA, contributes to the bacteremic step of *Escherichia coli* neonatal meningitis. Infect. Immun. *72*, 1216–1220.

Neilands, J.B. (1984). Siderophores of bacteria and fungi. Microbiol. Sci. *1*, 9–14.

Neilands, J.B. (1995). Siderophores: structure and function of microbial iron transport compounds. J. Biol. Chem. *270*, 26723–26726.

O'Brien, A.D., Tesh, V.L., Donohue-Rolfe, A., Jackson, M.P., Olsnes, S., Sandvig, K., Lindberg, A.A., and Keusch, G.T. (1992). Shiga toxin: biochemistry, genetics, mode of action, and role in pathogenesis. Curr. Top. Microbiol. Immunol. *180*, 65–94.

Oaks, E.V., Wingfield, M.E., and Formal, S.B. (1985). Plaque formation by virulent *Shigella flexneri*. Infect. Immun. *48*, 124–129.

Oglesby, A.G., Murphy, E.R., Iyer, V.R., and Payne, S.M. (2005). Fur regulates acid resistance in *Shigella flexneri* via RyhB and *ydeP*. Mol. Microbiol. *58*, 1354–1367.

Ouyang, Z., and Isaacson, R. (2006). Identification and characterization of a novel ABC iron transport system, *fit*, in *Escherichia coli*. Infect. Immun. *74*, 6949–6956.

Ouyang, Z., and Isaacson, R. (2008). Identification of a novel regulator for the *Escherichia coli fit* iron transport system. Open Microbiol. J. *2*, 94–99.

Payne, S.M., Wyckoff, E.E., Murphy, E.R., Oglesby, A.G., Boulette, M.L., and Davies, N.M. (2006). Iron and pathogenesis of *Shigella*: iron acquisition in the intracellular environment. Biometals *19*, 173–180.

Peigne, C., Bidet, P., Mahjoub-Messai, F., Plainvert, C., Barbe, V., Medigue, C., Frapy, E., Nassif, X., Denamur, E., Bingen, E., *et al.* (2009). The plasmid of *Escherichia coli* strain S88 (O45:K1:H7) that causes neonatal meningitis is closely related to avian pathogenic *E. coli* plasmids and is associated with high-level bacteremia in a neonatal rat meningitis model. Infect. Immun. *77*, 2272–2284.

Perry, R.D., Balbo, P.B., Jones, H.A., Fetherston, J.D., and DeMoll, E. (1999). Yersiniabactin from *Yersinia pestis*: biochemical characterization of the siderophore and its role in iron transport and regulation. Microbiology *145*, 1181–1190.

Postle, K., and Larsen, R.A. (2007). TonB-dependent energy transduction between outer and cytoplasmic membranes. Biometals *20*, 453–465.

Purdy, G.E., and Payne, S.M. (2001). The SHI-3 iron transport island of *Shigella boydii* 0–1392 carries the genes for aerobactin synthesis and transport. J. Bacteriol. *183*, 4176–4182.

Runyen-Janecky, L.J., and Payne, S.M. (2002). Identification of chromosomal *Shigella flexneri* genes induced by the eukaryotic intracellular environment. Infect. Immun. *70*, 4379–4388.

Runyen-Janecky, L.J., Reeves, S.A., Gonzales, E.G., and Payne, S.M. (2003). Contribution of the *Shigella flexneri* Sit, Iuc, and Feo iron acquisition systems to iron acquisition *in vitro* and in cultured cells. Infect. Immun. *71*, 1919–1928.

Russo, T.A., McFadden, C.D., Carlino-MacDonald, U.B., Beanan, J.M., Barnard, T.J., and Johnson, J.R. (2002). IroN functions as a siderophore receptor and is a urovirulence factor in an extraintestinal pathogenic isolate of *Escherichia coli*. Infect. Immun. *70*, 7156–7160.

Russo, T.A., McFadden, C.D., Carlino-MacDonald, U.B., Beanan, J.M., Olson, R., and Wilding, G.E. (2003). The siderophore receptor IroN of extraintestinal pathogenic *Escherichia coli* is a potential vaccine candidate. Infect. Immun. *71*, 7164–7169.

Sansonetti, P.J. (2006). The bacterial weaponry: lessons from *Shigella*. Ann. N Y Acad. Sci. *1072*, 307–312.

Sansonetti, P.J., d'Hauteville, H., Formal, S.B., and Toucas, M. (1982). Plasmid-mediated invasiveness of '*Shigella*-like' *Escherichia coli*. Ann. Microbiol. (Paris) *133*, 351–355.

Schmitt, M.P., and Payne, S.M. (1988). Genetics and regulation of enterobactin genes in *Shigella flexneri*. J. Bacteriol. *170*, 5579–5587.

Schmitt, M.P., and Payne, S.M. (1991). Genetic analysis of the enterobactin gene cluster in *Shigella flexneri*. J. Bacteriol. *173*, 816–825.

Schubert, S., Picard, B., Gouriou, S., Heesemann, J., and Denamur, E. (2002). *Yersinia* high-pathogenicity island contributes to virulence in *Escherichia coli* causing extraintestinal infections. Infect. Immun. *70*, 5335–5337.

Schubert, S., Rakin, A., Fischer, D., Sorsa, J., and Heesemann, J. (1999). Characterization of the integration site of *Yersinia* high-pathogenicity island in *Escherichia coli*. FEMS Microbiol. Lett. *179*, 409–414.

Schubert, S., Rakin, A., Karch, H., Carniel, E., and Heesemann, J. (1998). Prevalence of the 'high-pathogenicity island' of *Yersinia* species among *Escherichia coli* strains that are pathogenic to humans. Infect. Immun. *66*, 480–485.

Snyder, J.A., Haugen, B.J., Buckles, E.L., Lockatell, C.V., Johnson, D.E., Donnenberg, M.S., Welch, R.A., and Mobley, H.L. (2004). Transcriptome of uropathogenic *Escherichia coli* during urinary tract infection. Infect. Immun. *72*, 6373–6381.

Sturm, A., Schierhorn, A., Lindenstrauss, U., Lilie, H., and Bruser, T. (2006). YcdB from *Escherichia coli* reveals a novel class of Tat-dependently translocated hemoproteins. J. Biol. Chem. *281*, 13972–13978.

Torres, A.G., and Payne, S.M. (1997). Haem iron-transport system in enterohaemorrhagic *Escherichia coli* O157:H7. Mol. Microbiol. *23*, 825–833.

Torres, A.G., Redford, P., Welch, R.A., and Payne, S.M. (2001). TonB-dependent systems of uropathogenic *Escherichia coli:* aerobactin and heme transport and TonB are required for virulence in the mouse. Infect. Immun. *69*, 6179–6185.

Van Heyningen, W.E., and Gladstone, G.P. (1953). The neurotoxin of *Shigella shigae*. III. The effect of iron on production of the toxin. Br. J. Exp. Pathol. *34*, 221–229.

Vokes, S.A., Reeves, S.A., Torres, A.G., and Payne, S.M. (1999). The aerobactin iron transport system genes in *Shigella flexneri* are present within a pathogenicity island. Mol. Microbiol. *33*, 63–73.

Walters, M.S., and Mobley, H.L. (2009). Identification of uropathogenic *Escherichia coli* surface proteins by shotgun proteomics. J. Microbiol. Methods. *78*, 131–135.

Wilks, A. (2001). The ShuS protein of *Shigella dysenteriae* is a heme-sequestering protein that also binds DNA. Arch. Biochem. Biophys. *387*, 137–142.

Wyckoff, E.E., Duncan, D., Torres, A.G., Mills, M., Maase, K., and Payne, S.M. (1998). Structure of the *Shigella dysenteriae* haem transport locus and its phylogenetic distribution in enteric bacteria. Mol. Microbiol. *28*, 1139–1152.

Wyckoff, E.E., Lopreato, G.F., Tipton, K.A., and Payne, S.M. (2005). *Shigella dysenteriae* ShuS promotes utilization of heme as an iron source and protects against heme toxicity. J. Bacteriol. *187*, 5658–5664.

Yancey, R.J., Breeding, S.A., and Lankford, C.E. (1979). Enterochelin (enterobactin): virulence factor for *Salmonella typhimurium*. Infect. Immun. *24*, 174–180.

Yao, Y., Xie, Y., and Kim, K.S. (2006). Genomic comparison of *Escherichia coli* K1 strains isolated from the cerebrospinal fluid of patients with meningitis. Infect. Immun. *74*, 2196–2206.

Zhou, D., Hardt, W.D., and Galan, J.E. (1999). *Salmonella typhimurium* encodes a putative iron transport system within the centisome 63 pathogenicity island. Infect. Immun. *67*, 1974–1981.

Iron Uptake in Soft Rot *Erwinia*

6

Thierry Franza and Dominique Expert

Abstract

The critical role of iron in host–pathogen relationships has been elucidated in infectious diseases of mammals, where the importance of siderophores in microbial pathogenesis has been demonstrated. Our group has established the role of iron and its ligands in the virulence of the plant pathogenic bacteria *Dickeya dadantii* (*Erwinia chrysanthemi*) and *Erwinia amylovora*. The genomes of the two pectinolytic enterobacterial species *Pectobacterium atrosepticum* SCRI1043 and *D. dadantii* 3937 have been sequenced and annotated. This review focuses on the functions involved in iron acquisition in both species. Besides the production and utilization of siderophores, *P. atrosepticum* and *D. datantii* have the capacity to use other iron sources. Indeed, both species are able to use haem iron, whereas only *P. atrosepticum* can transport the ferric citrate complex and only *D. dadantii* can acquire ferrous iron. These different modes of iron capture indicate that these species have to cope with various environmental and ecological conditions during their pathogenic life cycle.

Erwinia, a 'messy' genus

First described in 1917 to comprise all members of the *Enterobacteriaceae* causing disease on plants (Winslow *et al.*, 1917), the genus *Erwinia* has long been exposed to many nomenclatural difficulties. This has led to the relocation of many species into other genera, thus depriving the genus of its most familiar members. For instance, the soft rot *Erwinia* spp. (*E. carotovora* subsp. *atroseptica*, *E. carotovora* subsp. *carotovora* and *E. chrysanthemi*), and *E. amylovora* were found to be too divergent phylogenetically to be included in a unique clade; therefore, the soft rot *Erwinia* spp. were moved to two new genera as *Pectobacterium* and *Dickeya* (Hauben *et al.*, 1998; Samson *et al.*, 2005; Ma *et al.*, 2007). *Pectobacterium* and *Dickeya* spp. are broad-host range pathogens, infecting potato and other economically important plants, vegetables and ornamentals. They have been isolated from many plant species and single strains can cause disease on various plant species under experimental conditions (Pérombelon, 2002). Their main weapon resides in the production of numerous cell-wall-degrading enzymes responsible for plant tissue maceration. In contrast, *E. amylovora* causes a necrotic disease called fire blight, affecting only apple, pear, and other rosaceous plants. Its pathogenicity is mainly due to the expression of a type three secretion system which delivers proteins able to trigger cell death in the host tissue (Boureau *et al.*, 2006).

In plants, as in human and animals, the availability of iron may be one of the factors limiting the growth of pathogenic microorganisms in the host. Although less documented than in animal infections, the role of iron in plant pathogenesis has been investigated in a number of bacterial and fungal diseases, with particular attention drawn to microbial siderophores as potential pathogenicity determinants. The assumption that ferric iron assimilation via siderophores could play a role in plant infection was experimentally confirmed with work on the pathogenicity of *Erwinia* spp. namely *Dickeya dadantii* and *E. amylovora* (Enard *et al.*, 1988; Franza *et al.*, 2005; Dellagi *et al.*, 1998, 1999). However, despite the apparent similarity in symptom elicitation to *D. dadantii*, *Pectobacterium carotovorum* does not require the production of siderophores for successful infection (Ishimaru and Loper, 1992). The possibility that this pathogen uses other high-affinity transport routes to acquire iron in the host is worth considering particularly if disease development requires substantial bacterial growth and movement.

Genome sequencing of diverse phytopathogens has yielded a wealth of information on candidate iron uptake mechanisms, which may contribute to improving our understanding of the lifestyle of plant-associated bacteria. In the case of soft rot *Erwinia*, the complete genome sequence of *D. dadantii* 3937 and *P. atrosepticum* SCRI1043 is available (https://asap.ahabs.wisc.edu/asap; Bell *et al.*, 2004). The two genomes have similar size (~ 5 Mb) but only 63% of their gene content is overlapping. It is likely that many genes potentially involved in pathogenesis and/or adaptation to different plant environments have been acquired by horizontal transfer. Interestingly, *P. atrosepticum* displays the original trait of causing blackleg, a specific disease of potato plants characterized by the bacterial development in the stem xylem vessels (Pérombelon, 2002). *D. dadantii* invade the intercellular spaces and walls of leaf tissues without moving throughout the vessels (Fagard *et al.*, 2007).

On the basis of functional studies and *in silico* analysis, we present a review of the *D. dadantii* and *P. atrosepticum* iron transport systems. This reveals common features and differences which we discuss with regard to the pathogenic behaviour of these two species representative of the soft rot enterobacteria.

Siderophore-mediated ferric iron acquisition

Several genetic approaches combined with the structural determination of the siderophores involved allowed the characterization in *D. dadantii* of two high-affinity iron uptake systems. Strain 3937 synthesizes two siderophores, achromobactin and chrysobactin which are produced in a sequential manner in culture supernatants of bacterial cells grown under iron limitation (Franza *et al.*, 2005). Achromobactin can be detected during the mid-exponential phase of growth while chrysobactin appears much later. As a catecholate readily identifiable using the Arnow test, chrysobactin (2,3-dihydroxybenzoyl-D-lysyl-L-serine), was the first siderophore of strain 3937 to be characterized (Persmark *et al.*, 1989) (Fig. 6.1A). Achromobactin is a hydroxycarboxylate siderophore derived from citrate to which are linked ethanolamine-α-ketoglutaric acid and 2,4-diaminobutyric-α-ketoglutaric acid (Fig. 6.1B). Achromobactin does not contain any catecholate or hydroxamate group (Münzinger *et al.*, 2000). This siderophore was uncovered in chrysobactin deficient mutants which are still able to form a halo of discoloration on chrome azurol S (CAS) agar medium, a universal assay for siderophore detection (Schwyn and Neilands, 1987). Chrysobactin-deficient mutants fail to grow in the presence of the strong ferric iron chelator ethylenediamine-*N,N′*-bis(2-hydroxyphenylacetic acid) (EDDHA) but the production of achromobactin enables them to thrive on a medium containing the ferrous iron chelator 2,2′-dipyridyl. Catechol siderophores as well as the hydroxamate aerobactin were also found to be produced by various field isolates of *E. carotovora* (syn. *P. carotovorum*) (Bull *et al.*, 1994). Purification of siderophores from one of these isolates, strain W3C105, led to the identification of chrysobactin (Barnes *et al.*, 1999). Production of achromobactin also occurs in *Pseudomonas syringae* pv. *syringae* B728a (Franza *et al.*, 2005), a pathogenic bacterium showing an epiphytic phase on plants.

The genetic organization of the *D. dadantii* chrysobactin system is reminiscent of that of the enterobactin region in *Escherichia coli* K-12. It comprises four Fur-regulated operons divergently transcribed from two bidirectional promoters. The *fct-cbsCEBA* operon (Fig. 6.1A) encodes the receptor Fct and the enzymes leading to the DHB moiety in chrysobactin biosynthesis (Franza and Expert, 1991). In addition, downstream from *cbsA*, we recently found a homologue of the *E. coli entH* gene (*cbsP*) which encodes a thioesterase necessary for optimal production of enterobactin in *E. coli* (Leduc *et al.*, 2007). By cleaving wrongly charged molecules on the carrier protein EntB, the EntH protein is predicted to play a proofreading role during biosynthesis of enterobactin *in vivo*. A similar role can be expected in *D. dadantii*. A second operon, *cbsHF*, encodes the chrysobactin synthase CbsF, which is predicted to be a non-ribosomal peptide synthetase (NRPS) with a multimodular structure allowing the assemblage of the three components, DHB, D-lysine and L-serine (Expert *et al.*, 2004) (Fig. 6.1A). The CbsH

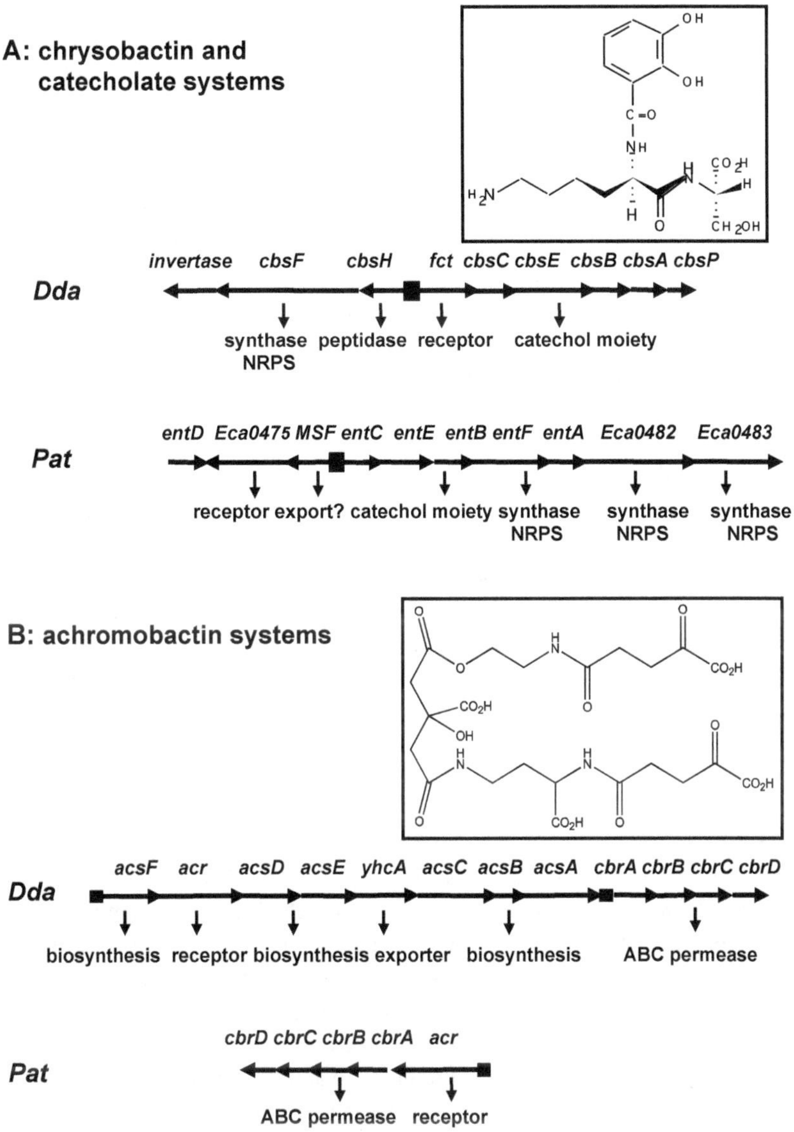

Figure 6.1 (A) Genetic organization of the operons involved in chrysobactin biosynthesis and uptake in *Dickeya dadantii* (*Dda*) and of the genes involved in biosynthesis and uptake of catecholate siderophore(s) in *Pectobacterium atrosepticum* (*Pat*). The structure of chrysobactin is boxed. (B) Genetic organization of the operons involved in achromobactin biosynthesis and transport in *D. dadantii* (*Dda*) and of the genes involved in achromobactin transport in *P. atrosepticum* (*Pat*). The structure of achromobactin is boxed.

protein is an oligopeptidase that degrades chrysobactin in the cytosol. Absence of this enzyme in a *cbsH*-negative mutant, results in bacterial growth inhibition caused by intracellular ferric iron chelation that follows chrysobactin accumulation (Rauscher *et al.*, 2002). Thus, this peptidase plays an important role in control of iron homeostasis in *D. dadantii*. A third cluster of genes located upstream of the *cbsHF* operon encodes the four components of ferric chrysobactin permease (CbuBCDG) and CbsS, a protein homologous to *E. coli* EntS, probably involved in the excretion of chrysobactin to the external medium (Expert *et al.*, 2004). Interestingly, the genome of *P. atrosepticum* possesses a gene cluster predicted to encode all the enzymes necessary for the biosynthesis of enterobactin, although quite different in organisation from those of *E. coli* and *D. dadantii* (Fig. 6.1A, Table 6.1). A relevant difference is the presence of two additional genes *Eca4082* and *Eca4083* (Table 6.2), besides *entF*, able to encode enzymes belonging to the NRPS family, with highest identities (55% and 54%, respectively) to those involved in the biosynthesis of amonabactins, the bis-catecholate siderophores from *Aeromonas hydrophila* (Barghouti *et al.*, 1989). Whether or not *P. atrosepticum* produces enterobactin and/or other catecholate siderophores has to be experimentally determined. In addition, the only ferric siderophore transport gene associated with this cluster is predicted to encode a protein sharing 38% identity with the *E. coli* CirA receptor (Fig. 6.1A). We found no potential ferric siderophore ABC type permease encoding gene in the vicinity of this biosynthetic gene cluster. On the other hand, in another part of the genome, we identified a *fepA*-like gene encoding a protein sharing 64% identity with FepA, the ferric enterobactin receptor from *E. coli* K-12. Thus, ferric enterobactin is likely to be transported by *P. atrosepticum*.

While chrysobactin is assembled by an NRPS-dependent pathway, achromobactin biosynthesis is NRPS-independent. It is considered as a model system for studying the substrate specificity, catalytic mechanism and structural biology of NIS (for NRPS-independent siderophore) synthetases (Schmelz *et al.*, 2009). Indeed, genetic studies showed that an operon of eight genes (*acsF acr acsD acsE yhcA acsC acsB acsA*) directs achromobactin production, its extracellular release and uptake of its ferric complex (Franza *et al.*, 2005) (Fig. 6.1B). The *acsF, acsE, acsD, acsC* and *acsA* genes are required for biosynthesis. The encoded proteins AcsD, AcsC and AcsA were proposed to belong to the NIS family of synthetases, AcsD and AcsC catalysing the condensation of citric acid with ethanolamine and 2,4-diaminobutyric acid, and AcsA catalysing acylation of the two resulting amino groups with α-ketoglutaric acid (Table 6.2). A structural and biochemical study of AcsD revealed that this protein represents a new member of adenylate-forming enzymes which catalyses stereospecific formation of an enzyme-bound-citryl-adenylate reacting with L-serine to form a likely achromobactin precursor (Schmelz *et al.*, 2009). The AcsE protein, predicted to be a pyridoxal-5′-phosphate (PLP)-dependent decarboxylase, is a good candidate to catalyse the removal of the serine-derived carboxyl group to provide a precursor to ethanolamine. Predicted to be a PLP-dependent enzyme of class III that participates in reactions of transamination, AcsF would be involved in the formation of diaminobutyrate, with glutamate as the amino group donor. More puzzling is AcsB, a putative aldolase the role of which might be to maintain the pool of carbohydrates necessary for the formation of achromobactin. Moreover, there is evidence for the existence of a physical interaction between the enzymes AcsDECBA and the inner membrane OutEFLM-forming platform of the *D. dadantii* type II secretion system, but the functional significance of this remains to be elucidated (Douet *et al.*, 2009). The *yhcA* gene encodes a protein with amino acid sequence characteristics of efflux pumps of the major facilitator superfamily (MFS), able to achieve extracellular release of achromobactin during biosynthesis and extrusion of iron internalized via achromobactin. The Acr protein was identified as the ferric achromobactin receptor. Downstream of this operon, a second operon (*cbrABCD*) encodes the ferric achromobactin permease (Fig. 6.1B). Interestingly, *cbr* negative mutants that do not produce a functional permease accumulate achromobactin in the medium and are derepressed for the production of chrysobactin. Both operons are controlled by iron availability, via direct Fur mediated metalloregulation. Interestingly, in the genome of *P. atrosepticum* we found a gene cluster sharing identity (> 62%) to the *cbrABCD* operon (Fig. 6.1B, Table 6.1). This cluster also contains a homologue of the *acr* achromobactin receptor gene (Fig. 6.1B). However, on the basis of sequence prediction, it is unlikely that this bacterium produces achromobactin. These data suggest that achromobactin could be used by *P. atrosepticum* as a xenosiderophore.

The role of iron in the pathogenicity of *D. dadantii* strain 3937 was established by investigating the virulence properties of mutants affected in siderophore production. Successful infection by strain 3937 on host plants including *Saintpaulia ionantha* and *Arabidopsis thaliana* results in a systemic disease that follows bacterial spreading from the leaf to the basis of the petiole. Chrysobactin defective mutants produce only localized symptoms on saintpaulia plants and are disadvantaged in their growth compared to a chrysobactin proficient strain (Enard *et al.*, 1988; Franza *et al.*, 2005). This decline coincides with the

Table 6.1 List of the common proteins involved in biosynthesis and transport of siderophores in *D. dadantii* 3937 (*Dda*) and *P. atrosepticum (Pat)* SCRI1043. Number of amino acids for each protein is indicated

Dda	Pat	Identity	Function
Catecholate biosynthetic proteins			
CbsA (252)	EntA (261)	66%	2,3-Dihydro-2,3-dihydroxy-benzoate dehydrogenase
CbsB (291)	EntB (291)	59%	Isochorismatase with phospho-pantetheine attachment site
CbsC (395)	EntC (398)	46%	Isochorismatehydroxymutase
CbsD (228)	EntD (238)	58%	Phosphopantetheinyltransferase component of synthetase
CbsE (542)	EntE (541)	62%	2,3-dihydroxybenzoate–AMP-ligase
CbsF (2864)	EntF (1047)	30%	Non-ribosomal peptide synthetase (NRPS)
CbsP (138)	Eca1857 (138)	68%	EntH (YbdB) thioesterase
Achromobactin biosynthetic and transport proteins			
AcsF (471)	Eca2243 (460)	66%	Aminobutyrate-oxoglutarate aminotransferase
Acr (704)	Acr (704)	61%	TonB-dependent ferric achromobactin receptor
CbrA (305)	CbrA (335)	63%	Achromobactin-binding periplasmic protein
CbrB (340)	CbrB (329)	60%	Achromobactin transport system, permease protein
CbrC (349)	CbrC (343)	62%	Achromobactin transport system permease protein
CbrD (269)	CbrD (279)	74%	Achromobactin transport ATP-binding protein

emergence of a necrotic border surrounding the lesion initiated by the mutant (Masclaux and Expert, 1995). Chrysobactin was detected in leaf intercellular fluids from plants inoculated with the wild type strain, suggesting that this compound can sequestrate the iron present in colonized tissues (Neema *et al.*, 1993). To further understand the biological function of chrysobactin, Tomisić *et al.* (2008) investigated the coordination properties of the different ferric complexes of this siderophore. They found that chrysobactin is a less effective ferric chelator than hexadentate siderophores, such as enterobactin or deferrioxamine B. However, chrysobactin exhibits a higher pFe value than citrate or malate (pFe of chrysobactin = 17.1 vs. pFe of citrate = 14.8) known to be the major ferric carriers in plants (Brown, 1978), and can effectively sequester the iron from such ferric complexes. Achromobactin-deficient mutants are also affected in the virulence but are more aggressive than the chrysobactin non-producers (Franza *et al.*, 2005). Double mutants deficient in achromobactin and chrysobactin production are impaired in symptom initiation, indicating that these two siderophores act in a complementary way to satisfy the bacterial iron needs within the plant. Although the coordination chemistry of achromobactin has not yet been investigated, it can be predicted that this ligand is less competitive than chrysobactin. It is noteworthy that the wild type cells but not the double mutant can survive for several days in intercellular spaces of host tissues without multiplying substantially. During the symptomatic phase, both strains can multiply, but the growth of the double mutant is 10-fold lower than that of the wild type strain. A study of the transcriptional expression of the *fct-cbsCEBA* operon during infection shows that the chrysobactin system is alternately up- and down-regulated in the leaf tissues (Masclaux and Expert, 1995). Using microarray technology to identify *D. dadantii* genes that are specifically regulated in the plant host, Okinaka *et al.* (2002) found that the *cbsA* gene was up-regulated after inoculation of African violets. In arabidopsis plants, infection by wild-type cells leads to the transcriptional activation of both the vacuolar metal transporter gene *AtNRAMP3* and the ferritin gene *AtFER1* with is followed by the

Table 6.2 List of proteins involved in biosynthesis and transport of siderophores present only in one species. Number of amino acids for each protein is indicated

Name	Function
In *D. dadantii* 3937	
Chrysobactin system proteins	
CbsH (390)	Chrysobactin oligopeptidase, 46% identical to the ferric enterobactin esterase Fes from *E. coli* K-12
CbsS (426)	430 kDa inner membrane protein, 50% identical to the enterobactin secretion machinery protein EntS from *E. coli* K-12
Achromobactin biosynthetic and transport proteins	
AcsA (647)	NRPS-independent siderophore (NIS) synthetase involved in achromobactin biosynthesis 65% identical to the protein AcsA from *Pseudomonas syringae* B728a
AcsB (258)	Achromobactin biosynthetic protein 74% identical to the HpcH/HpaI aldolase from *Pseudomonas syringae* B728a
AcsC (618)	NIS synthetase involved in achromobactin biosynthesis 70% identical to the protein AcsC from *Pseudomonas syringae* B728a
AcsD (620)	NIS synthetase involved in achromobactin biosynthesis 64% identical to the protein AcsD from *Pseudomonas syringae* B728a
AcsE (411)	Achromobactin biosynthetic protein 72% identical to the ornithine/diaminopimelate decarboxylase from *P syringae* B728a
YhcA (428)	Inner membrane protein involved in achromobactin secretion 70% identical to the EmrB/QacA family drug resistance transporter from *P. syringae* B728a
In *P. atrosepticum* SCRI1043	
Catecholate biosynthetic proteins	
Eca0482 (2116)	Non-ribosomal peptide synthetase 55% identical to a putative NRPS synthetase from *Aeromonas hydrophila*
Eca0483 (704)	NRPS synthetase with amino acid adenylation site (AMP binding site) 50% identical to a putative NRPS synthetase from *Aeromonas hydrophila*

accumulation of ferritin (Dellagi *et al.*, 2005; Segond *et al.*, 2009). These responses contribute to a basal level of host resistance to infection. Interestingly, similar responses are also observed after leaf infiltration with chrysobactin or other structurally unrelated siderophores, only if these molecules are free of iron (Dellagi *et al.*, 2009). These data indicate that infection by *D. dadantii* causes changes in plant iron homeostasis and suggest that these changes are due to the iron-binding property of chrysobactin.

Feo-mediated ferrous iron acquisition

Bacteria can also acquire the ferrous form of iron via the anaerobic-microaerophilic FeoAB pathway (Cartron *et al.*, 2006). The first evidence for the existence in *D. dadantii* 3937 of a ferrous iron transporter also comes from work of Okinaka *et al.* (2002) who showed that a homologue of the *E. coli feoB* gene was up-regulated during infection of African violets. Accordingly, a locus comprising two genes, *feoA* and *feoB*, organized as an operon was identified in the genome of strain 3937 (Fig. 6.2). 113 nucleotides upstream from the translation start of the *feoA* gene, we identified another gene encoding a FeoA-like protein of 79 amino acid residues (Fig. 6.2). The gene called *feoC*, associated with the *feo* operon in many other α-proteobacteobacteria including *E. coli*, is missing in *D. dadantii*. Also noteworthy is that a *feo* locus is

Feo system

ID-0017775	**79 AA peptide 41 % identical to the ferrous iron transporter protein A from *Rhodobacter sphaeroides***
FeoA	**75 AA peptide 44 % identical to the ferrous iron transporter protein A from *Escherichia coli* K12**
FeoB	**765 AA polypeptide 46 % identical to the G protein-coupled membrane peptide FeoB involved in ferrous iron uptake in *E. coli* K12**

EfeUOB system

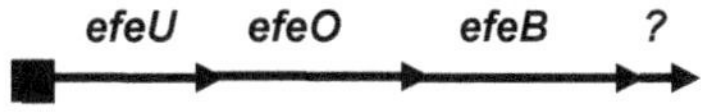

EfeU	**281 AA protein 75 % identical to the ferrous iron permease EfeU from *E. coli* O157:H7**
EfeO	**373 AA protein 74 % identical to the EfeO periplasmic protein containing a cupredoxin N-terminal domain from *E. coli* O157:H7**
EfeB	**437 AA protein 73 % identical to the periplasmic haem peroxidase-like protein EfeB from *E. coli* O157:H7**

Figure 6.2 Schematic organization of the *D. dadantii feoAB* and *efeUOB* operons involved in ferrous iron transport. '?' means the existence of a small ORF encoding an unknown protein.

absent from the genome of *P. atrosepticum*. Initially identified in *E. coli*, FeoB, the main component of this system is an integral membrane protein, the N-terminal domain of which contains a G-protein region having GTPase activity (Kammler *et al.*, 1993; Marlovits *et al.*, 2002). Although the function of FeoB has not been completely elucidated, it is assumed that this protein acts as the permease through which ferrous iron is transported into the cell. The G-protein region is essential for this transport function. The FeoA protein is present in 80% of Feo bacterial systems only. However, in 12% of the genomes containing a *feoB* gene, there is a *feoA–feoA* gene tandem (Cartron *et al.*, 2006). The function of the FeoA peptide is still unknown. It contains a SH3-domain, known to be involved in protein–protein interactions. It was suggested that FeoA interacts with its partner protein FeoB, perhaps to mediate FeoB-dependent ferrous iron uptake activity (Cartron *et al.*, 2006). FeoC is a small Fe–S protein assumed to be a transcriptional regulator.

The promoter region of the *D. dadantii feoAB* locus contains a Fur box but unlike *E. coli*, no obvious Fnr box could be identified. However, we found an Fnr box in the promoter region of the second *feoA* gene. A Fur box is also present in this promoter region. Thus, the *D. dadantii* FeoAB system may be functional in aerobic and anaerobic conditions. A *feoB*-negative mutant was constructed. In an achromobactin and chrysobactin negative background, the *feoB* mutation conferred a three-fold reduction in iron uptake under reducing conditions, compared with the *feoB*-positive strain. With respect to the pathogenicity on arabidopsis plants, there was no significant effect of the *feoB* mutation, when present in a wild type background. On the other hand, in a siderophore negative background, the *feoB* mutation results in a reduced number of systemic infections, which was twofold lower than with the siderophore non-producer and five times lower than with the wild-type strain. These data indicate that the Feo system contributes to the iron nutrition of bacterial cells in the host and are consistent with earlier results reported by Okinaka and collaborators (2002).

EfeU-mediated ferrous iron acquisition

The EfeUOB iron transport system, initially discovered in *E. coli* on the basis of similarities to the yeast oxidase-dependent ferrous iron transporter represents another bacterial iron transport mechanism (Grosse *et al.*, 2006; Cao *et al.*, 2007). The three genes encoding this system are organized in an operon (*efeUOB*) which is functional in several *E. coli* strains such as 0157:H7, but cryptic in strain K-12 due to the presence of a frameshift mutation altering the *efeU* gene expression. EfeU, a potential integral inner membrane protein, would act as a permease mediating the transfer of iron across the inner membrane, although the form of iron it transports, ferrous or ferric, is not clear. Indeed, ferrous iron proved to be the initial transport substrate, but it could be oxidized during its translocation process via the activity of EfeB, a periplasmic haem peroxidase-like protein (Cao *et al.*, 2007). The true role of EfeO is also unknown. The presence of a functional EfeUOB system in *E. coli* provides a growth advantage under iron-restriction and low-pH conditions, in aerobic and anaerobic environments. The *efeUOB* operon is repressed by Fe-Fur and pH-controlled by CpxAR, a two-component sensor regulator system involved in periplasmic stress response (Cao *et al.*, 2007).

Like FeoAB, the EfeUOB system exists only in *D. dadantii* (Fig. 6.2). The three proteins encoded are 70–75% identical to *E. coli* components and the operon gene structure is conserved. Furthermore, a CpxAR system exists in *D. dadantii*. As in *E. coli*, a CpxR binding site and a Fur box are present in the promoter region of the *D. dadantii efeUOB* operon, indicating that this operon could be induced under low iron and low pH conditions. No apparent frame shift mutation was identified in the *D. dadantii efeU* gene making likely that this system is functional in this bacterium. Indeed, we found that the *feoB* mutation does not completely abolish ferrous iron transport, suggesting that the EfeUOB could function under the conditions tested (see above). The role of the EfeUOB system in the pathogenicity would be interesting to investigate. External pH is an environmental signal involved in the differential expression of pectate lyase isoenzymes important for the virulence (Nachin and Barras, 2000). In chicory leaves, the pH of infected tissue varies from acidic to basic. Thus, it is possible that expression of the EfeUOB system could confer a growth advantage on bacteria at some stage of the disease process. The regulator(s) mediating the pH response of pectate lyase encoding genes has (have) not yet been identified. The possibility of a coordinate expression of iron transport and pectate lyase production in response to pH may be worth considering.

Iron ABC transport systems

In Gram-negative species, several iron acquisition systems based on the production of ABC permeases, were identified. In the absence of functional siderophore-mediated iron transport, the Yfe permease in *Yersinia pestis* (Bearden *et al.*, 1998) as well as the SitABC transporter in *Shigella flexneri* (Runyen-Janecky *et al.*, 2003), in *Salmonella enterica* (Kheres *et al.*, 2002) and other pathogens (Sabri *et al.*, 2006) can promote bacterial growth in the presence of ferrous or ferric iron chelators. However, the nature of the iron transported by these systems is poorly defined and they also act as manganese transporters. Interestingly, they play a role in bacterial iron nutrition during conditions of infection (Bearden and Perry, 1999; Runyen-Janecky *et al.*, 2003; Zaharik *et al.*, 2004). Similarly, there is a locus homologous to the *Y. pestis yfeABCD* operon in the genome of *D. dadantii* 3937 (Fig. 6.3). The *yfeA* gene was found to be up-regulated during infection of African violets and of spinach leaves (Okinaka *et al.*, 2002; Yang *et al.*, 2004). A similar locus with a different gene organization is also present in the genome of *P. atrosepticum* (Fig. 6.3). In *P. atrosepticum*, the *yfeE* gene encoding an inner membrane protein is situated in the close vicinity of the *yfeABCD* operon, whereas in *D. dadantii* this gene is present elsewhere in the genome (Fig. 6.3). In both species, the *yfeABCD* operons display a Fur binding site in their promoter region indicating that they can be iron-regulated.

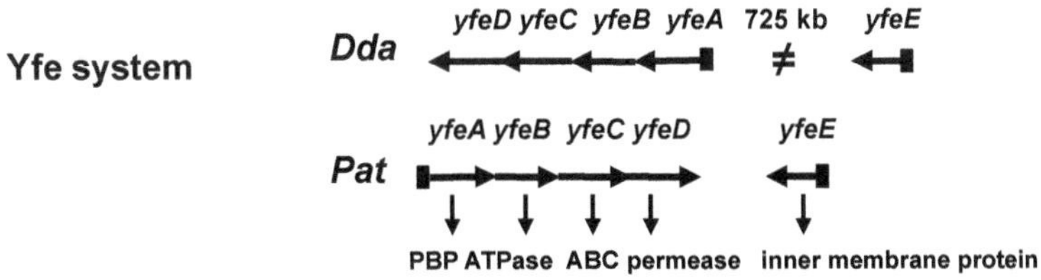

YfeA	316 AA protein 70 % identical to the periplasmic-binding protein YfeA from *Yersinia pestis*
YfeB	298 AA protein 72 % identical to the ATP-binding protein YfeB from *Y. pestis*
YfeC	284 AA protein 77 % identical to the integral membrane ABC permease protein YfeC from *Y. pestis*
YfeD	275 AA protein 66 % identical to the permease protein YfeD from *Y. pestis*
YfeE	182 AA protein 60 % identical to the inner membrane protein YfeE from *Y. pestis*

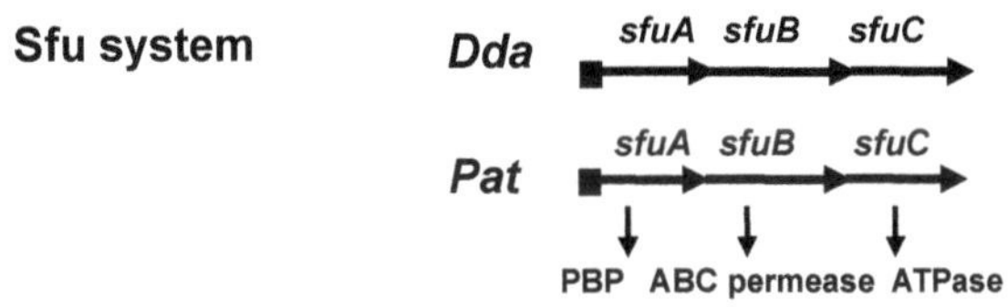

SfuA	338 AA protein 89 % identical to the iron periplasmic binding protein SfuA from *Serratia marcescens*
SfuB	539 AA protein 83 % identical to the ABC permease protein SfuB from *S. marcescens*
SfuC	344 AA protein 72 % identical to the ATP-binding protein SfuC from *S. marcescens*

Figure 6.3 Schematic organization of the genes from the Yfe and Sfu systems of *D. dadantii* (*Dda*) and *P. atrosepticum* (*Pat*).

The Sfu system of *Serratia marcescens*, encoded by the three *sfuA*, *sfuB* and *sfuC* genes, determines the transport of ferric iron via an ABC permease, without the requirement of a TonB-dependent outer membrane receptor (Zimmerman *et al.*, 1989; Angerer *et al.*, 1990). In what form ferric iron is transported via this system has not been elucidated. It was suggested that iron bound to diverse organic molecules which could traverse the outer membrane via non-specific porins is channelled into this cytoplasmic membrane transport pathway (Angerer *et al.*, 1990). We found that a locus similar to the *sfuABC* gene cluster is present in the genomes of *D. dadantii* and *P. atrosepticum* (Fig. 6.3). The proteins encoded share 72–90% identity with the *S. marcescens* Sfu components, indicating that a similar role could be fulfilled in these plant pathogens. In *D. dadantii* and *P. atrosepticum*, the *sfuABC* promoter regions have similarity to the consensus Fur-binding site suggesting that these genes are induced under low iron conditions.

Haem transport systems

First described in *Yersinia enterolitica*, the haem uptake system is present in many Gram-negative pathogens (Perry, 2004). This system comprises a TonB-dependent outer membrane receptor which takes up free or haemoprotein-bound haem from the medium, a periplasmic binding protein-dependent ABC permease and a cytoplasmic haem binding protein essential for prevention against toxicity of this substrate. In *Y. enterolitica*, these proteins (HmuR, HmuT-U-V and HmuS, respectively) are encoded by a Fur-regulated gene cluster forming the *hmuPRSTUV* operon. The function of the protein encoded by the first gene of this operon, *hmuP*, is unknown. In the genomes of both *D. dadantii* and *P. atrosepticum*, we found the existence of a gene cluster related to *hmuSTUV*, which could function as an operon (Fig. 6.4A). In *D. dadantii*, downstream of this cluster, there is a gene predicted to encode a TonB dependent outer membrane receptor (Fig. 6.4A). By contrast, no potential haem receptor gene was identified in the vicinity of the *P. atrosepticum hmuSTUV* gene cluster. Haem cross-feeding assay of a *D. dadantii* mutant impaired in chrysobactin and achromobactin biosynthesis/transport, showed that this putative haem uptake system could be functional under iron deficient conditions. Fur boxes are located in the promoter regions of the *hmuSTUV* operons of both species. The induction of *D. dadantii hmuSTUV* genes was

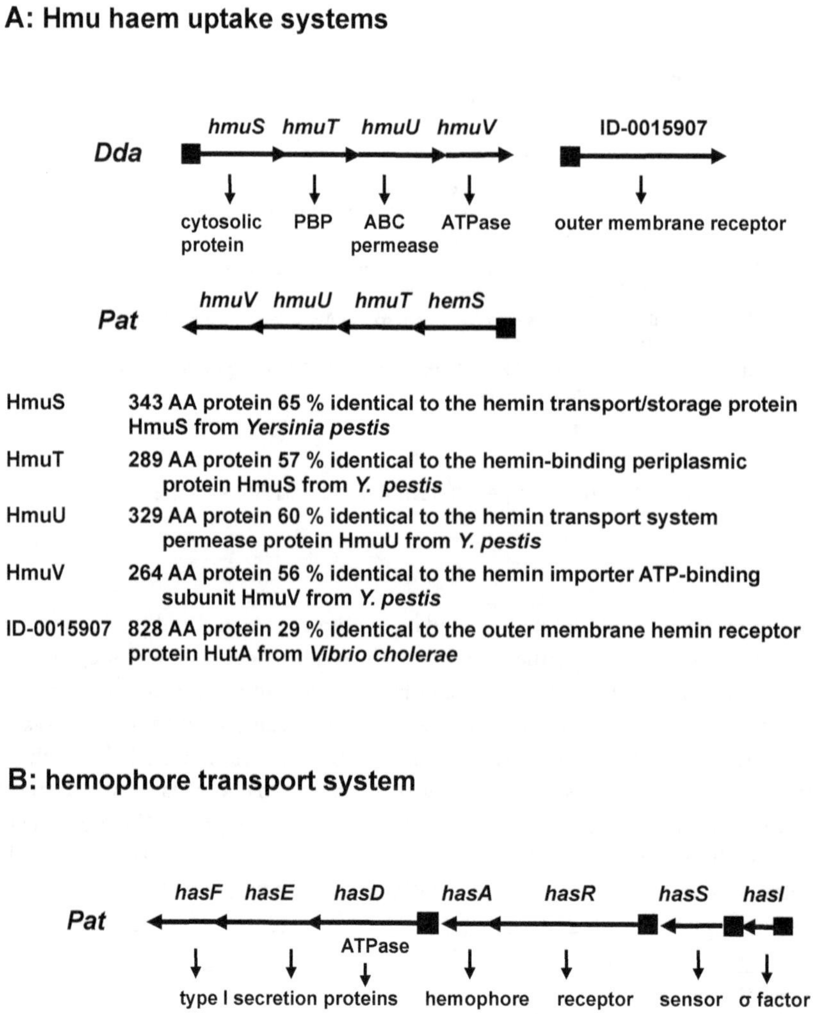

Figure 6.4 Genetic organization of the loci involved in haem uptake. A: genes from the Hmu system in *D. dadantii* (*Dda*) and in *P. atrosepticum* (*Pat*). (B) schematic representation of the *hasRA* and *hasDEF* operons from *P. atrosepticum* involved in transport and secretion of the HasA haemophore, respectively. The regulatory *hasIS* locus encodes the membrane anti-sigma sensor HasS and the ECF sigma factor HasI.

observed from a GFP-based *in vivo* expression technology (IVET) leaf array, using spinach as host plant (Yang *et al.* 2004). Okinaka *et al.* (2002) found the *hmuU* gene to be up-regulated during infection of saintpaulia plants.

The Has system involves a more elaborate mechanism for haem uptake (Wandersman and Delepelaire, 2004). It is based on the extracellular production of proteins called haemophores (HasA) which, due to their high affinity for haem, can extract this compound from various haemoproteins and convey it to specific TonB outer membrane receptors (HasR) that internalize only haem. The passage of haem in the cytoplasm is mediated by a periplasmic binding protein-dependent ABC permease. Initially characterized in *S. marcescens*, this system is also present in *Yersinia* and *Pseudomonas* species. In *S. marcescens*, the haemophore, the proteins involved in haemophore secretion (HasD, HasE and HasF), the outer membrane receptor and HasB, a TonB-like protein specific of the Has system (Létoffé *et al.*, 2004) are encoded by a Fur-regulated gene cluster constituting the *has* operon. This gene cluster is also positively regulated by a signalling process depending on a specific extracytoplasmic function (ECF) sigma HasI and its antisigma factor HasS encoded by the *hasIS* operon (Rossi *et al.*, 2003). In *P. atrosepticum*, a locus related to the *S. marcescens has* gene cluster was identified (Fig. 6.4B). However, the gene organization appears different. Two operons encoding HasR and HasA, and HasD, HasE and HasF, respectively can be predicted (Fig. 6.4B). A regulatory locus consisting of the *hasI and hasS* genes is situated just upstream from these two operons (Fig. 6.4B). A Fur box is present in the promoters of the *hasI* gene and the *hasRA* operon. As in *S. marcescens*, there is a HasI box in the promoter regions of the *hasS* and *hasR* genes (Biville *et al.*, 2004). No *tonB*-like gene is present in the vicinity of this cluster. Whether this Has-like system is functional requires investigation. By contrast, there is no evidence for the existence of this system in *D. dadantii*.

Citrate-mediated ferric iron acquisition

As an iron binding compound, citrate acts as an exogenous siderophore in bacteria harbouring a specific transport system for ferric citrate, called Fec. The mode of action of this system has been thoroughly investigated in *E. coli* K-12, although few other bacterial species including *Shigella* spp. also have the ability to transport ferric citrate via the Fec system (Braun *et al.*, 2006; Braun and Endriss, 2007). The specific TonB-dependent outer membrane receptor FecA and the ABC permease components FecBCDE required for internalization of ferric dicitrate in the cytoplasm are encoded by the *fecABCDE* operon. This operon is positively controlled by a signalling cascade that depends on FecA and the regulatory genes *fecI* and *fecR* also clustered in an operon located upstream of *fecABCDE.* By binding to FecA, ferric citrate triggers a signal which is transmitted to the cytoplasmic transmembrane protein FecR (Breidenstein *et al.*, 2006). The FecR-received signal is then transmitted to the cytoplasm, where the ECF sigma factor FecI is activated to bind RNA polymerase and specifically initiate transcription of the *fecABCDE* operon. Transcription of the *fecIR* operon is regulated by Fur. Interestingly, we found the existence of a complete *fec* system in the genome of *P. atrosepticum*, including the *fecIR* and *fecABCDE* operons. The proteins encoded share 70–80% identity with *E. coli* Fec counterparts. Although *D. dadantii* encodes a regulatory locus with low similarity to the *fecI* and *fecR* genes, we did not find the presence of ferric citrate transport genes in the genome. Accordingly, ferric citrate is not used as an iron source by this bacterium. It would be interesting to determine whether the *P. atrosepticum* Fec system is functional. As in plants, citrate is used to transport ferric iron to leaves by the xylem sap, expression of the Fec system by this bacterium which invades stem xylem vessels could be greatly beneficial. Fec mediated iron transport would be dispensable in a pathogen like *D. dadantii* which does not invade the xylem compartment.

Utilization of exogenous siderophores

Like many bacterial genera, *D. dadantii* 3937 has the capacity to utilize siderophores produced by diverse species of bacteria and fungi, including enterobactin and ferrichrome (Enard *et al.*, 1991). Potential receptors for the ferric complex of these two siderophores were identified in the outer membrane of *D. dadantii* cells grown under iron limitation. In addition, homologues of the *fhuA* and *pfeA* genes encoding the receptors for ferrichrome and ferric enterobactin in *E. coli* and *Pseudomonas aeruginosa*, respectively, were found to be up-regulated during infection of African violets (Okinaka *et al.*, 2002). The analysis of the *D. dadantii* genomic sequence revealed the presence of eight TonB-dependent outer membrane receptors including the five receptors already mentioned. The *D. dadantii tonB* gene encodes a product sharing at least 50% identity with TonB proteins of other enterobacteria. This TonB protein is involved in the transport of chrysobactin, achromobactin, enterobactin and ferrichrome (Enard and Expert, 2000). However, there are two additional *tonB* genes in this species. The one corresponding to ID-0019121 is clustered in an operon with the *exbB* and *exbD* genes. Another gene in reverse orientation, encoding a TonB-dependent outer membrane receptor, is associated with this cluster. The proteins encoded by this cluster display homology to those from *Pseudomonas* spp. suggesting that some siderophores from this genus are used as xenosiderophores by *D. dadantii*. Interestingly, the same 'TonB-ExbBD transport' locus is present in *P. atrosepticum*. The genome of *P. atrosepticum* reveals the presence of 17 TonB-dependent outer membrane receptors, one sharing 75% identity to the *E. coli* aerobactin receptor IutA. There are also six TonB homologues, three of them displaying identity to corresponding *D. dadantii* TonB proteins. Thus, it seems that *P. atrosepticum* is gifted with an important repertoire of proteins enabling it to utilize siderophores produced by other microorganisms or to transport non-iron compounds, such as large carbohydrates or vitamins (Schauer *et al.*, 2008).

Conclusion

Like other members of the *Enterobacteriaceae* interacting with animals and humans, pectinolytic species appear to contain multiple iron transport systems, illustrating the importance of iron acquisition to the fitness of these bacteria associated to plants. Experimental investigations are required to confirm the assumptions emerging from our genomic analysis of *D. dadantii* and *P. atrosepticum*. However, it seems likely that during evolution, these two closely related pathogens which display similar strategies of host attack and share common traits in their saprophytic life have developed different mechanisms for acquiring iron. Soft rot species have long been considered as opportunistic pathogens because in natural infections, their proliferation in the plant depends on a variety of environmental factors including temperature, humidity, oxygen tension and pH. These factors are known to influence iron availability. Outside the plant, their natural habitats are the rhizosphere and agricultural soils where many microbial populations interact with each other or with the plant through diverse relationships from mutualistic to competitive. In these environments, the possibility for these bacteria to express well-suited mechanisms for iron acquisition with the capacity to utilize xenosiderophores is likely to be essential for their survival. Globally, it seems that pectinolytic bacteria have in common the potentiality to produce siderophores of the catechol family, including chrysobactin and other related compounds of still unknown structure. The production of achromobactin is not a widely distributed trait among these species. However, achromobactin is produced by other phytopathogenic bacteria such as *P. syringae* and the fact that the corresponding genomic region has a similar G+C content to that of the *P. syringae* genome makes it possible that this region was inherited by lateral gene transfer. Another difference resides in the fact that *P. atrosepticum* has no ferrous iron acquisition system unlike *D. dadantii*, while it probably produces the machinery to acquire ferric citrate, the major iron source in the plant xylem vessels. Also, both bacteria have the ability to take up haem, but *D. dadantii* uses a direct transport route while in *P. atrosepticum* only a haemophore-dependent mechanism should be involved.

Acknowledgements

We wish to thank Professor Ching-Hong Yang from the University of Wisconsin-Milwaukee for the gift of the *D. dadantii feoB* mutant. D Expert is a researcher from the Centre National de la Recherche Scientifique (CNRS).

References

Angerer, A., Gaisser, S., and Braun, V. (1990). Nucleotide sequences of the *sfuA, sfuB,* and *sfuC* genes of *Serratia marcescens* suggest a periplasmic-binding-protein-dependent iron transport mechanism. J. Bacteriol. *172,* 572–578.

Barghouti, S., Young, R., Olson, M.O.J. Arceneaux, E.L., Clem, L.W., and Byers, B.R. (1989). Novel tryptophan- or phenylalanine-containing phenolate siderophore in *Aeromonas hydrophila.* J. Bacteriol. *171,* 1811–1816.

Barnes, H.H., and Ishimaru, C.A. (1999). Purification of catechol siderophores by boronate affinity chromatography: identification of chrysobactin from *Erwinia carotovora* subsp. carotovora. Biometals *12,* 83–87.

Bearden, S.W., Staggs, T.M., and Perry, R.D. (1998). An ABC transporter system of *Yersinia pestis* allows utilization of chelated iron by *Escherichia coli* SAB11. J. Bacteriol. *180,* 1135–1147.

Bearden, S.W., and Perry, R.D. (1999). The Yfe system of *Yersinia pestis* transports iron and manganese and is required for full virulence of plague. Mol. Microbiol. *32,* 403–414.

Bell, K.S., Sebaitha, M., Pritchard, L. Holden, M. T., Hymen, L. J., Holeva, M. C., Thompson, N. R., Bentley, S. D., Churcher, L. J., Mungall, K., *et al.* (2004). Genome sequence of the enterobacterial phytopathogen *Erwinia carotovora* subsp. atroseptica and characterization of virulence factors. Proc. Natl. Acad. Sci. U.S.A. *101,* 11105–11110.

Biville, F., Cwerman, H., Létoffé, S., Rossi, M.S., Drouet, V., Ghigo, J.M., and Wandersman, C. (2004). Haemophore-mediated signalling in *Serratia marcescens*: a new mode of regulation for an extra cytoplasmic function (ECF) sigma factor involved in haem acquisition. Mol. Microbiol. *53,* 1267–1277.

Boureau, T., ElMaarouf-Bouteau, H., Garnier, A., Brisset, M.-N., Périno, C., Pucheu, I., and Barny, M.A. (2006). DspA/E, a type III effector essential for *Erwinia amylovora* pathogenicity and growth in planta, induces cell death in host apple and nonhost tobacco plants. Mol. Plant Microbe Interact. *19,* 16–24.

Braun, V., and Endriss, F. (2007). Energy-coupled outer membrane transport proteins and regulatory proteins. Biometals *20,* 219–231.

Braun, V., Mahren, S., and Sauter, A. (2006). Gene regulation by transmembrane signaling. Biometals *19,* 103–113.

Breidenstein, E., Mahren, S., and Braun, V. (2006). Residues involved in FecR binding are localized on one side of the FecA signaling domain in *Escherichia coli.* J. Bacteriol. *188,* 6440–6442.

Brown, J.C. (1978). Mechanism of iron uptake by plants. Plant Cell Environ. *1,* 249–257.

Bull, C.T., Ishimaru, C.A., and Loper, J.E. (1994). Two genomic regions involved in catechol siderophore production by *Erwinia carotovora.* Appl. Environ. Microbiol. *60,* 662–669.

Cao, J., Woodhall, M.R., Alvarez, J., Cartron, M.L., and Andrews, S.C. (2007). EfeUOB (YcdNOB) is a tripartite, acid-induced and CpxAR-regulated, low-pH Fe^{2+} transporter that is cryptic in *Escherichia coli* K-12 but functional in *E. coli* O157:H7. Mol. Microbiol. *65,* 857–875.

Cartron, M.L., Maddocks, S., Gillingham, P., Craven, C.J., and Andrews, S.C. (2006). Feo transport of ferrous iron into bacteria. Biometals *19,* 143–157.

Dellagi, A., Brisset, M.N., Paulin, J.-P., and Expert, D. (1998). Dual role of desferrioxamine in *Erwinia amylovora* pathogenicity. Mol. Plant-Microbe Interact. *8,* 734–742.

Dellagi, A., Reis, D., Vian, B., and Expert, D. (1999). Expression of the ferrioxamine receptor gene of *Erwinia amylovora* CFBP 1430 during pathogenesis. Mol. Plant-Microbe Interact. *12,* 463–466.

Dellagi, A., Rigault, M., Segond, D., Roux, C., Kraepiel, Y., Cellier, F., Briat, J.F., Gaymard, F., and Expert, D. (2005). Siderophore-mediated up-regulation of Arabidopsis ferritin expression in response to *Erwinia chrysanthemi* infection. Plant J. *43,* 262–272.

Dellagi, A., Segond, D., Rigault, M., Fagard, M., Simon, C., Saindrenan, P., and Expert, D. (2009). Microbial siderophores exert a subtle role in Arabidopsis during infection by manipulating the immune response and the iron status. Plant Physiol. *150,* 1687–1696.

Douet, V., Expert, D., Barras, F., and Py, B. (2009). *Erwinia chrysanthemi* iron metabolism: the unexpected implication of the inner membrane platform within the type II secretion system. J. Bacteriol. *191,* 795–804.

Enard, C., Diolez, A., and Expert, D. (1988). Systemic virulence of *Erwinia chrysanthemi* 3937 requires a functional iron assimilation system. J. Bacteriol. *170,* 2419–2426.

Enard, C., Franza, T., Neema, C., Gill, P.R., Persmark, M., Neilands, J.B., and Expert, D. (1991). The requirement of chrysobactin dependant iron transport for virulence incited by *Erwinia chrysanthemi* on *Saintpaulia ionantha.* Plant Soil. *130,* 263–271

Enard, C., and Expert, D. (2000). Characterization of a *tonB* mutation in *Erwinia chrysanthemi* 3937: TonB(Ech) is a member of the enterobacterial TonB family. Microbiology *146,* 2051–2058.

Expert, D., Rauscher, L., and Franza, T. (2004). *Erwinia,* a plant pathogen. In Iron Transport in Bacteria: molecular genetics, biochemistry, microbial pathogenesis and ecology. J. Crosa and S. Payne, ed. (American Society of Microbiology Press Book), pp. 402–412.

Fagard, M., Dellagi, A., Roux, C., Périno, C., Rigault, M., Boucher, V., Shevchik, V., and Expert, D. (2007). *Arabidopsis thaliana* expresses multiple lines of defense to counter-attack *Erwinia chrysanthemi* Mol. Plant- Microbe Interact. *20,* 794–805.

Franza, T., and Expert, D. (1991). The virulence-associated chrysobactin iron uptake system of *Erwinia chrysanthemi* 3937 involves an operon encoding transport and biosynthetic functions. J. Bacteriol. *173,* 6874–6881.

Franza, T., Mahé, B., and Expert, D. (2005). *Erwinia chrysanthemi* requires a second iron transport route dependent of the siderophore achromobactin for extracellular growth and plant infection. Mol. Microbiol. *55,* 261–275.

Grosse, C., Scherer, J., Koch, D., Otto, M., Taudte, N., and Grass, G. (2006). A new ferrous iron-uptake transporter, EfeU (YcdN), from *Escherichia coli*. Mol. Microbiol. *62*, 120–131.

Hauben, L., Moore, E.R.B., Vauterin, L., Steenackers, M., Mergaert, J., Verdonck, L., and Swings, J. (1998). Phylogenetic position of phytopathogens within the *Enterobacteriaceae*. System. Appl. Microbiol. *21*, 384–397.

Ishimaru, C.A., and Loper, J.E. (1992). High-affinity iron uptake systems present in *Erwinia carotovora* subsp. carotovora include the hydroxamate siderophore aerobactin. J. Bacteriol. *174*, 2993–3003.

Kammler, M., Schon, C., and Hantke, K. (1993). Characterization of the ferrous iron uptake system of *Escherichia coli*. J. Bacteriol. *175*, 6212–6219.

Kehres, D.G., Janakiraman, A., Slauch, J.M., and Maguire, M.E. (2002). SitABCD is the alkaline Mn(2+) transporter of *Salmonella enterica* serovar Typhimurium. J. Bacteriol. *184*, 3159–3166.

Leduc, D., Battesti, A., and Bouveret, E. (2007). The hotdog thioesterase EntH (YbdB) plays a role *in vivo* in optimal enterobactin biosynthesis by interacting with the ArCP Domain of EntB. J. Bacteriol. *189*, 7112–7126

Létoffé, S., Delepelaire, P., and Wandersman, C. (2004). Free and hemophore-bound heme acquisitions through the outer membrane receptor HasR have different requirement for the TonB–ExbB-ExbD complex. J. Bacteriol. *186*, 4067–4074.

Ma, B., Hibbing, M.E., Kim, H.S., Reedy, R.M., Yedidia, I., Breuer, J., Breuer, J., Glasner, J.D., Perna, N.T., Kelman, A., and Charkowski, A.O. (2007). Host range and molecular phylogenies of the soft rot enterobacterial genera *Pectobacterium* and *Dickeya*. Phytopathology 97, 1150–1163.

Marlovits, T.C., Haase, W., Herrmann, C., Aller, S.G., and Unger, V.M. (2002). The membrane protein FeoB contains an intramolecular G protein essential for Fe(II) uptake in bacteria. Proc. Natl. Acad. Sci. U.S.A. *99*, 16243–16248.

Masclaux, C., and Expert, D. (1995). Signalling potential of iron in plant–microbe interactions: the pathogenic switch of iron transport in *Erwinia chrysanthemi*. Plant J. *7*, 121–128.

Münzinger, M., Budzikiewicz, H., Expert, D., Enard, C., and Meyer, J.M. (2000). Achromobactin, a new citrate siderophore of *Erwinia chrysanthemi*. Z. Naturforsch [C]. *55*, 328–332.

Nachin, L., and Barras, F. (2000). External pH: an environmental signal that helps to rationalize *pel* gene duplication in *Erwinia chrysanthemi*. Mol. Plant Microbe Interact. *13*, 882–886.

Neema, C., Laulhère, J.P., and Expert, D. (1993). Iron deficiency induced by chrysobactin in Saintpaulia leaves inoculated with *Erwinia chrysanthemi*. Plant Physiol. *102*, 967–973.

Okinaka, Y., Yang, C.H., Perna., N.T., and Keen., N.T. (2002). Microarray profiling of *Erwinia chrysanthemi* 3937 genes that are regulated during plant infection. Mol. Plant Microbe Interact. *15*, 619–629.

Pérombelon, M.C.M. (2002). Potato diseases caused by soft rot erwinias: an overview of pathogenesis. Plant Pathol. *51*, 1–12.

Perry, R.D. (2004). *Yersinia*. In Iron Transport in Bacteria: molecular genetics, biochemistry, microbial pathogenesis and ecology. J. Crosa and S. Payne, ed. (American Society of Microbiology Press Book), pp. 219–240.

Persmark, M., Expert, D., and Neilands, J.B. (1989). Isolation, characterization, and synthesis of chrysobactin, a compound with siderophore activity from *Erwinia chrysanthemi*. J. Biol. Chem. *264*, 3187–3193.

Rauscher, L., Expert, D., Matzanke, B.F., and Trautwein, A.X. (2002). Chrysobactin-dependent iron acquisition in *Erwinia chrysanthemi*. Functional study of a homolog of the *Escherichia coli* ferric enterobactin esterase. J. Biol. Chem. *277*, 2385–2395.

Rossi, M.S., Paquelin, J.M., Ghigo, J.M., and Wandersman, C. (2003). Hemophore-mediated signal transduction across the bacterial cell envelope in *Serratia marcescens*: the inducer and the transported substrate are different molecules. Mol. Microbiol. *48*, 1467–1480.

Runyen-Janecky, L.J., Reeves, S.A., Gonzales, E.G., and Payne, S.M. (2003). Contribution of the *Shigella flexneri* Sit, Iuc, and Feo iron acquisition systems to iron acquisition *in vitro* and in cultured cells. Infect. Immun. *71*, 1919–1928.

Sabri, M., Léveillé, S., and Dozois, C.M. (2006). A SitABCD homologue from an avian pathogenic *Escherichia coli* strain mediates transport of iron and manganese and resistance to hydrogen peroxide. Microbiology *152*, 745–758.

Samson, R., Legendre, J.B., Christen, R., Fischer-Le Saux, M., Achouak, W., and Gardan, L. (2005). Transfer of *Pectobacterium chrysanthemi* (Burkholder *et al.* 1953, Brenner *et al.* 1973) and *Brenneria paradisiaca* to the genus *Dickeya* gen. nov. as *Dickeya chrysanthemi* comb. nov. and *Dickeya paradisiaca* comb. nov. and delineation of four novel species, *Dickeya dadantii* sp. nov., *Dickeya dianthicola* sp. nov., *Dickeya dieffenbachiae* sp. nov., and *Dickeya zeae* sp. nov. Int. J. Syst. Evol. Microbiol. *55*, 1415–1427.

Schauer, K., Rodionov, D.A., and de Reuse, H. (2008) New substrates for TonB-dependenttransport: do we only see the 'tip of the iceberg'? Trends Biochem. Sci. *33*, 330–338.

Schmelz, S., Kadi, N., McMahon, S.A., Song, L., Oves-Costales, D., Oke, M., Liu, H., Johnson, K.A., Carter, L.G., Botting, C.H., White, M.F., Challis, G.L., and Naismith, J.H. (2009). AcsD catalyzes enantioselective citrate desymmetrization in siderophore biosynthesis. Nat. Chem. Biol. *5*, 174–182.

Schwyn, B., and Neilands, J.B. (1987). Universal chemical assay for the detection and determination of siderophores. Anal. Biochem. *106*, 47–56.

Segond, D., Dellagi, A., Lanquar, V., Rigault, M., Patrit, O., Thomine, S., and Expert, D. (2009). NRAMP genes function in *Arabidopsis thaliana* resistance to *Erwinia chrysanthemi* infection. Plant J. *58*, 195–207.

Tomisić, V., Blanc, S., Elhabiri, M., Expert, D., and Albrecht-Gary, A.M. (2008). Iron(III) uptake and release by chrysobactin, a siderophore of the phytophatogenic bacterium *Erwinia chrysanthemi*. Inorg. Chem. *47*, 9419–9430.

Wandersman, C., and Delepelaire, P. (2004). Bacterial iron sources: from siderophores to hemophores. Annu. Rev. Microbiol. *58*, 611–647.

Winslow, C.E. A., Broadhurst, J., Buchanan, R.E., Krumwiede, C., Rogers, L.A., and Smith, G.H. (1917). The families and genera of the bacteria. J. Bacteriol. *2*, 505–566.

Yang, S., Perna, N.T., Cooksey, D.A., Okinaka, Y., Lindow, S.E., Ibekwe, A.M., Keen, N.T., and Yang, C.H. (2004). Genome-wide identification of plant-up-regulated genes of *Erwinia chrysanthemi* 3937 using a GFP-based IVET leaf array. Mol. Plant Microbe Interact. *17*, 999–1008.

Zaharik, M.L., Cullen, V.L., Fung, A.M., Libby, S.J., Kujat Choy, S.L., Coburn, B., Kehres, D.G., Maguire, M.E., Fang, F.C., and Finlay, B.B. (2004). The *Salmonella enterica* serovar typhimurium divalent cation transport systems MntH and SitABCD are essential for virulence in an Nramp1G169 murine typhoid model. Infect. Immun. *72*, 5522–5525.

Zimmermann, L., Angerer, A., and Braun, V. (1989). Mechanistically novel iron(III) transport system in *Serratia marcescens*. J. Bacteriol. *171*, 238–243.

Iron Uptake in *Vibrio* and *Aeromonas*

7

Manuel L. Lemos and Carlos R. Osorio

Abstract

Vibrio and *Aeromonas* species are ubiquitous bacteria in aquatic environments worldwide. Many of the species are important pathogens for humans and/or aquatic animals. Several iron acquisition strategies have been developed by vibrios and aeromonads in order to get this essential element for surviving in their host and in aquatic habitats. All species studied so far have the ability to synthesize siderophores to sequester iron from the cell environment and transport it through their respective cognate outer membrane receptors. It has been demonstrated that this capacity is a relevant virulence factor for human and animal pathogens. Furthermore, all species studied can utilize exogenous siderophores, made by other bacteria. Another iron acquisition system described in both genera involves the use of haem as a source of iron, by a mechanism very well conserved among all species, which involves a haem transporter that includes a specific TonB-dependent outer membrane receptor(s) and an ABC-type inner membrane transporter. Alternative systems based on ferrous or ferric iron transporters have been reported in *V. cholerae.* How the different iron acquisition systems work together to supply iron to the cell and how they are used in the different environments where vibrios and aeromonads can be found is still an open question.

Introduction

The genus *Vibrio* contains a wide range of species many of them being important pathogens for humans and marine animals. A great number of species are encountered in marine and estuarine habitats as part of the bacterioplankton or associated with other organisms. Among the main human pathogens are *V. cholerae,* the etiological agent of cholera disease and *V. vulnificus,* which causes severe haemorrhagic septicaemias, and which is also an important pathogen for marine animals (Toranzo *et al.,* 2005). *V. parahaemolyticus* can cause gastroenteritis in humans, associated with the consumption of seafood and wound infections. Other important pathogens for marine animals, especially for fish, are *V. anguillarum* that causes a fatal haemorrhagic septicaemia in diverse fish species and it is one of the most important pathogens in the aquaculture industry, or *V. alginolyticus,* also a potential pathogen for fish and molluscs. Highly related to the vibrios are the members of the genus *Photobacterium,* also a group of marine bacteria that contains two important pathogens: *P. damselae* subsp. *damselae,* an opportunistic pathogen for humans and fish, and *P. damselae* subsp. *piscicida,* an important pathogen for a variety of fish species (Toranzo *et al.,* 2005).

The members of the family *Aeromonadaceae* are ubiquitous bacteria in aquatic environments worldwide, and have long been recognized as causative agents of infections in invertebrates and vertebrates, including fish, frogs, birds, domestic animals and humans (Joseph and Carnahan, 1994; Gosling, 1996). The genus *Aeromonas* has been historically divided into two groups: a group of psychrophilic, non-motile species, represented by *A. salmonicida,* and a second group of motile, mesophilic species associated with human disease, best represented by *A. hydrophila.* This is the only species known to cause disease in both fish and humans. In fish, it causes haemorrhagic septicaemia, and in humans it causes wound, soft tissue, and blood infections as well as acute gastroenteritis. *A. salmonicida* subsp. *salmonicida* is the causative agent of the so-called 'typical' furunculosis, which causes economically devastating losses in cultivated salmonids and non-salmonid fish, showing a widespread distribution (Toranzo *et al.*, 2005). The atypical strains of *A. salmonicida* are included within three subspecies, *masoucida, achromogenes* and *smithia,* and cause ulcerative diseases in a variety of fish species.

In order for iron to be used by these microorganisms, it must be obtained from the cell environment by specialized uptake mechanisms. Thus, different strategies to satisfy their iron demands have been developed by the members of *Vibrio* and *Aeromonas.* One of the main strategies, like in many other Gram-negative bacteria, consists of the synthesis of siderophores, although there are also siderophore-independent mechanisms, such as the uptake of iron from iron bearing substances that are found in the environment of the bacterium. In *Vibrio* and *Aeromonas* one of the main siderophore-independent mechanisms for obtaining iron sources is the utilization of haem and haem-containing proteins from the host. Furthermore, as most species of these two genera are bacteria related to aquatic environments which can survive in a free-living state, the importance of other yet poorly studied mechanisms, such as the uptake of iron by low-affinity mechanisms or other yet unknown systems, should not be disregarded. These diverse iron uptake mechanisms usually coexist in the same cell, which underscores the importance of iron in the cell metabolism.

In this chapter we try to give a global view of the current knowledge, as well as the future trends, about the iron uptake mechanisms of *Vibrio* and *Aeromonas* and their relationship with the virulence of the pathogenic species. Table 7.1 summarizes the diverse iron assimilation systems currently known in both genera.

Siderophore-mediated iron uptake in the genus *Vibrio*

Siderophore-mediated iron uptake in *V. cholerae*

The causative agent of cholera, *V. cholerae,* can acquire iron by multiple mechanisms (Wyckoff *et al.*, 2007). We describe below how this bacterium uses iron from haem or haemoglobin. Additional mechanisms include those for the transport of ferric and ferrous iron, and the transport of exogenous siderophores made by other organisms, as ferrichrome or enterobactin (Rogers *et al.*, 2000; Mey *et al.*, 2002, 2008; Wyckoff *et al.*, 2006).

Most *V. cholerae* strains synthesize a catecholate siderophore named vibriobactin (Griffiths *et al.*, 1984), which has been shown to contain three residues of 2,3-dihydroxybenzoic acid and two residues of threonine. Both threonine moieties are present in the form of oxazoline rings. Furthermore, the polyamine backbone of the molecule was proved to be norspermidine [N-(3-aminopropyl)-1,3-diaminopropane] (Fig. 7.1). Vibriobactin is synthesized and secreted into the environment. It binds ferric iron with high affinity and is then transported into the cell through ViuA, the specific outer membrane vibriobactin receptor (Butterton *et al.*, 1992).

The genes for biosynthesis and transport of vibriobactin are located in two separate gene clusters in chromosome I of *V. cholerae,* and each cluster contains both biosynthetic and utilization genes (Fig. 7.2). One of these clusters contains the gene *vibF* which encodes a high-molecular-weight non-ribosomal peptide synthetase, as well as the vibriobactin transport gene *viuA* (encodes the outer membrane receptor) and the vibriobactin utilization gene *viuB* (encodes a putative ferri-siderophore esterase). The second cluster contains the genes for the synthesis of DHBA from chorismate (*vibABC*), a gene for DHBA activation (*vibE*), a gene encoding a phosphopantetheinyl transferase (*vibD*), a gene for a NRPS (*vibH*), and genes for vibriobactin transport (*viuPDGC*) (Wyckoff *et al.*, 1997, 2001).

Biosynthesis and transport of vibriobactin

The mechanism of action of the different enzymes that participate in the biosynthesis of vibriobactin was elucidated by studying their biochemical properties, by *in vitro* synthesis of this siderophore molecule,

Table 7.1 Iron assimilation systems that have been described in species of *Vibrio* and *Aeromonas*

Species	Siderophore	Siderophore OM receptor	Exogenous siderophores used	Exogenous siderophores receptors	Haem uptake	Haem receptors	Other iron acquisition systems
Vibrio							
V. cholerae	Vibriobactin	ViuA	Ferrichrome Enterobactin Fluvibactin Agrobactin	FhuA IrgA, VctA	Yes	HutA HutR HasR	Feo FbpA VciA
V. anguillarum	Anguibactin Vanchrobactin	FatA FvtA	Ferrichrome Enterobactin Rhodotorulic acid		Yes	HuvA HuvS	Citrate
V. vulnificus	Vulnibactin	VuuA	Aerobactin	IutA	Yes	HupA	
V. parahaemolyticus	Vibrioferrin	PvuA	Ferrichrome Aerobactin	FhuA IutA	Yes	HutA	
V. alginolyticus	Vibrioferrin	PvuA	Ferrichrome		Yes		
V. mimicus	Aerobactin				Yes	MhuA	
V. hollisae	Aerobactin						
Aeromonas							
A. hydrophila	Amonabactins	FstC*			Yes		
A. salmonicida	Amonabactin? Anguibactin-like?	FstC* FstB*			Yes	HutA HutE	

*Putative outer membrane receptors deduced from its amino acid sequence.

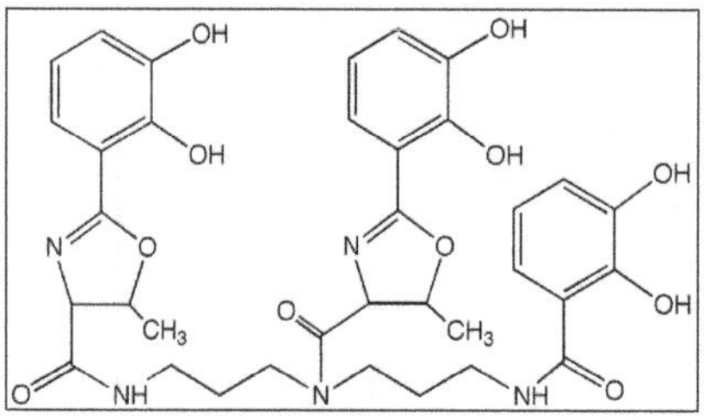

Vibriobactin
V. cholerae

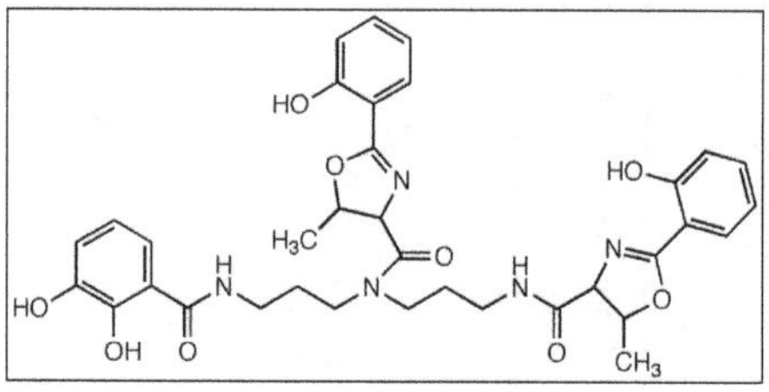

Vulnibactin
V. vulnificus

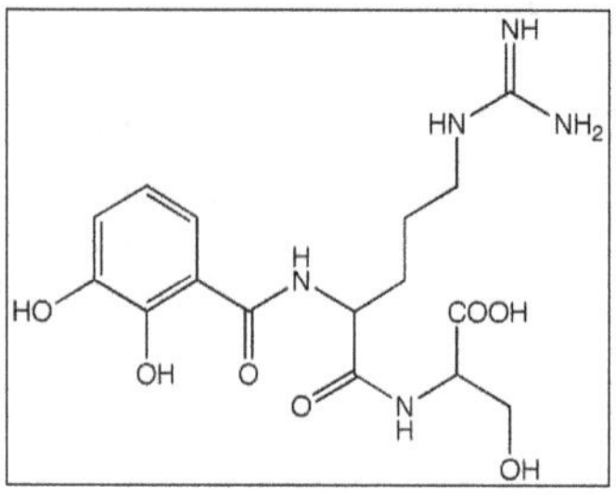

Vanchrobactin
V. anguillarum

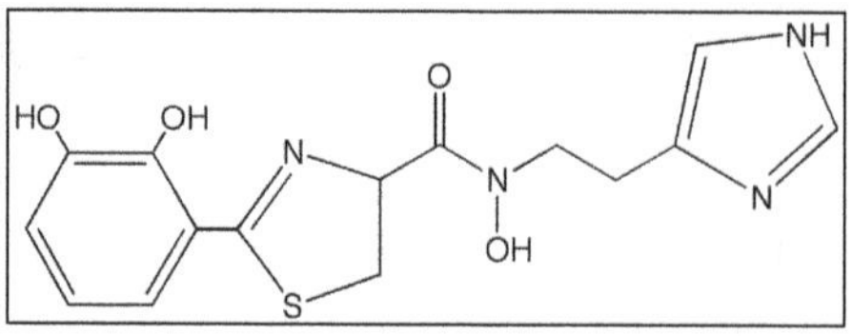

Anguibactin
V. anguillarum

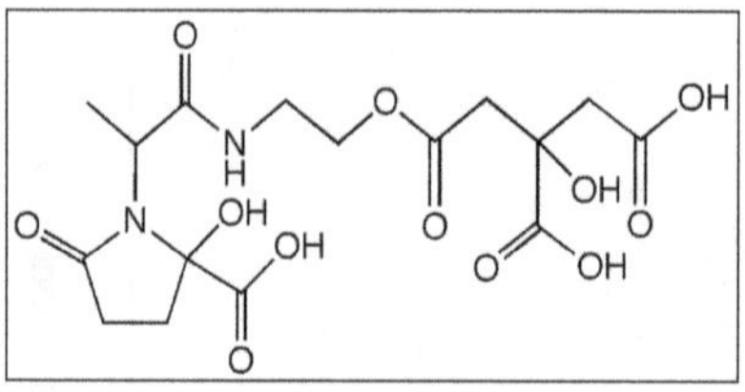

Vibrioferrin
V. parahaemolyticus
V. alginolyticus

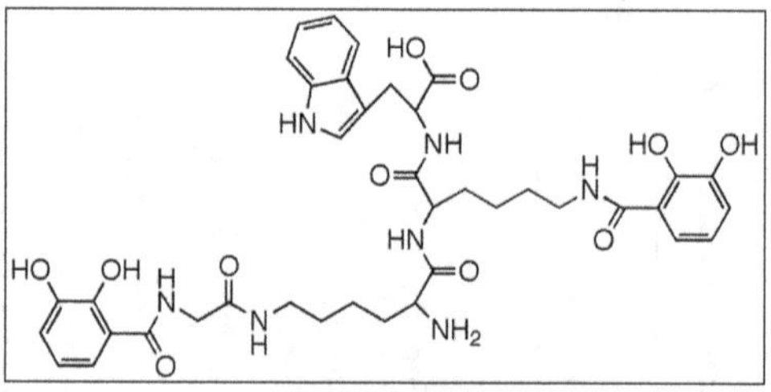

Amonabactin T
A. hydrophila
A. salmonicida

Figure 7.1 Structures of some representative siderophores synthesized by different species of *Vibrio* and *Aeromonas*.

and by studies with deletion mutants for biosynthetic genes. VibE and VibB are functional and structural analogs to the EntE and EntB enzymes of enterobactin synthesis, and to VabE and VabB enzymes described for *V. anguillarum* vanchrobactin biosynthesis. VibF is a multidomain NRPS with six predicted domains, Cy-Cy-A-C-PCP-C (Cy: Cyclization domain A: Adenylation domain; C: condensation; PCP: peptidyl carrier protein domain), and is the core of the vibriobactin assembly line. VibH is a NRPS that possesses only one condensation domain, and its sequence shows similarity with the first 452 amino acids of *E. coli* EntF. Hence, vibriobactin is synthesized following a typical non-ribosomal peptide synthetase-based mechanism that assembles monomers of DHBA, threonine and norspermidine into the final vibriobactin structure (Wyckoff *et al.*, 1997; Crosa and Walsh, 2002; Di Lorenzo *et al.*, 2004b). The vibriobactin molecule contains three 2,3-dihydroxybenzoyl residues. Two of them are linked to cyclized L-threonine moieties, generating dihydroxyphenyl oxazolinyl acyl groups (DHPO) which are linked to the norspermidine backbone. The third DHBA residue is directly linked to norspermidine (Fig. 7.1).

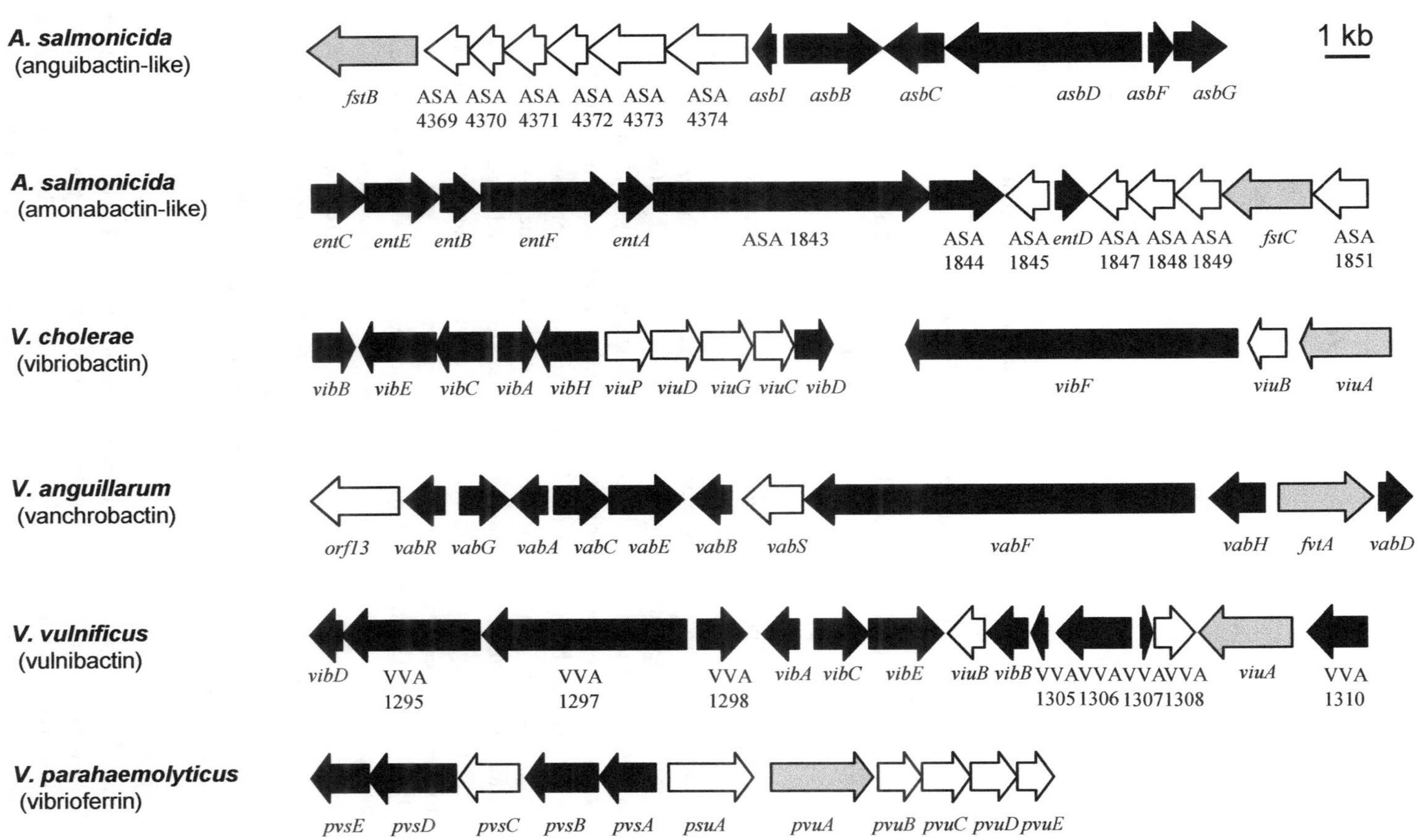

Figure 7.2 A scheme of siderophore biosynthesis and transport operons of *A. salmonicida*, *V. cholerae*, *V. anguillarum*, *V. vulnificus* and *V. parahaemolyticus*. Biosynthetic and regulatory genes are indicated by black arrows. White arrows denote genes involved in siderophore transport. The cognate siderophore outer membrane receptor genes are depicted in grey. Locus designations are used to identify unnamed *A. salmonicida* (ASA) and *V. vulnificus* (VVA) genes. Gene names for *V. cholerae* were used to label homologous loci in the *V. vulnificus* cluster.

The outer membrane receptor for vibriobactin, ViuA, is encoded in the same cluster as VibF and the vibriobactin utilization protein ViuB. ViuA is essential for vibriobactin-mediated iron uptake, since mutations in *viuA* result in a defective phenotype for vibriobactin transport. The other cluster contains four genes (*viuPDGC*) encoding additional proteins involved in the transport of vibriobactin. ViuP is homologous to the *E. coli* periplasmic binding protein FepB. ViuD and ViuG are predicted to be the cytoplasmic membrane permease proteins. ViuC is probably the ATPase component of an inner membrane ABC-transporter, as it contains typical domains of this protein family, namely Walker A and Walker B motifs. Contrary to *viuA* mutants, mutants in either *viuP* or *viuG* are not completely defective in vibriobactin transport, since their roles can be partially complemented by *vctP* and *vctG* genes, respectively, which are part of the enterobactin transport cluster (see below).

Besides vibriobactin, *V. cholerae* can use exogenous siderophores, made by other bacteria, like enterobactin, ferrichrome, agrobactin or fluvibactin (Wyckoff *et al.*, 2007). Enterobactin transport requires TonB and is independent of the vibriobactin receptor ViuA. Two receptors are used for enterobactin transport, IrgA and VctA. Mutation of either of these receptors does not significantly affect enterobactin utilization, but a strain defective in both receptors is unable to use enterobactin. Linked to the *vctA* gene is a set of genes, *vctPDGC*, which encode a periplasmic binding protein-dependent ABC transport system that functions in the utilization of both enterobactin and vibriobactin. Thus, unlike the outer membrane receptors, ViuPDGC and VctPDGC can transport both vibriobactin and enterobactin (Mey *et al.*, 2002; Wyckoff *et al.*, 2007).

Ferrichrome can be also transported by *V. cholerae* by means of the outer membrane receptor FhuA and the inner membrane transporters FhuBCD (Rogers *et al.*, 2000).

Siderophore-mediated iron uptake in *V. anguillarum*

V. anguillarum is the causative agent of vibriosis, a highly fatal haemorrhagic septicaemia affecting marine and freshwater fish species throughout the world, being one of the most important infectious disease in marine aquaculture industry. Up to 23 O serotypes are recognized in the European system, with O1 and O2 being the main ones implicated in disease worldwide (Toranzo *et al.*, 2005). Although the virulence mechanisms of *V. anguillarum* are not fully understood, it is known that the ability to scavenge iron through the utilization of siderophores contributes significantly to virulence. Two clearly different siderophore-mediated systems have been described in *V. anguillarum* strains belonging to serotypes O1 and O2. In most pathogenic strains of serotype O1 this system is mediated by the 65-kb plasmid pJM1, which harbours genes for the biosynthetic machinery and utilization of the siderophore anguibactin (Stork *et al.*, 2002; Fig. 7.1), although the biosynthesis of anguibactin needs chromosomally-encoded enzymes in addition to those encoded by pJM1 genes (Alice *et al.*, 2005).

On the other hand, all serotype O2 strains tested so far, and some plasmidless serotype O1 strains, produce a catecholate-type siderophore that is not related to the pJM1 plasmid-mediated system (Conchas *et al.*, 1991). This chromosome-encoded system produces a siderophore called vanchrobactin (Fig. 7.1), which has been recently isolated from a serotype O2 strain and chemically characterized. Its biosynthetic pathway has been also recently proposed (Balado *et al.*, 2006).

Other (exogenous) siderophores, like enterobactin, ferrichrome or rhodotorulic acid, can be also used as iron sources by as *V. anguillarum* by as yet unknown mechanisms.

Anguibactin-mediated iron uptake

Most *V. anguillarum* strains belonging to serotype O1 harbour the plasmid pJM1 (a ca. 65 kb characterized from strain 775) or other highly related plasmids. This type of plasmids confers the ability to invade the host through the synthesis of a highly efficient siderophore known as anguibactin. It has been demonstrated that the ability to produce anguibactin is a prerequisite for the virulence of plasmid-bearing strains. The siderophore is synthesized from 2,3-dihydroxybenzoic acid, L-cysteine, and N-hydroxyhistamine and has been characterized as ω-*N*-hydroxy-ω-*N*-{[2′-(2″,3″dihydroxyphenyl)-thiazolin-4′-yl] carboxy}-histamine (Jalal *et al.*, 1989; Fig. 7.1). The molecule can be classified as a hybrid between a

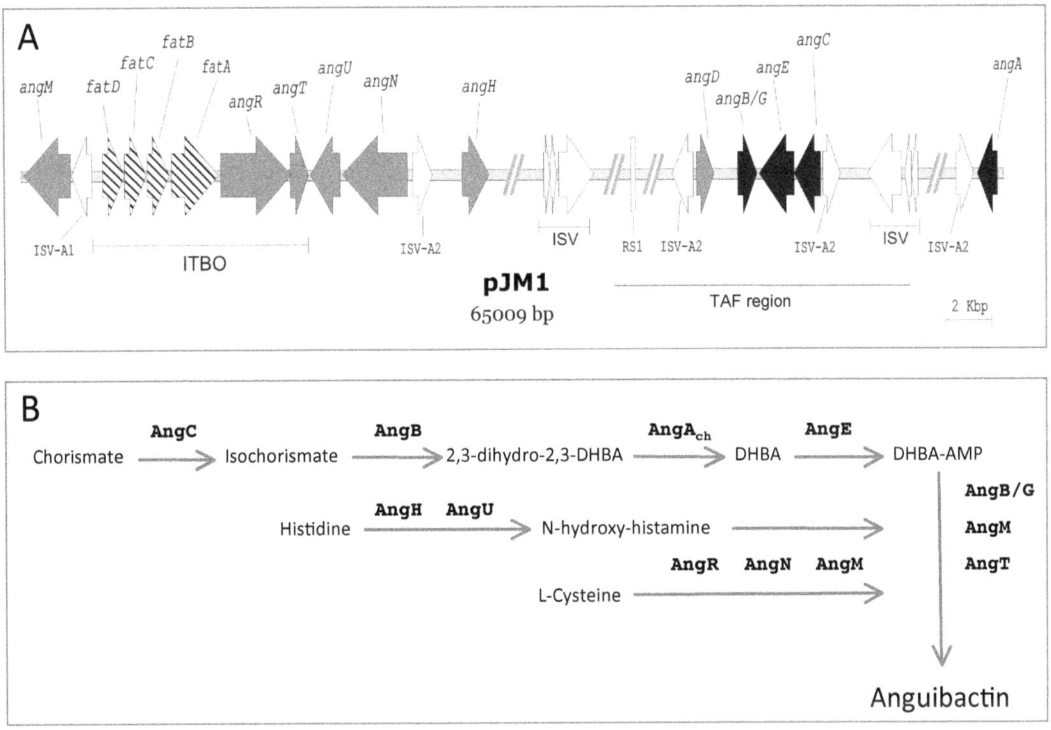

Figure 7.3 (A) Genetic organization of *V. anguillarum* anguibactin biosynthetic and transport genes contained in pJM1 plasmid. Some ORFs were omitted for clarity. The TAF region shown is approximate. ITBO, iron transport and biosynthesis operon. See Di Lorenzo *et al.* (2003) for details of the complete plasmid sequence. (B) Proposed simplified pathway for the synthesis of anguibactin.

catechol and a hydroxamate type of siderophore, since it contains a catechol group and the N-hydroxyl group of hydroxyhistamine.

Most biosynthetic genes are located in the pJM1 plasmid (Stork *et al.*, 2002; Fig. 7.3). The biosynthesis of anguibactin is accomplished by a complex of non-ribosomal peptide synthetases: AngB/G, AngM, AngN, AngR, and AngT. The biosynthesis genes *angR* and *angT* are included in an iron transport biosynthesis operon (ITBO) with *fatDCBA*, which are involved in the transport of ferric–anguibactin complexes. The adenylation domain of AngR activates L-cysteine and the condensation domain of AngM catalyses the formation of a peptidic bond between DHBA and cysteine. The AngM amino acid sequence shows regions of similarity to the consensus sequence for the peptidyl carrier protein (PCP) and the condensation (C) domains of NRPSs, but curiously, these two domains are not associated with an adenylation (A) domain (Di Lorenzo *et al.*, 2004a). One of the typical features of anguibactin is the presence of a thiazoline ring, which comes from the cyclation of cysteine. This reaction is catalysed by the two cyclation domains of AngN (Di Lorenzo *et al.*, 2008). Then, the dihydroxyphenyl-thiazoline is released from the PCP domain of AngM by nucleophilic attack of N-hydroxyhistamine, resulting in the formation of the final free anguibactin. N-hydroxyhistamine is synthesized by the monooxygenase AngU and AngH, a histidine decarboxylase (Di Lorenzo *et al.*, 2004b; Fig. 7.3).

The genes necessary for DHBA biosynthesis are present and expressed in plasmid pJM1 and also in the chromosome of the anguibactin-producing *V. anguillarum* O1 strain 775. The genes *angC*, *angB* and *angA* encode the enzymes that catalyse the synthesis of DHBA from chorismate, and *angE* encodes a 2,3-dihydroxybenzoate-AMP ligase that has a 60% identity with VibE from *V. cholerae*. Most of these genes have functional homologues in the pJM1 plasmid, except *angA* (coding for a putative 2,3-dihydro-2,3-dihydroxybenzoate dehydrogenase), for which only the chromosomal copy seems to be functional, and that it is essential for DHBA and anguibactin synthesis (Di Lorenzo *et al.*, 2003; Alice *et al.*, 2005).

Ferric anguibactin is transported by the FatABCD proteins, all encoded by the pJM1 plasmid. The genes that encode these proteins are transcribed as a single polycistronic mRNA, together with the biosynthetic genes *angR* and *angT* (ITB operon). FatA is an iron-regulated 86-kDa outer membrane receptor that is dependent on the energy transducer TonB2. Another protein necessary for ferric anguibactin transport is TtpC, which is part of the TonB2 system. FatB is the periplasmic binding protein and FatCD is the inner membrane permease (Di Lorenzo *et al.*, 2004b; Stork *et al.*, 2004; 2007).

Both the biosynthesis and the transport genes are under tight positive as well as negative control. Fur and a plasmid-mediated antisense RNA, RNAβ, act as transcriptional negative regulators, while positive regulation is facilitated by AngR and other factors encoded by the plasmid region named TAF (*trans*-acting factors). AngR is a bifunctional protein that participates in both the regulation and biosynthesis of anguibactin. A higher level of transcription of the ITB operon is observed when a certain non-contiguous region of pJM1 (TAF region) is present in conjunction to AngR. The TAF region was divided into two sub-regions, one involved in biosynthesis (TAFb) that harbours the gene *angB/G*, and the other in regulation (TAFr). The TAFr factor(s) are not very well known but since AngR has a leucine zipper domain it was suggested that AngR could interact with TAFr, and as a complex to activate transcription at the ITBO promoter (Stork *et al.*, 2002; Fig. 7.3).

The siderophore itself plays a positive role in the regulatory mechanism of the expression of both transport and biosynthesis genes. When anguibactin is present, even in absence of TAFr or AngR there is still transcription, although the highest level of transcription is achieved when all three positive regulators are acting synergistically (Stork *et al.*, 2002).

The anguibactin-mediated system seems to be restricted to serotype O1 strains (Conchas *et al.*, 1991) and was never found in any other serotypes. In this regard, Welch and Crosa (2005) found that the LPS O side chain present in serotype O1 strains seems to be necessary for the stability of FatA, the outer membrane receptor for anguibactin. So, it is possible that the anguibactin-mediated iron uptake system works properly only when associated with outer membranes bearing specific characteristics of serotype O1 strains. Interestingly, there are some highly virulent serotype O1 strains that do not harbour pJM1-like plasmids. Iron uptake in these strains are actually mediated by vanchrobactin (Naka *et al.*, 2008).

Vanchrobactin-mediated iron uptake

V. anguillarum strains which lack pJM1-like plasmids (mainly belonging to serotype O2) produce a catechol siderophore named vanchrobactin, which has been identified in strain RV22 as *N*-[*N*′-(2,3-dihydroxybenzoyl)-(D)-arginyl]-(L)-serine (Fig. 7.1) (Soengas *et al.*, 2006, 2007). The siderophore synthesis and transport system is encoded by a ca. 26-kb chromosomal region, containing 12 genes (Balado *et al.*, 2006). The main biosynthetic components are the *vabACBE* genes which specify the enzymes necessary for 2,3-dihydroxybenzoic acid biosynthesis and activation, and a gene (*vabF*) encoding a non-ribosomal peptide synthetase involved in the assembly of the siderophore components. The VabA, VabC, VabE proteins show an identity higher than 97% to AngA, AngC and AngE, respectively, which are encoded in the chromosome of *V. anguillarum* serotype O1 775 strain, and are involved in the synthesis and activation of DHBA, one of the precursors of anguibactin (Alice *et al.*, 2005). The same gene arrangement is observed for the RV22 and 775 homologues, and a putative Fur box predicted within the *vabA–vabC* intergenic region was found to be 100% identical to that reported between the equivalent *angA* and *angC* genes (Balado *et al.*, 2006).

VabF, the non-ribosomal peptide synthetase, shows a seven-domain structure, organized into two modules: the N-terminal module (module 1) follows the C–A–PCP (condensation–adenylation–peptidyl-carrier-protein) domain structure, whereas the C-terminal module (module 2) contains a C–A–PCP–TE (condensation–adenylation–peptidyl-carrier-protein–thioesterase) domain structure. Additional biosynthetic genes encode a DAHP (3-deoxy-D-arabino-heptulosonate-7-phosphate) synthetase (*vabG*), and a phosphopantheteinyl transferase (*vabD*). Although a biosynthetic general model is available (Balado *et al.*, 2006; Fig. 7.4), little is known yet about the secretion of vanchrobactin to the cell environment, and about the utilization of the siderophore after being internalized into the cytoplasm. In

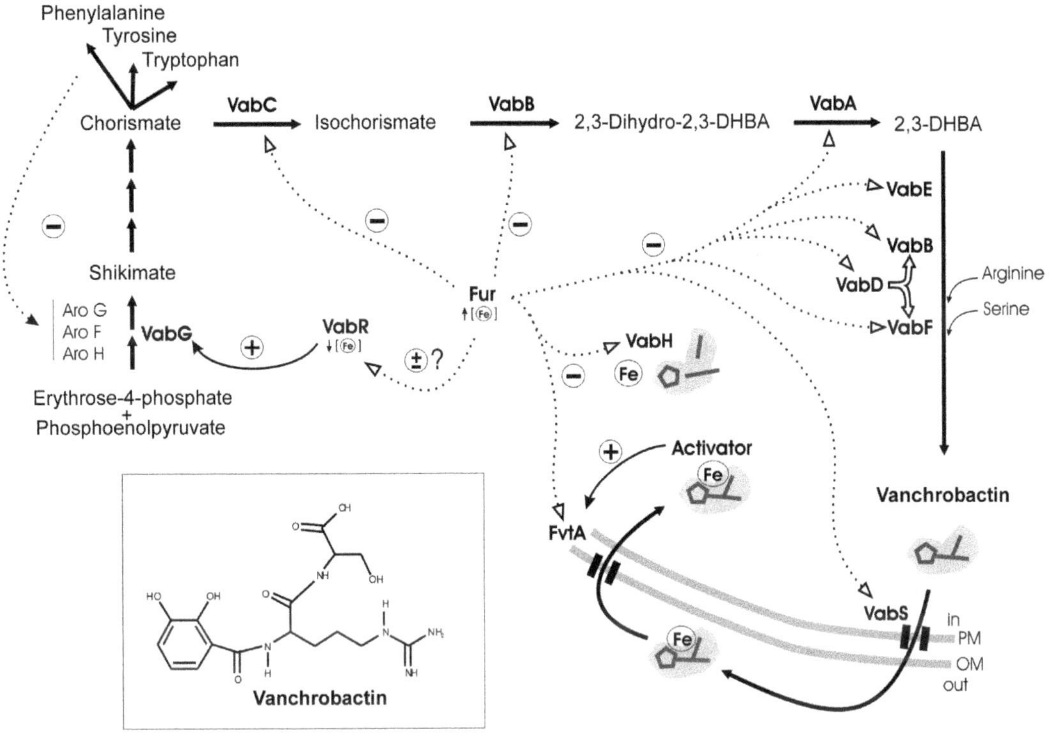

Figure 7.4 A proposed scheme for vanchrobactin biosynthesis and regulation. A circled plus symbol denotes activation and a circled minus symbol indicates repression. The Fur repressor protein, under conditions of high iron, represses transcription of the genes encoding VabCBAEDFHS and FvtA proteins (dashed lines ending in empty arrowheads). A dashed line ending in a filled arrowhead denotes negative regulation of genes, encoding AroGFH proteins, mediated by aromatic amino acids. The empty arrow arising from VabD shows the activation of the VabF peptide carrier protein (PCP) domains and VabB aryl carrier protein (ArCP) domain by VabD. Reproduced from Balado *et al.* (2008), Microbiology *154*, 1401, Copyright Society for General Microbiology (UK).

this regard, other genes that have been described are *vabS,* encoding a putative vanchrobactin exporter and *vabH,* encoding a putative siderophore esterase. It has been demonstrated that mutants lacking any of the biosynthetic genes are impaired for growth under conditions of iron limitation and do not produce vanchrobactin. Siderophore synthesis can be restored after transformation of the mutants with the wild type genes (Balado *et al.*, 2006; Balado *et al.*, 2008). This 12-gene cluster is organized into six iron-regulated transcriptional units. Transcriptional *lacZ* fusions demonstrated that the ferric uptake regulator (Fur) protein is the main iron-responsive regulator of these genes. Interestingly, the *vabG* gene is strongly iron-repressed, but Fur is not essential for this repression. Other regulatory components are a LysR-family transcriptional regulator (VabR) that positively regulates expression of *vabG* and acts as a suggested ferric-vanchrobactin-dependent activator that could play a positive regulatory role in transcription of the *fvtA-vabD* operon, a hypothesis reinforced by the presence of a predicted AraC box upstream of *fvtA* (Balado *et al.*, 2008).

Vanchrobactin transport to the cell is achieved by means of the TonB-dependent outer membrane receptor FvtA. This receptor seems to recognize primarily the catechol–iron centre since it is also the route of entry for several vanchrobactin analogues, in which one or the two amino acids have been changed. Another gene (*orf13*) that putatively encodes an outer membrane receptor was also identified as part of the vanchrobactin cluster, although it does not seem to be expressed (Balado *et al.*, 2009).

It is noteworthy that *fvtA* and most vanchrobactin biosynthesis genes are ubiquitous in both vanchrobactin- and anguibactin-producing *V. anguillarum* strains. In fact, the whole vanchrobactin biosynthesis and transport cluster is present in anguibactin-producing strains (Naka *et al.*, 2008; Balado *et al.*, 2009). In these strains vanchrobactin is not actually being produced since the *vabF* gene, coding for the NRPS that assembles the siderophore molecule, is inactivated by the presence of the insertion sequence RS1 (Naka *et al.*, 2008). When RS1 is removed, vanchrobactin synthesis can be detected in pJM1-bearing strains, although when both siderophores are produced by the same cell, anguibactin seems to have a higher affinity for iron than vanchrobactin, and hence will be the primary siderophore used for iron uptake. Furthermore, *fatDCBA*, the plasmid genes coding for the ferric anguibactin transport in plasmid pJM1-harbouring strains, are also present in the chromosome of plasmidless strains that produce vanchrobactin, although *fatA*, the gene coding for the outer membrane receptor, is inactivated by insertion of transposases (Balado *et al.*, 2009).

Moreover, there is evidence that indicate that most *V. anguillarum* strains (other ones than those belonging to serotype O1 or O2) which do not harbour pJM1-like plasmids synthesize vanchrobactin (M.L. Lemos and C. Jiménez, unpublished). In addition, vanchrobactin production was also observed in several *Vibrio* sp. strains isolated from the environment (M. Sandy, M. Haygood and A. Butler, personal communications).

From the data available, it seems reasonable to speculate that the vanchrobactin-dependent system represents the ancestral siderophore-dependent iron transport pathway within *V. anguillarum*. The anguibactin-mediated system could have been acquired later in the evolution of this species, probably by horizontal transfer, which led to the inactivation of the vanchrobactin system in strains harbouring pJM1-type plasmids, since anguibactin has a stronger affinity for iron than vanchrobactin.

Siderophore-mediated iron uptake in *V. vulnificus*

Vibrio vulnificus is a marine bacterial species which comprises two biotypes. Biotype 1 is an opportunistic human pathogen that has been isolated from clinical and environmental samples. Biotype 2 is mainly an eel pathogen whose importance as an opportunistic pathogen for humans has been also recognized (Amaro and Biosca, 1996). *V. vulnificus* has been reported to produce catechol and hydroxamate siderophores (Biosca *et al.*, 1996). Moreover, aerobactin can be used by *V. vulnificus* strains as an exogenous siderophore, and a gene cluster that mediates aerobactin utilization has been described (Tanabe *et al.*, 2005).

The catecholate siderophore vulnibactin (Fig. 7.1) seems to be the major siderophore in this pathogen. This siderophore is very similar to vibriobactin and consists of one residue of 2,3-dihydroxybenzoic acid as well as two residues of salicylic acid, both of which are involved in the formation of oxazoline rings with L-threonine bound to a norspermidine backbone. The structure was established as *N*-[3-(2,3-dihydroxybenzamido)propyl]-1, 3-bis[2-(2-hydroxy-phenyl)-*trans*-5-methyl-2oxazoline-4-carboxamido]propane (Okujo *et al.*, 1994). Although the genetics of siderophore biosynthesis in this species are not well understood, some genes with a role in vulnibactin-mediated iron acquisition have been identified so far. Thus, *venB* has been identified as encoding an isochorismate lyase that is involved in the synthesis of DHBA. V. *vulnificus* mutants in the *venB* gene are unable to produce catechol siderophores or to acquire iron from transferrin, and show reduced virulence in an infant mouse model (Litwin *et al.*, 1996). The gene *venB* (named *vibB* in Fig. 7.2) is part of a gene cluster that probably contains many of the genetic determinants for vulnibactin production and utilization. New insights into vulnibactin biosynthesis were recently gained after the study of two non-ribosomal peptide synthases encoded by genes *vvsA* and *vvsB* in *V. vulnificus* MO6–24/0 (Kim *et al.*, 2008), which are homologous to genes VVA_1295 and VVA_1297, respectively (Fig. 7.2). These authors demonstrated that mutation of either *vvsA* or *vvsB* caused a decreased production of catecholate siderophores in *V. vulnificus*, these two genes thus constituting the first report of multimodular NRPS genes involved in siderophore biosynthesis in this species. Interestingly enough, the amino acid sequences of VvsA and VvsB are similar to the C-terminus and N-terminus of *V. cholerae* VibF, respectively, which strongly suggests that *vvsA* and *vvsB* could have evolved from a pre-existing single gene.

The outer membrane receptor for ferric vulnibactin (VuuA) has been characterized as a 72-kDa iron-regulated outer membrane protein with 74% identity with the *V. cholerae* vibriobactin receptor, ViuA. Loss of this protein had a severe effect on virulence in a mouse model (Webster and Litwin, 2000).

Siderophore-mediated iron uptake in *V. parahaemolyticus*

V. parahaemolyticus produces the siderophore vibrioferrin (Fig. 7.1), which is a polyhydroxycarboxylate-type siderophore. It is synthesized from L-alanine, citric acid, ethanolamine and 2-ketoglutaric acid. The structure was determined as 1-{2-[2-(5-carboxy-5-hydroxy-2-oxo-1-pyrrolidinyl)propionamide]ethyl} citrate, which exists in two epimeric forms resulting from cyclization between an amidic nitrogen of the alanine residue and a keto group of the 2-ketoglutaric acid residue (Yamamoto *et al.*, 1994).

A gene cluster involved in the biosynthesis and transport of vibrioferrin has been identified in the genomic regions surrounding the first identified *pvuA* gene coding for the 78-kDa outer membrane protein that serves as the ferric vibrioferrin receptor (Funahashi *et al.*, 2002) (Fig. 7.2). Five of the genes, named *pvsABCDE,* constitute an operon that is expressed under iron-limiting conditions. Genes *pvsABDE* are implicated in the biosynthesis of the siderophore, and *pvsC* encodes an inner membrane exporter that is homologous to members of the major facilitator superfamily of multidrug efflux pumps. The genes named *pvuBCDE,* encode proteins strongly similar to *E. coli* FecBCDE, respectively, which are components of the ATP-binding cassette transporter system for ferric dicitrate. These transport genes are transcribed as a single mRNA with the upstream genes, *psuA* (a gene that encodes a putative receptor for an unknown siderophore) and *pvuA* (Tanabe *et al.*, 2003; Tanabe *et al.*, 2006).

Aerobactin, a citrate-based dihydroxamate siderophore, which is highly prevalent in species of the family *Enterobacteriaceae,* can also be efficiently used as an exogenous siderophore by *V. parahaemolyticus* (Funahashi *et al.*, 2003). The *V. parahaemolyticus* IutA protein shares 43% identity with IutA, the TonB-dependent outer membrane receptor for ferric aerobactin of *Escherichia coli.* The *iutA* gene from *V. parahaemolyticus* is transcribed in response to low-iron availability from a putative promoter overlapped with a sequence resembling a consensus *E. coli* Fur-binding sequence. The *V. parahaemolyticus iutA* homologue encodes an outer-membrane protein which functions as the receptor for ferric aerobactin since insertional inactivation of *iutA* impaired growth in the presence of aerobactin and incapacitated the outer-membrane fraction from iron-deficient cells for binding ^{55}Fe-labelled aerobactin. The *iutA* homologues seem to be widely distributed in clinical and environmental isolates of *V. parahaemolyticus* (Funahashi *et al.*, 2003).

Ferrichrome can also be used as an exogenous siderophore by *V. parahaemolyticus,* and this utilization requires the outer membrane receptor FhuA and the ABC transporter FhuCDB. The *fhuACDB* genes form an operon present in all clinical and environmental strains of *V. parahaemolyticus* tested. The ABC transport components FhuCDB are also required for the aerobactin transport to the cytosol, after being transported to the periplasm by the outer membrane receptor IutA (Funahashi *et al.*, 2009).

Iron uptake mediated by siderophores in other vibrios

Two compounds with siderophore activity were purified from cultures of *Vibrio fluvialis.* Norspermidine and 2,3-dihydroxybenzoic acid were identified as constituents common to both compounds, and L-threonine was identified in the hydrolysate of one compound, named fluvibactin. The structure of fluvibactin was established as N4-[2-(2,3-dihydroxy-phenyl)-*trans*-5-methyl-2-oxazoline-4-yl]carboxy-N1,N7-bis-(2,3-dihydroxybenzoyl)-norspermidine, and that of the other compound as N1,N7-bis-(2,3-dihydroxybenzoyl) norspermidine. Both of the purified compounds restored growth inhibition of the producer strain and also of *V. cholerae* non-O1 (Yamamoto *et al.*, 1993).

The presence of a gene cluster highly similar to the vibrioferrin biosynthesis and transport gene cluster from *V. parahaemolyticus,* was demonstrated in *V. alginolyticus* (Wang *et al.*, 2007), a pathogen for fish and molluscs. The 11-gene cluster consists of two divergently transcribed, Fur-regulated operons, *pvsABCDE* and *psuA-pvuABCDE,* sharing the same structural organization with the vibrioferrin locus (Fig. 7.2). The detection of this gene cluster in *V. alginolyticus* is reinforced by the fact that this species produced vibrioferrin and expressed an iron-repressible 78-kDa outer membrane protein, which strongly cross-reacted with the antiserum against PvuA, the ferric vibrioferrin receptor of *V. parahaemolyticus*. The cluster was not found in any other *Vibrio* species tested (Funahashi *et al.*, 2002).

Hydroxamate siderophores were purified from low-iron cultures of *V. hollisae* and *V. mimicus*. Both purified siderophores were identified as aerobactin. In response to iron limitation, strains from these species expressed two high molecular mass outer membrane proteins. The protein with an apparent molecular mass of 77 kDa, which was common to all strains examined, was suggested as the ferric aerobactin receptor (Okujo and Yamamoto, 1994).

Siderophore-mediated iron uptake in the genus *Aeromonas*

Within the species of aeromonads, *A. salmonicida* and *A. hydrophila* are the ones that have drawn more attention among microbiologists studying iron acquisition mechanisms. Thus, most of the current knowledge on iron uptake in the genus *Aeromonas* comes from studies with these two species and, predominantly, with *A. salmonicida* subsp. *salmonicida*.

Iron uptake mediated by siderophores in *A. hydrophila*

Most strains of *A. hydrophila* produce and use the catechol siderophore amonabactin, although some strains have been found to produce either enterobactin or yet uncharacterized phenolates (Barghouthi *et al.*, 1989; Massad *et al.*, 1991). Amonabactin can also be produced by other species of the genus *Aeromonas*, as *A. sobria* and *A. caviae* (Zywno *et al.*, 1992). Amonabactin is in fact a group of four different peptide-based bis-catecholate siderophores compounds, collectively known as the amonabactins. These siderophores are biscatecholates with the backbone composed of either tri or tetrapeptides in the sequence (gly)-(L)-lys-(L)-lys-(D)-aro, where glycine is the optional amino acid attached to the Nε of the N terminus lysine and aro is either tryptophan (amonabactin T) or phenylalanine (amonabactin P) (Fig. 7.1). The ligand units, 2,3-dihydroxybenzamide groups, are attached to the Nε amine of the C terminus lysine and either to glycine, if present, or to Nε amines of the N terminus lysine (Telford *et al.*, 1997). All four compounds are biologically active and all can supply iron to *A. hydrophila*. Addition of L-tryptophan to *A. hydrophila* cultures cause exclusive synthesis of amonabactin T, whereas addition of L-phenylalanine yields predominant production of amonabactin P (Barghouthi *et al.*, 1989).

Although very little is known about the genetic basis of amonabactin biosynthesis and transport, the completion of the *A. hydrophila* ATCC 7966^{T} genome sequence (Seshadri *et al.*, 2006) revealed the existence of a seven-gene cluster (*AHA2473–2479* in the genome sequence) that might contain the genetic determinants for biosynthesis and utilization of amonabactin. Of these seven genes, three appear to encode non-ribosomal peptide synthetases, and four genes encode enzymes with putative roles in the synthesis of 2,3-dihydroxybenzoic acid. One of these four genes, *amoA*, encodes a predicted isochorismate synthetase, and its role in amonabactin biosynthesis was demonstrated, its inactivation rendering *A. hydrophila* unable to produce amonabactin (Barghouthi *et al.*, 1991). Interestingly, the genome of *A. hydrophila* harbours additional genes related to siderophore biosynthesis and transport. Genes *AHA1964* and *AHA1966–1969* might encode membrane transporters involved in amonabactin uptake, although no experimental evidence on their actual function has been reported so far. Interestingly, genes annotated as *AHA3281* and *–3282* show similarity to *pvcAB* genes for siderophore pyoverdine synthesis in *Pseudomonas aeruginosa* although the role of these genes for the synthesis of pyoverdine has been put in doubt (Mossialos *et al.*, 2002).

It is known that amonabactin plays a key role in *A. hydrophila* utilization of host iron sources. It was early evidenced that *A. hydrophila* strains that produce amonabactin could grow in heat-inactivated serum and acquire iron from Fe-transferrin, whereas enterobactin-producing *A. hydrophila* strains could not (Massad *et al.*, 1991). Later, it was demonstrated that iron acquisition from transferrin and lactoferrin by *A. hydrophila* is indeed mediated by amonabactin (Stintzi and Raymond, 2000).

Iron uptake mediated by siderophores in *A. salmonicida*

Early studies by Chart and Trust (1983) evidenced that *A. salmonicida* subsp. *salmonicida* had the ability to sequester Fe^{3+} from the high-affinity iron chelators ethylenediamine-di-(*o*-hydroxyphenyl acetic acid) (EDDA), lactoferrin and transferrin, and that this species produces a low-molecular weight soluble siderophore, although nothing was then known about its chemical structure or the genetic basis for its synthesis and transport. Subsequent studies preliminarily characterized the siderophore as a catecholate (Hirst *et al.*, 1991). No production of siderophores was found in atypical *A. salmonicida* strains tested in those same studies, although most were able to grow in presence of high-affinity iron chelators which suggested that atypical strains utilized a siderophore-independent mechanism for iron acquisition.

It is not clear yet which siderophore(s) are produced and utilized by *A. salmonicida*, although all relevant published work report the production of a catechol-type siderophore (Chart and Trust, 1983; Hirst *et al.*, 1991; Fernandez *et al.*, 1998; Najimi *et al.*, 2008b). Hirst *et al.* (1991) suggested that most probably this siderophore was not amonabactin, since culture supernatants of an amonabactin-producing *A. hydrophila* strain failed to promote the growth of the *A. salmonicida* strains. Similarly, the strains tested by these authors did not utilize anguibactin or enterobactin, and the chromatographic properties of the *A. salmonicida* siderophore as analysed by thin-layer chromatography showed it to be different from those of anguibactin and amonabactin. Although amonabactin production was reported to occur in most of the DNA hybridization groups within the genus *Aeromonas* (Zywno *et al.*, 1992), to date amonabactin production by *A. salmonicida* subsp. *salmonicida* has not been demonstrated.

An analysis of the recently finished genome sequence of *A. salmonicida* A449 (Reith *et al.*, 2008) shows the presence of two gene clusters that could be involved in the biosynthesis and utilization of two different siderophores. A proteomic study of this same strain had previously reported the existence of two putative outer membrane ferric siderophore receptor proteins, FstC and FstB, expressed under iron-limiting conditions (Ebanks *et al.*, 2004). These two putative siderophore receptor genes are located in two different clusters on the A449 chromosome.

The first cluster (genes *ASA1838–1851*) includes the FstC receptor, as well as an homologue of *amoA*, an *A. hydrophila* gene encoding an isochorismate synthetase with a demonstrated role in the biosynthesis of siderophore amonabactin (Barghouthi *et al.*, 1991). This cluster altogether comprises three non-ribosomal peptide synthetase genes, a phosphopantetheinyl transferase gene (*entD*), four genes involved in the synthesis of 2,3-dihydroxybenzoic acid, one outer membrane receptor gene (*fstC*) and five genes encoding predicted ABC-type ferric transporters. This cluster might be responsible for the biosynthesis and utilization of amonabactin, or of an amonabactin-like siderophore (Fig. 7.2).

Interestingly, there is a second cluster (genes *ASA4368–4380*) in the chromosome of A449 that includes the gene for the FstB receptor, as well as 12 other genes encoding predicted siderophore biosynthesis and transport proteins with similarity to the *V. anguillarum* anguibactin and the *Acinetobacter baumannii* acinetobactin biosynthesis genes. A homologue of FstB, which was dubbed AsbE, was recently identified in *A. salmonicida* by a Fur titration assay (Najimi *et al.*, 2009). This second cluster is a candidate to encode the synthesis of an anguibactin-like catecholate siderophore, and has been preliminarily characterized in our laboratory in *A. salmonicida* subsp. *salmonicida* strain RSP74.1 isolated from turbot (Fig. 7.2) (Najimi *et al.*, 2008b; Najimi *et al.*, 2009).

Strain RSP74.1 produces a catechol-type siderophore and harbours a gene cluster identical to the second cluster reported in the A449 genome (genes *ASA4368–4380*). Interestingly, single-gene mutations within this cluster abolish siderophore biosynthesis, which suggests that this strain produces only one type of siderophore (Najimi *et al.*, 2008b). The genes encoded by loci *ASA4375–4380* were described as biosynthetic genes and named *asbIBCDFG* (Fig. 7.2). The *asb* cluster is organized in four transcriptional units: *asbG-asbF*, *asbD-asbC*, *asbB* and *asbI*, as confirmed by RT-PCR analyses. In addition, analysis of the expression of RNA transcripts showed that the four transcripts are weakly expressed under iron-rich conditions, and strongly expressed under iron limitation.

The *A. salmonicida* RSP74.1 siderophore might contain histamine

The proteins encoded by *asbG* and *asbC* show similarity to enzymes involved in the synthesis of siderophores containing histamine or its derivatives in their structures. Both genes are essential for growth of *A. salmonicida* RSP74.1 under iron limitation.

On the one hand, AsbG is a predicted histidine decarboxylase that catalyses decarboxylation of histidine to histamine. A Δ*asbG* mutant shows a clear decrease in siderophore production and is impaired for growth under iron-restricted conditions, whereas the levels of DHBA production remain at values similar to the wild type. However, addition of histamine complements the deficiency of a Δ*asbG* mutant for growth under iron limitation, which is consistent with histamine being the final product of the reaction catalysed by histidine decarboxylases. Thus, it is proposed that the product of *asbG* plays a crucial role in the biosynthesis of the *A. salmonicida* siderophore, and that histamine is a precursor in its biosynthesis. Other siderophores that contain histamine-derived moieties in their structures are anguibactin from *V. anguillarum*, pseudomonine from *Pseudomonas fluorescens* AH2, and acinetobactin from *Acinetobacter baumannii*. It is interesting to note that the proteins involved in histidine decarboxylation in these species also show similarity to *A. salmonicida* AsbG.

On the other hand, AsbC is a predicted histamine monooxygenase that shows high identity to *A. baumannii* BasC (70%) and *V. anguillarum* AngU (63%). Deletion of *asbC* causes impairment in the ability of *A. salmonicida* RSP74.1 to grow under iron-restricted conditions, but the addition of histamine to the culture medium does not enhance growth of the mutant. Chemical tests show that siderophore, but not DHBA production, is significantly reduced (more than a two-fold decrease) in an *asbC* mutant. Thus, AsbC is essential for growth of *A. salmonicida* RSP74.1 under iron-limiting conditions, is part of the siderophore biosynthesis pathway, and it is likely that it plays a role in yielding *N*-hydroxyhistamine or a derivative.

The *A. salmonicida* RSP74.1 catecholate siderophore is synthesized by a NRPS mechanism

AsbF is homologous to isochorismatases and 2,3-dihydroxybenzoate-AMP ligases, enzymes involved in the synthesis and/or activation of DHBA, a constituent of catecholate siderophores. However, the biosynthesis of DHBA from chorismate in bacteria requires the products of homologues of the *E. coli entC*, *entB* and *entA* genes, which are usually clustered together. It is interesting to note that although there are no *entC* and *entA* homologues in this cluster, these genes occur in the other siderophore cluster of *A. salmonicida*, the one that encodes a predicted amonabactin-like siderophore (Fig. 7.2). It is tempting to speculate that some of the functions encoded by the *ent* genes could be shared by the two siderophore clusters, although this hypothesis needs experimental demonstration.

A domain-prediction analysis shows that AsbB has a typical structure of non-ribosomal peptide synthetase, it contains a PCP domain for the attachment of a 4′-phosphopantetheine prosthetic group and also contains a condensation domain. Both the amino acid sequence and predicted domain structures of AsbB show similarity to the C-terminus of *V. cholerae* VibF. AsbB also shows similarity to *V. anguillarum* AngM and *A. baumannii* BasB.

AsbD is a predicted protein of 1459 amino acids with a calculated molecular mass of 160 kDa, and contains typical domains of non-ribosomal peptide synthetases involved in siderophore biosynthesis in bacteria: two condensation domains, and one adenylation domain (AMP-binding domain), and is similar to *V. cholerae* VibF and *A. baumannii* BasD. Mutation of *asbD* causes a severe impairment for growth under iron restricted conditions. Mutation of *asbD* would be expected to abolish the assembly of the complete siderophore molecule, but not the synthesis of DHBA, since domain analyses using the Pfam database do not predict domains for DHBA synthesis nor activation in AsbD. Indeed, analysis of the supernatants shows that its mutation significantly reduces siderophore production but not the production of DHBA.

It is interesting to note the sequence identity of *A. salmonicida* AsbD and AsbB with *A. baumannii* BasD and BasB respectively, and with *V. anguillarum* AngN and AngM. It is tempting to speculate that the siderophore produced by *A. salmonicida* RSP74.1 could also be structurally and functionally related to anguibactin and/or acinetobactin, although additional work is needed in order to support this hypothesis.

The catecholate siderophore produced by *A. salmonicida* RSP74.1 is necessary for the utilization of transferrin-bound iron. While the parental strain can grow at acceptable levels in minimal medium supplemented with apotransferrin (where most of the iron is expected to be bound to the protein), single mutations in *asbG, asbC* or *asbD* all cause impairment for growth under these conditions. It is noteworthy that the addition of histamine to the *asbG* mutant partially enhances growth in presence of transferrin. This suggests that *A. salmonicida* uses the catecholate siderophore for removal of iron from transferrin rather than relying on a receptor for this iron-binding protein.

One important question that remains to be answered satisfactorily is whether *A. salmonicida* subsp. *salmonicida* strains produce predominantly a unique type of siderophore or whether there is a certain degree of strain specificity in production of different molecular types. In this regard, studies carried out with cross-feeding assays yielded varied results. Hirst *et al.* (1991) found that the growth of *A. salmonicida* subsp. *salmonicida* strains was stimulated only by homologous iron-restricted supernatants, which suggested strain differences in the siderophore produced. However, Fernández *et al.* (1998) analysed 17 *A. salmonicida* strains isolated from salmonids in Spain and Scotland, and found that all of them produced a catecholate siderophore. Moreover, culture supernatants of those strains stimulated growth of all the homologous and most of the heterologous isolates, and those observations led to the proposal that *A. salmonicida* appears to be a homogeneous species with respect to siderophore production. Recently, we found that 18 *A. salmonicida* strains isolated from Spain and Portugal, all could cross-feed mutants for genes *asbD, asbG* and *asbC* under conditions of iron limitation (Najimi *et al.*, 2008b). This observation suggested that all the *A. salmonicida* strains used in that study produced a siderophore that could be efficiently used by the mutant strains. In addition, it was found that all those 18 strains harboured genes *asbD* and *asbE* (homologue of *fstB*) (Najimi *et al.*, 2008b; 2009), a result that points to a high degree of gene conservation for the anguibactin-like cluster in *A. salmonicida.*

Haem as iron source for *Vibrio* and *Aeromonas*

Haem is constituted by a protoporphyrin IX molecule with an iron atom. Formally, the term haem refers to the reduced form of the molecule (Fe^{2+}), the oxidized form (Fe^{3+}) being known as haemin, although 'haem' is widely used to refer to iron protoporphyrin IX in either oxidation state. The haem molecule is one of the most important enzyme cofactors, mediating oxygen transport and energy generation, among other functions. In fact, bacteria employ up to 10% of the total cell iron content to the biosynthesis of haem (Beale, 1996). However, haem has propensity to oxidize cellular constituents, and thus, in vertebrate hosts intracellular haem is mainly associated with haemoproteins, such as the abundant oxygen carriers haemoglobin and myoglobin, or haemopexin.

Haem and haem-containing proteins are a potential source of iron for pathogenic microorganisms. Usually, free haem is not encountered *in vivo* and thus it is expected that bacteria acquire haem mainly from haemoproteins (see Chapter 2). Thus, one of the main siderophore-independent mechanisms present in Gram-negative bacterial pathogens to obtain iron is the utilization of haem or haem-containing proteins from host tissues (Wandersman and Delepelaire, 2004). In this regard, *Vibrio* and *Aeromonas* species are not an exception. In fact, much of the current knowledge on haem uptake systems comes from work done with *Vibrio* species. Among *Vibrionaceae*, different species containing strains able to utilize haem iron have been studied so far. These include *Vibrio cholerae, V. anguillarum, V. vulnificus, V. parahaemolyticus, V. mimicus, V. alginolyticus, V. fluvialis*, and *Photobacterium damselae*. Haem uptake has been also demonstrated in two *Aeromonas* species: *A. hydrophila* and *A. salmonicida*.

Haem transport mechanism in vibrios: haem receptors and TonB

The mechanism of haem iron utilization in Gram-negative bacteria is believed to follow a very conserved general model, the differences between species being just a matter of gene organization rather than differences in the utilization mechanism. Haem or haemoproteins are recognized by iron regulated outer membrane protein receptors, and the haem moiety is unloaded and transported across the outer membrane. The haem receptors must be able to extract haem from different carrier proteins prior to the transport step.

Biochemical techniques, such as affinity chromatography and ligand binding have been used to identify haem-binding molecules. Genetic studies have also demonstrated the requirement in *Vibrionaceae* species for cell-surface proteins that serve to transport haem into the bacterial periplasm. Outer membrane proteins which function as haem receptors/transporters have been characterized in *V. cholerae* (HutA, HutR and HasR), *V. vulnificus* (HupA), *V. anguillarum* (HuvA and HuvS), and *V. mimicus* (MhuA). Protein sequence database searches indicate that similar receptors are also present in *V. alginolyticus*, *V. parahaemolyticus*, *V. splendidus*, *V. fischeri*, *V. harveyi* or *V. campbellii*. The molecular weights of these proteins range from 77.2 to 80.8 kDa, which indicates a conserved protein size of haem receptors in *Vibrionaceae*.

From the amino acid sequence alignment of four of these receptors, it is evident that some protein stretches are highly conserved. Sequence identity is high between *V. cholerae* HutA, *V. vulnificus* HupA and *V. anguillarum* HuvA (more than 50% sequence identity), whereas *V. mimicus* haem receptor MhuA (34% identity to HutA) and HuvS (38% identity to HuvA) show higher divergence. The overall similarity suggests that all these haem receptors share a common ancestor. Furthermore, these receptors share a very well conserved TonB-box, located near the amino terminus of the protein (consensus sequence: DEVVVSA, positions 28–34 of the mature protein), where the outer membrane receptors physically interact with the TonB protein. This sequence shares four out of seven residues with the TonB box consensus sequence (DTLVVTA) of the *E. coli* TonB-dependent outer membrane receptors (Postle and Larsen, 2007).

How the haem receptor interacts with the haem–protein complex to release the haem ligand is a process not yet fully understood. In *V. cholerae*, the HutA receptor is crucial for the ability of the bacterium to utilize haemoglobin as a haem source. Inactivation of HutA significantly reduces the ability of *V. cholerae* to use haemoglobin, although it can still use haemin, suggesting a direct interaction between HutA and haemoglobin (Henderson and Payne, 1994).

HutA is not the only haem receptor of *V. cholerae*. This species has at least two additional haem receptors, HutR and HasR, coexisting in the same cell (Mey and Payne, 2001). The mature HutR protein has a predicted molecular mass of 78.5 kDa, and the *hutR* gene is believed to be iron regulated. There is evidence that the HutA–TonB1 pair plays a crucial role in the ability of *V. cholerae* to utilize haemoglobin as an iron source, whereas HutR would function mainly in the utilization of haemin. Also, haem uptake through HasR requires TonB2. A *hutA*, *hutR* double mutant is significantly impaired in haemin utilization compared to wild-type strains, but can still form small colonies with haemin as sole iron source. However, a *hutA*, *hutR*, *hasR* triple mutant is completely deficient in haemin utilization. A single *hutR* mutant is not significantly affected in its ability to grow using haemin as sole iron source, while a *hutA* mutant has a significant defect in haemin utilization. It is possible that, though the three haem receptors of *V. cholerae* are efficient, HutA is required for optimal haem utilization. In fact, levels of HutA in the outer membranes of wild-type cells grown in low iron conditions are much higher than those of HutR, perhaps indicating a higher expression of *hutA* gene in iron-limiting conditions.

V. anguillarum possesses at least two different outer membrane haem receptors, HuvA and HuvS. Both receptors are functionally interchangeable but they do not coexist in the same cell. The utilization of one or the other seems to depend only of the strain (Mouriño *et al.*, 2005). HuvS shows significant similarity to other outer membrane haem receptors described in *Vibrionaceae*, but little similarity (39%) to HuvA. This haem receptor was present in 9 out of 11 of the *V. anguillarum* strains that tested negative for HuvA, which indicates that still other haem receptors may exist. The HuvS and HuvA sequences alignment, as well as the analysis of their respective upstream and downstream DNA sequences, suggest that horizontal transfer and recombination might be responsible for generating this genetic diversity (Mouriño *et al.*, 2005).

Several studies report that lipopolysaccharides of Gram-negative bacteria may also show haemin and/or haemoglobin binding activity (Belanger *et al.*, 1995; Grenier *et al.*, 1997; do Vale *et al.*, 2002). These interactions may facilitate further haem transport into the bacterial cell through the specific protein receptors.

Following a specific recognition, haem is transported across the outer membrane into the periplasmic space by a TonB-dependent process. However, it is not clear whether haem-binding proteins are always the actual haem transporter. This means that bacteria may have surface molecules with haem or haemoglobin-binding activity, but without transport functions. It is feasible that receptors comprising two polypeptide components exist, where one component binds the haem which is then transferred to the second component. In this regard, for some species the functional role of a single polypeptide component in haem uptake has been established, as is the case for the *V. cholerae* HutA receptor. However, in *V. anguillarum* it has been found that a *huvA* (which shows high similarity to *hutA*) mutant is unable to use haemin as sole iron source, but still retains the ability to bind it, thus suggesting that proteins with no haem transport function have haem binding activity (Mazoy *et al.*, 2003).

In several species, both the haemin and haemoglobin binding activity via protein receptors is constitutive (not iron regulated) since corresponding binding properties are exhibited at the same extent either in iron-restricted or in iron-enriched conditions. This is the case for species as *V. vulnificus* (Fouz *et al.*, 1996), *Photobacterium damselae* subsp. *damselae* (formerly *Vibrio damsela*) (Fouz *et al.*, 1994), and *V. anguillarum* (Mazoy and Lemos, 1996b). By contrast, in *V. cholerae* (Henderson and Payne, 1994) and *V. parahaemolyticus* (Yamamoto *et al.*, 1995), the haemin- or haemoglobin-binding activities are mediated by iron-regulated surface-accessible binding proteins.

One interesting question is whether *Vibrionaceae* species remove iron from the haem molecule and then transport only the iron into the cell, or if they transport the entire haem molecule into the cytoplasm. In this regard, results are available for *V. cholerae*, where it has been demonstrated that the entire haem molecule is transported into the bacterial cell (Henderson and Payne, 1993).

Haem transport is TonB dependent

Transport of haem across the outer membrane is an active process that uses energy of the cytoplasmic membrane to bring haem into the periplasm, and this energy is provided by the TonB system (Stojiljkovic and Hantke, 1992; Henderson and Payne, 1994). The TonB system consists of the cytoplasmic membrane proteins ExbB and ExbD, and the cytoplasmic membrane anchored protein TonB, that together are responsible for the transduction of the cytoplasmic membrane proton-motive force to the receptors. A conserved region in the outer membrane receptors known as the *TonB box*, is responsible for the interaction with TonB (Heller *et al.*, 1988). Interaction of TonB with outer-membrane receptors is believed to induce a conformational change of the receptors, resulting in the transport of bound ligands into the periplasmic space (Skare *et al.*, 1993).

All *Vibrio* species studied contain at least two sets of TonB systems known as TonB1 and TonB2. These two TonB systems are encoded by genes located in different chromosomes (TonB1 is located in chromosome II and TonB2 in chromosome I), at least in *V. cholerae*, *V. anguillarum* and *V. alginolyticus*. In these species, the TonB1 system is involved in haem transport and is part of the haem uptake cluster (Fig. 7.5) (Occhino *et al.*, 1998; Mouriño *et al.*, 2004; Wang *et al.*, 2008). The TonB2 system seems to be needed mainly for siderophore transport although it can be also used for haem transport. In fact, only the *tonB1 tonB2* double mutants are unable to use haem as a source of iron in *V. cholerae* and *V. anguillarum*. The presence of TonB along with the accessory proteins ExbB and ExbD is conserved across many Gram-negative species. In *V. anguillarum* however, a fourth protein, TtpC, which is encoded in the *tonB2* gene cluster, was recently discovered to be necessary for ferric-anguibactin transport mediated by the TonB2 system (Stork *et al.*, 2007). In addition, the TtpC homologue in *V. cholerae* is essential for enterobactin transport mediated by the *V. cholerae* TonB2 system. Although *V. anguillarum* has two TonB energy transduction systems, the TtpC protein is specific for the TonB2 system and not necessary

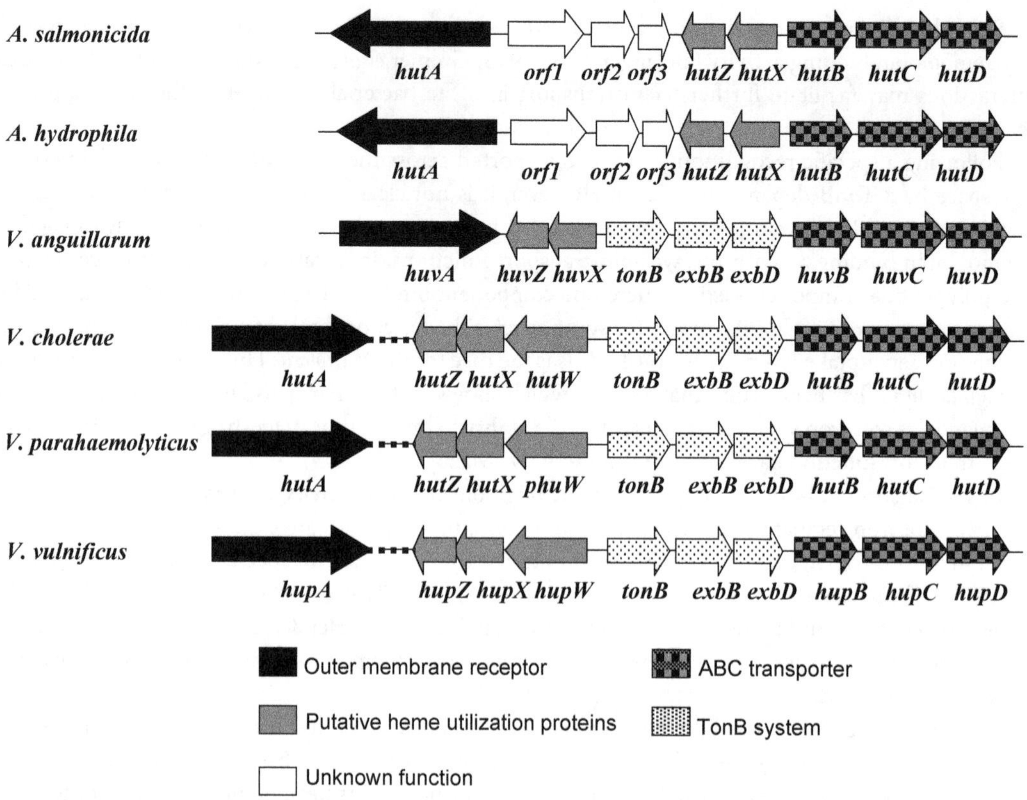

Figure 7.5 A scheme of the haem uptake gene clusters present in *A. salmonicida*, *A. hydrophila*, *V. cholerae*, *V. anguillarum*, *V. vulnificus* and *V. parahaemolyticus.* Function of each gene is indicated in the key.

for iron transport mediated by the TonB1 system. The same observation has been reported also for *V. alginolyticus* (Wang *et al.*, 2008).

Once haem enters the periplasmic space, a periplasmic binding protein is responsible for delivering it to a permease complex, which is located in the cytoplasmic membrane. Transport of haem into the cytoplasm is also ATP dependent. The genes involved in this process in *Vibrionaceae* were first described in *V. cholerae*. The *V. cholerae hutBCD* genes (Fig. 7.5) conform a specific transport system for haem, and it seems that they are not required for transport of other iron-containing molecules into the cytoplasm. The HutB protein is the periplasmic haemin-binding protein, whereas HutC and HutD form an ABC-transporter where HutC is the permease and HutD the ATP-binding protein responsible for the energy supply (Henderson and Payne, 1993; Occhino *et al.*, 1998). Homologous systems were also described in *V. anguillarum*, *V. parahaemolyticus* and *V. alginolyticus* (O'Malley *et al.*, 1999; Mouriño *et al.*, 2004).

Fate of haem in the cytoplasm

The fate of haem after entering the bacterial cytoplasm is not fully understood. Although some Gram-negative bacteria use haem oxygenase-like enzymes to acquire iron from haem (Zhu *et al.*, 2000b; Ratliff *et al.*, 2001), no haem oxygenases have been described to date in *Vibrionaceae* species. We can suppose that haem is degraded in bacterial cells in order to provide iron, since more than 90% of total cell iron is not associated with haem. However, it is not clear yet how haem is broken down. Some insights in this regard have been gained from studies with pathogenic neisseriae and from the Gram-positive

Corynebacterium diphtheriae (Schmitt, 1997; Zhu *et al.*, 2000a,b). These species encode haem oxygenases that may break down the haem and thus release the iron, and these enzymes have shown to be required for haem iron utilization in the two organisms. To date it is not known whether any *Vibrionaceae* uses a similar mechanism to remove the iron from haem. It was reported that the *V. cholerae* HutZ protein binds haem with high efficiency, and might act as a haem storage protein and as a shuttle protein that increases the efficiency of haem trafficking from the membrane to haem-containing proteins (Wyckoff *et al.*, 2004). Less is known about the role of HuvX and its homologues. Deletion of *huvX* did not cause a significant decrease in the ability of *V. anguillarum* to utilize haemin and haemoglobin as the unique sources of iron. Thus, it was speculated that it could have a regulatory role on transcription of the haem uptake cluster genes. However, the transcriptional activity of the *huvA*, *tonB* and *huvX* promoters did not show significant differences between a parental strain and a *huvX* mutant (Mouriño *et al.*, 2006).

Haem uptake in *A. salmonicida*

The first evidence that *A. salmonicida* subsp. *salmonicida* strains are able to internalize haem came from the studies of Ishiguro *et al.* (1986), who demonstrated that certain atypical *A. salmonicida* strains require haem as a porphyrin source for growth in laboratory media. However, these studies did not test the use of haem as a source of iron and, similarly, nothing was known about haem uptake in typical *A. salmonicida* subsp. *salmonicida* strains. Later, a study using a proteomic approach demonstrated that under iron-limited conditions *A. salmonicida* expresses three iron-regulated outer membrane receptors, and one of these receptors was proposed to be a haem receptor based on sequence similarity (Ebanks *et al.*, 2004).

Recently, in our laboratory we have demonstrated that *A. salmonicida* strains isolated from turbot utilize haem compounds as the sole iron source, and a nine-gene cluster (Fig. 7.5) containing some of the genes necessary for haem uptake was identified (Najimi *et al.*, 2008a). Six of the nine genes, *hutAZXBCD*, share significant degrees of similarity with haem transport and utilization genes in other bacteria. HutA is the *A. salmonicida* putative haem receptor, previously identified in a proteome analysis (Ebanks *et al.*, 2004). The roles of HutZ and HutX are not clear, neither are the roles of their homologues in other species. HutB is homologous to periplasmic haem-binding proteins, HutC has homology to ABC-type permease proteins, and HutD shows homology to the ATP-binding protein component of ABC transporters. It is known that deletion of *hutB* severely affects the ability to use haemin as the unique source of iron in *A. salmonicida*. A *hutB* mutant shows more than a fivefold reduction in growth levels after 24 h compared with the parental strain, when grown in an iron restricted medium plus haemin (Najimi *et al.*, 2008a). Thus, *hutB* is necessary, although probably not totally essential, for growth with haemin as sole iron source, and the residual growth of the *hutB* mutant could be attributed to other periplasmic iron-transporters encoded in the *A. salmonicida* genome.

Two Fur binding sites (Fur boxes) have been identified upstream of *hutA* and in the intergenic region between *hutX* and *hutB*, respectively. This cluster is organized into five transcriptional units, three are polycistronic (*orf2–orf3*, *hutX–hutZ* and *hutBCD*) and two are monocistronic (*hutA* and *orf1*), and all five would be expected to be transcribed from their own promoters. The spatial organization of the *A. salmonicida* haem uptake cluster shows some differences with respect to haem clusters in *Vibrio* species (Fig. 7.5). The gene encoding the outer membrane haem receptor is linked to the rest of the haem transport genes in *A. salmonicida* and *A. hydrophila*, a characteristic shared with *V. anguillarum*. However, in other *Vibrio* species the outer membrane receptor is instead located in another chromosomal location. In addition, the *A. salmonicida* cluster lacks a gene that codes for a putative coproporphyrinogen oxidase (HutW in *V. cholerae*). Another difference is the occurrence in *A. salmonicida* and *A. hydrophila* haem clusters of three genes (*orf1*, *orf2*, *orf3*; encoding a hypothetical iron regulated-protein, an isochorismatase-family protein and a putative lipoprotein, respectively, whose role in haem utilization is so far unknown) which are absent from the haem clusters of related Gram-negative bacteria. It is also noteworthy that no TonB system genes are linked to the haem uptake genes in *A. salmonicida*.

Multiplicity of outer membrane haem receptors in *A. salmonicida*

Although mutation of *hutB* causes a significant decrease in the ability to use haem as sole iron source, it is noteworthy that deletion of *hutA*, encoding a predicted outer membrane haem receptor, does not. A Δ*hutA* strain grown with 10 μM haemin in minimal medium with 2,2′-dipyridyl, shows initially some defect in its ability to grow during the first 12 h, but recovers growth levels similar to the parental strain after 24 h. Deletion of *hutA* does not completely abolish haem utilization, suggesting the existence of another haem receptor.

An *in silico* search of the genome sequence of *A. salmonicida* A449 shows that one of the ORFs (locus *ASA2695*) is annotated as a haemin receptor. Interestingly, this ORF encodes a predicted protein that shares 30% identity with the sequence of *A. salmonicida* HutA, and is 37% identical to *V. cholerae* HutR. This gene, encoding a 721 amino acid protein named HutE has been preliminarily characterized (Najimi *et al.*, 2009). Surprisingly, the deletion of *hutE* did not cause a significant defect in the ability to use haemin as sole iron source, which indicates that in *A. salmonicida* RSP74.1 neither *hutA* nor *hutE* are completely essential for this function. Both *hutA* and *hutE* are known to encode functional outer membrane haem receptors, as demonstrated by their ability to provide *E. coli* 101ESD with the ability to use haem iron (Najimi *et al.*, 2008a; Najimi *et al.*, 2009). Preliminary results with *hutA-hutE* double mutations showed that the double mutant still can use haemin as unique source of iron (C.R. Osorio, M. Najimi, M.L. Lemos, unpublished). Thus, it is likely that more than two functional outer membrane haem receptor genes coexist in the *A. salmonicida* genome, as is the case of *V. cholerae* where at least three functional haem receptors contribute to haem iron acquisition (see above).

Haem uptake in *A. hydrophila*

Most isolates of *A. hydrophila* utilize haem compounds independently of whether they produce the siderophore amonabactin or enterobactin (Massad *et al.*, 1991). Haem sources known to be utilized by this species include haemin, haemoglobin, and haemoglobin complexed to human haptoglobin. The release of haem-containing molecules from host tissues may be facilitated by the production of iron-regulated haemolysins in this species (Riddle *et al.*, 1981). A gene cluster in the *A. hydrophila* ATCC 7966^T genome (Seshadri *et al.*, 2006) (loci *AHA0964 to -0972*) contains the putative genetic determinants for haem utilization and is highly similar to the cluster described in *A. salmonicida* (Najimi *et al.*, 2008a) (Fig. 7.5). However, no reports exist on functional analysis of haem iron utilization genes in *A. hydrophila* to date.

Other iron assimilation systems

Besides iron acquisition mechanisms based on the synthesis of siderophores or haem uptake, other iron assimilation systems have been described in several bacteria. Among *Vibrio* and *Aeromonas* species these alternative systems are not well known yet. A homologue of the Feo system, a TonB-independent Fe^{2+} transporter, identified in *E. coli* and other bacteria (Cartron *et al.*, 2006), has been described in *V. cholerae* (Wyckoff *et al.*, 2006). *V. cholerae* FeoA and FeoB each have 40% amino acid identity with FeoA and FeoB of *E. coli*, respectively. The predicted *V. cholerae* FeoC protein is the same size as *E. coli* FeoC, but the two proteins have only 11% amino acid identity. Another transporter is encoded by the *vciAB* operon that encodes a predicted TonB-dependent outer membrane receptor, VciA, and a putative inner membrane protein, VciB, which functions by promoting iron uptake via a ferrous, but not ferric, iron transport system (Mey *et al.*, 2008).

Another ferric iron transporter identified in *V. cholerae* is termed FbpABC, which is an ABC transporter that consists of a periplasmic binding protein, a permease protein and an ATPase. The FbpABC proteins have about 50% amino acid identity with FbpABC of *Mannheimia haemolytica*, the most closely related system for which functional data are available (Wyckoff *et al.*, 2007). Unlike Feo, transport of iron by Fbp is inhibited by the reducing agent ascorbate, indicating that FbpABC is likely to be a ferric iron transporter (Wyckoff *et al.*, 2006). According to the genomic data available, the existence of additional ferric or ferrous iron transporters can not be ruled out. The presence of similar systems in other vibrios and aeromonads is unknown but it is likely that siderophore-independent iron transporters will be identified in other species in the following years.

Other mechanisms of uncertain significance have been also described in *Vibrio* species. For instance, *V. anguillarum* can use ferric dicitrate as an iron source (Mazoy *et al.*, 1997). *V. anguillarum* cells and isolated membranes can incorporate $^{55}Fe^{3+}$ complexed by citrate, without a difference between cells grown in the presence or absence of ferric dicitrate. Iron uptake from ferric dicitrate is an energy-dependent process not induced by the presence of iron or citrate in the medium, mediated by unknown membrane protein(s), which may require an iron reduction step to function. In fact, ferri-reductases have been detected in whole cells and cell fractions of *V. anguillarum* (Mazoy and Lemos, 1996a). In whole cells, the degree of reduction was dependent on the Fe^{3+} ligand and on the strain, with the greatest values corresponding to ferric dicitrate and serotype O1 strains, respectively. The ferric-reductase activity increases, over the basal levels, in presence of added ferric dicitrate. It is likely that this ferri-reductase activity is also involved in other iron acquisition mechanisms requiring an iron reduction step.

Conclusions and future trends

The haem internalization process is quite well understood in vibrios, and the roles of each of the transport genes have been identified quite satisfactorily. However, much less is known about the haem utilization proteins HutZXW. Although some progress has been made to understand the role of HutZ-like haem binding proteins, their actual mechanism has not been completely elucidated. Similarly, there is much room for future research on the function of HutX, HutW and their homologues. Of special interest are the three genes *orf123* which appear linked to other haem transport genes in *Aeromonas* species. What is the role in haem utilization of the lipoprotein encoded by *orf3*, and why a gene encoding an isochorismatase (*orf2*) occurs within a haem operon in Aeromonads, are two interesting questions that might also be answered in the future. Another basic question still unsolved about haem uptake is the relevance of this system in the infectious process. The available data do not permit to conclusively say that these systems are actually a virulence factor. However, as we have seen they are mechanisms extremely well conserved in all pathogenic species of *Vibrio* and *Aeromonas*, indicating relevance to the host invasion ability.

The regulatory mechanisms that control expression of siderophore biosynthesis and transport genes are far from being completely understood. We often read descriptions of new regulatory RNAs with a role in iron homeostasis, and it is expected that many more will be described in the following years. Identification of additional regulatory mechanisms is one of the topics for future research in iron uptake in *Vibrio* and *Aeromonas*. Recently, we reported that the iron-mediated repression of *vabG*, a gene encoding one of the components involved in vanchrobactin biosynthesis in *V. anguillarum*, is not due to the universal repressor Fur, but rather to a yet uncharacterized regulator (Balado *et al.*, 2008).

Genetic differences between isolates within a species are mainly due to the presence of horizontally-acquired DNA, and iron utilization gene clusters are not an exception. Actually, several siderophore biosynthesis and transport clusters have been described as forming part of horizontally-transmitted elements. The acquisition of a novel functional siderophore cluster through mobile DNA usually causes the implantation of the new system, and the inactivation of the pre-existing one, as we have seen in *V. anguillarum*. As an increasing number of strains from *Vibrio* and *Aeromonas* species are analysed, it is likely that new siderophore gene clusters will be identified, some of them being harboured in elements of mobile DNA.

Several different iron uptake systems have been identified in *V. cholerae*. However, it is reported that even after mutating three of these systems simultaneously (*vibB*, *feoB* and *fbpA*), *V. cholerae* can still obtain iron from an external iron chelator and remains virulent in a suckling mouse model. This suggests the existence of other iron transporters not yet identified (Wyckoff *et al.*, 2006). This is also applicable to all other vibrios, where the possible existence of ferrous or ferric iron transporters, as well as other siderophore-based systems, are still unknown. Furthermore, *V. cholerae* strains that were recently subjected to complete genome sequencing harbour gene clusters similar to that of the yersiniabactin cluster, which is part of the high pathogenicity island (HPI) described in pathogenic *Yersinia* species. This cluster was also identified and characterized in strains of the fish pathogen *Photobacterium damselae* subsp. *piscicida*, also a member of *Vibrionaceae*, and it was demonstrated that the siderophore produced by these strains has a strong functional relationship with yersiniabactin (Osorio *et al.*, 2006). It remains to be determined whether these clusters are functional in *V. cholerae* strains and whether they encode the synthesis of a yersiniabactin-like siderophore or not.

All these findings clearly suggest that the diversity in iron uptake systems in *Vibrio* and *Aeromonas* might be higher that initially expected and still much remains to be learned about the currently known systems and about other yet unknown mechanisms. The question of how the different iron acquisition systems work together to supply iron to the cell and how they are used in the different environments where vibrios and aeromonads can be found is still an open question. In summary, much more work is needed to completely understand the whole iron uptake process in these bacteria.

References

Alice, A.F., Lopez, C.S., and Crosa, J.H. (2005). Plasmid- and chromosome-encoded redundant and specific functions are involved in biosynthesis of the siderophore anguibactin in *Vibrio anguillarum* 775: a case of chance and necessity? J. Bacteriol. *187*, 2209–2214.

Amaro, C., and Biosca, E.G. (1996). *Vibrio vulnificus* biotype 2, pathogenic for eels, is also an opportunistic pathogen for humans. Appl. Environ. Microbiol. *62*, 1454–1457.

Balado, M., Osorio, C.R., and Lemos, M.L. (2006). A gene cluster involved in the biosynthesis of vanchrobactin, a chromosome-encoded siderophore produced by *Vibrio anguillarum*. Microbiology *152*, 3517–3528.

Balado, M., Osorio, C.R., and Lemos, M.L. (2008). Biosynthetic and regulatory elements involved in the production of the siderophore vanchrobactin in *Vibrio anguillarum*. Microbiology *154*, 1400–1413.

Balado, M., Osorio, C.R., and Lemos, M.L. (2009). FvtA is the receptor for the siderophore vanchrobactin in *Vibrio anguillarum*: utility as a route of entry for vanchrobactin analogues. Appl. Environ. Microbiol. *75*, 2775–2783.

Barghouthi, S., Payne, S.M., Arceneaux, J.E., and Byers, B.R. (1991). Cloning, mutagenesis, and nucleotide sequence of a siderophore biosynthetic gene (*amoA*) from *Aeromonas hydrophila*. J. Bacteriol. *173*, 5121–5128.

Barghouthi, S., Young, R., Olson, M.O., Arceneaux, J.E., Clem, L.W., and Byers, B.R. (1989). Amonabactin, a novel tryptophan- or phenylalanine-containing phenolate siderophore in *Aeromonas hydrophila*. J. Bacteriol. *171*, 1811–1816.

Beale, S.I. (1996). Biosynthesis of hemes. In *Escherichia coli* and *Salmonella*: cellular and molecular biology, F.C. Neidhardt, R. Curtiss III, J.L. Ingraham, E.C.C. Lin, K.B. Low, B. Magasanik, W.S. Reznikoff, M. Riley, M. Schaechter, and H.E. Umbarger, eds. (Washington, DC, USA: ASM Press), pp. 731.

Belanger, M., Begin, C., and Jacques, M. (1995). Lipopolysaccharides of *Actinobacillus pleuropneumoniae* bind pig hemoglobin. Infect. Immun. *63*, 656–662.

Biosca, E.G. Fouz, B., Alcaide, E., and Amaro, C. (1996). Siderophore-mediated iron acquisition mechanisms in *Vibrio vulnificus* biotype 2. Appl. Environ. Microbiol. *62*, 928–935.

Butterton, J.R., Stoebner, J.A., Payne, S.M., and Calderwood, S.B. (1992). Cloning, sequencing, and transcriptional regulation of *viuA*, the gene encoding the ferric vibriobactin receptor of *Vibrio cholerae*. J. Bacteriol. *174*, 3729–3738.

Cartron, M.L., Maddocks, S., Gillingham, P., Craven, C.J., and Andrews, S.C. (2006). Feo--transport of ferrous iron into bacteria. Biometals *19*, 143–157.

Chart, H., and Trust, T.J. (1983). Acquisition of iron by *Aeromonas salmonicida*. J. Bacteriol. *156*, 758–764.

Conchas, R.F., Lemos, M.L., Barja, J.L., and Toranzo, A.E. (1991). Distribution of plasmid- and chromosome-mediated iron uptake systems in *Vibrio anguillarum* strains of different origins. Appl. Environ. Microbiol. *57*, 2956–2962.

Crosa, J.H., and Walsh, C.T. (2002). Genetics and assembly line enzymology of siderophore biosynthesis in bacteria. Microbiol. Mol. Biol. Rev. *66*, 223–249.

Di Lorenzo, M., Poppelaars, S., Stork, M., Nagasawa, M., Tolmasky, M.E., and Crosa, J.H. (2004a). A nonribosomal peptide synthetase with a novel domain organization is essential for siderophore biosynthesis in *Vibrio anguillarum*. J. Bacteriol. *186*, 7327–7336.

Di Lorenzo, M., Stork, M., Alice, A.F., López, C.S., and Crosa, J.H. (2004b). *Vibrio*. In Iron transport in bacteria, J.H. Crosa, A.R. Mey, and S.M. Payne, eds. (Washington, DC, USA: ASM Press), pp. 241–255.

Di Lorenzo, M., Stork, M., Naka, H., Tolmasky, M.E., and Crosa, J.H. (2008). Tandem heterocyclization domains in a nonribosomal peptide synthetase essential for siderophore biosynthesis in *Vibrio anguillarum*. Biometals *21*, 635–648.

Di Lorenzo, M., Stork, M., Tolmasky, M.E., Actis, L.A., Farrell, D., Welch, T.J., Crosa, L.M., Wertheimer, A.M., Chen, Q., Salinas, P., Waldbeser, L., Crosa, J.H. (2003). Complete sequence of virulence plasmid pJM1 from the marine fish pathogen *Vibrio anguillarum* strain 775. J. Bacteriol. *185*, 5822–5830.

Do Vale, A., Magarinos, B., Romalde, J.L., Lemos, M.L., Ellis, A.E., and Toranzo, A.E. (2002). Binding of haemin by the fish pathogen *Photobacterium damselae* subsp. *piscicida*. Dis. Aquat. Organ. *48*, 109–115.

Ebanks, R.O., Dacanay, A., Goguen, M., Pinto, D.M., and Ross, N.W. (2004). Differential proteomic analysis of *Aeromonas salmonicida* outer membrane proteins in response to low iron and *in vivo* growth conditions. Proteomics. *4*, 1074–1085.

Fernandez, A.I., Fernandez, A.F., Perez, M.J., Nieto, T.P., and Ellis, A.E. (1998). Siderophore production by *Aeromonas salmonicida* subsp. *salmonicida*. Lack of strain specificity. Dis. Aquat. Organ. *33*, 87–92.

Fouz, B., Mazoy, R., Lemos, M.L., del Olmo, M.J., and Amaro, C. (1996). Utilization of hemin and hemoglobin by *Vibrio vulnificus* biotype 2. Appl. Environ. Microbiol. *62*, 2806–2810.

Fouz, B., Toranzo, A.E., Biosca, E.G. Mazoy, R., and Amaro, C. (1994). Role of iron in the pathogenicity of *Vibrio damsela* for fish and mammals. FEMS Microbiol. Lett. *121*, 181–188.

Funahashi, T., Moriya, K., Uemura, S., Miyoshi, S., Shinoda, S., Narimatsu, S., and Yamamoto, S. (2002). Identification and characterization of *pvuA*, a gene encoding the ferric vibrioferrin receptor protein in *Vibrio parahaemolyticus*. J. Bacteriol. *184*, 936–946.

Funahashi, T., Tanabe, T., Aso, H., Nakao, H., Fujii, Y., Okamoto, K., Narimatsu, S., and Yamamoto, S. (2003). An iron-regulated gene required for utilization of aerobactin as an exogenous siderophore in *Vibrio parahaemolyticus*. Microbiology *149*, 1217–1225.

Funahashi, T., Tanabe, T., Shiuchi, K., Nakao, H., and Yamamoto, S. (2009). Identification and characterization of genes required for utilization of desferri-ferrichrome and aerobactin in *Vibrio parahaemolyticus*. Biol. Pharm. Bull. *32*, 359–365.

Gosling, P.J. (1996). *Aeromonas* species in diseases of animals. In The genus *Aeromonas*, B. Austin, M. Altwegg, P.J. Gosling, and S.W. Joseph, eds. (Chichester: John Wiley & Sons,), pp. 121–175.

Grenier, D., Leduc, A., and Mayrand, D. (1997). Interaction between *Actinobacillus actinomycetemcomitans* lipopolysaccharides and human hemoglobin. FEMS Microbiol. Lett. *151*, 77–81.

Griffiths, G.L., Sigel, S.P., Payne, S.M., and Neilands, J.B. (1984). Vibriobactin, a siderophore from *Vibrio cholerae*. J. Biol. Chem. *259*, 383–385.

Heller, K.J., Kadner, R.J., and Gunther, K. (1988). Suppression of the *btuB*451 mutation by mutations in the *tonB* gene suggests a direct interaction between TonB and TonB-dependent receptor proteins in the outer membrane of *Escherichia coli*. Gene. *64*, 147–153.

Henderson, D.P., and Payne, S.M. (1993). Cloning and characterization of the *Vibrio cholerae* genes encoding the utilization of iron from haemin and haemoglobin. Mol. Microbiol. 7, 461–469.

Henderson, D.P., and Payne, S.M. (1994). Characterization of the *Vibrio cholerae* outer membrane heme transport protein HutA: sequence of the gene, regulation of expression, and homology to the family of TonB-dependent proteins. J. Bacteriol. *176*, 3269–3277.

Hirst, I.D., Hastings, T.S., and Ellis, A.E. (1991). Siderophore production by *Aeromonas salmonicida*. J. Gen. Microbiol. *137*, 1185–1192.

Ishiguro, E.E., Ainsworth, T., Kay, W.W., and Trust, T.J. (1986). Heme requirement for growth of fastidious atypical strains of *Aeromonas salmonicida*. Appl. Environ. Microbiol. *51*, 668–670.

Jalal, M.A.F., Hossain, M.B., van der Helm, D., Sanders-Loehr, J., Actis, L.A., and Crosa, J.H. (1989). Structure of anguibactin, a unique plasmid-related bacterial siderophore from the fish pathogen *Vibrio anguillarum*. J. Am. Chem. Soc. *111*, 292–296.

Joseph, S.W., and Carnahan, A.M. (1994). The isolation, identification, and systematics of the motile *Aeromonas* species. Annu. Rev. Fish Dis. *4*, 315–343.

Kim, I.H., Shim, J.I., Lee, K.E., Hwang, W., Kim, I.J., Choi, S.H., and Kim, K.S. (2008). Nonribosomal peptide synthase is responsible for the biosynthesis of siderophore in *Vibrio vulnificus* MO6–24/O. J. Microbiol. Biotechnol. *18*, 35–42.

Litwin, C.M., Rayback, T.W., and Skinner, J. (1996). Role of catechol siderophore synthesis in *Vibrio vulnificus* virulence. Infect. Immun. *64*, 2834–2838.

Massad, G., Arceneaux, J.E., and Byers, B.R. (1991). Acquisition of iron from host sources by mesophilic *Aeromonas* species. J. Gen. Microbiol. *137*, 237–241.

Mazoy, R., Botana, L.M., and Lemos, M.L. (1997). Iron uptake from ferric citrate by *Vibrio anguillarum*. FEMS Microbiol. Lett. *154*, 145–150.

Mazoy, R., and Lemos, M.L. (1996a). Ferric-reductase activities in whole cells and cell fractions of *Vibrio* (*Listonella*) *anguillarum*. Microbiology *142*, 3187–3193.

Mazoy, R., and Lemos, M.L. (1996b). Identification of heme-binding proteins in the cell membranes of *Vibrio anguillarum*. FEMS Microbiol. Lett. *135*, 265–270.

Mazoy, R., Osorio, C.R., Toranzo, A.E., and Lemos, M.L. (2003). Isolation of mutants of *Vibrio anguillarum* defective in haeme utilisation and cloning of *huvA*, a gene coding for an outer membrane protein involved in the use of haem as iron source. Arch. Microbiol. *179*, 329–338.

Mey, A.R., and Payne, S.M. (2001). Haem utilization in *Vibrio cholerae* involves multiple TonB-dependent haem receptors. Mol. Microbiol. *42*, 835–849.

Mey, A.R., Wyckoff, E.E., Hoover, L.A., Fisher, C.R., and Payne, S.M. (2008). *Vibrio cholerae* VciB promotes iron uptake via ferrous iron transporters. J. Bacteriol. *190*, 5953–5962.

Mey, A.R., Wyckoff, E.E., Oglesby, A.G., Rab, E., Taylor, R.K., and Payne, S.M. (2002). Identification of the *Vibrio cholerae* enterobactin receptors VctA and IrgA: IrgA is not required for virulence. Infect. Immun. *70*, 3419–3426.

Mossialos, D., Ochsner, U., Baysse, C., Chablain, P., Pirnay, J.P., Koedam, N., Budzikiewicz, H., Fernández, D.U., Schäfer, M., Ravel, J., and Cornelis, P. (2002). Identification of new, conserved, non-ribosomal peptide synthetases from fluorescent pseudomonads involved in the biosynthesis of the siderophore pyoverdine. Mol. Microbiol. *45*, 1673–1685.

Mouriño, S., Osorio, C.R., and Lemos, M.L. (2004). Characterization of heme uptake cluster genes in the fish pathogen *Vibrio anguillarum*. J. Bacteriol. *186*, 6159–6167.

Mouriño, S., Osorio, C.R., Lemos, M.L., and Crosa, J.H. (2006). Transcriptional organization and regulation of the *Vibrio anguillarum* heme uptake gene cluster. Gene. *374*, 68–76.

Mouriño, S., Rodriguez-Ares, I., Osorio, C.R., and Lemos, M.L. (2005). Genetic variability of the heme uptake system among different strains of the fish pathogen *Vibrio anguillarum*: identification of a new heme receptor. Appl. Environ. Microbiol. *71*, 8434–8441.

Najimi, M., Lemos, M.L., and Osorio, C.R. (2008a). Identification of heme uptake genes in the fish pathogen *Aeromonas salmonicida* subsp. *salmonicida*. Arch. Microbiol. *190*, 439–449.

Najimi, M., Lemos, M.L., and Osorio, C.R. (2008b). Identification of siderophore biosynthesis genes essential for growth of *Aeromonas salmonicida* under iron limitation conditions. Appl. Environ. Microbiol. *74*, 2341–2348.

Najimi, M., Lemos, M.L., and Osorio, C.R. (2009). Identification of iron regulated genes in the fish pathogen *Aeromonas salmonicida* subsp. *salmonicida*: genetic diversity and evidence of conserved iron uptake systems. Vet. Microbiol. *133*, 377–382.

Naka, H., Lopez, C.S., and Crosa, J.H. (2008). Reactivation of the vanchrobactin siderophore system of *Vibrio anguillarum* by removal of a chromosomal insertion sequence originated in plasmid pJM1 encoding the anguibactin siderophore system. Environ. Microbiol. *10*, 265–277.

Occhino, D.A., Wyckoff, E.E., Henderson, D.P., Wrona, T.J., and Payne, S.M. (1998). *Vibrio cholerae* iron transport: haem transport genes are linked to one of two sets of *tonB*, *exbB*, *exbD* genes. Mol. Microbiol. *29*, 1493–1507.

Okujo, N., Saito, M., Yamamoto, S., Yoshida, T., Miyoshi, S., and Shinoda, S. (1994). Structure of vulnibactin, a new polyamine-containing siderophore from *Vibrio vulnificus*. Biometals *7*, 109–116.

Okujo, N., and Yamamoto, S. (1994). Identification of the siderophores from *Vibrio hollisae* and *Vibrio mimicus* as aerobactin. FEMS Microbiol. Lett. *118*, 187–192.

O'Malley, S.M., Mouton, S.L., Occhino, D.A., Deanda, M.T., Rashidi, J.R., Fuson, K.L., Rashidi, C.E., Mora, M.Y., Payne, S.M., and Henderson, D.P. (1999). Comparison of the heme iron utilization systems of pathogenic Vibrios. J. Bacteriol. *181*, 3594–3598.

Osorio, C.R., Juiz-Rio, S., and Lemos, M.L. (2006). A siderophore biosynthesis gene cluster from the fish pathogen *Photobacterium damselae* subsp. *piscicida* is structurally and functionally related to the *Yersinia* high-pathogenicity island. Microbiology *152*, 3327–3341.

Postle, K., and Larsen, R.A. (2007). TonB-dependent energy transduction between outer and cytoplasmic membranes. Biometals *20*, 453–465.

Ratliff, M., Zhu, W., Deshmukh, R., Wilks, A., and Stojiljkovic, I. (2001). Homologues of neisserial heme oxygenase in gram-negative bacteria: degradation of heme by the product of the *pigA* gene of *Pseudomonas aeruginosa*. J. Bacteriol. *183*, 6394–6403.

Reith, M.E., Singh, R.K., Curtis, B., Boyd, J.M., Bouevitch, A., Kimball, J., Munholland, J., Murphy, C., Sarty, D., Williams, J. *et al.* (2008). The genome of *Aeromonas salmonicida* subsp. *salmonicida* A449: insights into the evolution of a fish pathogen. BMC Genomics. *9*, 427.

Riddle, L.M., Graham, T.E., and Amborski, R.L. (1981). Medium for the accumulation of extracellular hemolysin and protease by *Aeromonas hydrophila*. Infect. Immun. *33*, 728–733.

Rogers, M.B., Sexton, J.A., DeCastro, G.J., and Calderwood, S.B. (2000). Identification of an operon required for ferrichrome iron utilization in *Vibrio cholerae*. J. Bacteriol. *182*, 2350–2353.

Schmitt, M.P. (1997). Utilization of host iron sources by *Corynebacterium diphtheriae*: identification of a gene whose product is homologous to eukaryotic heme oxygenases and is required for acquisition of iron from heme and hemoglobin. J. Bacteriol. *179*, 838–845.

Seshadri, R., Joseph, S.W., Chopra, A.K., Sha, J., Shaw, J., Graf, J., Haft, D., Wu, M., Ren, Q., Rosovitz, M.J. *et al.* (2006). Genome sequence of *Aeromonas hydrophila* ATCC 7966T: jack of all trades. J. Bacteriol. *188*, 8272–8282.

Skare, J.T., Ahmer, B.M., Seachord, C.L., Darveau, R.P., and Postle, K. (1993). Energy transduction between membranes. TonB, a cytoplasmic membrane protein, can be chemically cross-linked *in vivo* to the outer membrane receptor FepA. J. Biol. Chem. *268*, 16302–16308.

Soengas, R.G., Anta, C., Espada, A., Nieto, R.M., Larrosa, M., Rodríguez, J., and Jiménez, C. (2007). Vanchrobactin: absolute configuration and total synthesis. Tetrahedron Lett. *48*, 3021–3024.

Soengas, R.G., Anta, C., Espada, A., Paz, V., Ares, I.R., Balado, M., Rodríguez, J., Lemos, M.L., and Jiménez, C. (2006). Structural characterization of vanchrobactin, a new catechol siderophore produced by the fish pathogen *Vibrio anguillarum* serotype O2. Tetrahedron Lett. *47*, 7113–7116.

Stintzi, A., and Raymond, K.N. (2000). Amonabactin-mediated iron acquisition from transferrin and lactoferrin by *Aeromonas hydrophila*: direct measurement of individual microscopic rate constants. J. Biol. Inorg. Chem. *5*, 57–66.

Stojiljkovic, I., and Hantke, K. (1992). Hemin uptake system of *Yersinia enterocolitica*: similarities with other TonB-dependent systems in gram-negative bacteria. EMBO J. *11*, 4359–4367.

Stork, M., Di Lorenzo, M., Mouriño, S., Osorio, C.R., Lemos, M.L., and Crosa, J.H. (2004). Two tonB systems function in iron transport in *Vibrio anguillarum*, but only one is essential for virulence. Infect. Immun. *72*, 7326–7329.

Stork, M., Di Lorenzo, M., Welch, T.J., Crosa, L.M., and Crosa, J.H. (2002). Plasmid-mediated iron uptake and virulence in *Vibrio anguillarum*. Plasmid. *48*, 222–228.

Stork, M., Otto, B.R., and Crosa, J.H. (2007). A novel protein, TtpC, is a required component of the TonB2 complex for specific iron transport in the pathogens *Vibrio anguillarum* and *Vibrio cholerae*. J. Bacteriol. *189*, 1803–1815.

Tanabe, T., Funahashi, T., Nakao, H., Miyoshi, S., Shinoda, S., and Yamamoto, S. (2003). Identification and characterization of genes required for biosynthesis and transport of the siderophore vibrioferrin in *Vibrio parahaemolyticus*. J. Bacteriol. *185*, 6938–6949.

Tanabe, T., Naka, A., Aso, H., Nakao, H., Narimatsu, S., Inoue, Y., Ono, T., and Yamamoto, S. (2005). A novel aerobactin utilization cluster in *Vibrio vulnificus* with a gene involved in the transcription regulation of the *iutA* homologue. Microbiol. Immunol. *49*, 823–834.

Tanabe, T., Nakao, H., Kuroda, T., Tsuchiya, T., and Yamamoto, S. (2006). Involvement of the *Vibrio parahaemolyticus* *pvsC* gene in export of the siderophore vibrioferrin. Microbiol. Immunol. *50*, 871–876.

Telford, J.R., and Raymond, K.N. (1997). Amonabactin: a family of novel siderophores from a pathogenic bacterium. J. Biol. Inorg. Chem. 2, 750–761.

Toranzo, A.E., Magariños, B., and Romalde, J.L. (2005). A review of the main bacterial fish diseases in mariculture systems. Aquaculture *246*, 37–61.

Wandersman, C., and Delepelaire, P. (2004). Bacterial iron sources: from siderophores to hemophores. Annu. Rev. Microbiol. *58*, 611–647.

Wang, Q., Liu, Q., Cao, X., Yang, M., and Zhang, Y. (2008). Characterization of two TonB systems in marine fish pathogen *Vibrio alginolyticus*: their roles in iron utilization and virulence. Arch. Microbiol. *190*, 595–603.

Wang, Q., Liu, Q., Ma, Y., Zhou, L., and Zhang, Y. (2007). Isolation, sequencing and characterization of cluster genes involved in the biosynthesis and utilization of the siderophore of marine fish pathogen *Vibrio alginolyticus*. Arch. Microbiol. *188*, 433–439.

Webster, A.C., and Litwin, C.M. (2000). Cloning and characterization of *vuuA*, a gene encoding the *Vibrio vulnificus* ferric vulnibactin receptor. Infect. Immun. *68*, 526–534.

Welch, T.J., and Crosa, J.H. (2005). Novel role of the lipopolysaccharide O1 side chain in ferric siderophore transport and virulence of *Vibrio anguillarum*. Infect. Immun. 73, 5864–5872.

Wyckoff, E.E., Mey, A.R., Leimbach, A., Fisher, C.F., and Payne, S.M. (2006). Characterization of ferric and ferrous iron transport systems in *Vibrio cholerae*. J. Bacteriol. *188*, 6515–6523.

Wyckoff, E.E., Mey, A.R., and Payne, S.M. (2007). Iron acquisition in *Vibrio cholerae*. Biometals *20*, 405–416.

Wyckoff, E.E., Schmitt, M., Wilks, A., and Payne, S.M. (2004). HutZ is required for efficient heme utilization in *Vibrio cholerae*. J. Bacteriol. *186*, 4142–4151.

Wyckoff, E.E., Smith, S.L., and Payne, S.M. (2001). VibD and VibH are required for late steps in vibriobactin biosynthesis in *Vibrio cholerae*. J. Bacteriol. *183*, 1830–1834.

Wyckoff, E.E., Stoebner, J.A., Reed, K.E., and Payne, S.M. (1997). Cloning of a *Vibrio cholerae* vibriobactin gene cluster: identification of genes required for early steps in siderophore biosynthesis. J. Bacteriol. *179*, 7055–7062.

Yamamoto, S., Hara, Y., Tomochika, K., and Shinoda, S. (1995). Utilization of hemin and hemoglobin as iron sources by *Vibrio parahaemolyticus* and identification of an iron-repressible hemin-binding protein. FEMS Microbiol. Lett. *128*, 195–200.

Yamamoto, S., Okujo, N., Fujita, Y., Saito, M., Yoshida, T., and Shinoda, S. (1993). Structures of two polyamine-containing catecholate siderophores from *Vibrio fluvialis*. J. Biochem. *113*, 538–544.

Yamamoto, S., Okujo, N., Yoshida, T., Matsuura, S., and Shinoda, S. (1994). Structure and iron transport activity of vibrioferrin, a new siderophore of *Vibrio parahaemolyticus*. J. Biochem. *115*, 868–874.

Zhu, W., Hunt, D.J., Richardson, A.R., and Stojiljkovic, I. (2000a). Use of heme compounds as iron sources by pathogenic neisseriae requires the product of the *hemO* gene. J. Bacteriol. *182*, 439–447.

Zhu, W., Wilks, A., and Stojiljkovic, I. (2000b). Degradation of heme in gram-negative bacteria: the product of the *hemO* gene of Neisseriae is a heme oxygenase. J. Bacteriol. *182*, 6783–6790.

Zywno, S.R., Arceneaux, J.E., Altwegg, M., and Byers, B.R. (1992). Siderophore production and DNA hybridization groups of *Aeromonas* spp. J. Clin. Microbiol. *30*, 619–622.

Iron Uptake in *Francisella*

8

Girija Ramakrishnan

Abstract

Francisella tularensis is unusual among Gram-negative bacteria in that its genome does not encode orthologues for TonB, ExbB and ExbD that typically energize the uptake of iron across the outer membrane. This organism secretes a siderophore similar in structure to rhizoferrin. The *fsl* operon of six genes encodes functions for biosynthesis and uptake of the siderophore. Two of these genes encode a siderophore synthetase belonging to the non-ribosomal peptide synthetase (**N**RPS)-**i**ndependent **s**ynthetase (NIS)-family and a protein belonging to the pyridoxyl phosphate-dependent decarboxylase family, and both are required for siderophore production. Siderophore utilization involves the product of the *fslE* gene, a protein unique to *Francisella* species that could function as a siderophore receptor. Additionally, genes related in sequence to *fslE* also play a role in siderophore acquisition. The mechanism for TonB-independent iron uptake in this microorganism remains to be elucidated.

Introduction

Francisella tularensis is a Gram-negative gamma-proteobacterium and is the causative agent of tularaemia, a zoonosis largely restricted to the Northern hemisphere (Sjostedt, 2007). Commonly transmitted by bites of arthropod vectors, tularaemia may also develop following direct contact with infected lagomorphs and rodents, by consumption of contaminated food or water or by inhalation of aerosolized bacteria.

F. tularensis forms a distinct evolutionary clade with a large number of endosymbionts and parasitic species, most of which remain undescribed (Keim *et al.*, 2007). It has evolved with a greatly reduced genome and lacks complete biochemical pathways for several essential metabolites, consistent with a parasitic lifestyle (Larsson *et al.*, 2005). The species comprises subspecies of varying virulence characteristics and geographic distribution. The most virulent subspecies *tularensis*, responsible for the most severe (Type A) form of tularaemia in humans, is restricted to North America. Isolates of the *holarctica* subspecies cause a less severe disease (Type B) and are prevalent in North America as well as in Europe. Strains of the *mediaasiatica* subspecies have distinct biochemical characteristics and have only been isolated in central Asia. The *novicida* subspecies is sometimes considered an independent species. While virulent in mice, it is avirulent in humans except for some immunocompromised individuals, and its genome is considered to be of a more ancestral lineage than the pathogenic subspecies. The *tularensis* subspecies of the organism is classified as a Category A select agent by the Centers for Diseases Control in the US because it is highly infectious and easily transmissible and may be exploited as a bioweapon (Dennis *et al.*, 2001).

F. tularensis is a facultative intracellular pathogen, capable of invading and replicating within a variety of cell types including macrophages, hepatocytes, fibroblasts, and endothelial cells. It has been shown to also replicate within arthropod cells and free-living amoebae. Macrophages and macrophage-like cell lines have been most commonly employed as a tissue culture infection model. Invading bacteria initially reside in membrane-bound phagosomes that become mildly acidified before the bacteria escape into the cytoplasm where they replicate (Clemens *et al.*, 2004; Golovliov *et al.*, 2003). There do not appear to be differences in the intracellular replication capability of the different subspecies, and while some studies have involved Type A strains, the majority of the work with tissue culture infection models has focused on the live vaccine strain (LVS, an attenuated *holarctica* derivative) or a *novicida* strain U112.

Iron and *F. tularensis*

Like most microorganisms, *F. tularensis* requires iron as a nutrient, and it is capable of utilizing ferrous and ferric forms of iron. As expected, iron availability impacts on the ability of *F. tularensis* to grow within host cells. Replication of live vaccine strain (LVS) within mouse peritoneal macrophages is inhibited by deferroxamine treatment or by blocking acidification of host cell endosomes with chloroquine and ammonium chloride (Bhatnagar *et al.*, 1995; Fortier *et al.*, 1995). Acidification releases necessary iron from iron–transferrin complexes necessary for the intracellular proliferation of the bacteria; the inhibition of growth by treatment with acidification inhibitors can be overcome by addition of supplemental inorganic ferric iron (Clemens *et al.*, 2009; Fortier *et al.*, 1995).

As with many pathogens, infection with *F. tularensis* affects host metabolism, and early studies reported the development of a typical iron-withholding response in infected patients (Pekarek *et al.*, 1969). Virulence may be regulated by iron availability; an LVS derived-strain was shown to have increased virulence in a mouse model after iron-limited growth (Bhatnagar *et al.*, 1995). Other studies suggest that host iron metabolism regulates susceptibility to *F. tularensis*. The *nramp1* gene encodes a proton-efflux pump that moves Fe^{2+} and Mn^{2+} ions from lysosomes and phagolysosomes into the cytosol, thereby regulating endosomal pH and iron availability within the phagosome (Forbes and Gros, 2001). The mouse *bcg(s)* mutation, which maps to the iron transporter gene *nramp1*, confers susceptibility to a variety of intracellular pathogens, including *Salmonella* and *Mycobacteria*. Paradoxically, the *bcg(s)* mutation renders mice resistant to infection with the LVS strain of *F. tularensis* (Kovarova *et al.*, 2000). LVS bacteria were shown to proliferate within *bcg(r)* macrophages significantly more than in *bcg(s)* macrophages (Kovarova *et al.*, 2002). The same report stated, although the data were not shown, that deferroxamine suppresses bacterial growth within *bcg(r)* macrophages and that provision of iron makes *bcg(s)* macrophages permissive for LVS growth. However, commonly used mice and mouse macrophage lines for studies with *F. tularensis* are genetically *bcg(s)*.

Iron uptake genes in *F. tularensis*

TonB is an inner membrane protein ubiquitous to Gram-negative bacteria that, in complex with ExbB and ExbD, transduces the energy from the proton motive force to outer membrane proteins that mediate uptake of ferric iron in siderophore complexes or in haem or transferrin conjugates. When the first *F. tularensis* genome was sequenced, it was recognized that the small 1.89-Mb genome does not encode a recognizable TonB orthologue (Larsson *et al.*, 2005). Multiple strains have since been sequenced, and it is clear that the organism lacks the classical TonB–ExbB–ExbD complex of proteins as also associated TonB-dependent receptors. The identification of siderophore production and utilization by *F. tularensis* strains established the existence of a TonB-independent siderophore uptake system in this organism (Deng *et al.*, 2006; Sullivan *et al.*, 2006). All of the subspecies encode a siderophore biosynthetic operon with a conserved Fur box in the promoter region, and this operon is located adjacent to the *fur* gene on the chromosome (Fig. 8.1). Studies of this operon have been conducted in different laboratories with representative strains of different subspecies: Schu S4 (*tularensis* subsp.), LVS (*holarctica* subsp.) and U112 (*novicida* subsp.).

While many of the features with regard to iron uptake are conserved across these subspecies, there are also underlying differences in their iron metabolism that are only partially recognized. For example, proteomics studies have indicated that strains of the *holarctica* subspecies possess higher levels of bacterioferritin than the *tularensis* subspecies (Hubalek *et al.*, 2003, 2004). We have observed differences in

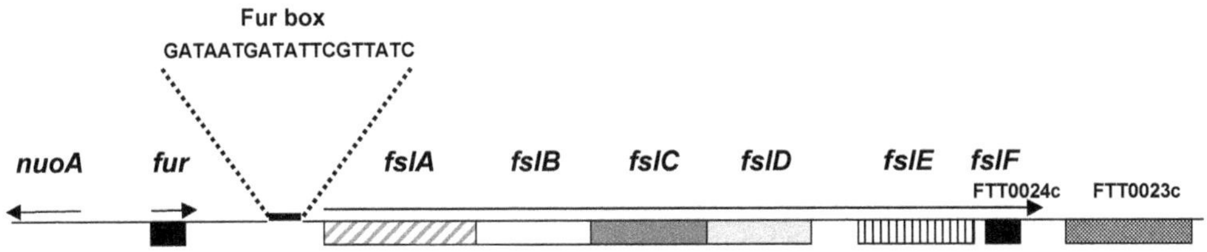

ORF	Alternate names	Closest annotated homolog in GenBank	Predicted siderophore-related function
FslA	FrgA FigA	YP_002305940.1 siderophore synthase (*Coxiella burnettii*) E=4e-95	siderophore synthase
FslB	FigB	YP_126628 multidrug resistance transporter (*Legionella pneumophila*)E=1e-39	transporter
FslC	FigC	PERMA_0884 diaminopimelate decarboxylase (*Persephonella marina*)E= 6e-54	siderophore biosynthetic enzyme
FslD	FigD	YP_001250062 drug resistance transporter Bcr/CflA (*L. pneumophila*) E=1e-57	?
FslE	FigE SrfA	-	Siderophore receptor
FslF	FigF	-YP_001140424.1lipase/ acyltransferase (*Aeromonas salmonicida*) in *novicida*	?

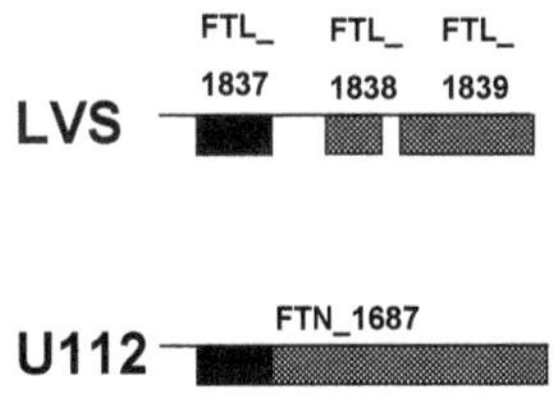

Figure 8.1 The *fsl* (*Francisella* siderophore locus) region of the *F. tularensis* chromosome. The *fsl* operon in the chromosome of Schu S4 (subsp. *tularensis*) is depicted on top. Variation in the *fslF* locus of LVS (subsp. *holarctica*) and U112 (subsp. *novicida*) is shown. Arrows indicate the direction of transcription of the genes. The inset table lists the closest homologues and the predicted function of the *fsl* genes.

the kinetics of growth under iron starvation of Schu S4 compared to the LVS (Sullivan *et al.*, 2006) and LVS and Schu S4 show different abilities to grow on iron-limiting plates (unpublished studies). Parallel studies in the different strains are therefore important for discerning differences in iron uptake and gaining an understanding of how they impact the physiology and influence virulence of the subspecies.

The *F. tularensis* siderophore

Culture supernatants of different *F. tularensis* strains grown under iron limitation produce a siderophore that can be detected using the Chrome azurol S (CAS) assay (Deng *et al.*, 2006; Sullivan *et al.*, 2006). This siderophore was purified from LVS and from Schu S4 by conventional anion-exchange chromatography followed by HPLC on an anion-exchange column (Sullivan *et al.*, 2006). ESI-MS analysis of the siderophore identified a 436 Da molecule, and the MS/MS spectra corresponded to a structure similar to, if not identical to, rhizoferrin. Rhizoferrin is a polycarboxylate siderophore produced by *Rhizopus* and other fungal species (Drechsel *et al.*, 1991; Thieken and Winkelmann, 1992). It has a relatively simple structure consisting of two citric acid moieties linked to the amino groups on a putrescine molecule by amide bonds (Fig. 8.2). The siderophore is acid labile, and exposure to acid results in conversion to dehydro and didehydro forms of the molecule.

In the nineteen sixties, Mager and colleagues published a series of papers on a 'growth initiation substance' (GIS) produced by *F. tularensis* that promoted growth of liquid cultures from small inocula of fewer than 10^5 organisms (Halmann *et al.*, 1967; Halmann and Mager, 1967; Mager, 1964). This GIS material behaved similarly to rhizoferrin by column chromatography and with regard to acid lability. Furthermore, a *Rhizopus* strain was also found capable of secreting GIS that could be utilized by *F. tularensis*. We believe that the GIS material is identical to rhizoferrin and the *Francisella* siderophore.

A strain of the bacterium *Ralstonia pickettii* also produces a siderophore similar to rhizoferrin, but whereas the citryl residues in fungal rhizoferrin are in the R,R configuration, the bacterial rhizoferrin adopts a S,S configuration (*enantio*-rhizoferrin) (Munzinger *et al.*, 1999). It has been proposed that bacterial carboxylate siderophores are synthesized by conserved biosynthetic pathways that result in molecules with an S, S-configuration (Drechsel and Winkelmann, 2005). If this is true, the *Francisella* siderophore would correspond to the S,S- form as well. Nevertheless, fungal rhizoferrin was found able to promote growth of strains of *F. tularensis* (Kiss *et al.*, 2008).

The fsl operon

A search of the Schu S4 genome sequence of *F. tularensis* for potential siderophore biosynthetic genes identified an open reading frame with homology to the superfamily of non-ribosomal peptide synthetase (NRPS)-independent synthetases (NIS) (Challis, 2005). We designated this gene *fslA* (for *Francisella* siderophore locus A) and it is present in all currently sequenced strains of the different *F. tularensis* subspecies. Other names for this gene are *figA* for 'iron induced gene A' (Deng *et al.*, 2006) and *frgA* for 'iron-regulated gene A' (Rohmer *et al.*, 2007). *fslA* is 253 nucleotides downstream of the *fur* homologue and is the first gene in an operon with six ORFs (Deng *et al.*, 2006; Kiss *et al.*, 2008; Milne *et al.*, 2007; Ramakrishnan *et al.*, 2008; Sullivan *et al.*, 2006) (Fig. 8.1). This region containing the *fsl* operon is highly conserved (98–99%) across the different strains and has a GC content of 32% similar to that of the whole genome (Larsson *et al.* 2005). A canonical Fur box is located 22 bp upstream of the *fslA* start codon. Based on sequence analysis and available experimental data, two of the genes *fslA* and *fslC* encode siderophore biosynthetic enzymes.

fslA encodes a 74-kDa protein that can be detected following growth under iron limitation (Deng *et al.*, 2006). The FslA sequence has 36% identity and 56% similarity to the *frgA* product of *Coxiella burnettii*, its closest homologue in the GenBank database. Another close homologue is LbtA of *Legionella pneumophila* (33% identity, 52% similarity), shown to be necessary for production of the siderophore legiobactin (Allard *et al.*, 2006). These NIS synthetases characteristically mediate the formation of amide or ester bonds between a substrate amine or alcohol and a carboxylic acid substrate (Challis, 2005). Phylogenetic analysis of NIS synthetases defines three classes (A, B and C) and it has been proposed that these classes reflect differences in substrate specificity of the enzymes. FslA clusters with the *C. burnettii* FrgA that is classified as B type enzyme.

The *fslC* ORF is annotated as a diaminopimelate decarboxylase based upon its closest homologue in GenBank. Diaminopimelate decarboxylase catalyses the last step in lysine biosynthesis. Since *F. tularensis* is a lysine auxotroph, it seems unlikely that FslC functions as this enzyme. It more likely is a pyridoxal-phosphate-dependent decarboxylase involved in siderophore biosynthesis and, consistent with this idea, an *fslC* mutant in U112 is deficient in siderophore production (Kiss *et al.*, 2008).

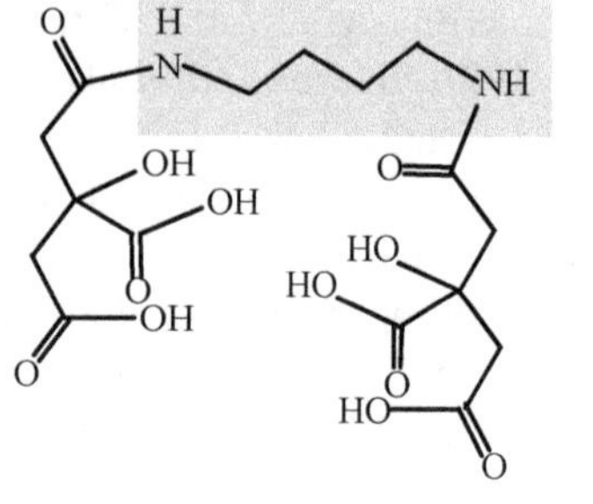

Figure 8.2 Structure of *F. tularensis* siderophore resembles rhizoferrin. The shaded part of the structure represents the putrescine part of the backbone that is linked by amide bonds to two citrate moieties. The structure was inferred on the basis of MS/MS analysis (Sullivan *et al.*, 2006).

The *fslB* and *fslD* ORFs encode proteins belonging to the major facilitator superfamily of transporters (MFS), suggesting that they could be involved in transport of the siderophore. Mutation of the *fslB* homologue in *novicida* strain U112 resulted in a deficiency in siderophore production and attenuation for growth under iron limitation (Kiss *et al.*, 2008). The same study reported that mutation of the *fslD* homologue resulted in reduced growth under iron-limitation, but this was not due to a deficiency in siderophore production or in siderophore utilization. It is not clear what role FslD may play in siderophore-mediated iron uptake.

The *fslE* gene lies 85 bp downstream of *fslD* and studies in Schu S4 and in U112 indicate that it is transcribed as part of the *fsl* operon (Kiss *et al*, 2008; Milne *et al.*, 2007; Ramakrishnan *et al.*, 2008). The sequence of the predicted protein is unique to *Francisella* species. The FslE protein of 509 amino acid residues is predicted to have a signal peptide and coiled coil domains (Larsson *et al.*, 2005), and is expressed as a protein of ~ 50 kDa (Ramakrishnan *et al.*, 2008). The protein, referred to as SrfA (for surface antigen A) has been identified in the outer membrane fraction of Schu S4 and LVS strains (Huntley *et al.*, 2007). Using the hidden Markov model-based beta-barrel prediction program PRED-TMBB (Bagos *et al.*, 2004), FslE can be modelled as a 14-stranded beta-barrel in the outer membrane with the amino-terminal third of the protein in the periplasm (Fig. 8.3). This structure is reminiscent of the typical 22-stranded TonB-dependent siderophore receptors from other Gram-negative bacteria (Chakraborty *et al.*, 2007).

A deletion mutant of *fslE* demonstrates reduced ability to grow under iron limitation both in Schu S4 and in U112 backgrounds (Kiss *et al.*, 2008; Ramakrishnan *et al.*, 2008). Their defect lays not in siderophore production, but in the ability to utilize siderophore: *fslE* mutants were unable to grow in a plate-based cross-feeding growth assay and they were also unable to utilize exogenously added siderophore in liquid culture (Kiss *et al.*, 2008; Ramakrishnan *et al.*, 2008). These studies suggest that FslE functions as a siderophore receptor although the mechanism for siderophore-mediated iron uptake remains to be elucidated. While uptake would appear to be TonB-independent, it is possible that there is an alternate protein functioning in the same capacity to transduce energy for the uptake process. Alternatively, iron from the siderophore complex may be taken up by a completely different mechanism.

The last ORF encoded by the *fsl* operon, FslF, shows variation due to shifts in the reading frame in the genomes of the different subspecies: *tularensis* (Schu S4), *holarctica* (LVS) and *novicida* (U112) (Larsson *et al.*, 2005; Rohmer *et al.*, 2007) (Fig. 8.1). *fslF* in Schu S4 (FTT0024c) encodes a protein

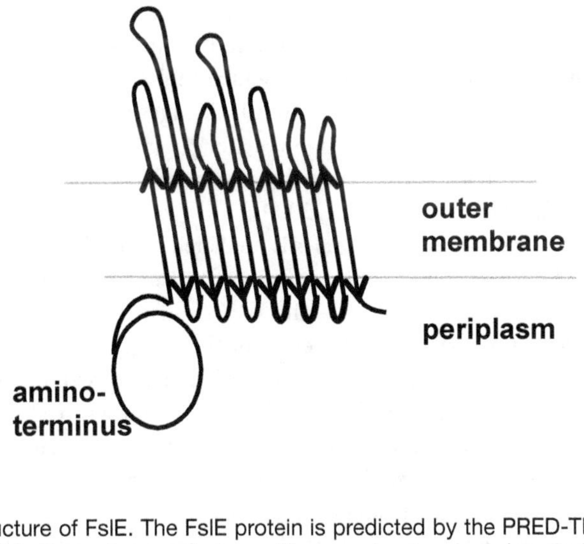

Figure 8.3 Predicted structure of FslE. The FslE protein is predicted by the PRED-TMBB program to form a 14 stranded beta-barrel in the outer membrane with an amino-terminal domain in the periplasm. This structure is reminiscent of a Gram-negative TonB-dependent siderophore receptor.

of 114 amino acid residues. Transcription of the operon in Schu S4 does not extend into the gene 257 bp downstream (locus FTT0023c encoding a lipase/acyltransferase) (Ramakrishnan *et al.*, 2008). In the LVS, the *fslF* ORF (locus FTL_1837) has 121 amino acid residues and is separated by a second short ORF from locus FTL_1839 encoding part of the lipase/acyltransferase. The *novicida* FslF protein FTN_1687 has 479 amino acid residues and includes sequences present both in the Schu S4 FslF and the lipase/acyltransferase as one fusion protein. The amino-terminal region conserved in all the strains is predicted by HMMTOP to include a *trans*-membrane domain (Tusnady and Simon, 1998; 2001), suggesting that they are localized in the cytoplasmic membrane. It is not currently clear as to whether any of these predicted proteins is involved in siderophore-mediated iron uptake.

Other iron uptake genes

Recent data suggest that FslE is not the sole mediator of siderophore-mediated iron acquisition in *Francisella* species. *fslE* is one member of a family of sequence-related genes present in *Francisella* species (Larsson *et al.*, 2005). These paralogues are conserved to various extents in the different subspecies and possess varying levels of similarity to *fslE*. The gene with the greatest similarity in strain Schu S4, *fupA* (locus FTT0918) is important for virulence and has recently been shown to be also important for siderophore-mediated iron uptake (Lindgren *et al.*, 2009). Studies in our laboratory indicate that *fslE* plays a redundant role in the LVS, unlike in Schu S4 and U112 strains (unpublished). This is probably because the *fslE*-related gene FTL_0439 in LVS has an iron acquisition function (unpublished observations). Thus siderophore-mediated iron acquisition may involve multiple sequence-related proteins.

Few other genes related to iron uptake have been identified. A ferrous iron transporter *feoB* gene is present and is not linked to the putative *feoA* homologue. There are no *mntH* homologues in the *F. tularensis* genome. An ABC transporter system specific for metals (which may include iron) has been identified on the basis of bioinformatics in Schu S4 (genes FTT0207c-0208c-0209c) (Atkins *et al.*, 2006). A gene annotated as encoding a proton-dependent oligopeptide transporter and with some homology to *iraB* of *Legionella pneumophila* has been implicated in energy-dependent iron uptake in U112 (FTN_1272) (Crosa *et al.*, 2009).

Fur and regulation of iron uptake genes

As expected, expression of the *fsl* genes is increased under iron limitation. This has been determined using different assays: direct analysis of RNA by qRT-PCR, use of microarrays and by the use of reporter *lacZ* fusions (Buchan *et al.*, 2008; Deng *et al.*, 2006; Ramakrishnan *et al.*, 2008; Sullivan *et al.*, 2006). In fact, the *fsl* genes are the most responsive to changes in iron levels. Additionally, microarray analysis and use of reporter fusions indicated that transcription of a number of other genes, including genes of the pathogenicity island (FPI) important for intracellular replication and virulence, is increased under iron limitation (Buchan *et al.*, 2008; Deng *et al.*, 2006). Changes in protein profiles following growth in iron-limiting media were noted several years ago (Bhatnagar *et al.*, 1995). A small number of proteins that change in response to iron levels have been characterized by proteomics approaches and by immunodetection (Lenco *et al.*, 2007; Schmerk *et al.*, 2009).

The iron-responsive expression of the *fsl* genes is mediated by the Fur repressor (Buchan *et al.*, 2008; Ramakrishnan *et al.*, 2008). *F. tularensis* Fur is a protein of 140 amino acid residues that is most similar to the *Neisseria gonorrhoeae* Fur protein (49% identity, 61% similarity). The *fur* gene itself is not regulated by iron levels (unpublished observations). Deletion of the *fur* gene in Schu S4 results in unregulated siderophore production by cells irrespective of iron levels in the growth medium and levels of the *fslA* transcript are increased several hundred-fold, while the *fslE* transcript is increased about 20-fold as determined by qRT-PCR (Ramakrishnan *et al.*, 2008). Similar results have been obtained also in the LVS background (unpublished). Deregulation of *fslA* gene transcription in a *fur* mutant has been documented in U112 using a reporter *lacZ* fusion (Buchan *et al.*, 2008). We have also observed a several-fold deregulation of the *feoB* gene in the *fur* mutant strains (unpublished), but not to the extent seen with the *fsl* genes.

As with many pathogens, iron acquisition is likely to be an important feature of pathogenesis. A recent study on the transcriptional profiling of Schu S4 within macrophages reported that the *fsl* genes were highly up-regulated in intracellularly replicating bacteria (Wehrly *et al.*, 2009). In a negative selection screen for virulence genes in mice using a collection of U112 transposon mutants, insertions in the *fslA-C* genes led to attenuation (Weiss *et al.*, 2007). An *fslE* deletion mutant was also found to be attenuated in mice (Milne *et al.*, 2007). The *fupA* gene, which plays a role in iron acquisition in Schu S4, surprisingly, does not appear to be iron regulated (Lindgren *et al.*, 2009). A *fupA* mutant is attenuated for virulence although it is not clear if the virulence defect is tied to the iron acquisition defect.

Conclusions

Interest in *F. tularensis* biology has undergone a revival in recent years. Iron is clearly a vital nutrient for this organism and the Fur-regulated siderophore-mediated iron uptake system encoded by the *fsl* genes represents a novel iron acquisition pathway. The *F. tularensis* siderophore of the organism has a rather simple structure resembling rhizoferrin, and two enzymes encoded by the *fsl* operon are required for siderophore production. It is not clear if additional biosynthetic loci are located elsewhere on the chromosome. *fslE* is a gene unique to *Francisella* strains and predicted to encode a beta-barrel protein in the outer membrane that is required for siderophore utilization and may function as a siderophore receptor. Additional members of the *fslE* family of genes may also play related roles in iron acquisition (Fig 8.4). In

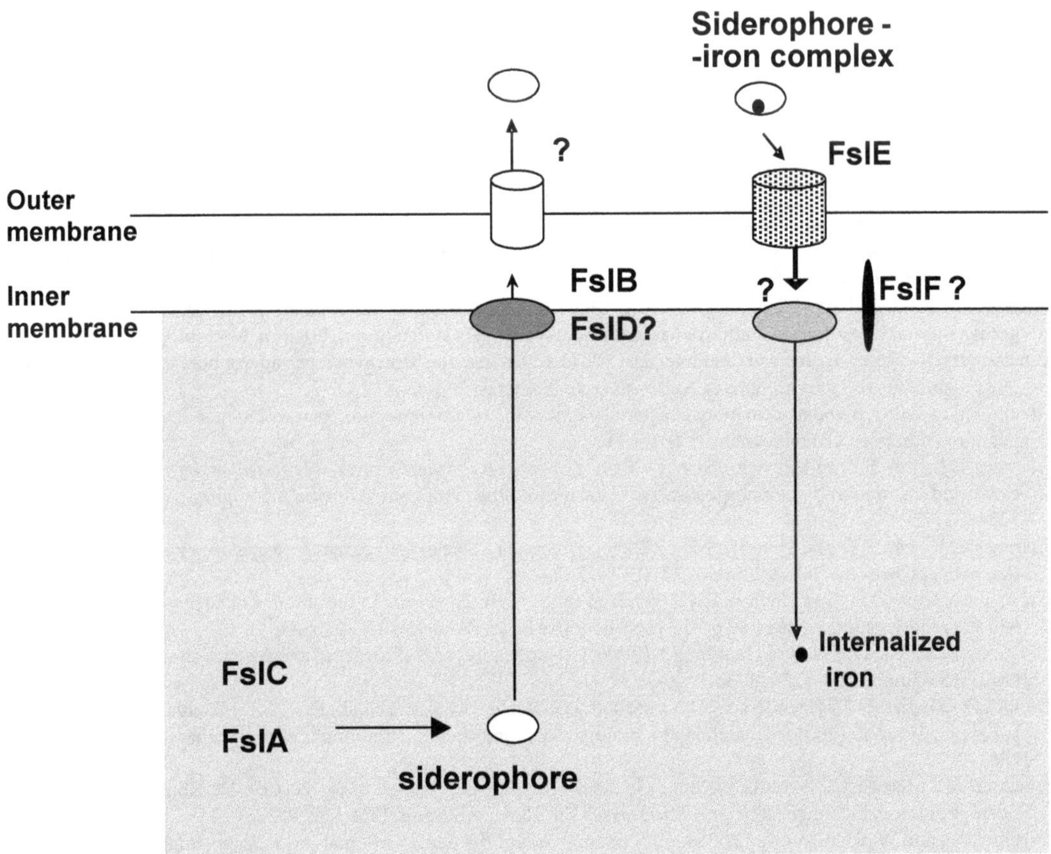

Figure 8.4 Model for role of *fsl* genes in siderophore-mediated iron uptake. The products of the *fslA* and *fslC* genes are enzymes involved in siderophore biosynthesis. The *fslB* (and possibly the *fslD*) gene products are likely to be involved in siderophore transport. Uptake involves FslE that may function as a receptor. The role of the *fslF* product, predicted to be in the inner membrane, is not clear.

the absence of identifiable *tonB*, *exbB* and *exbD* homologues, siderophore uptake is expected to involve a novel mechanism.

Future trends

These are still early years for the study of iron biology of *F. tularensis*, and there are currently more questions than answers. Efforts to understand the TonB-independent mechanism of siderophore uptake and/or utilization will probably be a major focus of study. Identification of other pathways for iron uptake will also be of interest. These studies need to be complemented by efforts to determine how host iron metabolism impacts the establishment and outcome of infection.

Web resources

BioHealthBase: NIAID Bioinformatics Resource Center (BRC) for *F. tularensis*
http://www.biohealthbase.org/GSearch/home.do?decorator=Francisella
FrantCyc: database of genome features and metabolic pathways of Schu S4 on the BiCyc server
http://biocyc.org/FRANT/organism-summary?object=FRANT
PEDANT 3 database: bioinformatics analysis of several *F. tularensis* genomes
http://pedant.gsf.de

Acknowledgements

Research in the author's laboratory is supported by grant AI067823 from the NIH (NIAID).

References

Allard, K.A., Viswanathan, V.K., and Cianciotto, N.P. (2006). *lbtA* and *lbtB* are required for production of the *Legionella pneumophila* siderophore legiobactin. J. Bacteriol. *188*, 1351–1363.

Atkins, H.S., Dassa, E., Walker, N.J., Griffin, K.F., Harland, D.N., Taylor, R.R., Duffield, M.L., and Titball, R.W. (2006). The identification and evaluation of ATP binding cassette systems in the intracellular bacterium *Francisella tularensis*. Res. Microbiol. *157*, 593–604.

Bagos, P.G., Liakopoulos, T.D., Spyropoulos, I.C., and Hamodrakas, S.J. (2004). PRED-TMBB: a web server for predicting the topology of beta-barrel outer membrane proteins. Nucleic Acids Res. *32*, W400–404.

Bhatnagar, N.B., Elkins, K.L., and Fortier, A.H. (1995). Heat stress alters the virulence of a rifampin-resistant mutant of *Francisella tularensis* LVS. Infect. Immun. *63*, 154–159.

Buchan, B.W., McLendon, M.K., and Jones, B.D. (2008). Identification of differentially regulated *Francisella tularensis* genes by use of a newly developed Tn5-based transposon delivery system. Appl. Environ. Microbiol. *74*, 2637–2645.

Chakraborty, R., Storey, E., and van der Helm, D. (2007). Molecular mechanism of ferricsiderophore passage through the outer membrane receptor proteins of *Escherichia coli*. Biometals *20*, 263–274.

Challis, G.L. (2005). A widely distributed bacterial pathway for siderophore biosynthesis independent of nonribosomal peptide synthetases. Chembiochem. *6*, 601–611.

Clemens, D.L., Lee, B.Y., and Horwitz, M.A. (2004). Virulent and avirulent strains of *Francisella tularensis* prevent acidification and maturation of their phagosomes and escape into the cytoplasm in human macrophages. Infect. Immun. *72*, 3204–3217.

Clemens, D.L., Lee, B.Y., and Horwitz, M.A. (2009). *Francisella tularensis* phagosomal escape does not require acidification of the phagosome. Infect. Immun. *77*, 1757–1773.

Crosa, L.M., Crosa, J.H., and Heffron, F. (2009). Iron transport in *Francisella* in the absence of a recognizable TonB protein still requires energy generated by the proton motive force. Biometals *22*, 337–344.

Deng, K., Blick, R.J., Liu, W., and Hansen, E.J. (2006). Identification of *Francisella tularensis* genes affected by iron limitation. Infect. Immun. *74*, 4224–4236.

Dennis, D.T., Inglesby, T.V., Henderson, D.A., Bartlett, J.G., Ascher, M.S., Eitzen, E., Fine, A.D., Friedlander, A.M., Hauer, J., Layton, M., *et al.* (2001). Tularemia as a biological weapon: medical and public health management. JAMA. *285*, 2763–2773.

Drechsel, H., Metzger, J., Freund, S., Jung, G., Boelaert, J.R., and Winkelmann, G. (1991). Rhizoferrin – a novel siderophore from the fungus *Rhizopus-Microsporus* Var Rhizopodiformis. Biol. Met. *4*, 238–243.

Drechsel, H., and Winkelmann, G. (2005). The configuration of the chiral carbon atoms in staphyloferrin A and analysis of the transport properties in *Staphylococcus aureus*. Biometals *18*, 75–81.

Forbes, J.R., and Gros, P. (2001). Divalent-metal transport by NRAMP proteins at the interface of host–pathogen interactions. Trends Microbiol. *9*, 397–403.

Fortier, A.H., Leiby, D.A., Narayanan, R.B., Asafoadjei, E., Crawford, R.M., Nacy, C.A., and Meltzer, M.S. (1995). Growth of *Francisella tularensis* LVS in macrophages: the acidic intracellular compartment provides essential iron required for growth. Infect. Immun. *63*, 1478–1483.

Golovliov, I., Baranov, V., Krocova, Z., Kovarova, H., and Sjostedt, A. (2003). An attenuated strain of the facultative intracellular bacterium *Francisella tularensis* can escape the phagosome of monocytic cells. Infect. Immun. *71*, 5940–5950.

Halmann, M., Benedict, M., and Mager, J. (1967). Nutritional requirements of *Pasteurella tularensis* for growth from small inocula. J. Gen. Microbiol. *49*, 451–460.

Halmann, M., and Mager, J. (1967). An endogenously produced substance essential for growth initiation of *Pasteurella tularensis*. J. Gen. Microbiol. *49*, 461–468.

Hubalek, M., Hernychova, L., Brychta, M., Lenco, J., Zechovska, J., and Stulik, J. (2004). Comparative proteome analysis of cellular proteins extracted from highly virulent *Francisella tularensis* ssp. tularensis and less virulent *F. tularensis* ssp. holarctica and *F. tularensis* ssp. mediaasiatica. Proteomics. *4*, 3048–3060.

Hubalek, M., Hernychova, L., Havlasova, J., Kasalova, I., Neubauerova, V., Stulik, J., Macela, A., Lundqvist, M., and Larsson, P. (2003). Towards proteome database of *Francisella tularensis*. J. Chromatogr. B Analyt. Technol. Biomed. Life Sci. *787*, 149–177.

Huntley, J.F., Conley, P.G., Hagman, K.E., and Norgard, M.V. (2007). Characterization of *Francisella tularensis* outer membrane proteins. J. Bacteriol. *189*, 561–574.

Keim, P., Johansson, A., and Wagner, D.M. (2007). Molecular epidemiology, evolution, and ecology of *Francisella*. Ann. NY Acad. Sci. *1105*, 30–66.

Kiss, K., Liu, W., Huntley, J.F., Norgard, M.V., and Hansen, E.J. (2008). Characterization of *fig* operon mutants of *Francisella novicida* U112. FEMS Microbiol. Lett. *285*, 270–277.

Kovarova, H., Halada, P., Man, P., Golovliov, I., Krocova, Z., Spacek, J., Porkertova, S., and Necasova, R. (2002). Proteome study of *Francisella tularensis* live vaccine strain-containing phagosome in Bcg/Nramp1 congenic macrophages: resistant allele contributes to permissive environment and susceptibility to infection. Proteomics. *2*, 85–93.

Kovarova, H., Hernychova, L., Hajduch, M., Sirova, M., and Macela, A. (2000). Influence of the *bcg* locus on natural resistance to primary infection with the facultative intracellular bacterium *Francisella tularensis* in mice. Infect. Immun. *68*, 1480–1484.

Larsson, P., Oyston, P.C., Chain, P., Chu, M.C., Duffield, M., Fuxelius, H.H., Garcia, E., Halltorp, G., Johansson, D., Isherwood, K.E., *et al.* (2005). The complete genome sequence of *Francisella tularensis*, the causative agent of tularemia. Nat. Genet. *37*, 153–159.

Lenco, J., Hubalek, M., Larsson, P., Fucikova, A., Brychta, M., Macela, A., and Stulik, J. (2007). Proteomics analysis of the *Francisella tularensis* LVS response to iron restriction: induction of the *F. tularensis* pathogenicity island proteins IglABC. FEMS Microbiol. Lett. *269*, 11–21.

Lindgren, H., Honn, M., Golovlev, I., Kadzhaev, K., Conlan, W., and Sjostedt, A. (2009). The 58-kilodalton major virulence factor of *Francisella tularensis* is required for efficient utilization of iron. Infect. Immun. *77*, 4429–4436.

Mager, J. (1964). A factor required for growth initiation of *Pasteurella tularensis*. Nature *203*, 898.

Milne, T.S., Michell, S.L., Diaper, H., Wikstrom, P., Svensson, K., Oyston, P.C., and Titball, R.W. (2007). A 55 kDa hypothetical membrane protein is an iron-regulated virulence factor of *Francisella tularensis* subsp. novicida U112. J. Med. Microbiol. *56*, 1268–1276.

Munzinger, M., Taraz, K., Budzikiewicz, H., Drechsel, H., Heymann, P., Winkelmann, G., and Meyer, J.M. (1999). S, S-rhizoferrin (enantio-rhizoferrin) – a siderophore of *Ralstonia* (*Pseudomonas*) *pickettii* DSM 6297 – the optical antipode of R, R-rhizoferrin isolated from fungi. Biometals *12*, 189–193.

Pekarek, R.S., Bostian, K.A., Bartelloni, P.J., Calia, F.M., and Beisel, W.R. (1969). The effects of *Francisella tularensis* infection on iron metabolism in man. Am. J. Med. Sci. *258*, 14–25.

Ramakrishnan, G., Meeker, A., and Dragulev, B. (2008). *fslE* is necessary for siderophore-mediated iron acquisition in *Francisella tularensis* Schu S4. J. Bacteriol. *190*, 5353–5361.

Rohmer, L., Fong, C., Abmayr, S., Wasnick, M., Larson Freeman, T.J., Radey, M., Guina, T., Svensson, K., Hayden, H.S., Jacobs, M., *et al.* (2007). Comparison of *Francisella tularensis* genomes reveals evolutionary events associated with the emergence of human pathogenic strains. Genome Biol. *8*, R102.

Schmerk, C.L., Duplantis, B.N., Wang, D., Burke, R.D., Chou, A.Y., Elkins, K.L., Ludu, J.S., and Nano, F.E. (2009). Characterization of the pathogenicity island protein PdpA and its role in the virulence of *Francisella novicida*. Microbiology *155*, 1489–1497.

Sjostedt, A. (2007). Tularemia: history, epidemiology, pathogen physiology, and clinical manifestations. Ann. NY Acad. Sci. *1105*, 1–29.

Sullivan, J.T., Jeffery, E.F., Shannon, J.D., and Ramakrishnan, G. (2006). Characterization of the siderophore of *Francisella tularensis* and role of *fslA* in siderophore production. J. Bacteriol. *188*, 3785–3795.

Thieken, A., and Winkelmann, G. (1992). Rhizoferrin: a complexone type siderophore of the Mucorales and entomophthorales (Zygomycetes). FEMS Microbiol. Lett. 73, 37–41.

Tusnady, G.E., and Simon, I. (1998). Principles governing amino acid composition of integral membrane proteins: application to topology prediction. J. Mol. Biol. *283*, 489–506.

Tusnady, G.E., and Simon, I. (2001). The HMMTOP transmembrane topology prediction server. Bioinformatics. *17*, 849–850.

Wehrly, T.D., Chong, A., Virtaneva, K., Sturdevant, D.E., Child, R., Edwards, J.A., Brouwer, D., Nair, V., Fischer, E.R., Wicke, L., *et al.* (2009). Intracellular biology and virulence determinants of *Francisella tularensis* revealed by transcriptional profiling inside macrophages. Cell. Microbiol. *11*, 1128–1150.

Weiss, D.S., Brotcke, A., Henry, T., Margolis, J.J., Chan, K., and Monack, D.M. (2007). *In vivo* negative selection screen identifies genes required for *Francisella* virulence. Proc. Natl. Acad. Sci. U.S.A. *104*, 6037–6042.

Haem and Iron Metabolism in *Bacteroides*

9

Edson R. Rocha and C. Jeffrey Smith

Abstract

Bacteroides spp. have an essential requirement for haem and non-haem iron. They cannot synthesize the tetrapyrrole macrocycle ring due to a lack of genes for the haem biosynthetic pathway. It is remarkable that haem-dependent organisms outnumber haem-independent organisms in the lower intestinal tract suggesting that haem biosynthesis is not essential for colonization of the colonic environment. However, this colonization advantage may be due to the fact that under anaerobic conditions in the presence of haem, *B. fragilis* can generate nearly double the amount of ATP than *Escherichia coli* per mol of glucose. This high energy yield is linked to a rudimentary haem-induced fumarate reductase and cytochrome *b*-dependent electron transport energy metabolism pathway which uses fumarate as the terminal electron acceptor. Moreover, *Bacteroides* spp. can incorporate iron-deuteroporphyrin and iron-mesoporphyrin into a functional type-*b* cytochrome. Haem can be demetalated without cleaving the tetrapyrrole ring releasing free iron and free protoporphyrin IX. The ability of the opportunistic human pathogen *B. fragilis* to cause infections seems to be due in part to its ability to scavenge haem and iron from host proteins. The in-frame translated intergenic region of the fused FeoAB proteins are exclusively present in gastro-intestinal colonizers belonging to the Bacteroidetes, Firmicutes and Actinobacteria phyla. Several members of the *Bacteroides* group have three orthologues of the mammalian-type bacterial ferritin gene, *ftnA*. FtnA may play an important role in protection against iron-induced oxidative stress in this group of highly aerotolerant anaerobes.

Introduction

Haem (iron-protoporphyrin IX) is essential for most living organisms and it is a prosthetic cofactor group in haemoproteins such as haemoglobin, myoglobin, cytochromes, catalases and peroxidases. Haem also is a preferred source of inorganic iron for many pathogenic bacteria. In the natural world, most organisms are able to synthesize haem from the precursor 5-γ-aminolaevulinic acid (5-ALA). Humans, fungi and the α-proteobacteria synthesize 5-ALA by conversion of succinate plus glycine through HemA (5-ALA synthase). Most bacteria, however, synthesize 5-ALA from an alternative pathway via L-glutamyl-tRNA to L-glutamate 1-semialdehyde through HemA (glutamyl-tRNA reductase). Subsequently, L-glutamate 1-semialdehyde is converted to 5-ALA by the action of HemL (glutamate-1-semialdehyde 2,1-aminomutase). 5-ALA is condensed to porphobilinogen and after several steps of oxidation and decarboxylations, this leads to the final precursor, the pyrrole macrocycle protoporphyrin IX (PpIX). Inorganic FeII is incorporated into the PpIX by ferrochelatase (HemH) in the final step of haem biosynthesis. Prokaryotes may modify haem to other derivative forms such as haem a, haem c and haem d (Ponka, 1999; Panek and O'Brian, 2002; Frankenberg *et al.*, 2003).

A few bacteria are known to have an essential requirement for exogenous haem or its precursors due to lack of genetic haem biosynthetic pathway. Among these bacteria are *Haemophilus*, *Bacteroides*,

Porphyromonas, Prevotella and *Bartonella* (Stojiljkovic and Perkins-Balding, 2002). Recent genomic analysis of prokaryotes, whose entire genome sequences have been completed, revealed that the number of organisms lacking part or entire haem biosynthetic pathway is higher than previously expected (O'Brian and Thony-Meyer, 2002). For more detailed consideration on the biochemistry and synthesis of haem in prokaryotes, the reader is directed to more complete sources (O'Brian and Thony-Meyer, 2002; Panek and O'Brian, 2002; Schobert and Jahn, 2002; Stojiljkovic, and Perkins-Balding, 2002; Frankenberg *et al.*, 2003). Because of the fact that *Bacteroides* are haem requiring organisms and the utilization of haem can provide a source of iron, we present in this chapter some aspects of haem utilization as a source of both the essential tetrapyrrole macrocycle ring and non-haem iron in the *Bacteroides* group that colonizes the GI tract of both ruminant and non-ruminant animals. The overwhelming majority of the studies done in this field have been carried out using the commensal and opportunistic human pathogen *B. fragilis* as a model strain and consequently, we will describe the aspects of iron and haem metabolism employed by this species.

Essential requirement for haem

Bacteroides species constitute approximately 20–40% of the total bacteria colonizing the lower intestinal tract with numbers reaching 10^{11} to 10^{12} CFU/gram of colonic content (Hooper *et al.*, 2002; Smith *et al.*, 2003; Eckburg *et al.*, 2005). *Bacteroides* species are known to have an absolute requirement for a source of exogenous haem for growth (Holdeman *et al.*, 1977; Sperry *et al.*, 1977; Rocha *et al.*, 1991). Nonetheless, when protoporphyrin IX plus a source of inorganic iron are supplied *in vitro, B. fragilis* is able to synthesize haem (Sperry *et al.*, 1977; Rocha *et al.*, 1991) even though a ferrochelatase (HemH) homologue has yet to be identified in this organism. The complete dependence of *Bacteroides* on exogenous haem is a paradox with its ability to outnumber and dominate the colonic microflora where it is predicted that 500 to a 1000 different bacterial species cohabit in a very nutritionally competitive niche (Hooper *et al.*, 2002; Smith *et al.*, 2003; Eckburg *et al.*, 2005; Ley *et al.*, 2005). It is intriguing that the predominant members of the human colon are unable to synthesize an essential nutrient such as haem. This characteristic should be an ecological disadvantage in view of the survival pressure to compete with organisms capable of the haem de novo synthesis. This ecological pressure would lead toward the selection of haem independent strains in the colonization of the intestinal tract (Fuller and Caldwell, 1982). However, it is the haem-independent bacteria in the intestinal tract such as *Escherichia coli* that are outnumbered by *Bacteroides* at 1000-fold. Similarly, other major intestinal colonizers such as *Bifidobacterium, Clostridium* and *Streptococcus faecalis*, whose complete genome sequences are available, lack complete haem biosynthesis pathway (Fig. 9.1). In this regard, it seems that lack of a tetrapyrrole macrocycle synthetic pathway is not a nutritional selective disadvantage to the colonic microflora competitiveness for colonization (Fuller and Caldwell, 1982). Fig. 9.1 is a schematic diagram comparing the genetic organization of gene homologues for the haem biosynthesis pathway in the Bacteroidetes phylum (also known as CFB group) and other intestinal colonizers. In the members of the Bacteroidetes whose genomic sequences have been completed, the aerobic free living genera, *Cytophaga, Flavobacterium* and *Salinibacter*, possess complete haem biosynthetic pathways. In contrast, the human gastrointestinal commensals *Bacteroides, Parabacteroides, Porphyromonas* and the acanthamoeba endocytobiont *Candidatus Amoebophilus asiaticus* (Horn *et al.*, 2001) lack nearly the entire pathway except for *Porphyromonas*, which possesses a *hemH* (ferrochelatase) homologue (Fig. 9.1). Though most genes required for haem biosynthesis are missing in these bacteria, one single component, HemN (oxygen-independent coproporphyrinogen III oxidase), is highly conserved among these organisms (Fig. 9.1). On the other hand, both HemN and HemF (oxygen-dependent coproporphyrinogen III oxidase) are present in the free-living aerobic Bacteroidetes, *Cytophaga, Flavobacterium* and *Salinobacter*. The significance of the conserved *hemN* locus in different phylogenetic groups of commensal intestinal colonizers is yet to be understood.

It is not clear whether the exogenous haem requirement is a contributing factor for the adaptation of *Bacteroides* to colonize host anaerobic niches such as the lower intestinal tract where the availability of haem and iron is provided by ingestion of red meat, epithelial desquamation, autolysing epithelial and

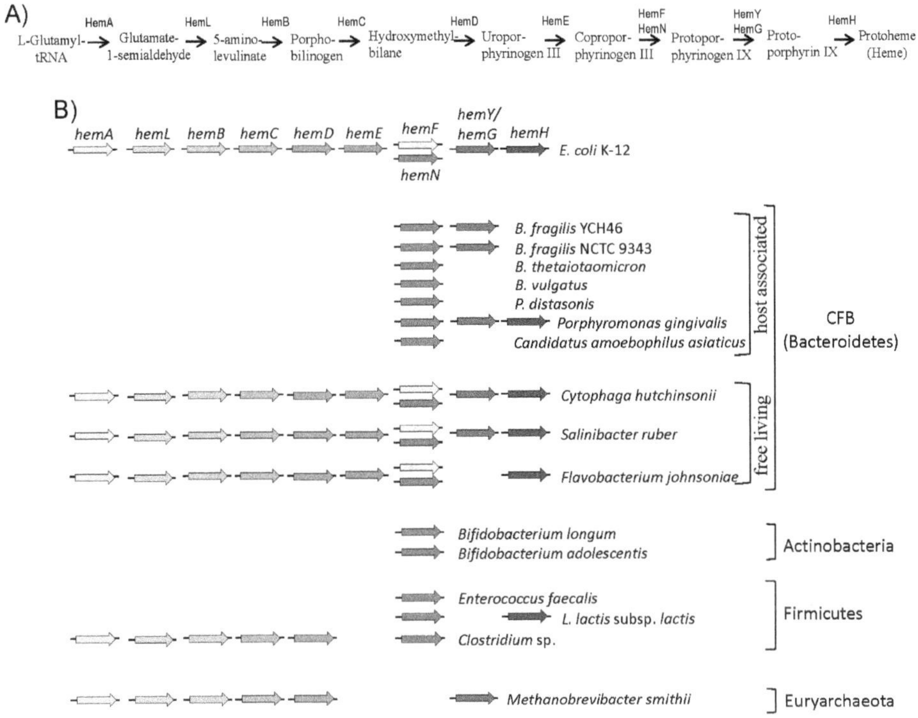

Figure 9.1 Conservation of the haem biosynthesis pathway gene loci in the genomes of Bacteroidetes, Firmicutes, Actinobacteria and Euryarchaeota groups of intestinal colonizers. Each gene locus shown is organized according to their biosynthetic pathway function and not by their genetic organization. *E. coli* K-12 (A) complete haem biosynthesis pathway with corresponding (B) gene loci designations are shown for the purpose of comparative analysis. Each gene is represented by an arrow with different shades of grey. Arrows of the same grey shade in a vertical column position are predicted to encode gene products with similarity respective to *E. coli* genes. Data presented were compiled from the following sources: http://www.ncbi.nlm.nih.gov/; http://www.sanger.ac.uk/Projects/B_fragilis/http://genome.wustl.edu/project_index.cgi; http://www.genome.ad.jp/kegg/pathway.html. HemA$^{Glu\text{-}tRNA}$, glutamyl-tRNA reductase (EC 1.2.1.70); HemB, porphobilinogen synthase (EC 4.2.1.24); HemC, hydroxymethylbilane synthase (EC 2.5.1.61); HemD, uroporphyrinogen-III synthase (EC:4.2.1.75); HemE: uroporphyrinogen decarboxylase (EC 4.1.1.37); HemF, coproporphyrinogen III oxidase (EC 1.3.3.3); HemG/HemY, protoporphyrinogen oxidase (EC 1.3.3.4), HemH, ferrochelatase (EC 4.99.1.1); HemL, glutamate-1-semialdehyde 2,1-aminomutase (EC 5.4.3.8); HemN, oxygen-independent coproporphyrinogen III oxidase (EC 1.3.99.22).The drawings are not to scale.

bacterial cells and a diversity of plant foods (Macy *et al.*, 1975; Young *et al.*, 1989; Fairweather-Tait *et al.*, 1999).

Haem availability plays an important role in the growth and fermentation products formed by saccharolytic *Bacteroides* species isolated from human large intestine, human clinical isolates and from the GI tract of ruminants (Caldwell *et al.*, 1965; Macy *et al.*, 1975; Sperry *et al.*, 1977; Chen and Wolin, 1981). *B. fragilis* ATCC 25285 did not grow without haem and required 1 ng/ml for growth while growth of *B. splanchnicus* and *B. eggerthii* had a slow growth without haem but growth was stimulated by the addition of 10 ng/ml and 100 ng/ml haem respectively (Chen and Wolin, 1981). Growth of *B. fragilis* ATCC 23745 on glucose was highly stimulated by the addition of 2 µg/ml haem in the medium (Macy *et al.*, 1975). The enterotoxigenic *B. fragilis* ATCC 43860 was unable to grow in the absence of haem or protoporphyrin IX, but protoporphyrin gave maximum growth at 0.01 µg/ml. Haem was also stimulatory but less effective than protoporphyrin IX as 0.5 μg/ml was required for maximum growth (Rocha *et al.*, 1991). When added into the culture media, the porphyrin precursors γ-aminolaevulinate or porphobilinogen are unable to replace haem (Sperry *et al.*, 1977). This is in agreement with the absence of genes involved in the metabolism and synthesis of haem in these organisms.

The major products of fermentation from glucose by *B. fragilis* ATCC 23745 in the absence of haem was fumarate, lactate and acetate with no detectable amounts of fumarate reductase and cytochromes (Macy *et al.*, 1975). In the presence of haem, succinate, propionate and acetate were the major products of fermentation. A *b*-type cytochrome, possibly a *c*-type cytochrome as well, and a very active fumarate reductase were induced by haem (Macy *et al.*, 1975). Chen and Wolin (1981) further expanded these findings using other strains of *Bacteroides*, the *B. fragilis* ATCC 25285, *B. splanchnicus* and *B. eggerthii*. The results showed that when haem was present at 1 ng/ml, malate instead of fumarate and lactate were the major product of glucose fermentation. When 1 μg/ml of haem was used, malate was not formed, lactate production decreased and succinate and acetate were the major products of fermentation. Haem is also required for the formation of butyrate by *B. splanchnicus*. It has also been demonstrated that vitamin B12 affects the fermentation products in a haem dependent manner (Chen and Wolin, 1981). In the presence of haem, *B. fragilis* can generate 4.5 mol of ATP per mol of glucose. In contrast, 1.7 mol ATP per mol of glucose is generated in the absence of haem as determined by molar growth yield (Macy *et al.*, 1975). In addition, studies by Baughn and Malamy (2002; 2003) have shown that in the presence of haem, the generation time for *B. fragilis* is reduced by three- to fourfold. This haem dependent anaerobic ATP generation during exponential growth is linked to a rudimentary cytochrome *b*-dependent electron transport system which uses fumarate as the terminal electron acceptor (Macy *et al.*, 1975; Harris and Reddy, 1977; Baughn and Malamy, 2003). The formation of succinate, the major glucose fermentation product of *Bacteroides* spp., is mediated by the haem-induced fumarate reductase (Macy *et al.*, 1975; Harris and Reddy, 1977; Baughn and Malamy, 2003).

Chen and Wolin (1981) have suggested that variations in haem availability in human lower intestinal tract may affect the growth of *Bacteroides* because of its effect on the malate or succinate fermentation pathways. This is particularly relevant regarding to the ability of *Bacteroides* to generate a high ATP yield from glucose in the presence of haem in the anaerobic lower intestinal tract where *Bacteroides* spp. outnumber eubacteria such as *E. coli*. This is probably because *E. coli* can make only 2.5 mol of ATP per mol of glucose by fermentation while *Bacteroides* can yield 4.5 mol of ATP per mol of glucose (Macy *et al.*, 1975). This highly efficient anaerobic haem-dependent energy system is an advantage for *Bacteroides* to grow and obtain maximal benefit of available haem in the colonic environment as co-colonizers such as *E. coli* K-12 cannot utilize exogenous haem or porphyrins and does not possesses any known uptake mechanisms for haem or porphyrin (Săsărman *et al.*, 1968; Tatsumi and Wachi, 2008).

Utilization of metallo-protoporphyrin IX side group derivatives

Utilization of tetrapyrroles with modified side groups in the macrocycle ring is a characteristic shared by different species of the *Bacteroides-Prevotella* group. One characteristic of the *Bacteroides* from human and ruminant origins is their ability to physiologically obtain and incorporate not only iron-protoporphyrin IX, but also iron-mesoprotoporphyrin IX (mesohaem) and iron-deuteroprotoporphyrin IX (deuterohaem), into functional cytochrome type-*b* (McCall and Caldwell, 1977; Fuller and Caldwell, 1982). Moreover, another unique characteristic of *Bacteroides* is its ability to demetalize Mg-protoporphyrin IX (Mg-PpIX), Mg-deuteroPpIX, Mg-mesoPpIX, Mn-PpIX, Mn-deuteroPpIX and Mn-mesoPpIX and then insert iron in the metal pocket through an unidentified putative ferrochelatase activity to form haem, mesohaem and deuterohaem respectively (Fig. 9.2). Either haem or side group modified haem are incorporated into unusual *b*-type cytochromes which, as mentioned above, are physiologically functional (McCall and Caldwell, 1977; Fuller and Caldwell, 1982).

This unique trait of *Bacteroides* is very relevant for the physiological metabolism of tetrapyrrole macrocycles in the intestinal tract. The unabsorbed haem that reaches the colon together with any haem associated with non-degraded haemoproteins, such as haemoglobin, myoglobin and catalases, that

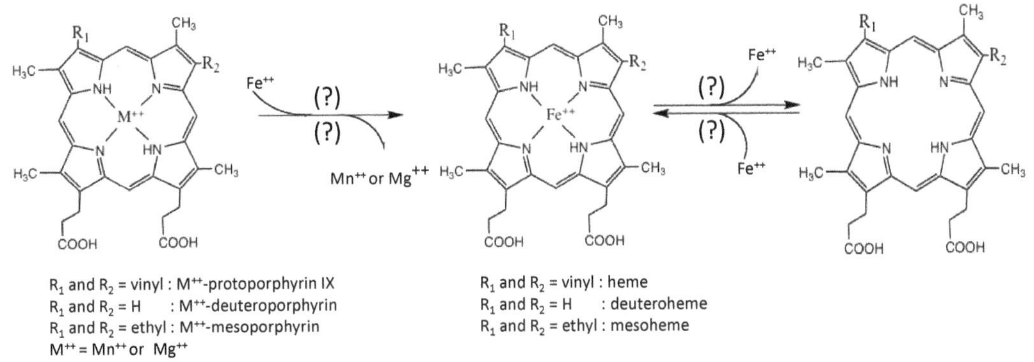

Figure 9.2 Reaction scheme for the demetalation of metalloporphyrin complexes and utilization of modified porphyrin side chain by *Bacteroides/Prevotella* group. *Bacteroides* spp. and *Prevotella* spp. can remove non-ferrous ion from the tetrapyrrole macrocycle and insert Fe^{++} into the tetrapyrrole ring mediated by a ferrochelatase-like activity. Iron-protoporphyrin can be demetalated to release free iron and free protoporphyrin. Iron-containing protoporphyrins are formed in cells grown with protoporphyrin plus inorganic iron though ferrochelatase (HemH) homologue is not present in the sequenced genomes of *Bacteroides* spp. (?): It is not known whether these activities are the action of multifunctional enzymes or the removal of non-ferrous iron and insertion of iron into the same moiety involves separate enzymes. Adapted from McCall and Caldwell (1977), Shah *et al.* (1979), Fuller and Caldwell (1982) and Brazier (1990).

find their ways into the colonic lumen are converted by bacteria to a range of haem-derived porphyrins (HDP) lacking iron (Young *et al.*, 1989; 1990). In addition to removal of iron from haem, modifications of the vinyl side chains also occur when haem is exposed to faecal bacteria. And at least four HDP may be found, protoporphyrin, deuteroporphyrin, mesoporphyrin and pemtoporphyrin (Young *et al.*, 1990). They appear in the faeces because of the bacterial action on the vinyl side chains of protoporphyrin (Beukeveld *et al.*, 1987). *In vivo*, the types of HDP are consistent within a given individual but differ in the amount and type between individuals. Factors such as faecal flora haem metabolism and redox potential may account for the differences in amount and type of HDP (Young *et al.*, 1990). The fluctuations in HDP content in the faeces of healthy persons depends entirely and exclusively on the anaerobic flora (Beukeveld *et al.*, 1987). Suppression of the subjects intestinal aerobic flora with polymixin E had no effect on faecal porphyrin excretion. In contrast, when selective suppression of anaerobic bacterial flora following the administration of cephaloridin or complete flora suppression with both polymixin E and cephaloridin were performed, it completely inhibited the transformation of protoporphyrin to pemto- and deuteroporphyrin and decreased mesoporphyrin contents in the faeces (Beukeveld *et al.*, 1987). The production of protoporphyrin IX by the demetalation of haem and modifications of the vinyl side chains also occur in the closely related members of the *Bacteroides* group, *Porphyromonas* and *Prevotella* (formerly black-pigmented *Bacteroides*) *in vitro* (Shah *et al.*, 1979) and *in vivo* from aspirated pus from various clinical sites including surgical patients (Brazier, 1990). This mechanism of demetalizing haem, which strips off iron from haem without breaking the macrocycle ring, releasing free PpIX and iron, is distinct from other prokaryotes which use haem oxygenases (HemO) and haem oxygenase-like enzymes (IsdG and IsdI) to cleave haem and release free iron, biliverdin and carbon monoxide (Zhu *et al.*, 2000; Skaar *et al.*, 2004).

Utilization of haem and iron from host proteins

In mammalian tissues, the iron withholding mechanisms maintain iron tightly bound to the prosthetic group haem, mainly in haemoglobin and myoglobin, and in iron binding proteins such as transferrin and lactoferrin and the iron storage protein, ferritin. Consequently, many aerobic bacteria and in particular, pathogens, will not grow in the low iron availability of host tissues, unless they possess specific iron acquisition mechanisms to scavenge iron from host iron-containing proteins (Otto *et al.*, 1992; Wandersman and Delepelaire, 2004). Although these mechanisms have been extensively studied in aerobic bacteria, iron uptake, transport, utilization and storage in the *Bacteroides* group have been largely overlooked. Our current knowledge in this field is relatively restricted to the studies reported with *B. fragilis*. The success of the opportunistic human pathogen *B. fragilis* to overcome host iron and haem withholding mechanisms seems to be due in part to its ability to scavenge haem and iron from host iron and haem-containing proteins. *B. fragilis* has an essential requirement for both haem iron and non-haem iron (Rocha *et al.*, 1991). Production of iron chelators by *B. fragilis* has not been found (Otto *et al.*, 1988), but it is able to utilize iron from haem, haemoglobin, haptoglobin–haemoglobin complex and transferrin under low iron conditions (Verweij-van Vught *et al.*, 1988; Otto *et al.*, 1994).

In response to iron restriction, *B. fragilis* strain BE1 synthesizes four iron-repressible OMP proteins (IROMPs) of 89, 49, 44 and 23 kDa (Otto *et al.*, 1988 and 1990). Evidence that the 44-kDa IROMP is expressed *in vivo* was provided by analysis of serum from patients post-infected with *B. fragilis* which showed antibodies reactive with the 44-kDa protein but not in serum from patients infected with *B. vulgatus*, *B. distasonis* and *B. cappilosus* (Otto *et al.*, 1991). This indicates that it is produced *in vivo* exclusively by *B. fragilis*. These findings may be related to the fact that *B. fragilis* is better adapted to utilize haem-containing compounds as iron source under low iron conditions compared to other members of the *B. fragilis* group (Verweij-van Vught *et al.*, 1988). *In vitro* experiments using anti-44-kDa protein antibodies inhibited growth of *B. fragilis* suggesting the involvement of this protein in haem uptake (Otto *et al.*, 1990). More recently, an α-enolase isolated from cytoplasm and inner membrane was also found to be iron-regulated in *B. fragilis* strain BE1, but in contrast to enolases from other organisms which are found on the cell surface and bind plasminogen, the *B. fragilis* enolase lacked this property (Sijbrandi *et al.*, 2005).

Among the *Bacteroides* group, *B. fragilis* is the only species able to grow on human serum or plasma possibly through its advantage in obtaining haem from haem-bound compounds and iron from transferrin (Verweij-van Vught *et al.*, 1988). The serum growth inhibitory effect on *B. vulgatus* was overcome by addition of haem and an unidentified growth factor present in tryptone (Verweij-van Vught *et al.*, 1986a,b). In addition, a contributing factor that may enable *B. fragilis* to overcome haem limitations imposed by the serum protein, haemopexin, is the ability to degrade the haemopexin-bound haem complex making haem available for growth (Rocha *et al.*, 2001). *Bacteroides* spp. also produce haemolysins which have been demonstrated to lyse erythrocytes *in vitro* (Robertson *et al.*, 2006). The two component haemolysins, HlyA and HlyB, are encoded by genes in an operon. *hlyBA* transcription is regulated by iron in a Fur-dependent manner suggesting that the presence of functional haemolysins in *B. fragilis* may contribute to its ability to obtain haem in extra-intestinal infections (Robertson *et al.*, 2006).

In intra-abdominal infections, the mutual molecular mechanisms of the bacterial synergism of *E. coli* and *B. fragilis* leading to development of abscess formation are not completely understood. However, Otto *et al.* (2002) have shown that *E. coli* haemoglobin protease (Hbp) autotransporter contributes to the pathogenic synergy of abscess formation in a murine infection model. This extracellular protein was initially isolated from culture supernatant of *E. coli* EB1 by affinity chromatography on a haem-agarose column (Otto *et al.*, 1998). The authors have shown that purified Hbp, a serine type protease, is able to deliver haem to *B. fragilis* strain BE1 which could partially explain the capacity of *B. fragilis* to seize iron from haem to overcome the low iron conditions in the host tissues (Otto *et al.*, 1998; Otto *et al.*, 2002). Using a mixed infection model with *E. coli* EB1 and *B. fragilis* BE1 strains, Otto *et al.* (2002) demonstrated that *E. coli* Hbp is able to supply haem to both *E. coli* and *B. fragilis* resulting in abscess formation. Only *E. coli* Hbp-positive strains were able to induce abscesses in mixed infections with *B. fragilis* BE1 in a mice model (Otto *et al.*, 2002).

Genetic organization of the feo system among intestinal colonizers

In Gram-positive and Gram-negative bacteria, the ferrous iron uptake (Feo) system primarily consists of two proteins, the small cytoplasmic, FeoA; and the large transmembrane protein, FeoB. In γ-proteobacteria, the Feo system also contains a third small protein component, FeoC (Cartron *et al.*, 2006). Analysis of genome sequences revealed the presence of a *feoAB* in the *B. fragilis* 638R and YCH46 strains and two *feoAB* homologous in the NCTC 9343 strain (http://www.sanger.ac.uk/Projects/B_fragilis/; Kuwahara *et al.*, 2004; Cerdeno-Tarraga *et al.*, 2005). In contrast to aerobic and facultative bacteria, intestinal anaerobes such as the *Bacteroides/Porphyromonas* group, *Eubacterium* and *Ruminococcus*, the *feoA* and *feoB* genes encode a single fused polypeptide containing both the FeoA and FeoB conserved domains (Fig. 9.3). Overall, very little is known about iron uptake and transport in the *Bacteroides* spp. Some studies have shown that both ferric and ferrous iron may be utilized by *B. fragilis* but the iron

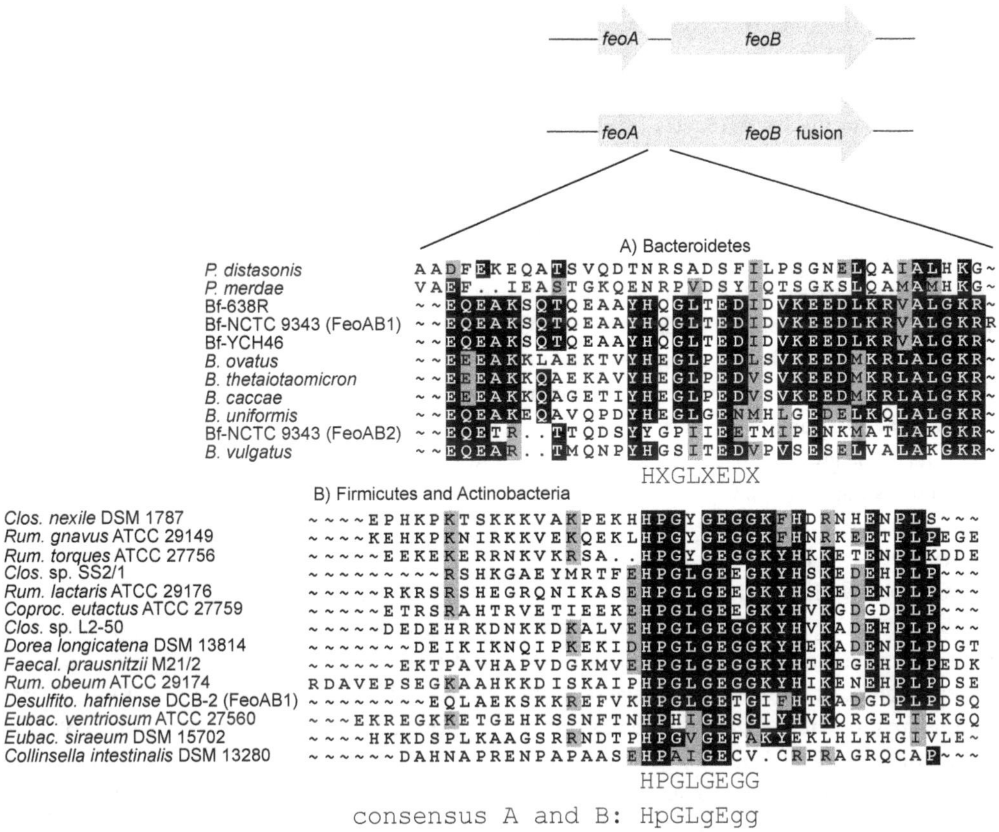

Figure 9.3 Multiple alignment of the deduced amino acid sequence for the in-frame translated intergenic region of the *feoAB* system in the intestinal colonizers (A) Bacteroidetes and (B) Firmicutes and Actinobacteria. *Bacteroides fragilis* NCTC 9343 FeoAB1 (CAH07068); *B. fragilis* NCTC 9343 FeoAB2 (CAH09534); *B. fragilis* YCH46 (BAD48168); *B. ovatus* ATCC 8483 (EDO09955); *B. thetaiotaomicron* VPI-5482 (AAO76699); *B. caccae* ATCC 43185 (EDM20115); *B. uniformis* ATCC 8492 (EDO51911); *B. vulgatus* ATCC 8482 (ABR40526); *Clostridium nexile* (EEA83224); *Clostridium* sp. SS2/1 (EDS21495); *Clostridium* sp. L2–50 (EDO57790); *Collinsella intestinalis* (EEA88022); *Coprococcus eutactus* (EDP26934); *Desulfitobacterium hafniense* (ACL18360); *Dorea longicatena* (EDM61700); *Eubacterium ventriosum* (EDM51574); *E. siraeum* (EDS00942); *Faecalibacterium prausnitzii* (EDP22159); *Parabacteroides distasonis* ATCC 8503 (ABR43218); *P. merdae* ATCC 43184 (EDN87569). *Ruminococcus gnavus* (EDN78735); *R. lactaris* (EDY32736); *R. obeum* (EDM86872); *R. torques* (EDK25291); Conserved amino acid residues (>50% identity) are labelled with black boxes. Semiconserved amino acid substitutions are depicted by grey boxes. Alignment of peptide sequences was performed using GCG programs PILEUP and BOX with the peptide scoring matrix default data file blosum62.cmp for the comparison of the amino acids substitution. A conserved sequence similarity between(A) Bacteroidetes and (B) Firmicutes and Actinobacteria is shown below the panel.

acquisition systems involved remain to be elucidated (Rocha *et al.*, 1991). A better understanding of the iron uptake in the Bacteroidetes group comes from the studies showing that *P. gingivalis* FeoB1 is essentially a specific ferrous iron high-affinity system transporter (Dashper *et al.*, 2005) while FeoB2 is implicated in manganese transport and resistance to oxidative stress generated by oxygen and H_2O_2 (Dashper *et al.*, 2005; He *et al.*, 2006). A *feoB1* mutant is avirulent in a mouse lesion model (Dashper *et al.*, 2005) and a *feoB2* mutant is less invasive to human umbilical vein endothelial cells (He *et al.*, 2006). The *B. thetaiotaomicron feoAB* has significantly higher expression levels in gnotobiotic mice during the suckling period compared to lower levels during the weaning period (Supplemental Table S2 in Bjursell *et al.*, 2006). This fact may have important implications in iron uptake by the human colonic flora since infants fed with added-iron formula have fewer babies colonized with *Bacteroides* and lower number of *Bacteroides* in the intestinal tract compared to non-added iron supplemented formula-fed infants (Balmer and Wharton, 1991).

In the GenBank database there are at least 30 organisms carrying the *feoA* and *feoB* genes fused in-frame belonging to the Bacteroidetes (CFB), the Gram-positive firmicutes and actinobacteria phyla of intestinal colonizers. For the FeoA–FeoB hybrids, the region between the FeoA and FeoB segments shows good sequence similarity when the firmicutes and actinobacteria are compared with each other, but not when these groups are compared to the Bacteroidetes (Fig. 9.3). When FeoA-FeoB from *Bacteroides* spp. were aligned with the amino acid sequences from the firmicutes and actinobacteria groups, the region between the FeoA and FeoB showed a conserved sequence similarity (HpGLgEgg). Within the *B. fragilis* group, Glu74, Glu76, Tyr86, Glu92 and Glu98 of the *B. fragilis* NCTC 9343 FeoAB1 amino acid sequence are highly conserved among all strains. The role this conserved region plays in the FeoAB ferrous iron transport system is not known. It is convenient to mention that genes encoding separate FeoA and FeoB polypeptides are also present in several species within the Bacteroidetes, firmicutes and actinobacteria group of bacteria (not shown) but, thus far, fused FeoAB peptides were not found in non-gastrointestinal isolates.

Iron storage and oxidative stress

In the microbial world iron is an essential nutrient but free iron forms a dangerous combination with oxygen by generating life-treating toxic cell-damaging oxygen radicals. As a result, intracellular iron is maintained at non-toxic levels by accumulating iron in a non-reactive insoluble inorganic form within the spherical globular shell of iron storage proteins (Andrews, 1998; Touati, 2000). Ferritins and bacterioferritins are thought to act as iron-storage proteins to provide a source of iron when iron is not readily available from exogenous sources, and to prevent excess iron from catalysing the production of toxic free radicals (Andrews, 1998; Touati, 2000). Both of these functions would seem to be superfluous in an organism which grows only in the absence of oxygen. The evidence that *B. fragilis* produces a ferritin (FtnA) is therefore of considerable interest, as it supports the suggestion that iron uptake in anaerobes is more complex than previously supposed. *B. fragilis* NCTC 4393, YCH46 and 638R strains possess both the typical non-haem ferritins with similarity to mammalian ferritin H-chain and haem-containing bacterioferritins (Rocha and Smith, 2004; Sund *et al.*, 2008). Interestingly, *B. thetaiotaomicron* and *B. vulgates*, the most common *Bacteroides* species isolated from fecal samples (Hooper *et al.*, 2002), possess three ferritins and a bacterioferritin. Though *B. uniformis*, *B. stercoris* and *P. merdae* also contain three FtnA homologues, they do not possess bacterioferritin. Comparison of *B. fragilis* NCTC 9343 FtnA with other *Bacteroides* ferritins, it showed100% identity with FtnA from *B. fragilis* 638R and YCH46 strains and approximately 89–56% identity to FtnA1 and FtnA2 from other *Bacteroides* species. However, *B. fragilis* NCTC 9343 FtnA has a lower percentage of amino acid identity when compared to FtnA3 (29% identity). A phylogenetic comparison between Bacteroidetes and other bacterial and eukaryotes ferritins showed that Bacteroidetes FtnA1 and FtnA2 are clustered together in a major group within bacterial ferritins (Fig. 9.4). In contrast, FtnA3 orthologues are branched together with Mycobacteria and eukaryotes. This extended number of ferritins in the Bacteroidetes group is quite unusual for bacteria since bacteria carry mostly two ferritins, two bacterioferritins or one of each while some lack any known

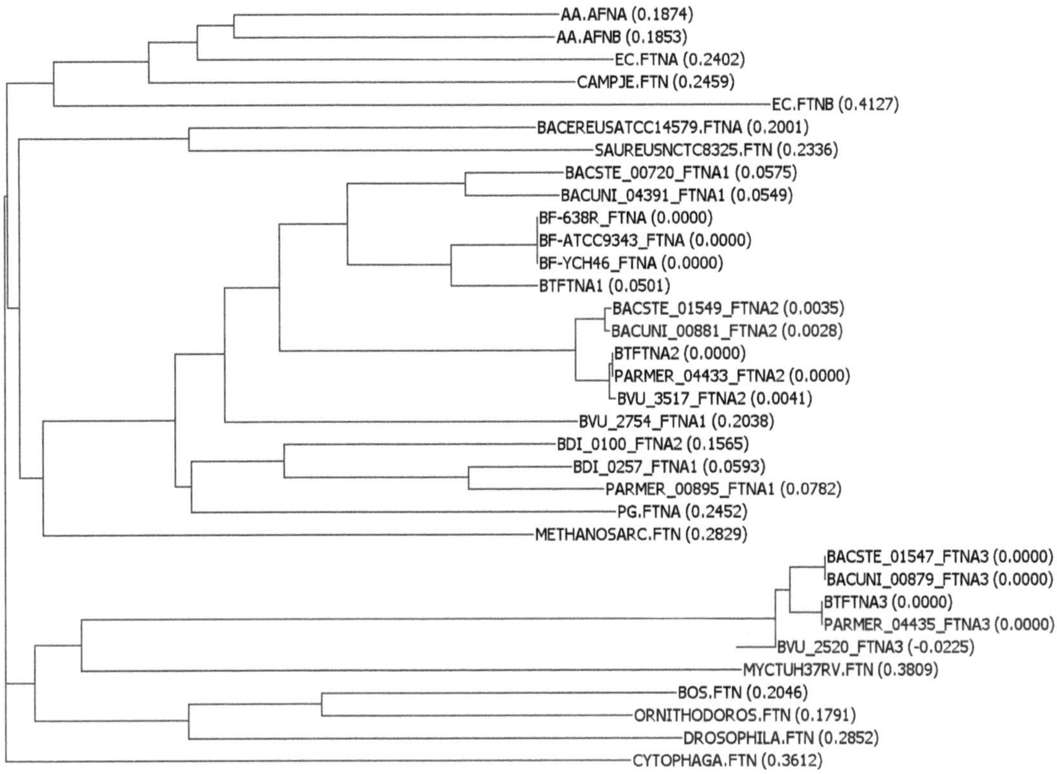

Figure 9.4 Phylogenetic comparison of 34 ferritins (Ftn) from diverse sources This unrooted tree was constructed from a multiple protein sequence alignment using GCG program PILEUP with peptide scoring matrix default data file blosom62.cmp for the comparison of amino acids substitution. A phylogenetic tree was calculated based on sequence distance utilizing the Neighbour Joining algorithm method built in the Vector NTI Advance 11 program AlignX. The following sequences with accession numbers were used: Aa, *Actinobacillus actinomycetemcomitans* FtnA (BAE53404) and FtnB (BAE53405); Bcereus, *Bacillus cereus* ATCC 14579 (NP_834718); Bf-638R, *Bacteroides fragilis* 638R (AAK29742); Bf-NCTC 9343, *B. fragilis* NCTC 9343 (YP_212498); Bf-YCH46, *B. fragilis* YCH46 (YP_100328); Bacste, *Bacteroides stercoris* ATCC 43183 FtnA1 (ZP_02434493), FtnA2 (ZP_02435306) and FtnA3 (ZP_02435304); Bt, *Bacteroides thetaiotaomicron* VPI-5482 FtnA1 (NP_810286), FtnA2 (NP_810020) and FtnA3 (NP_810022); Bacuni, *Bacteroides uniformis* ATCC 8492 FtnA1 (ZP_02072936), FtnA2 (ZP_02069467) and FtnA3 (ZP_02069465); Bvu, *Bacteroides vulgatus* ATCC 8482 FtnA1 (YP_001300026), FtnA2 (YP_001300763) and FtnA3 (YP_001300766); Bos, *Bos Taurus* ferritin heavy polypeptide (NP_776487); Camje, *Campylobacter jejuni* RM1221 (YP_178727); Cytophaga, *Cytophaga hutchinsonii* ATCC 33406 (YP_677424); Drosophila, *Drosophila melanogaster* ferritin 3 heavy chain (NP_572854); Ec, *Escherichia coli* K-12 MG1655 FtnA (NP_416418) and FtnB (NP_416415); Methanosarc, *Methanosarcina acetivorans* C2A (NP_615134); Mt: *Mycobacterium tuberculosis* H37Rv (NP_218358); Ornithodoros, *Ornithodoros moubata* (AAC19132); Parmer, *Parabacteroides merdae* ATCC 43184 FtnA1 (ZP_02030919), FtnA2 (ZP_02034381) and FtnA3 (ZP_02034383); Pdi, *Parabacteroides distasonis* ATCC 8503 FtnA1 (YP_001301664) and FtnA2 (YP_001301517); Saureus, *Staphylococcus aureus* subsp. *aureus* NCTC 8325 (YP_500600). Bold letters depict the non-haem iron-containing ferritin-like protein superfamily present in the Bacteroidetes (CFB) group.

iron-storage proteins (Andrews, 1998). The question why this group of intestinal communal anaerobes possess an abundant number of iron-storage proteins and the role they play in *Bacteroides* in relation to the intestinal ecosystem is not known.

The *Bacteroides* are highly aerotolerant and can survive for about 3 days in rich media while being shaken in a fully aerobic atmosphere (Rolfe *et al.*, 1977, 1978; Rocha *et al.*, 1996). These organisms also are highly resistant to H_2O_2; when pretreated with a sublethal concentration of H_2O_2 cultures retain high viability following 15 min exposure to 10 mM H_2O_2 (Rocha *et al.*, 1996). This extraordinary resistance to oxidative stress is due to an induced response (Rocha *et al.*, 1996). Protection against oxidative stress requires the strict management of iron pools as this and other transition metals participate in the Fenton reaction leading to hydroxyl radical formation. In this regard, three putative iron binding protein genes also are induced by oxidative stress, the non-specific DNA binding protein (*dps*), ferritin (*ftnA*) and bacterioferritin (*bfr*). Overall this is an extraordinary repertoire for an anaerobic organism not thought to need these detoxification enzymes during normal anaerobic growth.

To ensure that an appropriate physiological response is mounted for the level of stress encountered, there must be tight regulation of the oxidative stress response genes. In *B. fragilis* this is controlled in part by the LysR-family regulator, OxyR (Rocha *et al.*, 2000). There also are several genes that seem to be regulated indirectly by OxyR such as *ftnA* (Rocha *et al.*, 2003; Rocha and Smith, 2004). The rapid induction of the regulon is necessary for optimal survival during stress. From mutational studies it is known that *oxyR* mutants are more sensitive to H_2O_2 and during exposure to air the mutants discontinue growth much faster than wild-type. Consistent with this was the finding that survival of an *oxyR* mutant was similar to wild-type during extended aerobic exposure but when it was put in combination with *dps* and *ftnA* mutations there was a significant survival defect compared to the single mutations (Rocha and Smith, 2004). This supports the idea that other regulatory factors compensate for OxyR to induce expression of *dps* and *ftnA* in oxygen but these mechanisms are not known. Consistent with this idea, *bfr* expression was highly induced by air in an OxyR independent manner (Sund *et al.*, 2008).

Studies carried out with *B. fragilis* ferritin FtnA, which was the first mammalian-type ferritin isolated from bacteria (Rocha *et al.*, 1991), demonstrated that under anaerobic conditions, FtnA harbours few Fe atoms per molecule. This is probably due to the fact that anaerobically, iron is likely to be in the ferrous form and thus production of iron mediated oxygen radicals would not be a problem (Rocha *et al.*, 1991; Rocha and Smith, 2004). This is consistent with the fact that under anaerobic conditions, *B. fragilis ftnA* mRNA expression is not significantly altered either in high iron or low iron conditions. However, following oxygen exposure, *ftnA* mRNA expression levels are dependent upon both the iron content and the redox state of the culture media. Following oxidative stress, *ftnA* message increases about 10-fold in iron-replete medium compared to a four-fold increase under low-iron conditions (Rocha and Smith, 2004). Moreover, oxidation of the environment by oxidants such as diamide and potassium ferricyanide also induces the expression of *ftnA* in the absence of atmospheric oxygen *in vitro*. In addition, *B. fragilis ftnA* mRNA has two transcriptional start sites differentially regulated by iron and oxidants indicating that transcription of *ftnA* occurs in response to a change in environmental conditions (Rocha and Smith, 2004).

Conclusions and further perspectives

The utilization of exogenous haem by the commensal *Bacteroides* spp. allows high molar growth yields during fermentation which may be a competitive advantage for the colonization of the human lower intestinal tract. Haem also alters the formation of major short chain fatty acids produced as by-products of fermentation such as succinate, acetate, propionate and butyrate. In recent years, great interest has emerged in the ability of the host to harvest energy from microbial fermentation of dietary polysaccharides. Microbiome studies of the caecal contents in mice and human volunteers have shown that obesity is correlated with gut microbiota changes in the relative abundance of *Bacteroides* and firmicutes compared to lean individuals (Turnbaugh *et al.*, 2006). Higher levels of short-chain fatty acids are produced from fermentable carbohydrates in obesity-associated microbiomes than in lean controls (Turnbaugh *et al.*, 2006). It will be interesting to investigate the interrelationship between the micronutrients haem and iron and the effects they play in *Bacteroides* spp. metabolic fermentative pathways in the intestinal lumen. The unique ferrous iron uptake system in *Bacteroides* and other intestinal colonizers which encode a fused FeoA and FeoB peptides is an exciting area of future research in bacterial ferrous iron uptake in the intestinal tract. Moreover, the unusual high numbers of ferritin homologues in *Bacteroides* give this group of bacteria tremendous potential to store an abundant amount of iron. It still remains to be investigated if the mechanisms of iron uptake and storage by commensal intestinal microflora participate in their ability to inhibit colonization of unwanted colonizers and enteric pathogens. *B. fragilis* is the smallest component among the *Bacteroides* species found in the human colon reaching about 0.5–1.0% of the total flora, but is the most frequent anaerobe isolated from human infections (50–80%). Though its ability to obtain iron and haem from host tissues in extra-intestinal infections is an important advantage compared to other less pathogenic *Bacteroides* spp., the molecular, physiological and genetic mechanisms of iron and haem assimilation require further studies.

> *He who studies most deeply into the mysteries of nature will realize most fully his own ignorance and weakness. He will realize that there are depths and heights which he cannot reach, secrets which he cannot penetrate, vast fields of truth lying before him unentered. He will be ready to say, with Newton, 'I seem to myself to have been like a child on the seashore finding pebbles and shells, while the great ocean of truth lay undiscovered before me.*
>
> E. G. White, *Education*, 1903

Acknowledgements

This work was supported by the National Institutes of Health/NIAID grants AI068659 and AI079183 to Edson R. Rocha.

References

Andrews, S.C. (1998). Iron storage in bacteria. Adv. Microb. Physiol.:*40,* 281–351.

Balmer, S.E., and Wharton, B.A. (1991). Diet and faecal flora in the newborn: iron. Arch. Dis. Child. *66,* 1390–1394.

Baughn, A.D., and Malamy, M.H. (2002). A mitochondrial-like aconitase in the bacterium *Bacteroides fragilis*: implications for the evolution of the mitochondrial Krebs cycle. Proc. Natl. Acad. Sci. U.S.A. *99,* 4662–4667.

Baughn, A.D., and Malamy, M.H. (2003). The essential role of fumarate reductase in haem-dependent growth stimulation of *Bacteroides fragilis.* Microbiology *149,* 1551–1558.

Beukeveld. G.J., Wolthers, B.G., van Saene, J.J., de Haan, T.H., de Ruyter-Buitenhuis, L.W., and van Saene, R.H. (1987). Patterns of porphyrin excretion in feces as determined by liquid chromatography; reference values and the effect of flora suppression. Clin. Chem. 33, 2164–2170.

Bjursell, M.K., Martens, E.C., and Gordon, J.I. (2006). Functional genomic and metabolic studies of the adaptations of a prominent adult human gut symbiont, *Bacteroides thetaiotaomicron,* to the suckling period. J. Biol. Chem. *281,* 36269-36279.

Brazier, J.S. (1990). Analysis of the porphyrin content of fluorescent pus by absorption spectrophotometry and high performance liquid chromatography. J. Med. Microbiol. 33, 29–34.

Caldwell, D.R., White, D.C., Bryant, M.P., and Doetsch, R.N. (1965). Specificity of the heme requirement for growth of *Bacteroides ruminicola.* J. Bacteriol. 90, 1645–1654.

Cartron, M.L., Maddocks, S., Gillingham, P., Craven, C.J., and Andrews, S.C. (2006). Feo – transport of ferrous iron into bacteria. Biometals *19,* 143–157.

Cerdeno-Tarraga, A.M., Patrick, S., Crossman, L.C.,. Blakely, G., Abratt, V., Lennard, N., Poxton, I., Duerden, B., Harris, B., Quail, M.A., Barron, A., Clark, L., Corton, C., Doggett, J., Holden, M.T., Larke, N., Line, A., Lord, A., Norbertczak, H., Ormond, D., Price, C., Rabbinowitsch, E., Woodward, J., Barrell, B., and Parkhill, J. (2005). Extensive DNA inversions in the *Bacteroides fragilis* genome control variable gene expression. Science. *307,* 1463–1465.

Chen, M., and Wolin, M.J. (1981). Influence of heme and vitamin B12 on growth and fermentation of *Bacteroides* species. J. Bacteriol. *145,* 466–471.

Dashper, S.G., Butler, C.A., Lissel, J.P., Paolini, R.A., Hoffmann, B., Veith, P.D., O'Brien-Simpson, N.M., Snelgrove, S.L., Tsiros, J.T., and Reynolds, E.C. (2005). A novel *Porphyromonas gingivalis* FeoB plays a role in manganese accumulation. J. Biol. Chem. *280,* 28095–28102.

Eckburg, P.B., Bik, E.M., Bernstein, C.N., Purdom, E., Dethlefsen, L., Sargent, M., Gill, S.R., Nelson, K.E., and Relman, D.A. (2005). Diversity of the human intestinal microbial flora. Science. *308,* 1635–1638.

Fairweather-Tait, S.J., and Johnson, I.T. (1999). Bioavailability of minerals. In Colonic microbiota, nutrition and health, G.R. Gibson and M.B. Roberfroid, eds. (Dordrecht, The Netherlands: Kluwer Academic Publishers), pp. 233–244.

Frankenberg, N., Moser, J., and Jahn, D. (2003). Bacterial heme biosynthesis and its biotechnological application. Appl. Microbiol. Biotechnol. *63,* 115–127.

Fuller, M.D., and Caldwwell, D.R. (1982). Tetrapyrrole utilization by *Bacteroides fragilis.* Can. J. Microbiol. *28,* 1304–1310.

Harris, M.A., and Reddy, C.A. (1977). Hydrogenase activity and the H_2-fumarate electron transport system in *Bacteroides fragilis.* J. Bacteriol. *131,* 922–928.

He, J., Miyazaki, H., Anaya, C., Yu, F., Yeudall, W.A., and Lewis, J.P. (2006). Role of *Porphyromonas gingivalis* FeoB2 in metal uptake and oxidative stress protection. Infect. Immun. *74,* 4214–4223.

Holdeman, L.M., Cato, E.P., and Moore, W.E.C. (1977). Anaerobe laboratory manual, 4th ed. (Blacksburg: Anaerobe laboratory Virginia Polytechnic Institute and State University).

Hooper, L.V., Midtvedt, T., and Gordon. J.I. (2002). How host–microbial interactions shape the nutrient environment of the mammalian intestine. Annu. Rev. Nutr. *22*, 283–307.

Horn, M., Harzenetter, M.D., Linner, T., Schmid, E.N., Müller, K.D., Michel, R., and Wagner, M. (2001). Members of the *Cytophaga-Flavobacterium-Bacteroides* phylum as intracellular bacteria of acanthamoebae: proposal of '*Candidatus* Amoebophilus asiaticus'. Environ. Microbiol. 3, 440–449.

Kuwahara T., Yamashita, A., Hirakawa, H., Nakayama, H., Toh, H., Okada, N., Kuhara, S., Hattori, M., Hayashi, T., and Ohnishi, Y. (2004). Genomic analysis of *Bacteroides fragilis* reveals extensive DNA inversions regulating cell surface adaptation. Proc. Natl. Acad. Sci. U.S.A. *101*, 14919–14924.

Ley, R.E., Bäckhed, F., Turnbaugh, P., Lozupone, C.A., Knight, R.D., and Gordon, J.I. (2005). Obesity alters gut microbial ecology. Proc. Natl. Acad. Sci. U.S.A. *102*, 11070–11075.

Macy, J., Probst, I., and Gottschalk, G.(1975). Evidence for cytochrome involvement in fumarate reduction and adenosine 5′-triphosphate synthesis by *Bacteroides fragilis* grown in the presence of hemin. J. Bacteriol. *123*, 436–442.

McCall, D., and Caldwell, D.R. (1977). Tetrapyrrole utilization by *Bacteroides ruminicola*. J. Bacteriol. *131*, 809–814.

O'Brian, M.R., and Thöny-Meyer, L. (2002). Biochemistry, regulation and genomics of haem biosynthesis in prokaryotes. Adv. Microbiol. Physiol. *46*, 257–318.

Otto, B.R., Verweij-van Vught, A.M., van Doorn, J., and Maclaren, D.M. (1988). Outer membrane proteins of Bacteroides fragilis and Bacteroides vulgatus in relation to iron uptake and virulence. Microb. Pathog. *4*, 279–287.

Otto, B.R., Sparrius, M., Verweij-van Vught, A.M.J.J., and MacLaren, D.M. (1990). Iron regulated outer membrane protein of *Bacteroides fragilis* involved in heme uptake. Infect. Immun. *58*, 3954–3958.

Otto, B.R., Verweij, W.R., Sparrius, M., Verweij-van Vught, A.M., Nord, C.E., and MacLaren, D.M. (1991). Human immune response to an iron-repressible outer membrane protein of *Bacteroides fragilis*. Infect. Immun. *59*, 2999–3003.

Otto, B.R., Verweij-van Vught, A.M.J.J., and MacLaren, D.M. (1992). Transferrins and heme-compounds as iron sources for pathogenic bacteria. Crit. Rev. Microbiol. *18*, 217–233.

Otto, B.R., Sparrius, M., Wors, D.J., de Graaf, F.K., and MacLaren, D.M. (1994). Utilization of haem from the haptoglobin–haemoglobin complex by *Bacteroides fragilis*. Microb. Pathog. *17*, 137–147.

Otto, B.R., van Dooren, S.J., Nuijens, J.H., Luirink, J., and Oudega, B. (1998). Characterization of a hemoglobin protease secreted by the pathogenic *Escherichia coli* strain EB1. J. Exp. Med. *188*, 1091–1103.

Otto, B.R., van Dooren, S.J., Dozois, C.M., Luirink, J., and Oudega, B. (2002). *Escherichia coli* hemoglobin protease autotransporter contributes to synergistic abscess formation and heme-dependent growth of *Bacteroides fragilis*. Infect. Immun. *70*, 5–10.

Panek, H., and O'Brian, M.R. (2002). A whole genome view of prokaryotic haem biosynthesis. Microbiology *148*, 2273–2282.

Ponka, P. (1999). Cell biology of heme. Am. J. Med. Sci. *318*, 241–256.

Robertson, K.P., Smith, C.J., Gough, A.M., and Rocha, E.R. (2006). Characterization of *Bacteroides fragilis* hemolysins and regulation and synergistic interactions of HlyA and HlyB. Infect. Immun. *74*, 2304–2316.

Rocha, E.R., de Uzeda, M., and Brock, J.H. (1991). Effect of ferric and ferrous iron chelators on growth of Bacteroides fragilis under anaerobic conditions. FEMS Microbiol. Lett. *68*, 45–50.

Rocha, E.R., Selby, T., Coleman, J.P., and Smith, C.J. (1996). Oxidative stress response in an anaerobe, *Bacteroides fragilis*: a role for catalase in protection against hydrogen peroxide. J. Bacteriol. *178*, 6895–6903.

Rocha, E.R., Owens, Jr. G., and Smith, C.J. (2000). The redox-sensitive transcriptional activator OxyR regulates the peroxide response regulon in the obligate anaerobe *Bacteroides fragilis*. J. Bacteriol. *182*, 5059–5069.

Rocha, E.R., Smith, A., Smith, C.J., and Brock, J.H. (2001). Growth inhibition of *Bacteroides fragilis* by hemopexin: proteolytic degradation of hemopexin to overcome heme limitation. FEMS Microbiol. Lett. *199*, 73–78.

Rocha, E.R., Herren, C.D., Smalley, D.J., and Smith, C.J. (2003). The complex oxidative stress response of *Bacteroides fragilis*: the role of OxyR in control of gene expression. Anaerobe. *9*, 165–173.

Rocha, E.R., and Smith, C.J. (2004). Transcriptional regulation of the *Bacteroides fragilis* ferritin gene (*ftnA*) by redox stress. Microbiology *150*, 2125–2134.

Rolfe, R.D., Hentges, D.J., Barrett, J.T., and Campbell, B.J. (1977). Oxygen tolerance of human intestinal anaerobes. Am. J. Clin. Nutr. *30*, 1762–1769.

Rolfe, R.D., Hentges, D.J., Campbell, B.J., and Barrett, J.T. (1978). Factors related to the oxygen tolerance of anaerobic bacteria. Appl. Environ. Microbiol. *36*, 306–313.

Săsărman, A., Surdeanu, M., Szégli, G., Horodniceanu, T., Greceanu, V., and Dumitrescu, A. (1968). Hemin-deficient mutants of *Escherichia coli* K-12. J. Bacteriol. *96*, 570–572.

Schobert, M., and Jahn, D. (2002). Regulation of heme biosynthesis in non-phototrophic bacteria. J. Mol. Microbiol. Biotechnol. *4*, 287–294.

Shah, H.N., Bonnett, R., Mateen, B., and Williams, R.A. (1979). The porphyrin pigmentation of subspecies of *Bacteroides melaninogenicus*. Biochem. J. *180*, 45–50.

Sijbrandi, R., Den Blaauwen, T., Tame, J.R., Oudega, B., Luirink, J., and Otto, B.R. (2005). Characterization of an iron-regulated alpha-enolase of *Bacteroides fragilis*. Microbes Infect. 7, 9–18.

Skaar, E.P., Gaspar, A.H., and Schneewind, O. (2004). IsdG and IsdI, heme-degrading enzymes in the cytoplasm of *Staphylococcus aureus*. J. Biol. Chem. *279*, 436–443.

Smith, C.J., Rocha, E.R., and Paster, B.J. (2003). The medically important *Bacteroides* spp. in health and disease. In The Prokaryotes: An Evolving Electronic Resource for the Microbiological Community, 3rd edition, release 3.7, November 2, *2001*, M. Dworkin *et al.*, eds., (New York: Springer-Verlag, http://link.springer-ny.com/link/service/books/10125/).

Sperry, J.F., Appleman, M.D., and Wilkins, T.D. (1997). Requirement of heme for growth of *Bacteroides fragilis*. Appl. Environ. Microbiol. *34*, 386–390.

Stojiljkovic, I., and Perkins-Balding, D. (2002). Processing of heme and heme-containing proteins by bacteria. DNA Cell Biol. *21*, 281–295.

Sund, C.J., Rocha, E.R., Tzianabos, A.O., Wells, W.G., Gee, J.M., Reott, M.A., O'Rourke, D.P., and Smith, C.J. (2008). The *Bacteroides fragilis* transcriptome response to oxygen and H_2O_2: the role of OxyR and its effect on survival and virulence. Mol. Microbiol. *67*, 129–142.

Tatsumi, R., and Wachi, M. (2008). TolC-dependent exclusion of porphyrins in *Escherichia coli*. J. Bacteriol. *190*, 6228–6233.

Touati, D. (2000). Iron and oxidative stress in bacteria. Arch. Biochem. Biophys. *373*, 1–6.

Turnbaugh, P.J., Ley, R.E., Mahowald, M.A., Magrini, V., Mardis, E.R., and Gordon, J.I. (2006). An obesity-associated gut microbiome with increased capacity for energy harvest. Nature *444*, 1027–1031.

Verweij-van Vught, A.M., Namavar, F., Vel, W.A., Sparrius, M., and MacLaren, D.M. (1986a). Pathogenic synergy between *Escherichia coli* and *Bacteroides fragilis* or *B. vulgatus* in experimental infections: A non-specific phenomenon. J. Med. Microbiol. *21*, 43–47.

Verweij-van Vught, A.M., Sparrius, M., Namavar, F., Vel, W.A., and D.M. MacLaren, D.M. (1986b). Nutritional requirements of *Bacteroides fragilis* and *Bacteroides vulgatus* with reference to iron and virulence. FEMS Microbiol. Lett. *36*, 47–51.

Verweij-van Vught, A.M., Otto, B.R., Namavar, F., Sparrius, M., and MacLaren, D.M. (1988). Ability of *Bacteroides* species to obtain iron from iron salts, haem-compounds and transferring. FEMS Microbiol. Lett. *49*, 223–228.

Wandersman, C., and Delepelaire, P. (2004). Bacterial iron sources: from siderophores to hemophores. Annu. Rev. Microbiol. *58*, 611–647.

Young, G.P., Rose, I.S., and St John, D.J. (1989). Haem in the gut. I. Fate of haemoproteins and the absorption of haem. J. Gastroenterol. Hepatol. *4*, 537–545.

Young, G.P., Rose, I.S., St John, D.J., and Blake, D. (1990). Haem in the gut. Part II. Faecal excretion of haem and haem-derived porphyrins and their detection. J. Gastroenterol. Hepatol. *5*, 194–203.

Zhu, W., Wilks, A., and Stojiljkovic, I. (2000). Degradation of heme in gram-negative bacteria: the product of the *hemO* gene of *Neisseriae* is a heme oxygenase. J. Bacteriol. *182*, 6783–6790.

Campylobacter Fur and PerR Regulons

10

James Butcher, Annika Flint, Martin Stahl and Alain Stintzi

Abstract

Iron is known to catalyse a wide range of biochemical reactions essential for most living organisms, including *Campylobacter jejuni*. Paradoxically, this iron reactivity is also responsible for the generation of hydroxyl radicals (·OH), which are particularly biotoxic. In order to avoid iron toxicity, microorganisms must achieve an effective iron homeostasis by tightly regulating the expression of genes encoding the proteins involved in iron acquisition, metabolism and oxidative stress defences in response to iron availability. Interestingly, in addition to the classical ferric uptake regulator Fur, *C. jejuni* carries another member of the Fur family of metalloregulators, PerR. PerR is a peroxide-sensing regulator and typically regulates peroxide stress response in Gram-positive bacteria. Recent work indicates that the regulatory functions of Fur and PerR extend beyond their classically ascribed roles. These diverse functions include energy metabolism, protein glycosylation and flagella biogenesis. Moreover, the Fur and PerR regulons appear to overlap and co-regulate key genes at specific junctions. The current state of knowledge concerning *Campylobacter jejuni*'s response to iron, oxidative stressors and the roles of both Fur and PerR in controlling gene expression will be discussed and compared to those reported in other bacteria.

Introduction

The genus Campylobacter consists of Gram-negative, spirally curved, microaerophilic bacteria and contains 18 species (*Campylobacter fetus, C. hyointestinalis, C. laniena, C. sputorum, C. mucosalis, C. concisus, C. curvus, C. rectus, C. gracilis, C. showae, C. hominis, C. jejuni, C. coli, C. lari, C. insulaenigrae, C. canadensis, C. upsaliensis* and *C. helveticus*) (Debruyne *et al.*, 2008). *C. jejuni* and *C. coli* are the most frequently reported campylobacters associated with human infection (Blaser and Engberg, 2008).

Campylobacter jejuni is regarded as the most common cause of bacterial gastroenteritis, threatening human health worldwide (Butzler, 2004; Mead *et al.*, 1999; Moore *et al.*, 2005). *Campylobacter* infection is an acute diarrhoeal disease which ranges from a day of mild watery or bloody diarrhoea to severe abdominal pain that can last for several weeks (Zilbauer *et al.*, 2008). This pathogen is also involved in the development of Guillain–Barré syndrome, reactive arthritis and is suspected to be one of the possible aetiological factors in inflammatory bowel diseases (Blaser and Engberg, 2008; Boyanova *et al.*, 2004; Moore *et al.*, 2005). While most *Campylobacter* infections are self-limiting, antimicrobial therapies are recommended to treat severe cases and immunocompromised patients. Erythromycin and fluoroquinolones are the drugs of choice for treatment (Bos *et al.*, 2006). However, resistance to these classes of antibiotics has drastically increased during the last decade, compromising future treatments (Alfredson and Korolik, 2007).

C. jejuni colonizes the intestine of all avians, including chickens, turkeys, ducks, and wild birds (Lee and Newell, 2006). Although *C. jejuni* is also found in other livestock animals, 50–70% of *Campylobacter* illness is due to contaminated poultry and poultry products (Friedman *et al.*, 2000).

Campylobacter-positive chickens are typically colonized at up to 10^9 colony-forming-organisms per gram of intestinal content (Lee and Newell, 2006). During the slaughtering process, cecal contents can contaminate the carcass and the processing environment (Keener *et al.*, 2004). Given that 500–800 organisms are sufficient to induce disease in susceptible humans, even 1 µg of such material, an invisible amount, could be more than sufficient to pose a risk to the consumer (Black *et al.*, 1988; Robinson, 1981). This constitutes a substantial public health hazard that can be exacerbated by subsequently handling products inappropriately. Various types of poultry products have been implicated with human infections, including poultry or poultry liver, raw or rare chicken, cooked chicken, processed chicken, and prepared chicken meals (Humphrey *et al.*, 2007). A striking example of the role of poultry as a main source of human infection is the observed 40% reduction in the number of human *Campylobacter* cases reported in Belgium following the complete withdrawal of chicken and egg products from the market in 1999 in response to the dioxin crisis (Vellinga and Van Loock, 2002).

To date, the genome sequences of one *C. jejuni* subsp. *doylei* (*C. jejuni* 269.97), three human isolates of *C. jejuni* subsp. *jejuni* (*C. jejuni* NCTC 11168, 81–176, and 81116), and one chicken isolate (*C. jejuni* RM1221) have been completed (Fouts *et al.*, 2005; Hofreuter *et al.*, 2006; Parkhill *et al.*, 2000; Pearson *et al.*, 2007). In addition, the draft genome sequences of seven more human isolates of *C. jejuni* subsp. *jejuni* have been obtained (*C. jejuni* 260.94, 84–25, BH-01–0142, CF93–6, CG8421, CG8486, and HB93–13) (Poly and Guerry, 2008; Poly *et al.*, 2007). Despite this wealth of genomic information, the microbial factors governing *C. jejuni* virulence have not as of yet been identified. Moreover, the characterization of such virulence factors is confounded by the presence of unique genetic features which might confer distinctive metabolic and/or pathogenic properties to each strain. For example, the *cfrA* gene, which encodes the ferric-enterobactin receptor, is known to be required for the colonization of the chick gastrointestinal tract by *C. jejuni* NCTC 11168 (Palyada *et al.*, 2004); however it is missing in the genome of *C. jejuni* 81–176 (Hofreuter *et al.*, 2006). Consequently, since *C. jejuni* 81–176 efficiently colonizes chicks, this strain should possess alternative iron acquisition system(s). Similarly, two unique genes in *C. jejuni* 81–176, *dmsA* (a putative DMSO reductase) and *ggt* (a glutamyltranspeptidase), were found to significantly contribute to the virulence and tissue tropism of this particular strain, suggesting different virulence mechanisms in *C. jejuni* 81–176 from other human isolates such NCTC 11168, and CG8486 (Hofreuter *et al.*, 2006, 2008).

During its infection cycle, *C. jejuni* transits from contaminated food to the stomach, through the intestinal tract and finally to the faeces; allowing the transmission to a new host. During this stressful journey, it encounters life-threatening environmental conditions such as the acidic pH of the stomach (Merrell and Camilli, 2002), sudden temperature fluctuations (Konkel *et al.*, 1998; Konkel and Tilly, 2000), the high osmolarity of the gastrointestinal tract (Bremer and Kramer, 2000), intestinal gases (including hydrogen, methane and carbon dioxide) (Tannock, 2000), reactive oxygen and nitrogen compounds (Storz and Zheng, 2000), changes in nutrient availability (Hooper *et al.*, 1998, 2002), and low inorganic ion concentration (particularly iron starvation) (Hantke and Braun, 2000). In addition, most of these stresses vary between the external environment (e.g. food, water, and milk), different hosts and the different locations within the host. Undoubtedly, stress responses enable the bacterium to surmount these sudden environmental changes and successfully adapt to its new environment. Ultimately, a successful pathogenic colonization will lead to the establishment of infection and eventually the generation of a stress-free environment. These chemical and physical threats constitute key signals for pathogens to sense that they have invaded the host and also allows for discrimination between different locations. Transcriptional regulators orchestrate cellular responses to environmental stresses. Analysis of *C. jejuni* genome sequences has revealed the absence of many known key regulators of the stress defence systems found in other foodborne pathogens, such as *Salmonella* spp. and *Escherichia coli* (Parkhill *et al.*, 2000). Unlike most other Gram-negative bacteria, the well-characterized superoxide and peroxide sensing regulators, SoxRS and OxyR, are absent from the *C. jejuni* genomes. Interestingly, *C. jejuni* solely carries the peroxide-sensing regulator PerR, which typically regulates peroxide stress responses in Gram-positive bacteria (Mongkolsuk and Helmann, 2002). In addition to PerR, *C. jejuni* also contains a classical Fur (ferric uptake regulator) protein, which typically regulates iron uptake genes (Lee and Helmann, 2007). The presence of these two Fur-family orthologues, Fur and PerR, raises fundamental questions regarding the regulation of genes involved in iron metabolism and oxidative stress responses in *C. jejuni*. How are these genes regulated? Which Fur family orthologues regulate which genes? What is the nature of the Fur and PerR regulatory networks? At what level do these networks crosstalk?

Addressing the specific role of Fur and PerR could be rather difficult, as comparing the phenotypes of the wild-type strain with the respective mutant may not be very informative, due to the possibility of compensatory effects. However, with the technological advances of the post-genomic era, such as transcriptomics and proteomics, it is now possible to decode complex regulatory networks and to generate detailed analyses of bacterial stress responses. In this chapter, we will discuss the wealth of information on the Fur and PerR regulatory networks, how they are interconnected, how they orchestrate gene expression, and how they impact *C. jejuni* physiology and virulence. The nature of the iron and oxidative stress stimulons and their regulation by Fur and PerR will also be discussed. Furthermore, we will compare the iron and oxidative stress stimulons and the Fur and PerR regulons of *C. jejuni* with those of other bacteria. Note that most of the data discussed in this chapter were generated using *C. jejuni* NCTC 11168.

Chemistry of iron and reactive oxygen species

Iron is a transition metal absolutely required for most living organisms. In the cell, iron plays either a structural role by stabilizing the protein's tertiary or quaternary structure, or a functional role by contributing to the catalytic activity of enzymes or proteins involved in electron transfer (Harris, 2002). The oxidation state of iron varies from −2 to +6, with the lowest and highest oxidation states found in the $Fe(CO)_4^{2-}$ anion and the ferrate FeO_4^{2-} anion respectively (Harris, 2002). In biological systems, iron exists in three oxidation states: the reduced Fe^{2+} ferrous form, the oxidized Fe^{3+} ferric form, and the very reactive Fe^{4+} ferryl form (Harris, 2002). The ferryl species are formed during the catalytic cycle of various enzymes, including monooxygenase, cytochrome P450, and ribonucleotide reductase (Groves, 2006). Although this high-valent form of iron is essential for the function of these enzymes, there is no evidence that this highly reactive species participates in iron transport or storage. Thus, we will limit our discussion to the more stable ferrous and ferric forms of iron.

The redox reactivity of the Fe^{3+}/Fe^{2+} couple has been harnessed for catalysis by a wide range of proteins, enabling many major biological processes, such as nitrogen fixation, photosynthesis, oxygen transport, respiration and DNA synthesis. Iron's value relies almost entirely on its incorporation into enzymes and metalloproteins, either as a porphyrin or iron–sulphur cluster, or as a mononuclear, dinuclear or polynuclear complex (Andrews *et al.*, 2003). Paradoxically, while the physiochemical reactivity of iron is the raison d'être for its role as a catalytic or structural cofactor, it is also responsible for the generation of highly reactive hydroxyl radicals (Galaris and Pantopoulos, 2008; Imlay, 2003; Merkofer *et al.*, 2006). Hydroxyl radicals are extremely reactive oxidants capable of damaging DNA, proteins and lipids (Imlay, 2003). Iron reactivity with dioxygen species has received considerable attention and is described by the Fenton reaction (equation 10.1). In this reaction, ferrous ion reacts with hydrogen peroxide and produces hydroxyl radicals and ferric iron (Galaris and Pantopoulos, 2008). The ferric ion is reduced back to ferrous ion by superoxide anions, thus allowing iron to re-catalyse the Fenton reaction (equation 10.2). In biological systems, superoxide anions can be substituted to some extent by other reducing agents such as ascorbate.

$$Fe^{2+} + H_2O_2 + H^+ \rightarrow Fe^{3+} + HO^{\cdot} + H_2O \qquad (10.1)$$

$$Fe^{3+} + O_2^{\cdot -} \rightarrow Fe^{2+} + O_2 \qquad (10.2)$$

Hydrogen peroxide and superoxides are generated during aerobic bacterial metabolism by the auto-oxidation of the flavin cofactor of redox enzymes (Galaris and Pantopoulos, 2008). Under normal conditions, superoxide anions are removed by superoxide dismutase, which catalyses the dismutation of superoxides to hydrogen peroxide and oxygen (equation 10.3) (Imlay, 2008). Subsequently, hydrogen peroxide is converted to oxygen and water by peroxidases such as catalase (equation 10.4) (Imlay, 2008).

$$2O_2^{\cdot -} + 2H^+ \rightarrow H_2O_2 + O_2 \tag{10.3}$$

$$H_2O_2 + H_2O_2 \rightarrow 2H_2O + O_2 \tag{10.4}$$

Superoxides and hydrogen peroxide are moderately reactive (Winterbourn, 2008). Both oxidants interact with protein iron–sulphur clusters by binding to the iron atom and oxidizing the cluster (Winterbourn, 2008). The mechanism of protein inactivation by superoxide anions involves a one-electron oxidation of the iron–sulphur clusters and the release of hydrogen peroxide and ferrous ion (equations 10.5 and 10.6), which in turn could generate highly reactive hydroxyl radicals by Fenton chemistry (equation 10.1). Besides its role in the Fenton reaction, hydrogen peroxide also catalyses a two electron oxidation of iron–sulphur clusters, via the generation of a ferryl-like species, releasing ferric ion and the damaged $[3Fe\text{-}4S]^{1+}$ species (equations 10.7 and 10.8). Thus exposure to hydrogen peroxide inactivates proteins that contain iron–sulphur clusters.

$$[4Fe\text{–}4S]^{2+} + O_2^{\cdot -} + 2H^+ \rightarrow H_2O_2 + [4Fe\text{–}4S]^{3+} \tag{10.5}$$

$$[4Fe\text{–}4S]^{3+} \rightarrow [3Fe\text{–}4S]^{1+} + Fe^{2+} \tag{10.6}$$

$$[4Fe\text{–}4S]^{2+} + H_2O_2 \rightarrow [4Fe\text{–}4S/O]^{2+} + H_2O \tag{10.7}$$

$$[4Fe\text{–}4S/O]^{2+} + H^+ \rightarrow [3Fe\text{–}4S]^{1+} + Fe^{3+} + OH^- \tag{10.8}$$

Apart from the above described reactivity, hydrogen peroxide can also interact with the haem iron cluster of haem-containing proteins (Galaris and Pantopoulos, 2008). In this case, hydrogen peroxide catalyses the oxidation of Fe^{3+}-haem to a reactive oxo-Fe^{4+}-haem cation radical intermediate (Galaris and Korantzopoulos, 1997). Haem-containing proteins are reduced back to their initial form in the presence of antioxidants. In turn, this cycle contributes to the elimination of hydrogen peroxide (Galaris and Korantzopoulos, 1997). However, in the absence of adequate amounts of antioxidants, the oxo-Fe^{4+}-haem-containing proteins are further oxidized, leading to the liberation of the haem from the active site of the protein and the release of iron, which could then catalyse the Fenton reaction. Clearly, while the reactivity of superoxide anions and hydrogen peroxide is moderate, their main damaging effects are due to the secondary formation of hydroxyl radicals via the Fenton reaction (Merkofer *et al.*, 2006).

It is obvious that perturbation of the delicate cell balance between oxidants, antioxidants, and defence mechanisms will inevitably lead to deleterious effects, and thus create oxidative stress (Green and Paget, 2004; Jones, 2006). Moreover, the extent of cell damage is directly linked to the intracellular level of labile iron. Undoubtedly, the cellular iron status is closely linked to formation of oxidative stress via the catalysis of the Fenton reaction. It is therefore not surprising that iron homeostasis and oxidative stress defence systems are tightly regulated at both the transcriptional and post-transcriptional levels.

Iron sources available to *Campylobacter*

Campylobacter jejuni primarily colonizes the gastrointestinal tract of warm blooded animals, including avians and mammals (Lee and Newell, 2006). While the exact composition of the various iron sources available in the intestine remains incompletely described, haem, soluble complexes and salt forms of iron will be present (Conrad and Umbreit, 2000). Most of the iron originates from the diet, where it largely occurs as a ferric or haem species (Carpenter and Mahoney, 1992). Haem will be primarily derived from the breakdown of meat haemoglobin and myoglobin by pancreatic enzymes (Conrad *et al.*, 1967). Haem iron represents about one third of dietary iron in meat-eating Western countries, with the remainder consisting of ferric iron (Carpenter and Mahoney, 1992). While haem iron remains soluble throughout the gastrointestinal tract, ferric iron is insoluble under normal physiological conditions. However, it is solubilized during its transit through the low pH environment of the stomach (pH ~3) and then chelated as it enters less acidic duodenum. This chelation helps keep ferric iron in solution and is accomplished by intestinal mucins, organic acids (particularly, lactic, citric, succinic, malic, and pyruvic acids), bile acids, amino acids (particularly asparagine, glycine and histidine), membrane lipids (particularly fatty acids), and siderophores produced by the intestinal microflora (Conrad and Schade, 1968; Conrad and Umbreit, 2000; Conrad *et al.*, 1991; Ratledge, 2007; Ratledge and Dover, 2000; Weinberg, 2009). In contrast, certain food constituents (carbonates, phosphates, oxalates, phylates, and tannates) may precipitate ferric iron and form macromolecule complexes, sequestering ferric iron from endogenous or incoming microbes (Conrad and Umbreit, 2000). The stress hormone norepinephrine, which is secreted in the intestinal lumen, may also chelate iron and constitute an additional iron source for incoming microbes. Indeed, norepinephrine has been shown to stimulate *Campylobacter* growth under iron restricted conditions by acting in a siderophore-like manner (Cogan *et al.*, 2007). Importantly, a proportion of the dietary ferric iron might also be reduced to ferrous iron and maintained in that redox state by intestinal secretions or food constituents such as bile salts and ascorbic acid (Conrad and Schade, 1968). Other possible sources of iron could originate from the sloughing epithelial cells that are extruded from the mucosa into the intestinal lumen during normal mucosal cell proliferation, and from blood cells during inflammation. Finally, another potential source of iron is the non-haem iron binding protein lactoferrin that is present in the gastrointestinal fluid (Gonzalez-Chavez *et al.*, 2009). While lactoferrin tightly sequesters ferric iron from incoming pathogens, thus protecting the host against microbial infections, microbes have evolved specialized system(s) for iron acquisition from ferric-lactoferrin (Ratledge and Dover, 2000; Wandersman and Delepelaire, 2004).

In order to be a successful pathogen, *Campylobacter* must effectively acquire iron from any of these various potential iron sources and effectively compete for these sources with the normal gut microflora. Analysis of *Campylobacter jejuni* genomes reveals an absence of genes encoding enzymes involved in siderophore biosynthesis, suggesting the inability of *Campylobacter* to secrete these powerful iron chelators (Parkhill *et al.*, 2000). Although an early report suggested the presence of siderophore-like activity by some *Campylobacter* strains (Field *et al.*, 1986), further characterization of these strains failed to confirm this observation (Palyada *et al.*, 2009). Moreover, the lack of siderophore production by *Campylobacter* is supported by other studies (Baig *et al.*, 1986; Pickett *et al.*, 1992). Nevertheless, *C. jejuni* is able to acquire iron from exogenous siderophores and a ferric-enterobactin transport system has been fully characterized (Field *et al.*, 1986; Palyada *et al.*, 2004). The *cfrA* gene encodes the outer-membrane ferric-enterobactin receptor, the *ceuE* gene encodes the periplasmic binding protein, and *ceuBCD* encode the ABC transport component (Carswell *et al.*, 2008; Palyada *et al.*, 2004). Beside ferric-enterobactin, *C. jejuni* NCTC 11168 is thought to be able to acquire iron from the hydroxamate siderophore rhodotorulic acid via the p19 system (A. Stintzi, and J. Ketley, unpublished). This transport system consists of an ABC transporter (Cj1659–Cj1663), a periplasmic protein (p19), and a putative outer-membrane protein (Cj1658). Various microbial siderophores are discussed in Chapter 1. In addition to ferric-siderophores, *C. jejuni* NCTC 11168 is also able to utilize haem and the iron-binding glycoproteins transferrin and lactoferrin, via the ChuABCD and the Cj0175c-Cj0178 transport systems respectively (Miller *et al.*, 2008; Ridley *et al.*, 2006). Haem uptake and iron extraction are more comprehensively discussed in

Chapter 2. Finally, a ferrous iron transporter, FeoB, has been fully characterized in several strains of *C. jejuni* (*C. jejuni* NCTC 11168, 81–176 and ATCC 43431) (Naikare *et al.*, 2006).

Iron homeostasis and oxidative stress in *Campylobacter jejuni*

In order to avoid iron toxicity and the production of harmful hydroxyl radicals, microorganisms must achieve an effective iron homeostasis by tightly regulating the expression of genes encoding proteins involved in iron acquisition and metabolism in response to iron availability. In Gram-negative bacteria, the ferric uptake regulator Fur regulates the transcription of these genes. Fur was first described in 1981 as a repressor of iron acquisition systems in *E. coli* (Hantke, 1981). The ferric uptake regulator gene (*fur*) from *C. jejuni* NCTC 11168 has been recently characterized by proteomic and transcriptomic studies (Holmes *et al.*, 2005; Palyada *et al.*, 2004). *C. jejuni* also carries another member of the Fur family of metalloregulators, PerR, in addition to the *fur* gene. PerR is a peroxide-sensing regulator and typically regulates peroxide stress responses in Gram-positive bacteria (Mongkolsuk and Helmann, 2002). Its presence in *C. jejuni* might compensate for the lack of the well-characterized superoxide and peroxide sensing regulators of Gram-negative bacteria, SoxRS and OxyR (Imlay, 2008). Thus far, *C. jejuni* are the only Gram-negative bacteria that carry a PerR homologue. While a PerR protein has been annotated in *Neisseria gonorrhoeae*, this protein was recently shown to regulate manganese transport and thus should be reclassified as a Mur-like protein (Helmann *et al.*, 2007; Wu *et al.*, 2006). Inactivation of PerR in *C. jejuni* results in constitutive expression of KatA and AhpC, and enhanced resistance to peroxide stress (van Vliet *et al.*, 1999). Interestingly, both PerR and Fur appear to regulate the expression of catalase in an iron-dependent manner since both genes must be mutated to fully abolish *katA* iron regulation (van Vliet *et al.*, 1999; van Vliet *et al.*, 1998). This observation clearly indicates a close link between the oxidative stress and iron regulons in *C. jejuni* and suggests an overlap between the Fur and PerR regulons. The iron and oxidant stimulons, the Fur and PerR regulons, and the interlinks between these stimulons and regulons will be discussed in greater detail below.

The iron stimulon of *Campylobacter jejuni*

With the technological advances of the post-genomic era it is now possible to profile the effects of iron bioavailability on bacterial gene transcription at the genome level under carefully controlled *in vitro* growth conditions. Two independent studies have investigated the genome-wide transcriptional response of *Campylobacter jejuni* NCTC 11168 to the level of iron (Holmes *et al.*, 2005; Palyada *et al.*, 2004). Both studies compared the transcriptome profiles of iron-starved and iron-loaded *Campylobacter* cells grown to mid-log phase. Holmes and collaborators identified 147 iron responsive genes. The transcript abundance of 117 and 30 genes was found to be increased and decreased respectively in response to iron limitation. Palyada and collaborators identified 208 iron responsive genes, among which the transcript levels of 102 genes were found to be repressed by the addition of ferrous sulphate to the growth medium. The discrepancy in the total number of differentially expressed genes between these two studies probably results from the different growth conditions and experimental designs employed, and possibly from the use of two different clonal isolates of *C. jejuni* NCTC 11168. Nevertheless, 65 genes were found to be differentially expressed by both studies. About two thirds of these genes are predicted to encode iron acquisition systems, oxidative stress defences, and other probable transporter proteins. Notably, the transcript levels of all known or annotated iron acquisition systems were found to be iron responsive by both studies. On the other hand, most of the genes identified as iron responsive in one of the studies but not the other, encode proteins involved in general bacterial metabolism such as cofactor biosynthesis, electron transport, amino acid biosynthesis, and ribosomal protein synthesis and modification. The differential expression of this set of genes reflects the use of two different growth media, defined iron-restricted minimal growth medium (Palyada *et al.*, 2004) and rich broth supplemented with the iron chelator desferrioxamine (Holmes *et al.*, 2005), and the intrinsic link between the iron metabolism and general bacterial physiology. Accordingly, it is not surprising that the highest number of iron-responsive genes was identified by using a defined iron-restricted medium instead of a rich broth.

Interestingly, Palyada and collaborators also investigated the immediate transcriptional response of *C. jejuni* as a function of time following the addition of ferrous iron to iron starved cells (Palyada *et al.*, 2004). This study revealed the differential expression of 460 genes (with a minimum of 2-fold change in transcript abundance), representing 27% of the genome. The expression profile of these genes was shown to follow a strict chronological order, with the transcript level of some genes affected early and others later. About half of the 460 genes were still found to be differentially expressed at mid-log phase once adaptation to the new growth condition had occurred. The majority of these iron-responsive genes could be grouped into six main functional categories as described in further detail below and listed in Table 10.1.

Iron transport and storage

As expected, the transcript level of the genes encoding iron acquisition systems and iron storage was significantly reduced in response to the addition of ferrous sulphate. These genes encode for the haem uptake system (ChuABCD), the ferric-enterobactin receptor (CfrA), the ferric-enterobactin periplasmic binding protein (CeuE), the p19 iron uptake system (Cj1658, p19, and Cj1660–Cj1663), the ferric-transferrin or ferric-lactoferrin iron acquisition system (Cj0177, Cj0178, and Cj0173c–Cj0174c), the three outer membrane energy transduction systems (TonB1–ExbB1–ExbD1, TonB2–ExbB2–ExbD2, and TonB3), and the iron storage protein ferritin (Cft). Notably, the expression of all known or annotated iron acquisition systems was found to be iron responsive. While the ferrous ion uptake gene *feoB* was not reported as differentially expressed by Palyada *et al.*, the level of *feoB* transcript was 1.9-fold reduced at mid-log phase in an iron rich medium as compared to an iron restricted medium (Palyada *et al.*, 2004).

Oxidative stress defences

The transcript level of the genes encoding known or potential oxidative stress defences was decreased in iron loaded cells as compared to iron restricted cells. These genes code for alkyl hydroxyperoxide reductase (AhpC), thioredoxin and thioredoxin reductase (TrxA and TrxB), the iron-containing superoxide dismutase (SodB), catalase (KatA), an ankyrin-repeat containing protein (Cj1386), and a peroxiredoxin (Tpx). The alkyl hydroxyperoxide reductase (AhpC) catalyses the reduction of alkyl hydroperoxides to alcohols and has been shown to be important in oxidative stress defence against alkyl hydroperoxides in *C. jejuni* (Baillon *et al.*, 1999). In *E. coli* and many other bacteria, AhpC is recycled by the flavoprotein

Table 10.1 Selected members of the *Campylobacter jejuni* iron stimulon

Gene name and category	Proposed function	Iron
Iron transport and storage		
cfrA	Ferric-enterobactin uptake receptor	–
ceuBCDE	Ferric-enterobactin uptake complex	–
chuABCD	Haem transporter complex	–
Cj0178 (*ctuR/chaR*)	Putative transferrin/lactoferrin receptor	–
Cj0177 (*chaN*)	Putative iron transport protein	–
cfbpABC (Cj0173c-175c)	Putative iron uptake ABC transporter system	–
Cj1658-*p19*	Membrane proteins involved in iron transport	–
Cj1661-1663	Putative iron uptake ABC transporter system	–
exbB1-exbD1-tonB1	Outer-membrane energy transducer system	–
exbB2-exbD2-tonB2	Outer-membrane energy transducer system	–

Table 10.1 continued

Gene name and category	Proposed function	Iron
feoB	Ferrous iron transport protein	–
tonB3	TonB transport protein	–
cft	Ferritin	–
Oxidative stress defence		
ahpC	Alkyl hydroperoxide reductase	–
katA	Catalase	–
sodB	Superoxide dismutase (Fe)	–
tpx	Thiol peroxidase	–
trxA	Thioredoxin	–
trxB	Thioredoxin reductase	–
General protein glycosylation		
pglA	GalNAc transferase	+
pglC	Galactosyltransferase	+
pglE	UDP-4-keto-6-deoxy-GlcNAc C4 aminotransferase	+
pglF	UDP-GlcNAc C4,6 dehydratase	+
pglH	GalNAc transferase/polymerase	+
Energy metabolism		
ccoONPQ	cb-type cytochrome *c* oxidase	+
ldh	Lactate dehydrogenase	+
napABGH	Nitrate reductase	+
nrfA, *nrfH*	Nitrite reductase	+
nuoABCDEFGHIJKLMN	Complex I (NADH:ubiquione reductase)	+
oorDABC	2-Oxoglutarate:acceptoroxidoreductase	+
petABC	Cytochrome bc1 complex	+
sdhABC	Succinate dehydrogenase	+
Cj0020c	Putative cytochrome c551 peroxidase	–
Cj0358	Putative cytochrome c551 peroxidase	+
Cj0378c	Putative ferric reductase-like transmembrane protein	–
Cj0414, Cj0415	Gluconate dehydrogenase	–
Regulatory proteins		
cetA	Bipartate energy taxis response protein	–
perR	Peroxide stress regulator	–
Cj1000	Putative LysR family regulatory protein	–

Members of the iron stimulon mentioned in the text are highlighted, along with their proposed function and their transcriptional response to the addition of iron. The '+' and '–' symbols denote transcriptional responses at least 1.5-fold increased or decreased respectively.

AhpF, which acts as a dedicated disulphide reductase (Poole, 2005). While *C. jejuni* lacks an AhpF protein, the thioredoxin TrxA and thioredoxin reductase TrxB could exhibit an AhpF-like activity and recycle AhpC. The superoxide dismutase SodB converts superoxide anions to hydrogen peroxide, which is subsequently reduced to oxygen and water by the haem-containing catalase (KatA). Both AhpC and SodB have been shown to be required for *C. jejuni* survival in aerobic conditions (Baillon *et al.*, 1999; Purdy *et al.*, 1999). Moreover, *C. jejuni* Δ*katA* and Δ*sodB* mutants are affected in their survival ability within macrophages and INT407 cells respectively (Day *et al.*, 2000; Pesci *et al.*, 1994). The peroxiredoxin Tpx has been recently shown to protect *C. jejuni* against oxidative stress, particularly those caused by oxygen and hydrogen peroxide (Atack *et al.*, 2008). *C. jejuni* carries a second peroxiredoxin, the bacterioferritin comigratory protein (Bcp), which exhibits partially redundant antioxidant function with Tpx (Atack *et al.*, 2008). While the transcript level of *bcp* was not reported as significantly affected at mid-log phase by the iron concentration status, it transiently decreased up to threefold following the addition of ferrous sulphate to iron-restricted cells (Palyada *et al.*, 2004).

General protein glycosylation

C. jejuni possesses a pathway for general protein N-glycosylation consisting of 11 genes, named *pglA* to *pglK* (for protein glycosylation) (Nothaft *et al.*, 2008). Interestingly, the expression of at least five of these genes (*pglA, pglC, pglH, pglE* and *pglF*) was found to be significantly affected by the concentration of available iron in the growth medium (Palyada *et al.*, 2004). Moreover, the level of protein glycosylation was found to be enhanced under iron-limited conditions (Palyada *et al.*, 2004). A link between iron metabolism and the N-glycosylation pathway is further corroborated by the down-regulation of iron acquisition systems in *pgl* mutants, including the haem transporter (ChuABCD), the p19 system (p19-Cj1658 to Cj1664), and two of the three TonB energy transducing systems (TonB1-ExbB1-ExbD1 and TonB2-ExbB2-ExbD2) (Nothaft *et al.*, 2008). In contrast, the transcript of the ferrous ion transporter FeoB is increased in the *pgl* mutants (Nothaft *et al.*, 2008). A tight crosstalk between the N-glycosylation system and iron metabolism is in agreement with their co-requirement for epithelial cell invasion, host colonization and persistence (Poly and Guerry, 2008). Iron-restricted conditions, such as those encountered in the host, may strengthen *Campylobacter* virulence potential by increasing the level of protein glycosylation. The regulatory mechanism linking the expression of these two systems is unclear and requires further characterization.

Energy metabolism

Iron participates in a wide variety of critical biological functions, including the electron transport chain and central carbon metabolism. *Campylobacter jejuni* has a highly branched electron transport chain enabling *Campylobacter* to survive and grow under a wide range of environmental conditions (Kelly, 2008). Several of the key enzymes from this respiratory chain contain either haem or iron–sulphur ([Fe–S]) clusters as their catalytic centres (Fontecave, 2006). Consequently, the biogenesis of the electron transport chain is strongly dependent on the bioavailability of iron. It is therefore not surprising that the expression of the genes coding for these iron-containing enzymes is tightly regulated and repressed under iron-restricted conditions enabling an efficient and careful usage of iron. Moreover, given that the aerobic respiratory chain could be a significant source of endogenous reactive oxygen species (ROS), the route of electron transport used could significantly affect the amount of ROS generated. In *E. coli*, the main sites of endogenous ROS generation have been shown to be NADH dehydrogenase (from complex I), succinate dehydrogenase, and the cytochrome bc_1 complex of the respiratory chain (Imlay, 1995; Messner and Imlay, 1999; 2002; Naqui *et al.*, 1986), yielding up to 5 μM/s superoxide (Imlay, 1995, 2003) and 5 to 14 μM/s hydrogen peroxide (Seaver and Imlay, 2001). Interestingly, the genes encoding the *C. jejuni* complex I, also called NADH:ubiquinone oxidoreductase (*nuoABCDEFGHIJKLMN*), succinate dehydrogenase (*sdhABC*), and the cytochrome bc_1 complex (*petABC*) are repressed under iron-restricted conditions. While their preferential expression in presence of iron might reflect the need of this metal for their catalytic function, their down-regulation under iron-restriction might also restrict the generation of ROS to a minimum. The coordinate expression of these genes with the iron acquisition and oxidative stress systems might synergistically prepare cells for the anticipated influx of iron, preventing catalysis of the Fenton reaction and the formation of highly reactive hydroxyl radicals. Preventing the formation of ROS from the electron transfer chains under iron restriction might be of paramount importance for *C. jejuni* survival given the absence of manganese superoxide dismutase in *C. jejuni* and its complete reliance on its iron co-factored SodB for superoxide defence (Pesci *et al.*, 1994).

Complex I is the first enzyme in many respiratory chains and catalyses the transfer of electrons from NADH to the menaquinone pool, together with proton translocation across the bacterial membrane (Kelly, 2008). Interestingly, the *C. jejuni* complex I enzyme utilizes flavodoxin rather than NADH as an electron donor (Weerakoon and Olson, 2008). Flavodoxin is the electron acceptor for the tricarboxylic acid (TCA) cycle enzyme α-ketoglutarate dehydrogenase (encoded in *C. jejuni* by the 2-oxoglutarate oxidoreductase genes *oorDABC*), linking the respiratory chain to central carbon metabolism. The transcript levels of the *oorDABC* genes were similarly increased in response to the presence of iron, suggesting that *C. jejuni* preferentially respires using 2-oxoglutarate under iron-rich conditions. Other known electron donors to the menaquinone pool include hydrogen, formate, succinate, lactate, malate and proline (Kelly, 2008; Sellars *et al.*, 2002; Weerakoon *et al.*, 2009). The genes encoding lactate dehydrogenase (*ldh*) and succinate dehydrogenase (*sdhABCD*) were iron-induced. In contrast, the expression of the genes encoding malate dehydrogenase (*mdh*), proline dehydrogenase (*putA*), hydrogenase (*hypABCDE*), and formate dehydrogenase (*fdhABCD*) was either not responsive to the iron level or slightly induced under iron restriction (Palyada *et al.*, 2004). In *C. jejuni*, reduced ubiquinone can transfer its electrons to the cytochrome bc_1 complex (PetABC), which in turn reduces periplasmic cytochrome *c* (possibly Cj1153). Finally, cytochrome *c* is reoxidized by a *cb*-type cytochrome *c* oxidase (CcoONPQ), which uses oxygen as a terminal electron acceptor, or by one of two cytochrome *c* peroxidases (Cj0020c and Cj0358), which use hydrogen peroxide as a terminal electron acceptor. Other potential electron acceptors of the menaquinone pool include a nitrite reductase (NrfA and NrfH), a nitrate reductase (NapABGH), a cyanide-insensitive oxidase (CydAB), a dimethyl sulphoxide (DMSO) and trimethylamine N-oxide (TMAO) reductase, and possibly a *b*-type cytochrome (Cj0378) coupled to a molybdoprotein reductase (Cj0379) (Pittman and Kelly, 2005; Weingarten *et al.*, 2008). Finally, the respiratory chain could be bypassed with the entry of electrons directly via cytochrome *c* by respiration with gluconate (via gluconate dehydrogenase encoded by Cj0414 and Cj0415) or sulphite (via the sulphite oxidase encoded by *sorAB*) (Myers and Kelly, 2005; Pajaniappan *et al.*, 2008).

Interestingly, the expression of the genes encoding the above-mentioned proteins involved in the multiple electron transfer pathways was found to be iron responsive (Fig. 10.1). The avenues that appear to be favoured in the presence of iron are those requiring iron-rich respiratory complexes and utilizing oxygen or nitrate as a terminal electron acceptor. As noted above, these two electron transfer branches could generate significant amounts of reactive oxygen or nitrate species. These two branches are circumvented under iron-restricted conditions, with the preferential pathways being those utilizing hydrogen peroxide or TMAO/DMSO as terminal electron acceptor. This process would minimize the generation of hydroxyl radicals and would also contribute to the detoxification of hydrogen peroxide.

Proteins with regulatory function

The transcript levels of genes specifying several transcriptional regulators and two-component signal transduction systems were found to be significantly iron repressed (more than twofold), including the peroxide stress regulator *perR*, the putative LysR transcriptional regulator Cj1000, and the bipartite energy taxis system *cetA* and *cetB*. The regulation of *perR* expression by iron level reflects a tight interplay between the oxidative stress and iron stimuli. To date, the function of Cj1000 is uncharacterized. The CetA and CetB proteins constitute a bipartite energy taxis system in *C. jejuni* (Elliott and Dirita, 2008). This system is analogous to the energy taxis receptor Aer of *Escherichia coli*, which senses changes in the redox state of the electron transfer chain (Taylor, 2007). Consequently, the differential expression of *cetAB* in response to iron availability might reflect the concurrent shift in the electron pathway employed.

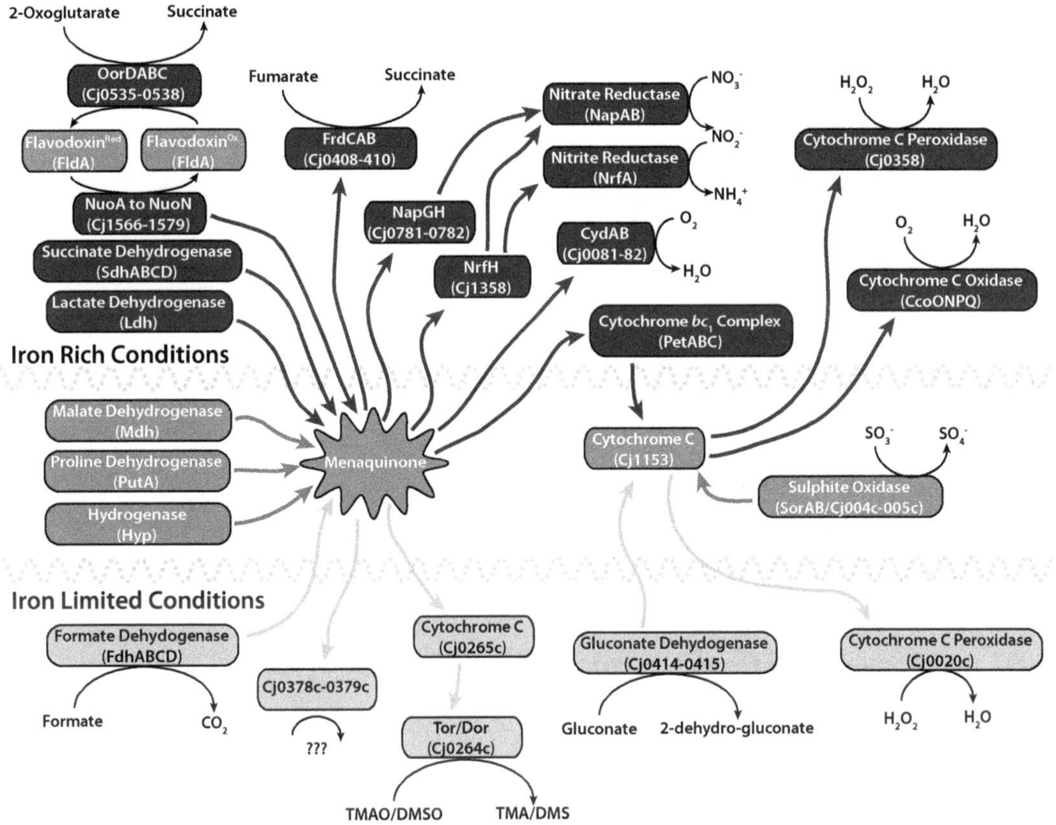

Figure 10.1 Iron-inducible electron transport pathways in *Campylobacter jejuni*. Major electron transport pathways are shown grouped according to their induction by iron. Dark boxes denote proteins induced under iron rich conditions while pale denotes those induced under iron restriction. Gray boxes indicate those proteins whose expression is not altered by iron levels. Figure adapted and updated from (Hendrixson, 2008; Palyada *et al.*, 2004; Weerakoon and Olson, 2008).

The oxidant stimulons of *Campylobacter*

Genes encoding proteins involved in oxidative stress defences are known to be induced in response to a disturbance in the oxidant-antioxidant balance in the favour of oxidants. Therefore, to characterize the oxidant stimulons of *C. jejuni*, we monitored gene expression under iron-restriction in response to 10 min exposures to three classes of oxidants: cumene hydroperoxide (an organic hydroperoxide), hydrogen peroxide, and menadione (a superoxide generator). There were 113 genes found to be at least 1.5-fold differentially expressed upon exposure of *C. jejuni* to at least one of the three different oxidants (Palyada *et al.*, 2009). Notably, several genes were found to be only inducible by certain oxidants, suggesting specific inducible responses to particular classes of oxidative agent. Specifically, of the 113 differentially expressed genes, 61, 83 and 71 genes have their transcript abundance affected by the addition of hydrogen peroxide, cumene hydroperoxide and menadione, respectively. The hydrogen peroxide stimulon consists of 29 up- and 32 down-regulated genes; the cumene hydroperoxide stimulon consists of 31 up- and 52 down-regulated genes; and the menadione stimulon is composed of 24 up- and 47 down-regulated genes. While the differences in gene content between the three stimulons might indicate specific transcriptional responses to each oxidant, it is also possible that these differences result from distinct levels of oxidative stress induced by each oxidant. The majority of the oxidant-responsive genes could be grouped into five functional categories as discussed below (and listed in Table 10.2).

Table 10.2 Selected members of the oxidant stimulons in *Campylobacter jejuni*

Gene name and category	Proposed function	Hydrogen peroxide	Cumene hydroperoxide	Menadione
Oxidative stress defence				
ahpC	Alkyl hydroperoxide reductase	+	+	
cft	Ferritin	+		
dps	Putative bacterioferritin			+
katA	Catalase	+	+	
sodB	Superoxide dismutase	+		+
Cj0358	Putative cytochrome C551 peroxidase			+
Heat shock				
grpE	**Heat shock** protein		+	
hrcA	Putative heat shock regulator		+	
Haem acquisition				
chuB	Haem uptake system permease protein	-	-	
chuC	Haem uptake system ATP-binding protein	-	-	-
chuD	Haem uptake system periplasmic haemin-binding protein		-	
DNA replication				
dnaA	Chromosomal replication initiator protein		-	
dnaG	DNA primase		-	-
dnaX	Putative DNA polymerase III subunit gamma		-	
purD	Phosphoribosylamine–glycine ligase		-	-
pyrE	Putative orotate phosphoribosyltransferase		-	
pyrF	Orotidine 5′-phosphate decarboxylase		-	
Surface structures				
accA	Acetyl-coenzyme A carboxylase carboxyltransferase subunit α		-	-
accB	Putative biotin carboxyl carrier protein of acetyl-CoA carboxylase		-	
fabH	Putative 3-oxoacyl-[acyl-carrier-protein] synthase		-	-
flhB	Flagellar biosynthetic protein	-		-
glf	UDP-galactopyranose mutase	-	-	-
mraY	Phospho-N-acetylmuramoyl-pentapeptide-transferase		-	
neuC2	Putative N-acetylglucosamine-6-phosphate2-epimerase/N-acetylglucosamine-6-phosphatase		-	
pglH	GalNAc transferase/polymerase	-		-

Table 10.2 continued

Gene name and category	Proposed function	Hydrogen peroxide	Cumene hydroperoxide	Menadione
plsX	Putative fatty acid/phospholipid synthesis protein	-	-	
Cj1305c, Cj1310c	Hypothetical proteins (617 family)		-	
Cj1423c	Putative D-glycero-D-manno-heptose 1-phosphate guanosyltransferase	-	-	
Cj1432c	Putative sugar transferase		-	
Cj1436c	Aminotransferase	-		
Cj1437c	Aminotransferase	-	-	

Members of the various oxidant stimulons mentioned in this chapter are highlighted, along with their annotated function and transcriptional response to each oxidant. The '+' and '-' symbols denote transcriptional responses at least 1.5-fold increased or decreased respectively.

Genes involved in oxidative stress defence

The analysis of the *C. jejuni* genome reveals the presence of enzymes and proteins that are thought or known to protect *Campylobacter* against the damaging effects of oxidative stress, including SodB (Pesci *et al.*, 1994), KatA (Day *et al.*, 2000), AhpC (Baillon *et al.*, 1999), Cft (Wai *et al.*, 1996), FdxA (van Vliet *et al.*, 2001), Dps (Ishikawa *et al.*, 2003), and Tpx (Atack *et al.*, 2008). Previous reports have shown that AhpC and SodB are required for *C. jejuni* survival in aerobic conditions and that *C. jejuni* Δ*katA* and Δ*sodB* mutants are affected in their survival ability within macrophages and INT407 cells respectively (Day *et al.*, 2000; Purdy *et al.*, 1999). Inactivation of *katA* resulted in severe sensitivity toward hydrogen peroxide and a slight sensitivity towards cumene hydroperoxide, suggesting that KatA might also exhibit an organic-hydroperoxidase activity (Palyada *et al.* 2009). Nevertheless, AhpC is the major enzyme in organic-hydroperoxide detoxification in *C. jejuni*, as demonstrated by an increased sensitivity of a *C. jejuni* NCTC 11168 Δ*ahpC* mutant toward cumene hydroperoxide. Interestingly, inactivation of *ahpC* increased hydrogen peroxide resistance suggesting a compensatory effect, probably through the over-expression of genes that deal with this particular stress (such as *katA*, *tpx* or *bcp*) (Palyada *et al.* 2009). Similarly, mutation of *ahpC* in *Staphylococcus aureus*, *Bacillus subtilis*, and *Xanthomonas campestris* was also shown to increase H_2O_2 resistance (Bsat *et al.*, 1996; Charoenlap *et al.*, 2005; Cosgrove *et al.*, 2007). Inactivation of *sodB* led to impaired resistance toward H_2O_2, cumene hydroperoxide and menadione (Palyada *et al.* 2009). Menadione is known to be a superoxide generator. Consequently, the menadione sensitivity of the *sodB* mutant is in agreement with the function of SodB as an enzyme that dismutases superoxide to hydrogen peroxide. Interestingly, the *sodB* mutation also reduced the ability of *C. jejuni* to resist H_2O_2 and cumene hydroxyperoxide exposure (Palyada *et al.* 2009); however these two oxidants are not directly detoxified by SodB. Based on this observation, it is tempting to speculate that the increase of endogenous superoxide anions produced during normal metabolism in the Δ*sodB* mutant will catalyse the release of ferrous iron from iron–sulphur clusters. In turn, a larger pool of labile iron will be available to catalyse the Fenton reaction in presence of hydrogen peroxide and produce hydroxyl radicals. In the presence of cumene hydroperoxide, this pool of iron might catalyse the generation of alkoxyl radicals which will initiate lipid peroxidation. As a consequence, the Δ*sodB* mutant might be sensitive to H_2O_2 and cumene hydroperoxide due to increased metal-catalysed ROS production. Both *C. jejuni* Δ*cft* and Δ*dps* deletion mutants were found to be more sensitive to H_2O_2 than their parent strains (Ishikawa *et al.*, 2003; Wai *et al.*, 1996). The lack of Cft may result in increased levels of labile iron in the cell, which would contribute to hydrogen peroxide toxicity by leading to increased hydroxyl radical formation. A similar mechanism of protection has also been proposed for Dps's role in protection against oxidative stress in *C. jejuni* (Ishikawa *et al.*, 2003). Although Dps proteins in other bacteria bind DNA and are believed protect it from oxidative damage; *C. jejuni* Dps does not appear to have this ability and may solely confer resistance to H_2O_2 by sequestering iron (Ishikawa *et al.*, 2003). A *C. jejuni* Δ*fdxA* deletion mutant was significantly affected in its aerotolerance but not in its resistance to hydrogen peroxide and cumene hydroperoxide (van Vliet *et al.*, 2001). The specific function of FdxA in oxidative stress defence remains to be elucidated. It has been shown that a Δ*tpx/dcp* double mutant is affected in aerotolerance and is hypersensitive to oxidant exposure (Atack *et al.*, 2008). Given that the single mutants were not significantly affected, it has been proposed that Tpx and Bcp are partially redundant antioxidant enzymes (Atack *et al.*, 2008). Finally, while the two cytochrome *c* peroxidases, Cj0020c and Cj0358, were expected to contribute to hydrogen peroxide resistance, they were not differentially expressed upon H_2O_2 exposure (Palyada *et al.* 2009). Moreover, the peroxidase activities of these two proteins do not appear to contribute to resistance to hydrogen peroxide (Bingham-Ramos and Hendrixson, 2008). Nevertheless, the transcript level of Cj0358 was increased upon menadione exposure suggesting a potential role in resistance to menadione-related stress.

Genes encoding proteins involved in heat shock

HrcA and *grpE* are two genes which are part of the cumene hydroperoxide stimulon and are involved in heat shock response in *C. jejuni* (Stintzi, 2003). The putative heat shock regulator HrcA, which is thought to repress the *groEL–groES* and *hrcA-grpE–dnaK* operons, was up-regulated in response to cumene hydroperoxide. GrpE is a nucleotide exchange factor involved in the exchange of ADP for ATP on DnaK, which along with DnaJ, helps prevent the formation of aggregates of denatured proteins within the cell (Harrison, 2003). The induction of these genes in response to oxidative stress in *C. jejuni* may function to prevent harmful aggregations forming from damaged oxidized proteins.

Genes involved in haem transport

The transcript levels of the genes involved in haem transport, *chuABCD*, were down-regulated in response to oxidant exposure in *C. jejuni*. Inactivation of the outer-membrane receptor ChuA led to an inability to grow in the presence of haem or haemoglobin as an iron source (Ridley *et al.*, 2006). The genes *chuBCD* encode for a haem uptake system permease protein, ATP-binding protein and periplasmic haem-binding protein respectively (Ridley *et al.*, 2006). The down-regulation of haem acquisition genes might indirectly contribute to cell survival by preventing further production of reactive oxygen species catalysed by the uptake of additional exogenous iron. In addition, the repression of these genes might also be an indirect consequence of iron released from protein iron–sulphur clusters (such as aconitase and fumarase) upon oxidative stress, which in turn would result in Fe^{2+}-Fur repression (Imlay, 2008).

Genes encoding proteins involved in DNA replication

Several genes involved in DNA replication were down-regulated in response to exposure to cumene hydroperoxide. These genes include *pyrE*, *pyrF*, *purD*, *dnaX*, *dnaA*, and *dnaG*. The genes *pyrE* and *pyrF* encode for a putative orotate phosphoribosyltransferase and an orotidine 5′-phosphate decarboxylase respectively. Both enzymes are involved in pyrimidine base biosynthesis whereas the gene *purD* encodes a phosphoribosylamine-glycine ligase involved in purine biosynthesis. *dnaX* encodes a putative DNA polymerase III gamma subunit, *dnaA* encodes a chromosomal replication initiator protein, and *dnaG* codes for DNA primase. The down-regulation of these genes in response to oxidative stress seems to indicate a growth arrest; possibly to allow for repair of damaged DNA, and thus avoiding the accumulation of mutations.

Genes encoding proteins involved in surface structures

The transcript levels of a number of genes encoding proteins involved in capsule and membrane biogenesis were found to be significantly decreased upon oxidant exposure suggesting a remodelling of the outer-membrane and cellular structure in response to oxidative stress. In particular, several genes involved in biogenesis of the lipids (*accA, accB, fabH,* and *plsX*), the peptidoglycan (*mraY*), the capsule (Cj1436c, Cj1432c, *glf*, Cj1423c, and Cj1437c), the O-linked glycosylated flagellum (Cj1305c, Cj1310c, *neuC2*, and *flhB*), and the N-linked glycosyl group (*pglH*) were found to be repressed. The extent and significance of this membrane structure remodelling remains to be investigated. Nevertheless, a similar response was also observed in *Rhodobacter sphaeroides* cells exposed to hydrogen peroxide (Zeller *et al.*, 2005).

The iron and oxidant stimulons of *Campylobacter*: how common and how unique from other bacteria?

A number of studies have employed microarray analysis to identify iron-regulated genes in various microbes, including *Helicobacter pylori* (Ernst *et al.*, 2005a; Merrell *et al.*, 2003), *Actinobacillus pleuropneumoniae* (Deslandes *et al.*, 2007), *Vibrio cholera* (Mey *et al.*, 2005), *Vibrio vulnificus* (Alice *et al.*, 2008), *Francisella tularensis* (Deng *et al.*, 2006), *Escherichia coli* (McHugh *et al.*, 2003), *Synechococcus elongates* (Nodop *et al.*, 2008), *Yersinia pestis* (Zhou *et al.*, 2006), *Pseudomonas aeruginosa* (Ochsner *et al.*, 2002; Palma *et al.*, 2003), *Pseudomonas syringae* (Bronstein *et al.*, 2008), *Neisseria gonorrhoeae* (Ducey *et al.*, 2005), *Neisseria meningitidis* (Basler *et al.*, 2006), *Haemophilus influenzae* (Whitby *et al.*, 2009) and *Bacillus subtilis* (Baichoo *et al.*, 2002). While comparison of the various microarray data could be confounded by differences in growth medium and microarray experimental design, comparative analysis could still be very instructive by allowing the identification of commonalities and discordances. As expected, genes encoding proteins involved in iron acquisition systems are systematically induced by iron restriction in all analysed bacteria. In general, the transcript level of genes belonging to general metabolism and respiration pathways are significantly affected by iron availability, likely to ensure optimal bacterial metabolism under iron restricted or iron rich conditions. In particular, genes encoding iron-containing proteins, such as iron–sulphur containing enzymes involved in electron transfer chains, were usually found to be preferentially expressed under iron rich conditions. This allows the bacterium to avoid the depletion of intracellular iron by these iron-intensive proteins under iron scarce conditions. While the transcriptional response of *C. jejuni* to iron availability is very similar to the response from other microbes, the expression profile of several genes was found to be rather distinctive to *Campylobacter jejuni*, and probably delineates a unique regulatory network. In particular, the expression of the *C. jejuni* ferritin gene (*cft*) was found to be up-regulated under iron-restricted conditions (Palyada *et al.*, 2004). To the best of our knowledge, the gene encoding ferritin is typically induced by iron, in agreement with its function as an iron storage protein. Therefore, the up-regulation of the ferritin gene under iron restricted conditions in *C. jejuni* is to some extent surprising. A possible explanation is that in contrast to the ferritin from other bacteria, *C. jejuni* ferritin has been found to contribute to the defence against iron-mediated oxidative stress (Wai *et al.*, 1996). In turn, the up-regulation of ferritin under iron restriction might protect *C. jejuni* from the anticipated accumulation of iron which would induce oxidative stress through the Fenton reaction. Whereas the genes encoding the oxidative stress defence systems in *C. jejuni* were found to be induced under iron restriction, in most bacteria (with the exception of *Bacillus subtilis*) the expression of this class of genes is not directly responsive to iron availability but rather to oxidant exposure. The co-regulation of the oxidative stress defence genes by iron and oxidants in *C. jejuni* is most likely the direct consequence of the absence of the transcriptional regulators OxyR and SoxRS, which are widespread among bacteria and activate genes in response to peroxide stress and superoxide radicals respectively (Mongkolsuk and Helmann, 2002). Instead, *C. jejuni* carries a PerR homologue which senses hydrogen peroxide through oxidation of its iron-coordinating histidyl residues (Lee and Helmann, 2006b). When iron is in short supply, the demetallated apo-PerR is unable to repress gene expression, resulting in the induction of the PerR-repressed genes even in the absence of oxidative stress. A similar situation is observed in *B. subtilis*, which also lacks an OxyR homologue and uses a PerR transcriptional repressor in its place. However, in contrast to *C. jejuni*, *B. subtilis* carries two additional transcription factors OhrR and σ^B, which also co-ordinate the transcriptional response to peroxide stress (Helmann *et al.*, 2003). In turn, *B. subtilis* appears to be better equipped than *C. jejuni* to co-ordinate gene expression as a function of changing environmental conditions. While the induction of oxidative stress defence genes in the absence of oxidants in *C. jejuni* could appear as a waste of energy, such a mechanism should prime the bacterium for the anticipated acquisition of iron. Moreover, iron-limited conditions, as those encountered in the host, would predispose *C. jejuni* for an efficient defence against oxidative stress induced by the host immune system during inflammation.

With respect to the specific genes belonging to the oxidant stimulons, there are some interesting similarities and differences present in *C. jejuni* when compared with other bacterial species, in particular with regard to the three main antioxidant enzymes KatA, AhpC and SodB. *C. jejuni* carries a single monofunctional catalase encoded by the *katA* gene (Day *et al.*, 2000). In contrast to this, *E. coli* contains two catalases: hydroperoxidase I (KatG) and hydroperoxidase II (KatE) (Zamocky and Koller, 1999). Interestingly, KatE is primarily expressed during stationary phase or during conditions which invoke slow growth in the bacteria, whereas KatG is part of the hydrogen peroxide stimulon and is induced upon exposure to this oxidant (Zheng *et al.*, 2001). These two different catalases are under the regulation of two different transcription factors. KatG is regulated by OxyR whereas KatE is regulated by the stationary phase sigma factor, σ^s. This accounts for KatG belonging to the hydrogen peroxide stimulon unlike KatE (Zamocky *et al.*, 2008; Zheng *et al.*, 2001). In *P. aeruginosa*, there are three monofunctional catalases present: KatA, KatB and KatC (Ochsner *et al.*, 2000). KatA is the major catalase present in *P. aeruginosa. KatA*'s expression is induced by both iron and quorum sensing, with its expression greatest during stationary growth (Hassett *et al.*, 1999). Both *katA* and *katB* are part of the hydrogen peroxide stimulon. However while *katB* is regulated by OxyR, *katA* is stimulated by an OxyR independent mechanism (Ochsner *et al.*, 2000). The microaerophilic bacterium *H. pylori* contains a single monofunctional catalase, KatA, similar to *C. jejuni* (Wang *et al.*, 2006). However, in contrast to *C. jejuni, H. pylori katA* appears to be activated by Fur in the presence of iron rather than repressed (Harris *et al.*, 2002). It is not yet known whether KatA in *H. pylori* is part of the hydrogen peroxide stimulon. The *C. jejuni* genome contains a single iron containing superoxide dismutase encoded by the *sodB* gene, which is part of the hydrogen peroxide stimulon. Likewise, *H. pylori* contains a single *sodB* gene, which also contains an iron co-factor (Wang *et al.*, 2006). In *H. pylori, sodB* is regulated by apo-Fur and its expression is repressed when iron levels are low (Ernst *et al.*, 2005a,b). This is significantly different from the situation in *C. jejuni,* where the *sodB* gene is induced under iron restriction (Palyada *et al.*, 2004). This regulation of *sodB* expression in *C. jejuni* is paradoxical given that *sodB* requires iron as a cofactor and thus SodB would probably not be functional when iron is in short supply. Other bacteria have dealt with this problem by over-expressing a manganese-cofactored superoxide dismutase, SodA, under iron-limited conditions. The rationale behind *sodB* expression in *C. jejuni* is unclear.

The Fur protein and mechanisms of Fur regulation

In most bacteria, including *C. jejuni,* the primary regulator of iron levels is the ferric uptake repressor (Fur) protein (Lee and Helmann, 2007). Although Fur is one of the most extensively studied metalloregulators, there is still considerable controversy over its mechanisms of binding DNA and affecting gene transcription. Fur is a homodimer that binds upstream in the promoter region of iron acquisition genes and blocks transcription during iron replete conditions. Fur requires a metal co-factor to bind DNA and this cofactor can be one of several divalent metals, including Fe, Mn, Zn, or Co (Mills and Marletta, 2005). However, only iron is believed to be physiologically present at levels sufficient to cause gene repression. Many Fur proteins (such as *E. coli* Fur) also contain a high affinity zinc binding site that is believed to be important for protein folding and stability (Lee and Helmann, 2007). The crystal structure of *P. aeruginosa* Fur (Fur_{PA}) has been solved and contains two metal binding sites that are both occupied by Zn (Pohl *et al.*, 2003). Only one site was able to exchange the zinc ion with iron, causing this site to be designated as the metal binding site and the other as the structural zinc site. However mutational studies have suggested that the designated zinc binding site may actually be the regulatory metal site and that the Fur_{PA} may not contain a structural zinc ion (Lewin *et al.*, 2002). This is contrary to the situation in *E. coli*, where all the residues involved in zinc binding where found to be essential for Fur_{EC} function (Coy *et al.*, 1994). Fur_{CJ} contains homologous resides to all the putative ligands for the metal sites identified in both Fur_{EC} and Fur_{PA}, however the functional requirement of these ligand binding sites and the presence of a structural Zn ion has not yet been characterized.

The binding sequence and specificity of Fur has also been subject to controversy. The most compelling model for Fur binding comes from studies in *E. coli* and *B. subtilis*. Initial studies in *E. coli* identified a 19-bp consensus sequence (GATAATGATAATCATTAC) for Fur binding (Lavrrar and McIntosh, 2003). However, this sequence appeared to be too large for a small dimeric repressor such as Fur. Additional studies with synthetic DNA oligomers suggested that Fur may recognize an array of three hexameric repeats, with the minimal recognition site being GATAAT (Escolar *et al.*, 1998). This model postulates that a minimum of three Fur monomers binding to this site, with one Fur monomer each interacting with one hexameric repeat. However, as Fur exists as a dimer, it is uncertain as to whether this site would interact with monomeric or dimeric Fur. Further studies in *B. subtilis* refined the 19-bp consensus to be two overlapping 7–1–7 binding motifs (Baichoo and Helmann, 2002). In this model each 7–1–7 sequence would bind one dimer of Fur and the 19 bp consensus would have two Fur dimers bound. A recent bioinformatic study in *E. coli* has landed support to this model, although it proposes that the flanking two nucleotides are also important for Fur_{EC} binding (Chen *et al.*, 2007). However another bioinformatic study in *P. aeruginosa* found that the Fur_{PA} dimer could be modelled to bind a 9–1–9 consensus sequence and that the Fur_{PA} dimers may bind cooperatively downstream or upstream from the initial binding site (van Oeffelen *et al.*, 2008). In *C. jejuni*, a Fur_{CJ} consensus sequence has been proposed and was found to be quite different from that of *E. coli* (Palyada *et al.*, 2004). This was somewhat surprising as Fur_{CJ} has been shown to partially complement a Δ*fur* mutant in *E. coli* (Wooldridge *et al.*, 1994).

The precise binding of Fur to DNA is further obscured by the fact that monomeric Fur is capable of binding Fur boxes, albeit at much lower affinities (D'Autreaux *et al.*, 2007). However, it should be noted that this lower affinity of monomeric Fur vs. dimeric Fur DNA binding is based on comparing their affinity to the same sequence. Should monomeric Fur have a DNA binding role *in vivo*, it may bind to a different sequence and have distinct roles from dimeric Fur. As well, it has been suggested that the regulatory metal bound by dimeric Fur could also influence its binding kinetics and binding preferences (Mills and Marletta, 2005). Fur is also known to form oligomers on DNA and protect DNA sequences from degradation well beyond its classical binding sites. This oligomerization has been shown to be both Fur concentration and pH dependent (D'Autreaux *et al.*, 2007). These features suggest that Fur may be capable of regulating gene expression beyond its traditional role in metal regulation. Indeed, Fur has been implicated in the pH stress response of *H. pylori* (Gancz et *al.*, 2006) and *Salmonella typhimurium* (Foster and Hall, 1992). This regulation does not require the presence of an activating metal and can be genetically separated from its metal sensing function (Hall and Foster, 1996). Recent work in *H. pylori* suggests that Fur_{HP} may form more stable dimers than Fur_{EC} and contains additional residues that may contribute to oligomerization (Vitale *et al.*, 2009). In addition, apo-Fur_{HP} has been shown to directly bind and repress gene expression, most notably in the case of *H. pylori sodB* (Carpenter *et al.*, 2009; Ernst *et al.*, 2005b). These reports provide additional support that apo-Fur participates in gene regulation, although most studies cannot definitively distinguish between apo-Fur and Fe^{2+}-Fur binding. It should be noted however that in *Bradyrhizobium japonicum*, the Fur-like protein Irr has been shown to bind DNA without the presence of an activating metal and both repress and activate gene transcription (Lee and Helmann, 2007). Further details concerning iron regulation in the rhizobia can be found in Chapter 3.

Δ*fur* mutants commonly show several genes that are repressed rather than activated. This suggests that Fur is able to act as an activator of gene transcription as well as a repressor (Lee and Helmann, 2007). It was later discovered that much of this gene activation was an indirect consequence of *fur* deletion. Fur was found to repress a small regulator RNA, RyhB in *E. coli* and two small RNAs in *P. aeruginosa*, prrF1 and prrF2 (Massé and Gottesman, 2002; Massé *et al.*, 2007; Wilderman *et al.*, 2004). These RNAs post-transcriptionally repress their target genes by binding their mRNAs and enhancing degradation. This degradation coordinates the iron sparing response and decreases the translation of iron-containing proteins during times of iron starvation. RyhB also regulates acid resistance genes in *Shigella flexneri* (Oglesby *et al.*, 2005). There has been no *ryhB* homologue identified in *C. jejuni* to date, although its presence should not be ruled out *a priori*. Notably, succinate dehydrogenase, from which *ryhB* is also transcribed, is regulated differently in *E. coli* as compared to *C. jejuni* (Massé and Gottesman, 2002; Palyada *et al.*, 2004). Recent work in *H. pylori* has revealed that Fur_{HP} is capable of directly binding genes which are known to be to be Fur activated (Carpenter *et al.*, 2009; Danielli *et al.*, 2006; Ernst *et al.*, 2005b). To date this is the best published evidence of direct Fur gene activation. Preliminary ChIP-on-chip data indicates that direct gene activation by Fur is also present in *C. jejuni*, although the mechanism by which this activation occurs is currently unknown (A. Stintzi and J. Butcher, unpublished).

In summary, five different mechanisms of Fur regulation have been described: (1) repression of transcription by Fe^{2+}-Fur proteins; (2) activation of gene expression by Fe^{2+}-Fur proteins; (3) indirect repression of gene expression by Fe^{2+}-Fur proteins via small RNA molecules (e.g. *ryhB* in *E. coli*); (4) repression of gene expression by the iron-free form of Fur proteins (apo-Fur); and (5) activation of gene expression by apo-Fur. Importantly, not all five types of regulation have been described in the same organism (Fig. 10.2), with only one or two types of regulation typically present in a particular bacterium (Lee and Helmann, 2007). Moreover, although Fur is ubiquitous in Gram-negative bacteria, its functions and regulons significantly differ among them.

The Fur regulon

The *C. jejuni* Fur regulon was characterized by two independent studies (Holmes *et al.*, 2005; Palyada *et al.*, 2004). Both studies compared the transcriptome profile of the wild-type strain with a *fur* deletion mutant grown in iron-rich and iron-limited conditions. While the majority of the Fur regulated genes appear to be Fe^{2+}-Fur repressed, analysis of the transcriptome data suggests other types of Fur regulation, including apo-Fur repression, and both apo-Fur and Fe^{2+}-Fur activation (Table 10.3). Importantly, these studies cannot distinguish direct from indirect regulation. Consequently, not all of these four modes of Fur regulation might truly take place. Interestingly, unlike the *fur* gene from other bacteria, the expression of the *fur* gene from *C. jejuni* is not auto-regulated (Palyada *et al.*, 2004).

In agreement with the classical Fur regulation paradigm, all the genes encoding the iron acquisition systems were found to be Fe^{2+}-Fur repressed (Table 10.3). These genes encode the ferric-enterobactin transport system (*cfrA-ceuBCE*), the haem transporter complex (*chuABCD*), the ferric-transferrin/ferric-lactoferrin iron acquisition system (Cj0177–0178, and *cfbpABC*), the p19 system (*p19*, Cj1658 and Cj1660–1663), and several components of the three TonB–ExbB-ExbD complexes (*tonB1*, *tonB3*, *exbB1*, *exbD1*, *exbB2* and *exbD2*). Another category of genes whose expression is Fe^{2+}-Fur repressed are involved in oxidative stress defence. These genes include catalase (*katA*), alkyl-hydroperoxidase reductase (*ahpC*), a thioredoxin reductase (*trxB*), an ankyrin-repeat containing protein (Cj1386), and a probable oxidoreductase (Cj0559). Unexpectedly, these studies have identified the ferric uptake regulator Fur as a master regulator in *C. jejuni*. Besides genes encoding proteins with a clear role in iron acquisition or oxidant detoxification, Fur directly or indirectly regulates genes involved in flagella biogenesis, energy metabolism, fatty acid biosynthesis, other non-iron transporters, and numerous hypothetical proteins (Holmes *et al.*, 2005; Palyada *et al.*, 2004). While the expression of all Fur regulated genes identified by Palyada *et al.* was iron responsive, not all iron responsive genes were found to be Fur regulated. This observation suggests the presence of additional levels of gene regulation in response to iron availability. In fact, the responsiveness of several of the genes with respect to iron availability could not be explained by the mode of Fur regulation. For example, the transcript levels of Cj0403 (hypothetical protein), Cj0569 (hypothetical protein), Cj0011c (a DNA binding protein involved in DNA transformation; Jeon and Zhang, 2007) and Cj0012c (also named rrc, an rbo/Rbr-like protein; Yamasaki *et al.*, 2004) are Fe^{2+}-Fur activated and yet their gene expression is repressed under iron-rich conditions (Palyada *et al.*, 2004). The expression of these genes is probably under the control of an additional transcriptional regulator and probably not directly regulated by Fur.

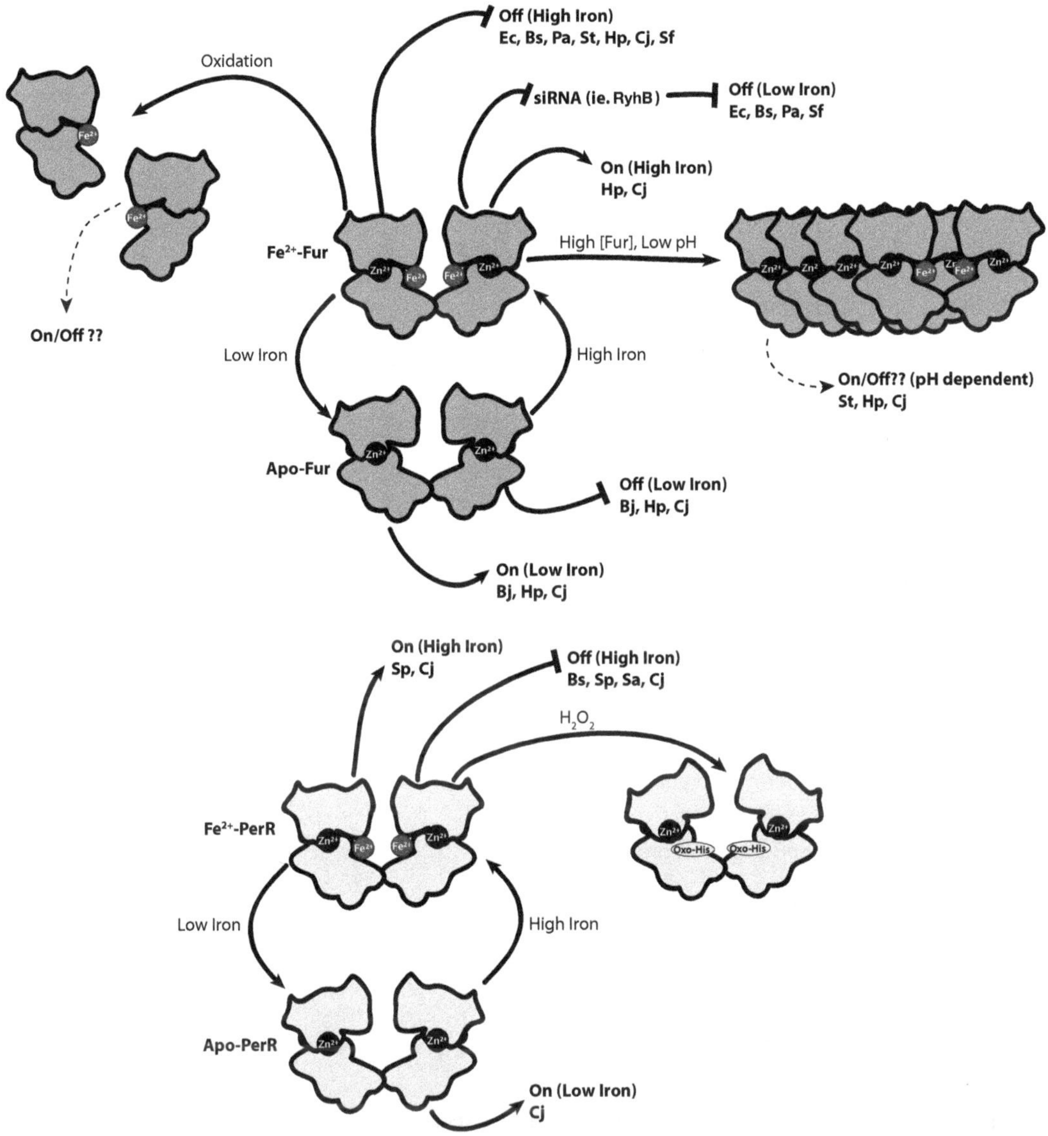

Figure 10.2 Methods of gene regulation by the Fur and PerR proteins. Top: The known modes of Fur regulation are shown along with bacterial species where this type of regulation has been identified. Figure adapted and updated from (D'Autreaux *et al.*, 2007; Lee and Helmann, 2007). Bottom: The known modes of PerR regulation are shown along with bacterial species where this type of regulation has been identified (Gryllos *et al.*, 2008; Helmann *et al.*, 2003; Horsburgh *et al.*, 2001). Ec, *Escherichia coli*; Bs, *Bacillus subtilis*; Pa, *Pseudomonas aeruginosa*; St, *Salmonella typhimurim*; Hp, *Helicobater pylori*; Cj, *Campylobacter jejuni*; Bj, *Bradyrhizobium japonicum;* Sp, *Streptococcus pyrogenes*; Sa, *Staphylococcus aureus*; Sf, *Shigella flexneri.*

Fur regulation beyond iron acquisition and metabolism

In addition to the classical role of Fur in the regulation of iron metabolism, Fur has been linked to oxidative, nitrosative, and acid stress responses in several bacteria (Baichoo *et al.*, 2002; Ernst *et al.*, 2005b; Gao *et al.*, 2008; Holmes *et al.*, 2005; Mey *et al.*, 2005; Palyada *et al.*, 2004; Yang *et al.*, 2006; Zhou *et al.*, 2006). Transcriptomic studies have revealed that Fur regulates between 50 and 100 genes depending on the species. Interestingly, these genes not only code for iron acquisition systems, iron storage and

Table 10.3 The Fur and PerR regulons in *Campylobacter jejuni*

Gene name and category	Proposed function	Fur –Fe	Fur +Fe	Iron	PerR –Fe	PerR +Fe	PerR +Fe +H_2O_2
Iron transport							
cfrA	Ferric-enterobactin uptake receptor		–	–			
ceuBCDE	Ferric-enterobactin uptake complex		–	–	–		
chuABCD	Haem transporter complex		–	–	–		
chuZ (Cj1613c)	Putative pyridoxamine 5′-phosphate oxidase		–	–	–		
Cj0178 (*ctuR*/*chaR*)	Putative transferrin/lactoferrin receptor		–	–			
Cj0177 (*chaN*)	Puative iron transport protein		–	–	–		
cfbpABC	Putative iron uptake ABC transporter system		–	–	–		
Cj1658-*p19*	Membrane proteins involved in iron transport		–	–	–		
Cj1661–1663	Putative iron uptake ABC transporter system		–	–	–		
exbB1-exbD1-tonB1	Outer-membrane energy transducer system		–	–			
exbB2-exbD2-tonB2	Outer-membrane energy transducer system		–	–	–		
tonB3	TonB transport protein		–	–	–		
Oxidative stress defence							
ahpC	Alkyl hydroperoxidase reductase		–	–	–		
cft	Ferritin			–			
fdxA	Ferredoxin	–		+			
katA	Catalase		–	–	–	–	
rrc (Cj0012c)	Rubredoxin oxidoreductase/rubrerythrin homologue	–	+	+		–	
trxB	Thioredoxin reductase		–	–	–	–	
Cj0559	Probable oxidoreductase		–	–			
Cj1386	Ankyrin-repeat containing protein		–	–			
Cj1545c	MdaB protein homologue				–		
Flagella biosynthesis							
flaD	Putative flagellin	+		+	–	–	–
flgB	Putative flagellar basal-body rod protein		+		–	–	–
flgC	Flagellar basal-body rod protein					–	
flgD	Putative flagellar hook assembly protein		+		–	–	–
flgE	Flagellar hook protein			+	–	–	–
flgE2	Flagellar hook subunit protein	+		–	–	–	–

Table 10.3 continued

Gene name and category	Proposed function	Fur −Fe	Fur +Fe	Iron	PerR −Fe	PerR +Fe	PerR +Fe $+H_2O_2$
flgG, *flgG2*	Flagellar basal-body rod proteins	+		+	−	−	−
flgH	Putative flagellar L-ring protein precursor	+		+	−	−	−
flgI	Flagellar P-ring protein	+		+	−	−	−
flgK	Putative flagellar hook-associated protein		+	+	−	−	−
flgR	Sigma-54-associated transcriptional activator				−	−	−
flhB	Flagellar biosynthetic protein			+			
fliK	Putative flagellar hook-length control protein			+	−	−	−
fliP	Flagellar biosynthesis protein		−				
Energy metabolism							
fba	Fructose-bisphosphate aldolase				+		
fldA	Flavodoxin		−	−			
fumC	Fumarate hydratase		+	+			
nuoN	NADH dehydrogenase I chain N		+	+			
napA	Periplasmic nitrate reductase			+			
napGH	Putative ferredoxins			+	+		
oorAB	2-Oxoglutarate:acceptoroxidoreducta se subunits			+			
sdhA	Succinate dehydrogenase flavoprotein subunit	−		+			
Cj0264c	Molybdopterin-containing oxidoreductase`	+		−			
Cj0265c	Putative cytochrome C-type haem-binding protein	+	+	−			
Cj0378c	Putative integral membrane protein		−	−			
Cj0379c	Putative molybdopterin-containing oxidoreductase		−				
Cj0414–0415	Putative oxidoreductase subunits	+	+	−			
Cj1377c	Putative ferredoxin		−	−			
Cj1664–1665	Possible thiredoxins		−	−			
Other transport proteins							
pstC	Putative phosphate transport system permease protein	+					
pstS	Possible periplasmic phosphate binding protein				−		
Cj0045c	Putative iron-binding protein				−		−

Table 10.3 continued

Gene name and category	Proposed function	Fur −Fe	Fur +Fe	Iron	PerR −Fe	PerR +Fe	PerR +Fe $+H_2O_2$
Cj0204	Putative oligopeptide transporter, OPT family			–			
Cj0444	Pseudogene (TonB-dependent OM receptor)		–	–			
Cj0850c	Transmembrane transport protein						
Cj0948c	Putative transmembrane transport protein	–		+			
Cj1582c	Putative peptide ABC-transport permease protein			+			

The members of the Fur and PerR regulons are grouped according to functional category. Each gene's transcript was up- or down-regulated at least 2-fold in at least one experimental condition with a *p* value $<10^{-4}$. Iron restricted conditions are denoted as '–Fe' while iron replete conditions are denoted as '+Fe'. The '+' and '–' symbols denote activation or repression of gene expression by Fur, PerR or iron.

metabolism, but also for proteins involved in oxidative stress defence, virulence and acid resistance as observed in *C. jejuni*. However, it is important to re-emphasize that transcriptomic studies do not allow one to differentiate between direct and indirect regulation. As such, the Fur regulation of some of these genes might be indirect or due to the activation of a regulatory cascade. In fact, it is clear that positive regulation by Fur is often indirect via the repression of a non-coding small RNA (*ryhB* in *E. coli*) by Fe^{2+}-Fur (Massé *et al.*, 2007). In addition, Fur has been shown to regulate in an iron-independent manner the expression of genes involved in acid survival in *Salmonella enterica* serovar Typhimurium (Hall and Foster, 1996). More recently, transcriptome analyses have revealed the importance of Fur in the regulation of pH homeostasis in *H. pylori* (Gancz *et al.*, 2006). Indeed, Fur was found to be required for growth at low pH and up to 93 genes were found to exhibit altered expression in a Fur-deficient strain in response to pH shock. Interestingly, acidic pH was recently shown to significantly favour the oligomerization of Fur, leading to different DNA binding properties, and thus providing a possible mechanism for Fur-mediated pH regulation (D'Autreaux *et al.*, 2007). A Δ*fur* deletion mutant of *C. jejuni* has been demonstrated to be impaired in its ability to grow at pH 5.5 (Reid *et al.*, 2008), and preliminary experiments indicate that the Δ*fur* mutant is severely affected in its ability to survive acid shock at pH 3 (Palyada *et al.*, 2009). These observations indicate that Fur plays a role in acid resistance in *C. jejuni*. Up to now, the transcriptomic studies of the Δ*fur* mutant have aimed at identifying iron and Fur regulated genes but not Fur regulated and acid induced genes. Acid responsive genes would not be differentially expressed between the Δ*fur* mutant and the wild-type strain under normal growth conditions. Consequently, the reported Fur regulon is likely to be incomplete. More in depth studies of the transcriptional responses of the Δ*fur* mutant to pH stress would be required to generate a comprehensive list of the members from the Fur regulon.

The Fur regulon of *Campylobacter*: how common and how unique from other bacteria?

While the specific members of the Fur regulon differ between individual bacteria, the biochemical pathways regulated by Fur are relatively similar. Genes encoding proteins involved in iron acquisition systems are classically Fe^{2+}-Fur repressed in most bacteria, including *C. jejuni*. *B. japonicum* and other related rhizobial species might be an exception to this rule as it contains another Fur-like protein Irr, that is involved in iron homeostasis (Lee and Helmann, 2007). *B. japonicum* Fur has been shown to exhibit only a modest role repressing iron regulated in genes under iron rich conditions (Zhou *et al.*,

2006). Genome-wide transcriptomic studies of the Δ*fur* mutant from numerous bacteria have revealed additional categories of genes that are Fur regulated. In *E. coli*, Fur also regulates genes encoding iron-containing proteins and proteins involved in the tricarboxylic acid cycle, energy metabolism, and redox stress resistance (McHugh *et al.*, 2003). In *Y. pestis*, Fur regulates, in addition to the classical iron metabolism genes, genes encoding proteins involved in a variety of cellular processes including virulence factors, pesticin activity, oxidative stress defence and electron transport (Gao *et al.*, 2008; Zhou *et al.*, 2006). In *B. subtilis*, the Fur regulon is primarily composed of genes involved in siderophore biosynthesis and iron acquisition (Baichoo *et al.*, 2002). Further discussion of iron uptake in *B. subtilis* can be found in Chapter 12 by Gaballa and Helmann. However, the *B. subtilis* Fur regulon also includes genes implicated in the utilization of lichenan, flagella biogenesis, amino-acid metabolism, redox activity and nitrite reduction. In *H. pylori*, Fur-regulated genes were shown to be involved metal metabolism, nitrogen metabolism, motility, cell wall synthesis and cofactor synthesis (Danielli *et al.*, 2006; Ernst *et al.*, 2005a). It is apparent from these studies that Fur is a global regulator in most bacteria, including *C. jejuni*, by controlling the expression of genes implicated in biological functions beyond classical iron metabolism.

The Fur regulon and its effect on colonization and virulence

It has long been established that microorganisms must be able to acquire iron and to tightly regulate iron homeostasis in order to establish and maintain successful host colonization and infection (Ratledge, 2007; Weinberg, 2009). *C. jejuni* is no exception to the rule. Deletion mutants in various iron acquisition systems and the *fur* regulator were shown to be significantly affected in their ability to colonize the host (Palyada *et al.*, 2004). A Δ*cfrA* mutant, which is defective in enterobactin-mediated iron acquisition, was found to be unable to colonize the gastrointestinal tract of chickens (Palyada *et al.*, 2004). Given that enterobactin is not synthesized by *C. jejuni*, this observation suggests that *C. jejuni* gut colonization relies on the production of enterobactin or enterobactin-like molecules by the gut microflora. Deletion mutants in the Cj0178 and *ceuE* genes, shown to be implicated in ferric-transferrin or ferric-lactoferrin mediated iron acquisition (Miller *et al.*, 2008) and ferric-enterobactin transport respectively, were also significantly affected in chick colonization (Palyada *et al.*, 2004). Moreover, the ΔCj0178 mutant was also shown to be attenuated in its ability to survive in the rabbit ileal loop (Stintzi *et al.*, 2005). Other members of the Fur regulon shown to be required for chick gut colonization include *tonB1* and *tonB3* (A. Stintzi, unpublished), the oxidative stress defence genes *ahpC* and *katA* (A. Stintzi, unpublished), flagellum genes (Hendrixson and DiRita, 2004), the NADH dehydogenase genes (*nuo*) (Weerakoon *et al.*, 2009), and the gluconate dehydrogenase Cj0415 (Pajaniappan *et al.*, 2008; Stintzi *et al.*, 2005). In addition, mutagenesis of the *fur* gene also significantly affected the organism's ability to colonize the gastrointestinal tract of chicks, demonstrating the importance of iron homeostasis *in vivo* (Palyada *et al.*, 2004). The effect of Fur on the colonization potential of *C. jejuni* is likely to be multifactorial. First, maintaining a balanced iron level is undoubtedly essential to avoid oxidative stress due to the over-accumulation of iron. Second, the constitutive expression of outer-membrane iron acquisition systems in the Δ*fur* mutant could be attractive targets for chick antibodies, triggering the immune clearance of the Δ*fur* mutant. Thirdly, several of the Fur activated genes under iron restriction are known to be required for gut colonization. In particular, the genes encoding proteins involved in flagellum biogenesis are crucial for chick colonization (Bingham-Ramos and Hendrixson, 2008; Hendrixson, 2008; Hendrixson and DiRita, 2004). Although the above mentioned iron acquisition systems and Fur-regulated genes are of paramount importance for gastrointestinal tract colonization, the haem transporter ChuA and the rhodotorulic mediated iron acquisition system p19 appear to not be required for gut colonization (A. Stintzi, unpublished). This observation suggests that *C. jejuni* acquires iron from different sources other than haem and rhodotorulic acid in the gastrointestinal tract. In fact, in addition to the ferric-enterobactin (CfrA related iron acquisition) and ferric-lactoferrin (Cj0178 related iron acquisition) transporters, the ferrous ion acquisition system (FeoB) also plays an important role in gut colonization (Naikare *et al.*, 2006). Altogether, these data indicate the requirement of three different iron acquisition systems for efficient gut colonization, highlighting the significance of iron requirement and metabolism *in vivo*.

The PerR protein and mechanisms of PerR regulation

Most Gram-negative bacteria sense and respond to reactive oxygen species (ROS) by the OxyR and SoxRS systems, which recognize hydrogen peroxide and superoxide stresses respectively (Mongkolsuk and Helmann, 2002). *C. jejuni* does not contain homologues of either of these proteins and it solely carries the peroxide-sensing regulator PerR. As a member of the Fur family of transcriptional repressors, PerR is thought to be a zinc-containing metalloprotein that utilizes either iron or manganese as a regulatory metal ion. PerR regulates the oxidative stress response in several bacteria, including *Clostridia, Streptococcus, Bacillus* and *Staphylococcus* (Gryllos *et al.*, 2008; Helmann *et al.*, 2003; Hillmann *et al.*, 2008; Horsburgh *et al.*, 2001). The primary role of PerR is to sense H_2O_2 and most of its functional characterization has been carried out in *B. subtilis* using both crystallographic and biochemical approaches (Lee and Helmann, 2006a,b; Traore *et al.*, 2006; 2009). OxyR and PerR use two different mechanisms to sense H_2O_2. OxyR senses the presence of H_2O_2 by the oxidation of two cysteines (Imlay, 2008). These oxidized cysteines subsequently form an intramolecular disulphide bond that causes an alteration in its regulatory domain, triggering the transcription of OxyR's target genes. In contrast, PerR senses hydrogen peroxide by Fe^{2+}-catalysed oxidation of its regulatory metal binding site, which in turn leads to gene derepression (Lee and Helmann, 2006b). Crystal structure studies have revealed that PerR binds its regulatory iron metal through a coordination complex that includes three histidines (H37, H91, H93) and two aspartates (D85, D104) (Traore *et al.*, 2006; 2009). When Fe^{2+} is bound to its regulatory metal site, the PerR dimer is able to bind DNA and repress gene transcription. The presence of H_2O_2 causes the oxidation of either H37 or H91; presumably via the Fenton reaction between H_2O_2 and the bound iron (Lee and Helmann, 2006b). This metal-catalysed oxidation results in the formation of either oxo-H37 or oxo-H91 PerR species; weakening the affinity of the regulatory metal binding site for iron and eliciting gene derepression (Traore *et al.*, 2009). There is currently no known mechanism by which the oxidized histidine could be reduced to enable PerR recycling (Lee and Helmann, 2006b). Manganese is also able to compete with iron for binding to the regulatory metal site in PerR (Herbig and Helmann, 2001). However, when bound to Mn^{2+}, the Mn^{2+}-PerR form is insensitive to oxidative inactivation (at physiological levels) and remains associated with DNA even in the presence of H_2O_2 (Lee and Helmann, 2006a).

PerR is believed to repress its target transcripts by binding upstream of the transcriptional start site to DNA sequences termed PerR-boxes. The PerR-boxes from *B. subtilis* and *Streptococcus pyogenes* have been thoroughly characterized and are believed to be composed of a 7–1–7 inverted repeat similarly to Fur boxes (Baichoo and Helmann, 2002; Gryllos *et al.*, 2008). The extreme similarity between the *B. subtilis* Fur and PerR boxes (1 bp difference) makes it difficult to distinguish which genes are being regulated by each protein. Despite this similarity, Fur_{BS} and $PerR_{BS}$ do not recognize each other's consensus sequences, although artificial sequences recognized by both proteins can be synthesized (Baichoo and Helmann, 2002). Interestingly, in *C. jejuni* the expression of *katA* is regulated by both Fur and PerR, suggesting the presence of both Fur and PerR boxes or of a composite Fur/PerR box recognized by both regulators (van Vliet *et al.*, 1999; 2002). Currently there is no defined PerR box in *C. jejuni*, although bioinformatics analysis of the *ahpC* and *katA* promoter regions have revealed a Fur/PerR box-like sequence that is also present in other genes implicated in oxidative stress defence (van Vliet *et al.*, 2002). However, it is currently unclear whether these putative binding sites are bound by either Fur and/or PerR, and whether Fur_{CJ} and $PerR_{CJ}$ are able to recognize each other's binding sequence.

PerR has not yet been demonstrated to exploit the vast repertoire of mechanisms employed by Fur to regulate gene expression (Fig. 10.2). Currently, repression of genes by Fe^{2+}-PerR is the only experimentally confirmed mode of direct regulation (Lee and Helmann, 2007). While there are several reports of PerR activating genes in various bacteria, these studies were not able to differentiate between indirect or direct regulation (Gryllos *et al.*, 2008; Helmann *et al.*, 2003). However, recent work using ChIP-on-chip analysis suggests that $PerR_{CJ}$ may be able to directly activate gene expression (A. Stintzi and J. Butcher, unpublished).

The PerR regulon

PerR has previously been shown to regulate the expression of two genes; *katA* and *ahpC* (van Vliet *et al.*, 1999). We have recently further defined the PerR regulon using a transcriptomics approach (Palyada *et al.*, 2009). A *C. jejuni* NCTC 11168 isogenic deletion mutant of *perR* was constructed and its transcriptome was compared to the wild type strain in the presence and absence of both iron and H_2O_2. The PerR regulon consists of at least 66 genes and these genes appear to be regulated by PerR both dependently and independently of iron and H_2O_2. Several of the most striking gene categories are detailed below, and the entire regulon is listed in Table 10.3.

Genes encoding oxidative stress defences

Notably, all the genes belonging to the hydrogen peroxide stimulon that were more than 2-fold up-regulated in response to H_2O_2 exposure were found to be PerR regulated. This observation indicates that PerR is the primary regulator of H_2O_2 stress in *C. jejuni*. As expected, many of the genes found in the PerR regulon have been previously shown or are believed to be implicated in oxidative stress defence. These genes include *katA* (catalase), *ahpC* (alkyl hydroxyperoxide reductase), *cft* (ferritin), *trxB* (thioredoxin), Cj1545c and *perR* itself. The role of KatA, AhpC, Cft and TrxB in oxidative stress defence has already been discussed above. Cj1545c encodes an MdaB homologue, which is an NAD(P)H quinolone reductase. MdaB reduces quinones to quinols using a two-electron transfer mechanism to transfer electrons from NAD(P)H to quinone. In *H. pylori*, MdaB has been shown to contribute to oxidative stress defence by presumably competing with the quinone one-electron reduction pathway, and thus preventing the production of semiquinone radicals, which in turn are able to form superoxide radicals by reacting with molecular oxygen (Wang and Maier, 2004). Interestingly, PerR appears to directly or indirectly activate its own transcription in the presence of iron and H_2O_2. This auto-activation of PerR is paradoxical as it would consequently result in the down-regulation of oxidative stress defences at a time when they are the most needed. In fact, the transcript levels of *katA* and *ahpC* were found to be lower in the presence of both iron and hydrogen peroxide than their levels in the presence of either hydrogen peroxide or iron alone. While the rationale for this mode of regulation is unclear, it might be a direct consequence of the overlap between the PerR and Fur regulons.

Genes implicated in flagella biogenesis

Several members from the PerR regulon are implicated in the biogenesis of the flagellum. The σ^{54}-associated transcription factor FlgR is regulated by PerR, as well as a component of the putative flagellar export apparatus FlhB and the hook-length control protein FliK (Hendrixson and DiRita, 2003; Kamal *et al.*, 2007). The FliK and FlgR regulons have recently been defined using transcriptomic approaches (Kamal *et al.*, 2007). These studies found that deletion of *flgR* resulted in the down-regulation of several basal body components [*flgBCG(G2)HI* genes] and flagellar hook components [*flgDE(E2)K* genes]. In contrast, the Δ*fliK* deletion mutant exhibited up-regulation of these components as well as up-regulation of *fliE* and the σ^{28} factor *fliA*. Interestingly, the PerR regulon includes most of the flagellar components that are regulated by both FliK and FlgR such as *flgBCDE(E2)G(G2)HIK*. Intriguingly, these genes, including *fliK* and *flgR* are all up-regulated in the *perR* mutant, despite the opposing regulatory effects of FliK and FlgR on the expression of many of these genes. In fact, preliminary ChIP-on-chip analysis has revealed that many of these flagellar genes are direct binding targets of PerR and not simply up-regulated due to altered *flgR* or *fliK* expression (A. Stintzi and J. Butcher, unpublished). The significance of PerR's regulation of flagella biosynthesis is currently unclear and merits further study.

Genes encoding iron acquisitions systems.

Interestingly, the transcripts of genes encoding components of several iron acquisition systems, including *chuABCD*, *cfrA*, and Cj1658-*p19*, were more abundant in the Δ*perR* mutant than the wild-type strain under iron restricted conditions. Given that these genes are also known to be Fe-Fur repressed, their expression profiles suggest that the Δ*perR* mutant might be more iron-starved than the wild-type strain, leading to their over-expression in the Δ*perR* mutant. Indeed, the Δ*perR* mutant might be significantly affected in its iron homeostasis due to the over-expression of catalase which will deplete haem pools from the cell (Helmann *et al.*, 2003). Alternatively, these genes might be directly regulated by apo-PerR. While this alternative could not be excluded, the rationale for an apo-PerR repression of the iron acquisition genes is puzzling.

Genes encoding proteins implicated in energy metabolism

The PerR regulon also contains several members which are involved in energy metabolism, including *napA, napG,* and *napH*. The *napA* gene encodes a nitrite reductase, while *napG* and *napH* code for ferredoxins. All of these proteins contain [4Fe–4S] clusters and their transcript levels are decreased in the Δ*perR* mutant as compared with the wild-type strain under iron-restricted conditions. As previously mentioned, this regulation could be the indirect consequence of the *perR* mutation through enhanced iron scarcity. Indeed, these genes belong to the iron stimulon and have been shown to be preferentially expressed under iron replete conditions (Table 10.1). Nevertheless, the direct regulation of these genes by PerR could not be excluded and in fact their down-regulation might prevent the formation of endogenous hydroxyl radicals.

Genes encoding regulators

Beside PerR itself, there are several regulator-encoding genes present in the PerR regulon, including FlgR, CheA, and CprS. As stated above, FlgR controls flagellar gene expression. CheA is a histidine kinase that is involved in chemotaxis. Upon activation, CheA undergoes an autophosphorylation and subsequently phosphorylates CheY (Korolik and Ketley, 2008). CheY diffuses towards the flagella and modulates flagellar motion through FliM (Baker *et al.*, 2006). Recent studies have identified the CheA adaptor protein CheW as being involved in controlling a highly efficient mechanism of *C. jejuni* cellular invasion (van Alphen *et al.*, 2008). CprS is a recently characterized histidine kinase that is involved in the regulation of biofilm formation (Svensson *et al.*, 2009). Interestingly, characterization of the Δ*cprS* proteome revealed the up-regulation of several oxidative stress proteins including KatA, AhpC and TrxB. Intriguingly, the Δ*cprS* mutant was more sensitive to H_2O_2 despite the up-regulation of both *katA* and *ahpC*. The regulation of CprS, FlgR and CheA by PerR indicates a significant overlap between these other regulatory networks, including flagellum biogenesis and biofilm formation. Thus PerR might be considered a master regulator in *C. jejuni*.

The PerR regulon of *Campylobacter*: how common and how different from other bacteria?

The PerR regulon has been characterized by genome-wide analysis in *B. subtilis* and *S. pyogenes* (Gryllos *et al.*, 2008; Helmann *et al.*, 2003). In addition, several members of the *S. aureus* PerR regulon have also been determined (Horsburgh *et al.*, 2001).

The *B. subtilis* PerR regulon is composed of approximately 75 members and some of the key genes include *katA, ahpC, perR,* and genes associated with competence (Helmann *et al.*, 2003). Similarly to *B. subtilis*, the *katA, ahpC* and *perR* genes also belong to the PerR regulon of *C. jejuni*. Moreover, a gene known to contribute to natural transformation in *C. jejuni* (Jeon and Zhang, 2007), Cj0011c, was also found to be PerR repressed (Palyada *et al.*, 2009). Interestingly, in *B. subtilis* $PerR_{BS}$ was found to repress its own transcription (Helmann *et al.*, 2003), while in *C. jejuni* $PerR_{CJ}$ appears to directly or indirectly activate its own transcription under conditions of oxidative stress (Palyada *et al.*, 2009).

The *S. pyogenes* PerR regulon contains 42 members (Gryllos *et al.*, 2008). *S. pyogenes* does not contain a catalase, although its alkyl hydroperoxidase (AhpC) was found to be regulated by PerR as is the case in *C. jejuni*. The most striking feature of the *S. pyogenes* PerR regulon is the high representation of genes involved in sugar metabolism, with 40% of the regulon belonging to this functional category. $PerR_{CJ}$ does not appear to regulate sugar metabolism in *C. jejuni* to the same extent, although this may be due to the fact that *C. jejuni*'s primary carbon sources are derived from peptides rather than carbohydrates (Wright *et al.*, 2009).

The PerR regulon in *S. aureus* has not yet been fully defined using genome-wide approaches; however several members of its PerR regulon have been characterized (Horsburgh *et al.*, 2001). These include *katA, ahpC, fur, perR, ftn* and *bcp*. As previously mentioned, $PerR_{CJ}$ regulates the expression of *katA, ahpC* and *perR*, but not *fur*. The gene *ftn* encodes a ferritin that is involved in storing iron within the cell. The homologous protein in *C. jejuni* is *cft* and was not identified as being regulated by $PerR_{cJ}$. Instead, *cft* was found to be Fur regulated (Palyada *et al.*, 2004). As previously mentioned, Bcp is a thioredoxin that is involved in oxidative stress defence. While Bcp has recently been shown to be important for aerotolerance in *C. jejuni* (Atack *et al.*, 2008), it was not identified in the PerR regulon. In contrast with $PerR_{CJ}$, $PerR_{SA}$ represses its own expression as seen in *B. subtilis* (Baichoo *et al.*, 2002; Horsburgh *et al.*, 2001).

It is worth noting that none of the other characterized PerR regulons include members of the flagella biosynthesis pathways. In addition, with exception of *fur* and *perR*, the other PerR regulons do not appear to contain genes with regulatory functions as seen with *cheA*, *flgR* and *cprS* in *C. jejuni* (Palyada *et al.*, 2009). The observed integrated network in *C. jejuni* could be the consequence of its relatively small genome and its paucity of transcriptional regulators. In turn, PerR may play more of a global regulatory role in *C. jejuni* as compared to other bacteria.

PerR regulation beyond oxidative stress defences

The characterization of the *C. jejuni* PerR regulon has revealed its involvement in diverse processes beyond its traditional role in oxidative stress defence (Palyada *et al.*, 2009). Genes involved in various processes including flagella biosynthesis, energy metabolism, surface structures, and both amino acid and fatty acid biosynthesis are all present in the PerR regulon. In addition, PerR appears to be a master regulator; regulating other regulators that are involved in chemotaxis, motility, and biofilm formation. This degree of regulatory interconnectivity suggests that gene expression in *C. jejuni* can be modulated to respond to a variety of environmental factors, without the need for an array of specialized transcription factors to sense and process environmental inputs. By combining its few transcriptional regulators into tightly interconnected networks, *C. jejuni* may be able to respond to environmental cues as efficiently as microorganisms with much larger genomes. PerR may therefore play an important and thus far unappreciated role in ensuring that these networks respond in synchrony to environmental factors.

The PerR regulon and its effect on virulence

Successful regulation of the PerR regulon has been demonstrated to be important for virulence in several pathogenic bacteria. In *S. pyogenes*, a Δ*perR* deletion mutant displayed reduced resistance to murine macrophages *in vitro* (Gryllos *et al.*, 2008). This increased sensitivity was abolished in the presence oxidative burst inhibitors, suggesting that the defect in survival was related to oxidative stress management. This deletion mutant was also attenuated in an *in vivo* model of pharyngeal infection (Gryllos *et al.*, 2008).

In *S. aureus*, a Δ*perRSA* deletion mutant was found to be attenuated in an *in vivo* mouse abscess model of infection (Horsburgh *et al.*, 2001). Curiously, a Δ*katA* deletion mutant was not affected, suggesting that the defects in the Δ*perRSA* deletion mutant extend beyond oxidative stress defence (Horsburgh *et al.*, 2001).

In *C. jejuni*, a Δ*perR* deletion mutant displayed reduced colonization levels (10^3-fold reduction) in the chick cecal model of colonization, despite exhibiting extreme resistance to H_2O_2 (Palyada *et al.*, 2009). In addition, a Δ*katA* deletion mutant was completely unable to colonize the chick caecum, while a Δ*ahpC* deletion mutant was severely attenuated in colonization (10^4-fold reduction) (Palyada *et al.*, 2009). The *in vivo* attenuation of the Δ*perR* mutant suggests that PerR's role in colonization may also extend beyond the classical oxidative stress defence systems. One possible role for PerR in colonization could be in modulating flagella biosynthesis, as the Δ*perR* deletion mutant was also affected in its motility, and proper flagella function has been shown to be essential for *C. jejuni* colonization (Zilbauer *et al.*, 2008).

Conclusion and future trends

There is little doubt as to the importance of properly co-regulating iron acquisition and oxidative stress defence for the well-being of *C. jejuni*, as disruption or over-expression of iron-related molecules has deleterious consequences. Fur and PerR play vital roles in modulating this regulation. Moreover, their importance extends beyond iron acquisition and oxidative stress defence and encompasses a variety of factors that contribute to *C. jejuni*'s ability to colonize multiple hosts and survive in various hostile environments. Overlap in the genes that are controlled by these two transcriptional regulators allows a robust response to oxidative stress by minimizing the catalysis of the Fenton reaction and enhancing oxidative stress defence systems (Fig. 10.3).

However there are still many unanswered questions concerning Fur, PerR and their regulatory functions. Most significantly, the difficulty encountered when attempting to define the Fur and PerR consensus binding sequences is intriguing and might suggest the presence of additional regulatory component(s). Moreover, it is currently unclear how these repressor proteins are able to directly activate gene expression in *C. jejuni* and other bacteria. This activation could be due to competition with other DNA binding proteins or through interactions with other transcriptional regulators. There is also increasing evidence that Fur family proteins may be able to modulate gene expression in the absence of an activating metal. How this regulation occurs and its physiological relevance are important questions that must be answered to truly understand the roles that these regulators play *in vivo*.

Most of the research on iron and oxidative stress regulation in *C. jejuni* has focused on transcriptional regulation. No study has yet reported the presence of post-transcriptional regulation in *C. jejuni*. Interestingly, the post-transcriptional regulator CsrA has been recently found to contribute to oxidative stress defence in *C. jejuni* (Fields and Thompson, 2008). In other bacteria, the CsrA protein interacts with target mRNA transcripts and affects their stability and/or translation and thus plays a central role in the pathogenesis of various bacteria (Lucchetti-Miganeh *et al.*, 2008). Although *C. jejuni* CsrA's regulatory targets were not identified, a *csrA* mutant was attenuated in its ability to survive oxidative stress (Fields and Thompson, 2008). This observation suggests the presence of an additional level of regulation for iron and oxidative stress-related proteins. This initial work warrants more in-depth studies to characterize the CsrA post-transcriptional regulatory system in *C. jejuni*.

In addition, the Fur and PerR regulons intersect with those of several other transcriptional regulators. These include the nitrosative stress regulator NssR, the planktonic growth sensor kinase CprS, and the flagella biogenesis σ^{54} factor FlgR. NssR regulation has previously been shown to be influenced by Fur regulation, and Δ*fur* mutants are more resistant to nitrosative stress (Monk *et al.*, 2008). As well, several members of the nitrosative stress response are members of the NssR, Fur and PerR regulons. Moreover, this co-regulation appears to be an adaptive response to nitrosative stress, and not an indirect consequence of nitrosylation agents (Monk *et al.*, 2008). The Δ*cprS* mutant displayed increased levels of several oxidative stress proteins, including KatA, AhpC and SodB. Curiously, despite this up-regulation, the Δ*cprS* mutant was more sensitive to both H_2O_2 and the organic peroxide t-BuOOH (Svensson *et al.*, 2009). The Δ*cprS* mutant also displayed altered levels of the orphan two component regulator Cj0355c. This protein has been shown to be responsive to paraquat exposure (Garenaux *et al.*, 2008) and preliminary results indicate that it is a direct binding target of PerR (A. Stintzi and J. Butcher, unpublished). In addition to FlgR itself, many components of flagella biogenesis are regulated by both Fur and PerR (Table 10.3), although PerR regulation appears to be more important.

It is apparent that the regulatory networks of many transcriptional regulators in *C. jejuni* are both intertwined and interconnected. This degree of complexity will only be revealed as additional transcriptional and post transcriptional regulatory networks are defined and characterized within *C. jejuni*. Delineating these regulatory networks and obtaining a unified vision of *C. jejuni* gene regulation *in vivo* is the next challenge that exists within the field and would yield invaluable insight into *C. jejuni*'s ability to infect and cause disease in human populations.

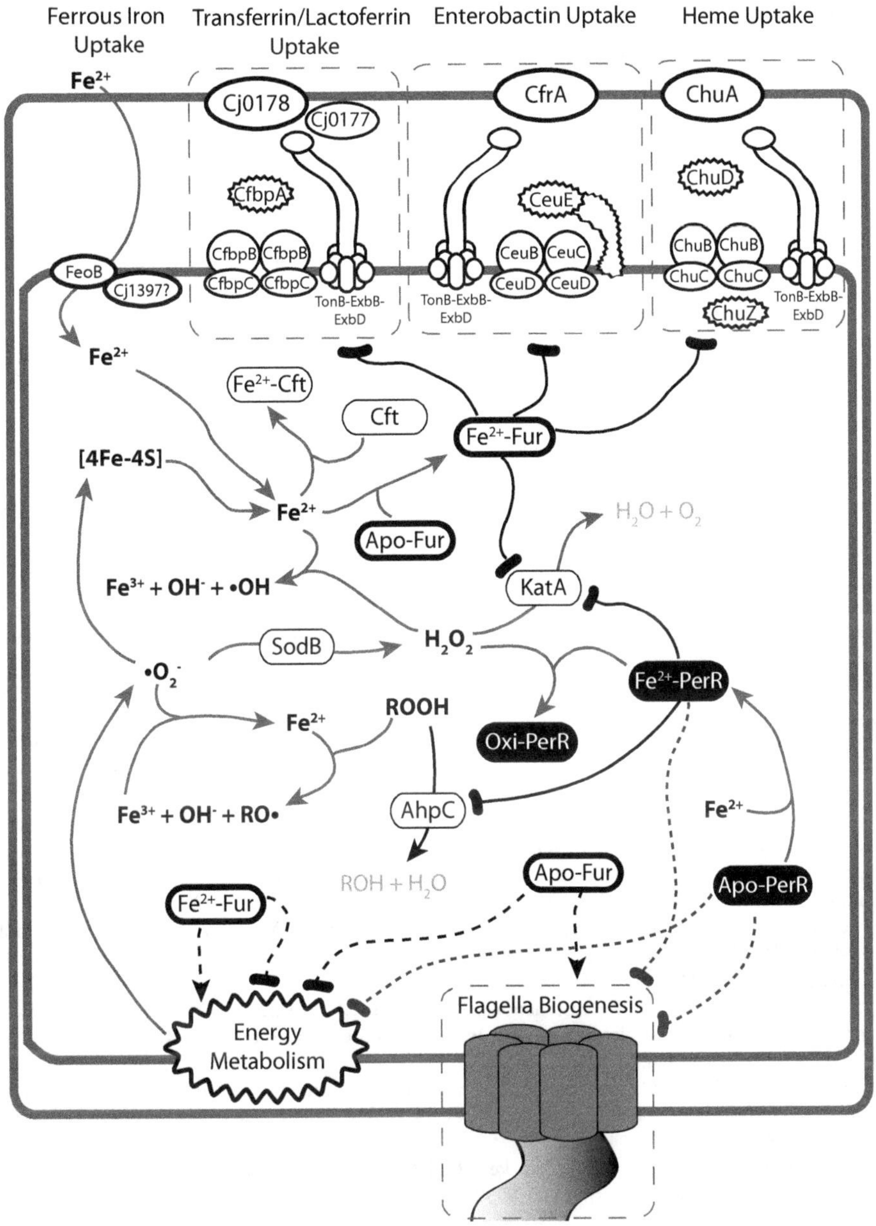

Figure 10.3 Cross-talk between iron, oxidative stress, Fur and PerR in *Campylobacter jejuni.* Superoxide anions are generated from energy metabolism and detoxified by SodB to H_2O_2. Superoxide anions can also react with iron sulphur complexes within proteins and release free ferrous ion. Hydrogen peroxide can be detoxified through KatA or can react with ferrous ion through the Fenton reaction and produce ferric ion and hydroxyl radicals. Hydrogen peroxide can also oxidize and inactivate Fe^{2+}-PerR, causing de-repression of oxidative defence genes. Organic peroxides are detoxified by AhpC and result in the corresponding organic alcohols and H_2O. Organic peroxides can also react with free ferrous ion through the Fenton reaction. Ferric ion generated from initial Fenton reactions can be reduced by superoxide anions or other reducing agents present in the cell to reform ferrous ion. Excess ferrous ion is sequestered with Cft to prevent additional oxidative stress through Fenton chemistry. Ferrous ion can also be taken up through FeoB and complexes with Apo-PerR and Apo-Fur to form Fe^{2+}-PerR and Fe^{2+}-Fur. Fe^{2+}-Fur represses genes involved in iron acquisition pathways and partially represses KatA. Fe^{2+}-Fur and apo-Fur also modulate the expression of energy metabolism genes according to iron availably and this regulation may be indirect or direct. Fe^{2+}-PerR represses the expression of KatA and AhpC. Apo -PerR also appears to regulate genes involved in iron acquision, flagella biogenesis and energy metabolism, although it is unclear whether this is direct regulation by PerR or an indirect consequence of *perR* deletion.

Acknowledgements

Research in Dr Stintzi's laboratory has been supported by the Canadian Institutes of Health Research and the National Institutes of Health (grant AI055612).

References

Alfredson, D.A., and Korolik, V. (2007). Antibiotic resistance and resistance mechanisms in *Campylobacter jejuni* and *Campylobacter coli*. FEMS Microbiol. Lett. *277*, 123–132.

Alice, A.F., Naka, H., and Crosa, J.H. (2008). Global gene expression as a function of the iron status of the bacterial cell: influence of differentially expressed genes in the virulence of the human pathogen *Vibrio vulnificus*. Infect. Immun. *76*, 4019–4037.

Andrews, S.C., Robinson, A.K., and Rodriguez-Quinones, F. (2003). Bacterial iron homeostasis. FEMS Microbiol. Rev. *27*, 215–237.

Atack, J.M., Harvey, P., Jones, M.A., and Kelly, D.J. (2008). The *Campylobacter jejuni* thiol peroxidases Tpx and Bcp both contribute to aerotolerance and peroxide-mediated stress resistance but have distinct substrate specificities. J. Bacteriol. *190*, 5279–5290.

Baichoo, N., and Helmann, J.D. (2002). Recognition of DNA by Fur: a reinterpretation of the Fur box consensus sequence. J. Bacteriol. *184*, 5826–5832.

Baichoo, N., Wang, T., Ye, R., and Helmann, J.D. (2002). Global analysis of the *Bacillus subtilis* Fur regulon and the iron starvation stimulon. Mol. Microbiol. *45*, 1613–1629.

Baig, B.H., Wachsmuth, I.K., and Morris, G.K. (1986). Utilization of exogenous siderophores by *Campylobacter* species. J. Clin. Microbiol. 23, 431–433.

Baillon, M.L., van Vliet, A.H., Ketley, J.M., Constantinidou, C., and Penn, C.W. (1999). An iron-regulated alkyl hydroperoxide reductase (AhpC) confers aerotolerance and oxidative stress resistance to the microaerophilic pathogen *Campylobacter jejuni*. J. Bacteriol. *181*, 4798–4804.

Baker, M.D., Wolanin, P.M., and Stock, J.B. (2006). Signal transduction in bacterial chemotaxis. Bioessays *28*, 9–22.

Basler, M., Linhartova, I., Halada, P., Novotna, J., Bezouskova, S., Osicka, R., Weiser, J., Vohradsky, J., and Sebo, P. (2006). The iron-regulated transcriptome and proteome of *Neisseria meningitidis* serogroup C. Proteomics *6*, 6194–6206.

Bingham-Ramos, L.K., and Hendrixson, D.R. (2008). Characterization of two putative cytochrome *c* peroxidases of *Campylobacter jejuni* involved in promoting commensal colonization of poultry. Infect. Immun. *76*, 1105–1114.

Black, R.E., Levine, M.M., Clements, M.L., Hughes, T.P., and Blaser, M.J. (1988). Experimental *Campylobacter jejuni* infection in humans. J. Infect. Dis. *157*, 472–479.

Blaser, M.J., and Engberg, J. (2008). Clinical aspects of *Campylobacter jejuni* and *Campylobacter coli*. In Campylobacter, I. Nachamkin, C. Szymanski, and M.J. Blaser, eds. (Washington, ASM Press), pp. 99–121.

Bos, J.C., Schultsz, C., Vandenbroucke-Grauls, C.M., Speelman, P., and Prins, J.M. (2006). [Optimising antibiotic policies in the Netherlands. IX. SWAB guidelines for antimicrobial therapy in adults with acute infectious diarrhoea]. Ned Tijdschr Geneeskd *150*, 1116–1122.

Boyanova, L., Gergova, G., Spassova, Z., Koumanova, R., Yaneva, P., Mitov, I., Derejian, S., and Krastev, Z. (2004). *Campylobacter* infection in 682 bulgarian patients with acute enterocolitis, inflammatory bowel disease, and other chronic intestinal diseases. Diagn. Microbiol. Infect. Dis. *49*, 71–74.

Bremer, E., and Kramer, R. (2000). Coping with osmotic challenges: osmoregulation through accumulation and release of compatible solutes in bacteria. In Bacterial Stress Responses, G. Storz, and H. Hengee-Aronis, eds. (Washington, D.C., ASM Press), pp. 79–97.

Bronstein, P.A., Filiatrault, M.J., Myers, C.R., Rutzke, M., Schneider, D.J., and Cartinhour, S.W. (2008). Global transcriptional responses of *Pseudomonas syringae* DC3000 to changes in iron bioavailability *in vitro*. BMC Microbiol. *8*, 209–223.

Bsat, N., Chen, L., and Helmann, J.D. (1996). Mutation of the *Bacillus subtilis* alkyl hydroperoxide reductase (*ahpCF*) operon reveals compensatory interactions among hydrogen peroxide stress genes. J. Bacteriol. *178*, 6579–6586.

Butzler, J.P. (2004). *Campylobacter*, from obscurity to celebrity. Clin. Microbiol. Infect. *10*, 868–876.

Carpenter, B.M., Gancz, H., Gonzalez-Nieves, R.P., West, A.L., Whitmire, J.M., Michel, S.L., and Merrell, D.S. (2009). A single nucleotide change affects *fur*-dependent regulation of *sodB* in *H. pylori*. PLoS ONE *4*, e5369.

Carpenter, C.E., and Mahoney, A.W. (1992). Contributions of heme and nonheme iron to human nutrition. Crit. Rev. Food Sci. Nutr. *31*, 333–367.

Carswell, C.L., Rigden, M.D., and Baenziger, J.E. (2008). Expression, purification, and structural characterization of CfrA, a putative iron transporter from *Campylobacter jejuni*. J. Bacteriol. *190*, 5650–5662.

Charoenlap, N., Eiamphungporn, W., Chauvatcharin, N., Utamapongchai, S., Vattanaviboon, P., and Mongkolsuk, S. (2005). OxyR mediated compensatory expression between *ahpC* and *katA* and the significance of *ahpC* in protection from hydrogen peroxide in *Xanthomonas campestris*. FEMS Microbiol. Lett. *249*, 73–78.

Chen, Z., Lewis, K.A., Shultzaberger, R.K., Lyakhov, I.G., Zheng, M., Doan, B., Storz, G., and Schneider, T.D. (2007). Discovery of Fur binding site clusters in *Escherichia coli* by information theory models. Nucleic Acids Res. *35*, 6762–6777.

Cogan, T.A., Thomas, A.O., Rees, L.E., Taylor, A.H., Jepson, M.A., Williams, P.H., Ketley, J., and Humphrey, T.J. (2007). Norepinephrine increases the pathogenic potential of *Campylobacter jejuni*. Gut *56*, 1060–1065.

Conrad, M.E., Benjamin, B.I., Williams, H.L., and Foy, A.L. (1967). Human absorption of hemoglobin-iron. Gastroenterology *53*, 5–10.

Conrad, M.E., and Schade, S.G. (1968). Ascorbic acid chelates in iron absorption: a role for hydrochloric acid and bile. Gastroenterology *55*, 35–45.

Conrad, M.E., and Umbreit, J.N. (2000). Iron absorption and transport-an update. Am. J. Hematol. *64*, 287–298.

Conrad, M.E., Umbreit, J.N., and Moore, E.G. (1991). A role for mucin in the absorption of inorganic iron and other metal cations. A study in rats. Gastroenterology *100*, 129–136.

Cosgrove, K., Coutts, G., Jonsson, I.M., Tarkowski, A., Kokai-Kun, J.F., Mond, J.J., and Foster, S.J. (2007). Catalase (KatA) and alkyl hydroperoxide reductase (AhpC) have compensatory roles in peroxide stress resistance and are required for survival, persistence, and nasal colonization in *Staphylococcus aureus*. J. Bacteriol. *189*, 1025–1035.

Coy, M., Doyle, C., Besser, J., and Neilands, J.B. (1994). Site-directed mutagenesis of the ferric uptake regulation gene of *Escherichia coli*. Biometals *7*, 292–298.

D'Autreaux, B., Pecqueur, L., Gonzalez de Peredo, A., Diederix, R.E., Caux-Thang, C., Tabet, L., Bersch, B., Forest, E., and Michaud-Soret, I. (2007). Reversible redox- and zinc-dependent dimerization of the *Escherichia coli* Fur protein. Biochemistry *46*, 1329–1342.

Danielli, A., Roncarati, D., Delany, I., Chiarini, V., Rappuoli, R., and Scarlato, V. (2006). *In vivo* dissection of the *Helicobacter pylori* Fur regulatory circuit by genome-wide location analysis. J. Bacteriol. *188*, 4654–4662.

Day, W.A., Jr., Sajecki, J.L., Pitts, T.M., and Joens, L.A. (2000). Role of catalase in *Campylobacter jejuni* intracellular survival. Infect. Immun. *68*, 6337–6345.

Debruyne, L., Gevers, D., and Vandamme, P. (2008). Taxonomy of the family Campylobacteraceae. In Campylobacter, I. Nachamkin, C. Szymanski, and M.J. Blaser, eds. (Washington, ASM Press), pp. 3–25.

Deng, K., Blick, R.J., Liu, W., and Hansen, E.J. (2006). Identification of *Francisella tularensis* genes affected by iron limitation. Infect. Immun. *74*, 4224–4236.

Deslandes, V., Nash, J.H., Harel, J., Coulton, J.W., and Jacques, M. (2007). Transcriptional profiling of *Actinobacillus pleuropneumoniae* under iron-restricted conditions. BMC Genomics *8*, 72–91.

Ducey, T.F., Carson, M.B., Orvis, J., Stintzi, A.P., and Dyer, D.W. (2005). Identification of the iron-responsive genes of *Neisseria gonorrhoeae* by microarray analysis in defined medium. J. Bacteriol. *187*, 4865–4874.

Elliott, K.T., and Dirita, V.J. (2008). Characterization of CetA and CetB, a bipartite energy taxis system in *Campylobacter jejuni*. Mol. Microbiol. *69*, 1091–1103.

Ernst, F.D., Bereswill, S., Waidner, B., Stoof, J., Mader, U., Kusters, J.G., Kuipers, E.J., Kist, M., van Vliet, A.H., and Homuth, G. (2005a). Transcriptional profiling of *Helicobacter pylori* Fur- and iron-regulated gene expression. Microbiology *151*, 533–546.

Ernst, F.D., Homuth, G., Stoof, J., Mader, U., Waidner, B., Kuipers, E.J., Kist, M., Kusters, J.G., Bereswill, S., and van Vliet, A.H. (2005b). Iron-responsive regulation of the *Helicobacter pylori* iron-cofactored superoxide dismutase SodB is mediated by Fur. J. Bacteriol. *187*, 3687–3692.

Escolar, L., Perez-Martin, J., and de Lorenzo, V. (1998). Binding of the Fur (ferric uptake regulator) repressor of *Escherichia coli* to arrays of the GATAAT sequence. J. Mol. Biol. *283*, 537–547.

Field, L.H., Headley, V.L., Payne, S.M., and Berry, L.J. (1986). Influence of iron on growth, morphology, outer membrane protein composition, and synthesis of siderophores in *Campylobacter jejuni*. Infect. Immun. *54*, 126–132.

Fields, J.A., and Thompson, S.A. (2008). *Campylobacter jejuni* CsrA mediates oxidative stress responses, biofilm formation, and host cell invasion. J. Bacteriol. *190*, 3411–3416.

Fontecave, M. (2006). Iron-sulfur clusters: ever-expanding roles. Nat. Chem. Biol. *2*, 171–174.

Foster, J.W., and Hall, H.K. (1992). Effect of *Salmonella typhimurium* ferric uptake regulator (*fur*) mutations on iron- and pH-regulated protein synthesis. J. Bacteriol. *174*, 4317–4323.

Fouts, D.E., Mongodin, E.F., Mandrell, R.E., Miller, W.G., Rasko, D.A., Ravel, J., Brinkac, L.M., DeBoy, R.T., Parker, C.T., Daugherty, S.C., *et al.* (2005). Major structural differences and novel potential virulence mechanisms from the genomes of multiple *Campylobacter* species. PLoS Biol. *3*, e15.

Friedman, C.R., Neimann, J., Wegener, H.C., and Tauxe, R.V. (2000). Epidemiology of *C. jejuni* infections in the United States and other industrialized nations. In Campylobacter, I. Nachamkin, and M.J. Blaser, eds. (Washington, American Society for Microbiology), pp. 121–138.

Galaris, D., and Korantzopoulos, P. (1997). On the molecular mechanism of metmyoglobin-catalyzed reduction of hydrogen peroxide by ascorbate. Free Radic. Biol. Med. *22*, 657–667.

Galaris, D., and Pantopoulos, K. (2008). Oxidative stress and iron homeostasis: mechanistic and health aspects. Crit. Rev. Clin. Lab. Sci. *45*, 1–23.

Gancz, H., Censini, S., and Merrell, D.S. (2006). Iron and pH homeostasis intersect at the level of Fur regulation in the gastric pathogen *Helicobacter pylori*. Infect. Immun. *74*, 602–614.

Gao, H., Zhou, D., Li, Y., Guo, Z., Han, Y., Song, Y., Zhai, J., Du, Z., Wang, X., Lu, J., *et al.* (2008). The iron-responsive Fur regulon in *Yersinia pestis*. J. Bacteriol. *190*, 3063–3075.

Garenaux, A., Guillou, S., Ermel, G., Wren, B., Federighi, M., and Ritz, M. (2008). Role of the Cj1371 periplasmic protein and the Cj0355c two-component regulator in the *Campylobacter jejuni* NCTC 11168 response to oxidative stress caused by paraquat. Res. Microbiol. *159*, 718–726.

Gonzalez-Chavez, S.A., Arevalo-Gallegos, S., and Rascon-Cruz, Q. (2009). Lactoferrin: structure, function and applications. Int. J. Antimicrob. Agents 33, *301*, e301–308.

Green, J., and Paget, M.S. (2004). Bacterial redox sensors. Nat. Rev. Microbiol. *2*, 954–966.

Groves, J.T. (2006). High-valent iron in chemical and biological oxidations. J. Inorg. Biochem. *100*, 434–447.

Gryllos, I., Grifantini, R., Colaprico, A., Cary, M.E., Hakansson, A., Carey, D.W., Suarez-Chavez, M., Kalish, L.A., Mitchell, P.D., White, G.L., *et al.* (2008). PerR confers phagocytic killing resistance and allows pharyngeal colonization by group A *Streptococcus*. PLoS Pathog. *4*, e1000145.

Hall, H.K., and Foster, J.W. (1996). The role of Fur in the acid tolerance response of *Salmonella typhimurium* is physiologically and genetically separable from its role in iron acquisition. J. Bacteriol. *178*, 5683–5691.

Hantke, K. (1981). Regulation of ferric iron transport in *Escherichia coli* K12: isolation of a constitutive mutant. Mol. Gen. Genet. *182*, 288–292.

Hantke, K., and Braun, V. (2000). The art of keeping low and high iron concentration in balance. In Bacterial Stress Responses, G. Storz, and H. Hengee-Aronis, eds. (Washington, D.C., ASM Press), pp. 275–288.

Harris, A.G., Hinds, F.E., Beckhouse, A.G., Kolesnikow, T., and Hazell, S.L. (2002). Resistance to hydrogen peroxide in *Helicobacter pylori*: role of catalase (KatA) and Fur, and functional analysis of a novel gene product designated 'KatA-associated protein', KapA (HP0874). Microbiology *148*, 3813–3825.

Harris, W.R. (2002). Iron chemistry. In Molecular and Cellular iron transport, D.M. Templeton, ed. (New York, Marcel Dekker, Inc.), pp. 2–35.

Harrison, C. (2003). GrpE, a nucleotide exchange factor for DnaK. Cell Stress Chaperones *8*, 218–224.

Hassett, D.J., Ma, J.F., Elkins, J.G., McDermott, T.R., Ochsner, U.A., West, S.E., Huang, C.T., Fredericks, J., Burnett, S., Stewart, P.S., *et al.* (1999). Quorum sensing in *Pseudomonas aeruginosa* controls expression of catalase and superoxide dismutase genes and mediates biofilm susceptibility to hydrogen peroxide. Mol. Microbiol. *34*, 1082–1093.

Helmann, J.D., Soonsanga, S., and Gabriel, S. (2007). Metalloregulators: Arbiters of Metal Sufficiency. In Molecular Microbiology of Heavy Metals, D.H. Nies, and S. Silver, eds. (Springer Berlin/Heidelberg), pp. 37–71.

Helmann, J.D., Wu, M.F., Gaballa, A., Kobel, P.A., Morshedi, M.M., Fawcett, P., and Paddon, C. (2003). The global transcriptional response of *Bacillus subtilis* to peroxide stress is coordinated by three transcription factors. J. Bacteriol. *185*, 243–253.

Hendrixson, D.R. (2008). Regulation of flagellar gene expression and assembly. In *Campylobacter*, I. Nachamkin, C. Szymanski, and M.J. Blaser, eds. (American Society for Microbiology), pp. 545–559.

Hendrixson, D.R., and DiRita, V.J. (2003). Transcription of sigma54-dependent but not sigma28-dependent flagellar genes in *Campylobacter jejuni* is associated with formation of the flagellar secretory apparatus. Mol. Microbiol. *50*, 687–702.

Hendrixson, D.R., and DiRita, V.J. (2004). Identification of *Campylobacter jejuni* genes involved in commensal colonization of the chick gastrointestinal tract. Mol. Microbiol. *52*, 471–484.

Herbig, A.F., and Helmann, J.D. (2001). Roles of metal ions and hydrogen peroxide in modulating the interaction of the *Bacillus subtilis* PerR peroxide regulon repressor with operator DNA. Mol. Microbiol. *41*, 849–859.

Hillmann, F., Fischer, R.J., Saint-Prix, F., Girbal, L., and Bahl, H. (2008). PerR acts as a switch for oxygen tolerance in the strict anaerobe *Clostridium acetobutylicum*. Mol. Microbiol. *68*, 848–860.

Hofreuter, D., Novik, V., and Galan, J.E. (2008). Metabolic diversity in *Campylobacter jejuni* enhances specific tissue colonization. Cell Host Microbe 4, 425–433.

Hofreuter, D., Tsai, J., Watson, R.O., Novik, V., Altman, B., Benitez, M., Clark, C., Perbost, C., Jarvie, T., Du, L., *et al.* (2006). Unique features of a highly pathogenic *Campylobacter jejuni* strain. Infect. Immun. *74*, 4694–4707.

Holmes, K., Mulholland, F., Pearson, B.M., Pin, C., McNicholl-Kennedy, J., Ketley, J.M., and Wells, J.M. (2005). *Campylobacter jejuni* gene expression in response to iron limitation and the role of Fur. Microbiology *151*, 243–257.

Hooper, L.V., Bry, L., Falk, P.G., and Gordon, J.I. (1998). Host-microbial symbiosis in the mammalian intestine: exploring an internal ecosystem. Bioessays *20*(4), 336–343.

Hooper, L.V., Midtvedt, T., and Gordon, J.I. (2002). How host–microbial interactions shape the nutrient environment of the Mammalian intestine. Annu. Rev. Nutr. 22, 283–307.

Horsburgh, M.J., Clements, M.O., Crossley, H., Ingham, E., and Foster, S.J. (2001). PerR controls oxidative stress resistance and iron storage proteins and is required for virulence in *Staphylococcus aureus*. Infect. Immun. *69*, 3744–3754.

Humphrey, T., O'Brien, S., and Madsen, M. (2007). *Campylobacters* as zoonotic pathogens: a food production perspective. Int. J. Food Microbiol. *117*, 237–257.

Imlay, J.A. (1995). A metabolic enzyme that rapidly produces superoxide, fumarate reductase of *Escherichia coli*. J. Biol. Chem. *270*, 19767–19777.

Imlay, J.A. (2003). Pathways of oxidative damage. Annu. Rev. Microbiol. *57*, 395–418.

Imlay, J.A. (2008). Cellular defenses against superoxide and hydrogen peroxide. Annu. Rev. Biochem. *77*, 755–776.

Ishikawa, T., Mizunoe, Y., Kawabata, S., Takade, A., Harada, M., Wai, S.N., and Yoshida, S. (2003). The iron-binding protein Dps confers hydrogen peroxide stress resistance to *Campylobacter jejuni*. J. Bacteriol. *185*, 1010–1017.

Jeon, B., and Zhang, Q. (2007). Cj0011c, a periplasmic single- and double-stranded DNA-binding protein, contributes to natural transformation in *Campylobacter jejuni*. J. Bacteriol. *189*, 7399–7407.

Jones, D.P. (2006). Redefining oxidative stress. Antioxid. Redox. Signal. *8*, 1865–1879.

Kamal, N., Dorrell, N., Jagannathan, A., Turner, S.M., Constantinidou, C., Studholme, D.J., Marsden, G., Hinds, J., Laing, K.G., Wren, B.W., *et al.* (2007). Deletion of a previously uncharacterized flagellar-hook-length control gene *fliK* modulates the sigma54-dependent regulon in *Campylobacter jejuni*. Microbiology *153*, 3099–3111.

Keener, K.M., Bashor, M.P., Curtis, P.A., Sheldon, B.W., and Kathariou, S. (2004). Comprehensive review of *Campylobacter* and poultry processing. Comprehensive Reviews in Food Science and Food Safety 3, 105–116.

Kelly, D.J. (2008). Complexity and versatility in the physiology and metabolism of *Campylobacter jejuni*. In Campylobacter, I. Nachamkin, C. Szymanski, and M.J. Blaser, eds. (Washington ASM Press), pp. 41–61.

Konkel, M.E., Kim, B.J., Klena, J.D., Young, C.R., and Ziprin, R. (1998). Characterization of the thermal stress response of *Campylobacter jejuni*. Infect. Immun. *66*, 3666–3672.

Konkel, M.E., and Tilly, K. (2000). Temperature-regulated expression of bacterial virulence genes. Microbes Infect. *2*, 157–166.

Korolik, V., and Ketley, J. (2008). Chemosensory signal transduction pathway of *Campylobacter jejuni*. In Campylobacter, I. Nachamkin, C. Szymanski, and M.J. Blaser, eds. (American Society of Microbiology), pp. 351–366.

Lavrrar, J.L., and McIntosh, M.A. (2003). Architecture of a Fur binding site: a comparative analysis. J. Bacteriol. *185*, 2194–2202.

Lee, J.W., and Helmann, J.D. (2006a). Biochemical characterization of the structural Zn^{2+} site in the *Bacillus subtilis* peroxide sensor PerR. J. Biol. Chem. *281*, 23567–23578.

Lee, J.W., and Helmann, J.D. (2006b). The PerR transcription factor senses H_2O_2 by metal-catalysed histidine oxidation. Nature *440*, 363–367.

Lee, J.W., and Helmann, J.D. (2007). Functional specialization within the Fur family of metalloregulators. Biometals *20*, 485–499.

Lee, M.D., and Newell, D.G. (2006). *Campylobacter* in poultry: filling an ecological niche. Avian Dis. *50*, 1–9.

Lewin, A.C., Doughty, P.A., Flegg, L., Moore, G.R., and Spiro, S. (2002). The ferric uptake regulator of *Pseudomonas aeruginosa* has no essential cysteine residues and does not contain a structural zinc ion. Microbiology *148*, 2449–2456.

Lucchetti-Miganeh, C., Burrowes, E., Baysse, C., and Ermel, G. (2008). The post-transcriptional regulator CsrA plays a central role in the adaptation of bacterial pathogens to different stages of infection in animal hosts. Microbiology *154*, 16–29.

Massé, E., and Gottesman, S. (2002). A small RNA regulates the expression of genes involved in iron metabolism in *Escherichia coli*. Proc. Natl. Acad. Sci. U.S.A. *99*, 4620–4625.

Massé, E., Salvail, H., Desnoyers, G., and Arguin, M. (2007). Small RNAs controlling iron metabolism. Curr. Opin. Microbiol. *10*, 140–145.

McHugh, J.P., Rodriguez-Quinones, F., Abdul-Tehrani, H., Svistunenko, D.A., Poole, R.K., Cooper, C.E., and Andrews, S.C. (2003). Global iron-dependent gene regulation in *Escherichia coli*. A new mechanism for iron homeostasis. J. Biol. Chem. *278*, 29478–29486.

Mead, P.S., Slutsker, L., Dietz, V., McCaig, L.F., Bresee, J.S., Shapiro, C., Griffin, P.M., and Tauxe, R.V. (1999). Food-related illness and death in the United States. Emerg. Infect. Dis. *5*, 607–625.

Merkofer, M., Kissner, R., Hider, R.C., Brunk, U.T., and Koppenol, W.H. (2006). Fenton chemistry and iron chelation under physiologically relevant conditions: Electrochemistry and kinetics. Chem. Res. Toxicol. *19*, 1263–1269.

Merrell, D.S., and Camilli, A. (2002). Acid tolerance of gastrointestinal pathogens. Curr. Opin. Microbiol. *5*, 51–55.

Merrell, D.S., Thompson, L.J., Kim, C.C., Mitchell, H., Tompkins, L.S., Lee, A., and Falkow, S. (2003). Growth phase-dependent response of *Helicobacter pylori* to iron starvation. Infect. Immun. *71*, 6510–6525.

Messner, K.R., and Imlay, J.A. (1999). The identification of primary sites of superoxide and hydrogen peroxide formation in the aerobic respiratory chain and sulfite reductase complex of *Escherichia coli*. J. Biol. Chem. *274*, 10119–10128.

Messner, K.R., and Imlay, J.A. (2002). Mechanism of superoxide and hydrogen peroxide formation by fumarate reductase, succinate dehydrogenase, and aspartate oxidase. J. Biol. Chem. *277*, 42563–42571.

Mey, A.R., Wyckoff, E.E., Kanukurthy, V., Fisher, C.R., and Payne, S.M. (2005). Iron and *fur* regulation in *Vibrio cholerae* and the role of *fur* in virulence. Infect. Immun. *73*, 8167–8178.

Miller, C.E., Rock, J.D., Ridley, K.A., Williams, P.H., and Ketley, J.M. (2008). Utilization of lactoferrin-bound and transferrin-bound iron by *Campylobacter jejuni*. J. Bacteriol. *190*, 1900–1911.

Mills, S.A., and Marletta, M.A. (2005). Metal binding characteristics and role of iron oxidation in the ferric uptake regulator from *Escherichia coli*. Biochemistry *44*, 13553–13559.

Mongkolsuk, S., and Helmann, J.D. (2002). Regulation of inducible peroxide stress responses. Mol. Microbiol. *45*, 9–15.

Monk, C.E., Pearson, B.M., Mulholland, F., Smith, H.K., and Poole, R.K. (2008). Oxygen- and NssR-dependent globin expression and enhanced iron acquisition in the response of *Campylobacter* to nitrosative stress. J. Biol. Chem. *283*, 28413–28425.

Moore, J.E., Corcoran, D., Dooley, J.S., Fanning, S., Lucey, B., Matsuda, M., McDowell, D.A., Megraud, F., Millar, B.C., O'Mahony, R., *et al.* (2005). *Campylobacter*. Vet. Res. *36*, 351–382.

Myers, J.D., and Kelly, D.J. (2005). A sulphite respiration system in the chemoheterotrophic human pathogen *Campylobacter jejuni*. Microbiology *151*, 233–242.

Naikare, H., Palyada, K., Panciera, R., Marlow, D., and Stintzi, A. (2006). Major role for FeoB in *Campylobacter jejuni* ferrous iron acquisition, gut colonization, and intracellular survival. Infect. Immun. *74*, 5433–5444.

Naqui, A., Chance, B., and Cadenas, E. (1986). Reactive oxygen intermediates in biochemistry. Annu. Rev. Biochem. *55*, 137–166.

Nodop, A., Pietsch, D., Hocker, R., Becker, A., Pistorius, E.K., Forchhammer, K., and Michel, K.P. (2008). Transcript profiling reveals new insights into the acclimation of the mesophilic fresh-water cyanobacterium *Synechococcus elongatus* PCC 7942 to iron starvation. Plant Physiol. *147*, 747–763.

Nothaft, H., Amber, S., Aebi, M., and Szymanski, C. (2008). N-linked protein glycosylation in *Campylobacter*. In Campylobacter, I. Nachamkin, C. Szymanski, and M.J. Blaser, eds. (American Society for Microbiology), pp. 447–470.

Ochsner, U.A., Vasil, M.L., Alsabbagh, E., Parvatiyar, K., and Hassett, D.J. (2000). Role of the *Pseudomonas aeruginosa oxyR-recG* operon in oxidative stress defense and DNA repair: OxyR-dependent regulation of *katB-ankB, ahpB,* and *ahpC-ahpF*. J. Bacteriol. *182*, 4533–4544.

Ochsner, U.A., Wilderman, P.J., Vasil, A.I., and Vasil, M.L. (2002). GeneChip expression analysis of the iron starvation response in *Pseudomonas aeruginosa*: identification of novel pyoverdine biosynthesis genes. Mol. Microbiol. *45*, 1277–1287.

Oglesby, A.G., Murphy, E.R., Iyer, V.R., and Payne, S.M. (2005). Fur regulates acid resistance in *Shigella flexneri* via RyhB and *ydeP*. Mol. Microbiol. *58*, 1354–1367.

Pajaniappan, M., Hall, J.E., Cawthraw, S.A., Newell, D.G., Gaynor, E.C., Fields, J.A., Rathbun, K.M., Agee, W.A., Burns, C.M., Hall, S.J., *et al.* (2008). A temperature-regulated *Campylobacter jejuni* gluconate dehydrogenase is involved in respiration-dependent energy conservation and chicken colonization. Mol. Microbiol. *68*, 474–491.

Palma, M., Worgall, S., and Quadri, L.E. (2003). Transcriptome analysis of the *Pseudomonas aeruginosa* response to iron. Arch. Microbiol. *180*, 374–379.

Palyada, K., Threadgill, D., and Stintzi, A. (2004). Iron acquisition and regulation in *Campylobacter jejuni*. J. Bacteriol. *186*, 4714–4729.

Palyada, K., Sun, Y. Q., Flint, A., Butcher, J., Naikare, H., and Stintzi, A. (2009). Characterization of the oxidative stress stimulon and PerR regulon of *Campylobacter jejuni*. BMC Genomics *10*, 481–499.

Parkhill, J., Wren, B.W., Mungall, K., Ketley, J.M., Churcher, C., Basham, D., Chillingworth, T., Davies, R.M., Feltwell, T., Holroyd, S., *et al.* (2000). The genome sequence of the food-borne pathogen *Campylobacter jejuni* reveals hypervariable sequences. Nature *403*, 665–668.

Pearson, B.M., Gaskin, D.J., Segers, R.P., Wells, J.M., Nuijten, P.J., and van Vliet, A.H. (2007). The complete genome sequence of *Campylobacter jejuni* strain 81116 (NCTC11828). J. Bacteriol. *189*, 8402–8403.

Pesci, E.C., Cottle, D.L., and Pickett, C.L. (1994). Genetic, enzymatic, and pathogenic studies of the iron superoxide dismutase of *Campylobacter jejuni*. Infect. Immun. *62*, 2687–2694.

Pickett, C.L., Auffenberg, T., Pesci, E.C., Sheen, V.L., and Jusuf, S.S. (1992). Iron acquisition and hemolysin production by *Campylobacter jejuni*. Infect. Immun. *60*, 3872–3877.

Pittman, M.S., and Kelly, D.J. (2005). Electron transport through nitrate and nitrite reductases in *Campylobacter jejuni*. Biochem. Soc. Trans. *33*, 190–192.

Pohl, E., Haller, J.C., Mijovilovich, A., Meyer-Klaucke, W., Garman, E., and Vasil, M.L. (2003). Architecture of a protein central to iron homeostasis: crystal structure and spectroscopic analysis of the ferric uptake regulator. Mol. Microbiol. *47*, 903–915.

Poly, F., and Guerry, P. (2008). Pathogenesis of *Campylobacter*. Curr. Opin. Gastroenterol. *24*, 27–31.

Poly, F., Read, T., Tribble, D.R., Baqar, S., Lorenzo, M., and Guerry, P. (2007). Genome sequence of a clinical isolate of *Campylobacter jejuni* from Thailand. Infect. Immun. *75*, 3425–3433.

Poole, L.B. (2005). Bacterial defenses against oxidants: mechanistic features of cysteine-based peroxidases and their flavoprotein reductases. Arch. Biochem. Biophys. *433*, 240–254.

Purdy, D., Cawthraw, S., Dickinson, J.H., Newell, D.G., and Park, S.F. (1999). Generation of a superoxide dismutase (SOD)-deficient mutant of *Campylobacter coli*: evidence for the significance of SOD in *Campylobacter* survival and colonization. Appl. Environ. Microbiol. *65*, 2540–2546.

Ratledge, C. (2007). Iron metabolism and infection. Food Nutr. Bull. *28*, S515–523.

Ratledge, C., and Dover, L.G. (2000). Iron metabolism in pathogenic bacteria. Annu. Rev. Microbiol. *54*, 881–941.

Reid, A.N., Pandey, R., Palyada, K., Naikare, H., and Stintzi, A. (2008). Identification of *Campylobacter jejuni* genes involved in the response to acidic pH and stomach transit. Appl. Environ. Microbiol. *74*, 1583–1597.

Ridley, K.A., Rock, J.D., Li, Y., and Ketley, J.M. (2006). Heme utilization in *Campylobacter jejuni*. J. Bacteriol. *188*, 7862–7875.

Robinson, D.A. (1981). *Campylobacter* infection. R. Soc. Health J. *101*, 138–140.

Seaver, L.C., and Imlay, J.A. (2001). Alkyl hydroperoxide reductase is the primary scavenger of endogenous hydrogen peroxide in *Escherichia coli*. J. Bacteriol. *183*, 7173–7181.

Sellars, M.J., Hall, S.J., and Kelly, D.J. (2002). Growth of *Campylobacter jejuni* supported by respiration of fumarate, nitrate, nitrite, trimethylamine-N-oxide, or dimethyl sulfoxide requires oxygen. J. Bacteriol. *184*, 4187–4196.

Stintzi, A. (2003). Gene expression profile of *Campylobacter jejuni* in response to growth temperature variation. J. Bacteriol. *185*, 2009–2016.

Stintzi, A., Marlow, D., Palyada, K., Naikare, H., Panciera, R., Whitworth, L., and Clarke, C. (2005). Use of genome-wide expression profiling and mutagenesis to study the intestinal lifestyle of *Campylobacter jejuni*. Infect. Immun. 73, 1797–1810.

Storz, G., and Zheng, M. (2000). Oxidative stress. In Bacterial stress responses, G. Storz, and H. Hengee-Aronis, eds. (Washington, D.C., ASM Press), pp. 47–59.

Svensson, S.L., Davis, L.M., MacKichan, J.K., Allan, B.J., Pajaniappan, M., Thompson, S.A., and Gaynor, E.C. (2009). The CprS sensor kinase of the zoonotic pathogen Campylobacter jejuni influences biofilm formation and is required for optimal chick colonization. Mol. Microbiol. *71*, 253–272.

Tannock, G.W. (2000). The intestinal microflora: potentially fertile ground for microbial physiologists. In Advances in Microbial Physiology, P.R.K., ed. (Sheffield, UK, Academic Press), pp. 25–46.

Taylor, B.L. (2007). Aer on the inside looking out: paradigm for a PAS-HAMP role in sensing oxygen, redox and energy. Mol. Microbiol. *65*, 1415–1424.

Traore, D.A., El Ghazouani, A., Ilango, S., Dupuy, J., Jacquamet, L., Ferrer, J.L., Caux-Thang, C., Duarte, V., and Latour, J.M. (2006). Crystal structure of the apo-PerR-Zn protein from *Bacillus subtilis*. Mol. Microbiol. *61*, 1211–1219.

Traore, D.A., Ghazouani, A.E., Jacquamet, L., Borel, F., Ferrer, J.L., Lascoux, D., Ravanat, J.L., Jaquinod, M., Blondin, G., Caux-Thang, C., *et al.* (2009). Structural and functional characterization of 2-oxo-histidine in oxidized PerR protein. Nat. Chem. Biol. *5*, 53–59.

van Alphen, L.B., Bleumink-Pluym, N.M., Rochat, K.D., van Balkom, B.W., Wosten, M.M., and van Putten, J.P. (2008). Active migration into the subcellular space precedes *Campylobacter jejuni* invasion of epithelial cells. Cell Microbiol. *10*, 53–66.

van Oeffelen, L., Cornelis, P., Van Delm, W., De Ridder, F., De Moor, B., and Moreau, Y. (2008). Detecting cis-regulatory binding sites for cooperatively binding proteins. Nucleic Acids Res. *36*, e46.

van Vliet, A.H., Baillon, M.A., Penn, C.W., and Ketley, J.M. (2001). The iron-induced ferredoxin FdxA of *Campylobacter jejuni* is involved in aerotolerance. FEMS Microbiol. Lett. *196*, 189–193.

van Vliet, A.H., Baillon, M.L., Penn, C.W., and Ketley, J.M. (1999). *Campylobacter jejuni* contains two fur homologs: characterization of iron-responsive regulation of peroxide stress defense genes by the PerR repressor. J. Bacteriol. *181*, 6371–6376.

van Vliet, A.H., Ketley, J.M., Park, S.F., and Penn, C.W. (2002). The role of iron in *Campylobacter* gene regulation, metabolism and oxidative stress defense. FEMS Microbiol. Rev. *26*, 173–186.

van Vliet, A.H., Wooldridge, K.G., and Ketley, J.M. (1998). Iron-responsive gene regulation in a *Campylobacter jejuni fur* mutant. J. Bacteriol. *180*, 5291–5298.

Vellinga, A., and Van Loock, F. (2002). The dioxin crisis as experiment to determine poultry-related *campylobacter* enteritis. Emerg. Infect. Dis. *8*, 19–22.

Vitale, S., Fauquant, C., Lascoux, D., Schauer, K., Saint-Pierre, C., and Michaud-Soret, I. (2009). A ZnS_4 structural zinc site in the *Helicobacter pylori* ferric uptake regulator. Biochemistry *48*, 5582–5591.

Wai, S.N., Nakayama, K., Umene, K., Moriya, T., and Amako, K. (1996). Construction of a ferritin-deficient mutant of *Campylobacter jejuni*: contribution of ferritin to iron storage and protection against oxidative stress. Mol. Microbiol. *20*, 1127–1134.

Wandersman, C., and Delepelaire, P. (2004). Bacterial iron sources: from siderophores to hemophores. Annu. Rev. Microbiol. *58*, 611–647.

Wang, G., Alamuri, P., and Maier, R.J. (2006). The diverse antioxidant systems of *Helicobacter pylori*. Mol. Microbiol. *61*, 847–860.

Wang, G., and Maier, R.J. (2004). An NADPH quinone reductase of *Helicobacter pylori* plays an important role in oxidative stress resistance and host colonization. Infect. Immun. 72, 1391–1396.

Weerakoon, D.R., Borden, N.J., Goodson, C.M., Grimes, J., and Olson, J.W. (2009). The role of respiratory donor enzymes in *Campylobacter jejuni* host colonization and physiology. Microb. Pathog. *47*, 8–15.

Weerakoon, D.R., and Olson, J.W. (2008). The *Campylobacter jejuni* NADH:ubiquinone oxidoreductase (complex I) utilizes flavodoxin rather than NADH. J. Bacteriol. *190*, 915–925.

Weinberg, E.D. (2009). Iron availability and infection. Biochim. Biophys. Acta. *1790*, 600–605.

Weingarten, R.A., Grimes, J.L., and Olson, J.W. (2008). Role of *Campylobacter jejuni* respiratory oxidases and reductases in host colonization. Appl. Environ. Microbiol. *74*, 1367–1375.

Whitby, P.W., Seale, T.W., VanWagoner, T.M., Morton, D.J., and Stull, T.L. (2009). The iron/heme regulated genes of *Haemophilus influenzae*: comparative transcriptional profiling as a tool to define the species core modulon. BMC Genomics *10*, 6–24.

Wilderman, P.J., Sowa, N.A., FitzGerald, D.J., FitzGerald, P.C., Gottesman, S., Ochsner, U.A., and Vasil, M.L. (2004). Identification of tandem duplicate regulatory small RNAs in *Pseudomonas aeruginosa* involved in iron homeostasis. Proc. Natl. Acad. Sci. U.S.A. *101*, 9792–9797.

Winterbourn, C.C. (2008). Reconciling the chemistry and biology of reactive oxygen species. Nat. Chem. Biol. *4*, 278–286.

Wooldridge, K.G., Williams, P.H., and Ketley, J.M. (1994). Iron-responsive genetic regulation in *Campylobacter jejuni*: cloning and characterization of a *fur* homolog. J. Bacteriol. *176*, 5852–5856.

Wright, J.A., Grant, A.J., Hurd, D., Harrison, M., Guccione, E.J., Kelly, D.J., and Maskell, D.J. (2009). Metabolite and transcriptome analysis of *Campylobacter jejuni in vitro* growth reveals a stationary-phase physiological switch. Microbiology *155*, 80–94.

Wu, H.J., Seib, K.L., Srikhanta, Y.N., Kidd, S.P., Edwards, J.L., Maguire, T.L., Grimmond, S.M., Apicella, M.A., McEwan, A.G., and Jennings, M.P. (2006). PerR controls Mn-dependent resistance to oxidative stress in *Neisseria gonorrhoeae*. Mol. Microbiol. *60*, 401–416.

Yamasaki, M., Igimi, S., Katayama, Y., Yamamoto, S., and Amano, F. (2004). Identification of an oxidative stress-sensitive protein from *Campylobacter jejuni*, homologous to rubredoxin oxidoreductase/rubrerythrin. FEMS Microbiol. Lett. *235*, 57–63.

Yang, J., Sangwan, I., and O'Brian M.R. (2006). The *Bradyrhizobium japonicum* Fur protein is an iron-responsive regulator *in vivo*. Mol. Genet. Genomics *276*, 555–564.

Zamocky, M., Furtmuller, P.G., and Obinger, C. (2008). Evolution of catalases from bacteria to humans. Antioxid. Redox. Signal. *10*, 1527–1548.

Zamocky, M., and Koller, F. (1999). Understanding the structure and function of catalases: clues from molecular evolution and *in vitro* mutagenesis. Prog. Biophys. Mol. Biol. *72*, 19–66.

Zeller, T., Moskvin, O.V., Li, K., Klug, G., and Gomelsky, M. (2005). Transcriptome and physiological responses to hydrogen peroxide of the facultatively phototrophic bacterium *Rhodobacter sphaeroides*. J. Bacteriol. *187*, 7232–7242.

Zheng, M., Wang, X., Templeton, L.J., Smulski, D.R., LaRossa, R.A., and Storz, G. (2001). DNA microarray-mediated transcriptional profiling of the *Escherichia coli* response to hydrogen peroxide. J. Bacteriol. *183*, 4562–4570.

Zhou, D., Qin, L., Han, Y., Qiu, J., Chen, Z., Li, B., Song, Y., Wang, J., Guo, Z., Zhai, J., *et al.* (2006). Global analysis of iron assimilation and *fur* regulation in *Yersinia pestis*. FEMS Microbiol. Lett. *258*, 9–17.

Zilbauer, M., Dorrell, N., Wren, B.W., and Bajaj-Elliott, M. (2008). *Campylobacter jejuni*-mediated disease pathogenesis: an update. Trans. R. Soc. Trop. Med. Hyg. *102*, 123–129.

Iron Dependency of, and Transport by, Cyanobacteria 11

Kerstin Nicolaisen and Enrico Schleiff

Abstract

Cyanobacteria are dependent on but can also be compromised by metals such as iron. On the one hand the demand for iron for photosystem functionality represents a challenge for the iron uptake machinery in iron limiting environments. On the other hand intoxication by iron causes a severe problem for growth and reproduction. To overcome this dilemma cyanobacteria have developed a regulatory network controlling iron uptake. They produce siderophores, which are distinct from that of other bacteria. Furthermore, the iron metabolism is linked to the nitrogen metabolism as documented for example in *Anabaena* sp. PCC 7120. In this chapter we focus on three aspects of this regulatory system. We describe the physiological consequence of iron starvation or iron intoxication in cyanobacteria. Especially molecular, morphological and physiological changes will be discussed. In the second part we elucidate the regulatory network sensing the iron status within the cell with focus on the Fur mediated transcriptional network and its cross-talk to other regulatory systems. In the final section we focus on the molecular components involved in uptake of iron by cyanobacteria.

Introduction

More than 3 billion years ago cyanobacteria evolved, which were able to produce oxygen (Sleep and Bird, 2008). The development of oxygen production was dependent on the availability of nutrients such as phosphorus, which in turn was largely removed from the environment by precipitation in the form of iron oxides (Bejerrum and Canfield, 2002). At the same time iron is a redox component required by virtually all organisms and may have catalysed reactions that constituted the first steps in the origin of life (Cody *et al.*, 2000; Russell and Martin, 2004). Iron is also a key element for the synthesis of several cellular compounds, e.g. ribonucleotides, haem and chromophores (Straus, 1994). Particularly oxygenic photosynthesis is iron-depended since photosystem I (PSI) contains 12 iron atoms in three 4Fe–4S clusters (Jordan *et al.*, 2001) and ferredoxin is needed in the electron transport chain (Laudenbach *et al.*, 1988).

In terrestrial environments the iron concentration can be rather high. It has been even hypothesized that within the precambian springs, rivers, lakes and oceans the iron concentration could have approached 100 mg/kg (Brown *et al.*, 2005). Nowadays, such conditions can only be found at thermal springs and vents. A possible intoxication by high iron concentrations is often discussed, but to the best of our knowledge it was never experimentally confirmed. Owing to the formation of insoluble hydroxides from ferric iron (Fe^{3+}) under oxygenic, aqueous conditions the biological availability of free Fe^{3+} in the environment became less than 10^{-18} M. This in turn is too low to allow growth of aerobic microorganisms (Braun *et al.*, 1990; Morel and Price, 2003; Schalk *et al.*, 2004). Therefore iron scavenging mechanisms may have developed subsequently in response to diminishing concentrations of soluble iron (Brown *et al.*, 2005). As consequence of this historical change from high to low availability of free environmental iron, it is hypothesized that iron limitation might have been one of the selective forces during evolution of cyanobacteria, leading to the survival of cyanobacteria with the most efficient iron uptake systems (Ting *et al.*, 2002).

Today, cellular metal homeostasis is dynamic and changes in response to availability, annual cycles and growth phase of photosynthetic organisms (Shcolnick and Keren, 2006). For uptake of iron many bacteria produce iron specific chelators, so called siderophores (Neilands, 1995; Katoh *et al.*, 2000; read also Chapter 1). To date, more than 500 different siderophores are known (Winkelmann, 2001). Nevertheless, these low molecular weight compounds (500–1500 Da) can be divided into three large groups based on chemical composition: (i) catechols, (ii) hydroxycarboxylates and (iii) hydroxamates (Ferguson and Deisenhofer, 2002). Siderophore production is induced upon iron starvation (Goldman *et al.*, 1983; Clarke *et al.*, 1987; Ferguson and Deisenhofer, 2002; Andrews *et al.*, 2003). To the best of our knowledge, it is not yet known how siderophores are intracellularly produced and how they are secreted from cyanobacteria. Indeed, there are so far only three structurally described siderophores, which are known to be secreted by cyanobacteria, namely schizokinen (Simpson and Neiland, 1976), anachelin (Beiderbeck *et al.*, 2000; Ito *et al.*, 2001) and synechobactin (Ito and Butler, 2005).

The iron loaded siderophores are taken up by the cell via TonB-dependent iron transporters present in the outer membrane (Schalk *et al.*, 2004; Faraldo-Gómez and Sansom, 2003). Subsequently, the iron is transported by a periplasmic 'shuttle' protein and a plasma membrane inserted machinery and can be stored in bacterioferritins within the cell (Bertani *et al.* 1997; Keren *et al.*, 2004). It was also observed that iron-free siderophores bind to outer membrane transporters (OMTs). These might play a role in the regulation of the iron uptake mechanism, but it is not yet described for the transport systems of cyanobacteria and the biological significance of this interaction remains elusive to date (Schalk *et al.*, 2004).

In this chapter, we focus on the influence of iron on the physiology of cyanobacteria, the regulation of iron metabolism, the response to iron starvation and on the existing information on transport systems, and we will compare this knowledge to that available for proteobacteria.

The importance of iron for cyanobacteria

Cyanobacterial organisms of the phytoplankton generally require higher Fe:C quotas than other eukaryotic phytoplankton (Kustka *et al.*, 2002). This observation can be extended such that the overall metal quota is much higher in photosynthetic than in non-photosynthetic organisms. For example, the iron content of *Synechocystis* cells, which was determined by atomic absorption spectroscopy, is found to be one order of magnitude higher in comparison to *Escherichia coli* (Table 11.1; Keren *et al.*, 2004). This results from the iron demand not only for respiratory redox enzymes as in non-photosynthetic bacteria, but also for the photosynthetic machinery and nitrogen fixation apparatus of cyanobacteria (Shi *et al.*, 2007).

Iron containing cofactors are, for example, iron–sulphur clusters, cytochromes and non-haem iron. Non-haem iron is positioned between Q_A and Q_B quinone molecules in photosystem II (PSII) and it is assumed that it is facilitating the electron transfer between the two (Shcolnick and Keren, 2006). In consequence, more than 25% of the iron quota of cyanobacteria is present in the PSI, which has been supported by atomic absorption spectroscopy of isolated thylakoid membranes (Keren *et al.*, 2004). In all cyanobacteria, the functional photosynthetic apparatus requires 22–23 Fe atoms, depending on whether cytochrome c553 or plastocyanin is used as an electron donor to PSI (Ferreira and Strauss, 1994). Hence, iron is a very important, potentially limiting nutrient metal for photoautotrophic growth, as iron starvation for example causes oxidative damage not observed in heterotrophic bacteria (Latifi *et al.*, 2005).

Not only photosynthesis requires an enhanced iron uptake by cyanobacteria. Diazotrophy imposes an additional demand of iron for growth. NO_3^- or NH_4^+ grown diazotrophs require less iron in culture media than those grown on N_2, presumably due to the down-regulated nitrogenase activity (Kustka *et al.*, 2002). For *Trichodesmium erythraeum* IMS101 it is even described that nitrogen fixation is far more susceptible to iron limitation than photosynthesis (Shi *et al.*, 2007). *T. erythraeum* IMS101 is a filamentous,

Table 11.1 Metal content of different cyanobacteria in comparison with *Escherichia coli*

	E. coli				
	mM	mmol/kg (dry wt)	*Anabaena variabilis* mmol/kg (dry wt)	*Nostoc* sp. mmol/kg (dry wt)	*Anabaena* sp. PCC 7120 mmol/kg (dry wt)
Mg	26.5	205 ± 10.0	450	400	375
K	74.1	576 ± 34.6	310	330	240
Ca	1.21×10^{-1}	1.5 ± 0.6	300	330	250
Mn	1.70×10^{-2}		24	27	17
Fe	2.21×10^{-1}		61	71	48
Cu	3.13×10^{-2}		0.6	0.6	1.1
Zn	1.46×10^{-1}		2	1.4	1.7
Na		401 ± 22.2	0.07	0.09	0.04
P			410	440	370
Al			12	70	50
	Outten and O'Haloran (2001)	Chang *et al.* (1986)	Roger *et al.* (1986)		

Given are the intracellular concentrations of elements indicated in the first column either in mM (based on cell volume) or mmol per kg dry weight. Please note that the reported concentration of copper is twice as high in *Anabaena* sp. PCC 7120 as compared to the other two cyanobacteria and that the concentration of aluminium is 6 to 10-fold higher in *Anabaena* sp. PCC 7120 and *Nostoc* sp. as compared to *Anabaena variabilis*.

non-heterocyst-forming cyanobacterium, which lives in tropical and subtropical surface water. To protect the nitrogenase, photosynthetically evolved O_2 is removed by a linear photosynthetic electron transport and an increased respiration rate (Shi *et al.*, 2007). The expression of nitrogenase, PSII and PSI is temporally separated by a diurnal cycle (Shi *et al.*, 2007). However, the nitrogenase contains 19 iron atoms per heterodimeric protein complex moiety [one [4Fe–4S] cluster in the nitrogenase reductase and one P cluster ([8Fe–7S])/one FeMoCo cluster ([7Fe–9S–Mo]) in the nitrogenase], but has a lower iron usage efficiency as compared to other iron-containing enzymes involved in nitrogen metabolism, such as nitrate reductase or nitrite reductase (Raven, 1988). The minimal iron requirement for the nitrogenase activity was estimated to be 10–30 mol Fe per mol fixed N (Kustka *et al.*, 2002). Therefore, iron is essential for many aspects of cyanobacterial growth and physiology.

Influence of iron on cyanobacterial growth, development and morphology

The importance of iron was initially analysed by describing phenotypes evoked by iron starvation. Here, for example, electron microscopic pictures or the growth behaviour of cells cultured under normal or iron starvation conditions were compared.

For coccoid cyanobacteria like *Synechococcus* sp. PCC 7942 (*Anacystis nidulans* R2, Sherman and Sherman, 1983) and *Synechococcus* sp. PCC 7002 (*Agmenellum quadruplicatum*, Hardie *et al.*, 1983) chlorosis was observed as the most obvious manifestation of iron limitation (Boyer *et al.*, 1987). Furthermore, iron starvation of the cyanobacterium *Synechococcus* sp. PCC 7942 results in phycobilisome replacement by glycogen containing granules (Sherman and Sherman, 1983). However, Boussiba and Richmond (1980) suggested that the degradation of cellular phycocyanin (an accessory pigment found in phycobilisomes) could be only a secondary effect of iron limitation. Nitrogen deficiency caused by reduced levels of iron-containing enzymes required for nitrogen assimilation may also cause the cells to degenerate phycocyanin pools and use them as a source of nitrogen (Wilhelm, 1995). Another effect of iron limitation described for *Synechococcus* sp. PCC 7942 is the detachment of most thylakoid membrane stacks and a reduction of the number of carboxysomes (Sherman and Sherman, 1983). A similar disaggregation of thylakoid membranes was observed for iron starved cyanobacteria of the species *Acaryochloris marina* (Swingley *et al.*, 2005) and *Anabaena* sp. PCC 7120. In the latter, however, the carboxysomes are still present after iron starvation. In *Synechococcus* sp. PCC 7002 three to six pairs of thylakoid membranes are equally spaced and concentric along most of the cell body. In iron-starved cells this packing is loose and some mesosome-like structures are formed. After a prolonged time of iron starvation (60–100 h) the thylakoid system is rearranged and a separation between the cytoplasmic membrane and the peptidoglycan layer of the cell envelope occurred (Hardie *et al.*, 1983). After further extension of iron starvation to 200 h the ribosome content is depleted, the number of polyphosphate bodies is decreased and peripheral lipid bodies disappear (Hardie *et al.*, 1983), possibly indicating the death of the cell.

When the filamentous cyanobacterium *T. erythraeum* IMS101 was studied under conditions of iron starvation, filament fragmentation and shortening was observed and a lower cell proliferation rate (Shi *et al.*, 2007) which stopped after about 5 to 8 days of iron depletion (Küpper *et al.*, 2008). In contrast, the growth rate of *Synechococcus* sp. PCC 7942 remained constant under iron limiting conditions, but the cells were only one-half or two-thirds of the size normally observed (Sherman and Sherman, 1983). However, subsequent studies demonstrated that the growth defects may be influenced by the amount of iron remaining as impurities of solutions and in some species the growth rate can even 'recover' (increase) after a prolonged phase of iron starvation (Wilhelm and Trick, 1994). The growth rate (μ_{max}) of several cyanobacteria, e.g. *Synechococcus* sp. PCC 7942, *Synechococcus* sp. PCC 7002, *Anabaena variables* and *Oscillatora tenuis* is reduced as the availability of iron decreases (Kerry *et al.*, 1988; Brown and Trick, 1992; Wilhelm, 1995; Wilhelm *et al.*, 1996). Further, it became obvious that the growth rate of *O. tenuis* at an 'intermediate' iron concentration was even lower than that at the lowest iron concentration tested (Kerry *et al.*, 1988). This does not hold true for *Plectonema boryanum* (Kerry *et al.*, 1988), where the growth rate directly correlates with the iron content of the medium. As *P. boryanum* does not secrete any siderophores, Wilhelm (1995) concluded from this observation that the iron concentration for which the lowest growth rate is observed is just above the threshold of the iron concentration inducing siderophore synthesis. Based on this observation the author concluded that above the limiting concentration a low affinity iron transport system mediates iron uptake, whereas below the critical concentration a high affinity system dependent on the siderophores produced by the individual cyanobacterium takes over.

Brown *et al.* (2005) described a novel cyanobacterium, *Chroogloeocystis siderophila*, isolated from an iron-depositing hot spring. For normal growth it needs higher iron concentrations than those provided by widely used growth media, more than 40 μM. This suggests that *C. siderophila* employs a low affinity mechanism of iron uptake rather than the one that is known for example for *Synechocystis* sp. PCC 6803, *Trichodesmium*, *Cyanothece* or *Anabaena*, which would only require 1–2 μM iron (Brown *et al.*, 2005; Berman-Frank *et al.*, 2007). However, the concentration of 40 μM iron is not perilous for other cyanobacteria. For example, for *Anabaena* spp., *Nostoc* spp. and *Nodularia* spp. a concentration of 200 μM iron did not result in lethality (Okmen *et al.*, 2007).

In general it can be concluded that iron starved cyanobacteria are reduced in their growth rate and that their thylakoid systems are disorganized. Further effects seem to be rather secondary, since iron as cofactor for nitrogen fixation and photosynthesis is missing upon iron starvation, which also leads to starvation effects. Iron intoxication, however, appears not to be a significant problem for the cyanobacteria existing today.

Influence of iron on cyanobacterial physiology

The overall changes of cyanobacterial physiology upon iron starvation are manifold. For example the phycocyanin and chlorophyll content is reduced, the number of thylakoid membranes declines, the occurrence of the iron starvation induced PsbC homologue CP43′ causes a characteristic blue shift of chlorophyll absorption and leads to dramatic alterations in the 77 K chlorophyll fluorescence emission spectrum because the capacity for state transitions is reduced and the photosystem is locked in state I (Guikema and Sherman, 1983; Sandmann *et al.*, 1990; Sandström *et al.*, 2002; Ivanov *et al.*, 2006). A further effect is the reduced photochemistry of PSII and PSI. Especially PSI, the most iron–rich complex of the photosynthetic apparatus, is highly susceptible to iron limitation and its activity decreases more dramatically than PSII during iron starvation (Sherman and Sherman, 1983; Guikema and Sherman, 1983; Ivanov *et al.*, 2001; Sandström *et al.*, 2002). The transition of PSI from its trimeric to a monomeric state seems to be one of the major effects of iron starvation, which was described for *Synechococcus* sp. PCC 7942 and *Synechocystis* sp. PCC 6803 (Ivanov *et al.*, 2006). Consistently, microarray analysis of *Synechocystis* sp. PCC 6803 unravelled that iron deficiency induces down-regulation of the PsaL subunit, which is required for the trimerization of PSI, and the PsaI subunit, which is important for the stability of PsaL (Singh *et al.*, 2003).

The IsiA protein

In many cyanobacteria, such as *Synechococcus elongatus* PCC 7942, *Synechocystis* sp. PCC 6803 and *Anabaena* sp. PCC 7120, expression of the iron stress induced gene A (*isiA*) in response to iron starvation was observed (Fig. 11.1, Latifi *et al.*, 2005; Kouril *et al.*, 2005), which is a result of Fur mediated derepression of transcription (see below; Ghassemian and Straus, 1996). The *isiA* gene product CP43′ forms a chlorophyll–protein complex of 18 subunits leading to an antenna ring around trimeric and monomeric PSI (Latifi *et al.*, 2005; Kouril *et al.*, 2005), as exemplified by electron microscopy for *Synechocystis* sp. PCC 6803 (Boekema *et al.*, 2001; Bibby *et al.*, 2001a; Kouril *et al.*, 2005). The association of IsiA with PSI results in an increase of the antenna size by about 70% (Bibby *et al.*, 2001b) and of the PSI absorption cross-section by 2-fold (Andrizhiyevskaya *et al.*, 2002). On the basis of these observations a function of CP43′ as an alternative antenna complex for PSI is discussed. This complex is supposed to increase the functional absorption cross-section for PSI by efficient energetic coupling between the CP43′ ring and the PSI reaction centre core. Alternatively, CP43′ is discussed as quencher for protecting the PSII from photo-inhibitory damage. This was suggested because iron stressed cells are locked in state I and show a characteristic increase of PSII fluorescence at 685 nm and a decrease of PSI fluorescence peak due to adjustments in the energy distribution between the two photosystems (Ivanov *et al.*, 2006).

Alternatively it was discussed that *isiA* induction is a rather general stress response than specific for iron starvation. This notion was based on the enhanced *isiA* expression observed in response to oxidative stress (Michel and Pistorius, 2004). However, iron limitation causes the accumulation of reactive oxygen species (ROS) by about 10-fold in comparison to normal growth as shown in *Anabaena* sp. PCC 7120 and *Synechocystis* sp. PCC 6803 (Latifi *et al.*, 2005). Therefore, the induction of *isiA* expression by other means might reflect a function in photosystem protection in general.

Flavodoxin can replace ferredoxin

In times of iron starvation the iron-rich soluble electron transfer protein ferredoxin can be replaced by flavodoxin as demonstrated, for example, in *S. elongatus* PCC 7942 and *Synechocystis* sp. PCC 6803 (Latifi *et al.*, 2005; Kouril *et al.*, 2005). This protein encoded by *isiB* contains a flavin that does not require iron as cofactor. Flavodoxin transfers electrons from PSI to ferredoxin-$NADP^+$ reductase as efficient as ferredoxin because the oxidation/reduction potential determined for the semiquinone/hydroquinone pair in different flavodoxins is comparable to those of ferredoxin (Fillat *et al.*, 1988, 1990). The ratio of flavodoxin to ferredoxin is proposed as a marker for the iron nutritional status of a cyanobacterial cell (Doucette *et al.*, 1996). The repression of ferredoxin formation leaves iron available for the synthesis of other iron containing proteins. Noticeably, flavodoxin is constitutively expressed in heterocysts, independent of the iron concentration, which emphasizes the role of this protein in the transport of electrons to the nitrogenase (Sandmann *et al.*, 1990). However, some distinctions in the mechanism of protection by flavodoxin exist. While in *Anabaena* ATCC 29413 flavodoxin can replace ferredoxin in times of iron

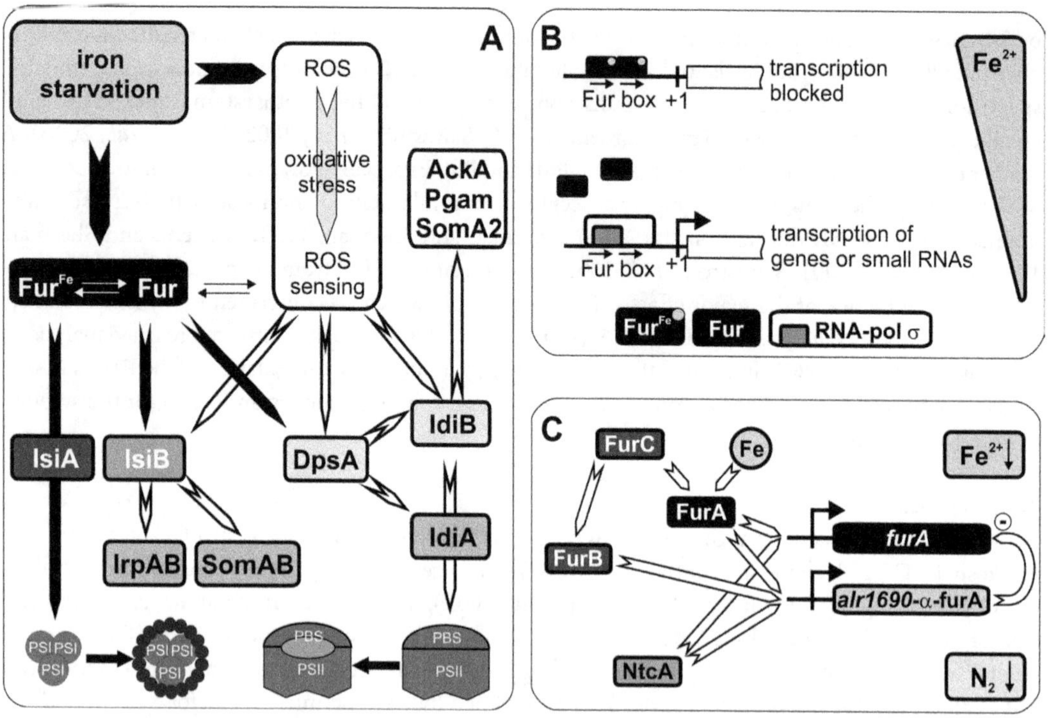

Figure 11.1 The regulatory network acting in response to iron deficiency. Iron deficiency can be sensed by two alternative routes. On the one hand, iron loaded Fur in most cases represses the expression of Fur dependent genes. Iron depletion therefore leads to a de-repression activating the transcription of *isiA*, *isiB*, *dpsA* and other genes (black arrows). On the other hand, iron starvation leads to the accumulation of radical oxygen species which in turn activate a signal transduction cascade leading to the expression of *isiB*, *dpsA*, *idiB* and other genes (white arrows). DpsA and IdiB act as transcriptional regulators for instance for *idiA*. The physiological consequence of the expression of *isiA* and *idiA* is the protection of the photosystems, as indicated. (B) Fur (black) is associated with the Fur-Box when iron (grey) is loaded and inhibits the association of the RNA polymerase (white). Under low iron concentrations, Fur dissociates and transcription can be initiated. (C) The regulatory Fur network identified in *Anabaena* sp. PCC 7120 is shown. FurC influences the activity of FurB and FurA. FurA functions self-regulatory on its own promoter. FurB and FurA together regulate the *alr1690-α-furA* antisense RNA gene, which in turn regulates the transcription of *furA*. *Alr1690*-α-*furA* and *furA* are also targeted by NtcA, the master regulator for nitrogen sensing. This interaction represents one link between iron and nitrogen starvation.

starvation, *Anabaena* ATCC 29211 generally lacks the potential to synthesize flavodoxin (Sandmann *et al.*, 1990). In the latter only the ferredoxin content is decreased at very low iron concentrations (below 1 μM), at which ferredoxin is already replaced by flavodoxin in *Anabaena* ATCC 29413.

The genomic occurrence of isiA and isiB

In *Prochlorococcus marinus* SS120 which belongs to the order of the *Prochlorales, S. elongatus* PCC 7942 and *Synechocystis* sp. PCC 6803 belonging to the order of *Chroococcales* and in *A. marina* MBIC11017 classified as uncharacterized cyanobacteria, *isiA* is in an operon also containing *isiB* (Fig. 11.2). In the last three cyanobacteria a third gene within the same operon was identified, namely *isiC*. *isiC* encodes a hydrolase-like protein which shows 54.3% similarity to the esterase Fes from *Escherichia coli* (Nodop *et al.*, 2008). In *E. coli* Fes hydrolyses ester bonds of internalized ferri-enterobactin siderophores (Andrews *et al.*, 2003) and therefore, the gene product might be involved in the recycling of iron containing molecules in times of iron starvation.

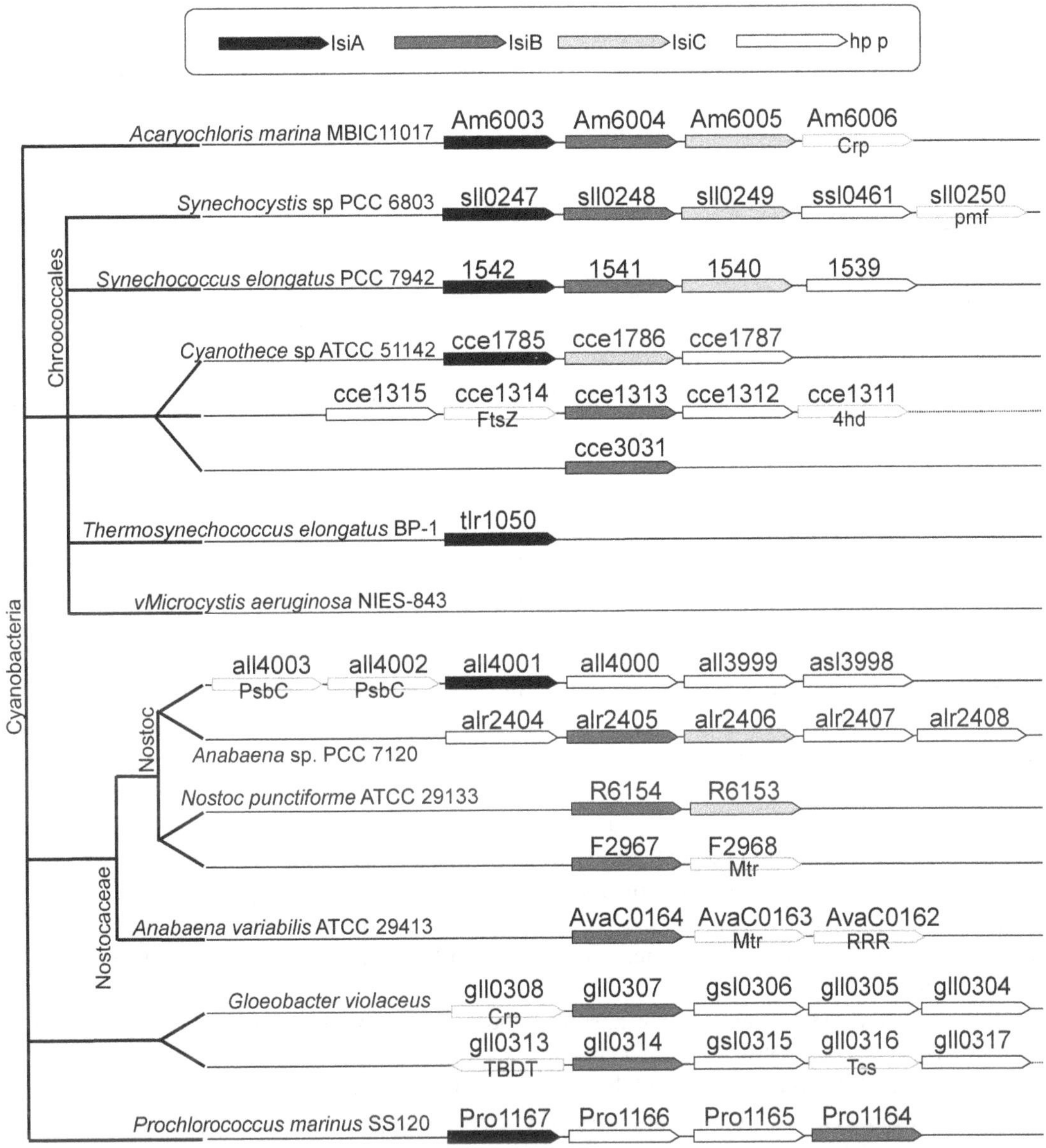

Figure 11.2 The genomic structure of *isiA*, *isiB* and *isiC* is shown for selected cyanobacteria. The IsiA, IsiB and IsiC amino acid sequence of *Synechocystis* sp PCC 6803 and *Synechococcus elongatus* PCC 7942 and *Anabaena* sp. PCC 7120 was used as bait to perform a forward and reverse Blast search on CYANOBASE (Nakamura *et al.*, 1998; 2000) and only double positives are shown. On the left side the relation of the cyanobacteria according to the taxonomy browser at www.ncbi.nlm.nih.gov (Wheeler *et al.*, 2000) is shown. The gene accession numbers according to cyanobase are indicated (at the top), *isiA* (black), *isiB* (dark grey), *isiC* (grey) and other genes (white). Other genes represent hypothetical open reading frames, if not indicated otherwise (pmf, pantothenate metabolism flavoprotein; 4hd, 4-hydroxyphenylpyruvate dioxygenase; Tcs, tetracenomycin C synthesis protein homologue; Crp, transcriptional regulator; MTr, multicomponent transcriptional regulator; RRR, Response regulator receiver domain protein).

However, the operonic arrangement of these genes is not conserved in all cyanobacterial species. In *Cyanothece* sp. ATCC 51142, belonging to the *Chroococcales isiA* is in an operon with *isiC*, but not with *isiB*. Interestingly, in *Cyanothece* sp. ATCC 51142, as well as in *Nostoc punctiforme* ATCC 29133 and *Gloeobacter violaceus* two genes coding for *isiB* can be identified (Fig. 11.2). In the more distant

Chroococcales, *Thermosynechococcus elongatus* BP-1 and *Microcystis aeruginosa* NIES-843, no *isiB* and *isiC* were recognized. This, however, has to be taken with care at least for *isiB*, as the evolutionary distance might be the reason why the corresponding flavodoxin cannot be identified by sequence comparison. The same holds true for the *isiA* in *M. aeruginosa* NIES-843. As *isiA* shares a significant similarity to *psbC* we were unable to assign the correct gene, which is also the case for *G. violaceus*, *A. variabilis* ATCC 29413 and *N. punctiforme* ATCC 29133. Hence, future work has to elucidate whether the protective mechanism involving *isiA* and *isiB* is indeed conserved in all cyanobacteria.

In the family of *Nostocaceae* an interesting phenomenon can be observed. On the one hand, *isiB* and *isiC* are found in close genomic proximity in the cyanobacteria of the genus *Nostoc* (Fig. 11.2). On the other hand, in *Anabaena variabilis* ATCC 29413 and *Nostoc punctiforme* ATCC 29133 one *isiB* is found upstream of a transcriptional regulator. A similar ensemble, but in opposite order, is found for one *isiB* of *G. violaceus*. The diversity of the genomic organisation suggests that cyanobacteria might have adopted various strategies to escape iron starvation.

Additional players involved in the response to iron starvation

Under iron or manganese deficiency, the expression of the iron deficiency-induced protein A (IdiA) is strongly enhanced, as found for example in *S. elongatus* PCC 7942 (Fig. 11.1, Michel *et al.*, 1998, 2004). However, in contrast to the transcript of *isiAB*, which disappears 30 minutes after iron repletion, the transcript of *idiA* remains detectable even ten hours after iron repletion (Yousef *et al.*, 2003). In deletion mutants of *idiA* in *Synechocystis* sp. PCC 6803, PSII is more susceptible to inactivation and the formation of the PSI–IsiA complex is more advanced under mild iron limitation (Tölle *et al.*, 2002). Consistently, IdiA is a predominantly thylakoid membrane-associated protein (Michel *et al.*, 1998) and interacts with the PSII (Tölle *et al.*, 2002). The observed increased degradation of a subunit of the photosystem, D1, in *isiA* mutants (Michel *et al.*, 1996; Exss-Sonne *et al.*, 2000) and higher resistance towards the herbicide bentazone in *idiA* expression strains (Bagchi *et al.*, 2003) is consistent with a model in which IdiA shields and protects the acceptor side of PSII that is exposed to cytoplasm after phycobilisome degradation (Nodop *et al.*, 2008). Indeed, electron microscopic analysis of single PSII particles indicated that IdiA directly interacts with CP43 and D1 (Lax *et al.*, 2007). However, this protection mechanism is specific for cyanobacteria, since IdiA-like proteins have not been found in green algae and higher plants (Michel and Pistorius, 2004).

The expression of *idiA* is controlled by the transcriptional activator IdiB (Fig. 11.1, Michel *et al.*, 1999; 2001) and *idiB* expression itself is induced upon iron starvation (Yousef *et al.*, 2003). Similar to *isiAB*, but in contrast to *idiA*, the transcript of *idiB* disappears shortly after iron repletion of starved cells (Yousef *et al.*, 2003). IdiB is a helix-turn-helix transcriptional regulator of the of the Crp/Fnr family (Michel and Pistorius, 2004). Interestingly, a similar gene is found in the same genomic region as the *isiABC* operon in *A. marina* MBIC11017 and as *isiB* encoded by *gll0307* in *G. violaceus* (Fig. 11.2). A third iron-regulated gene is *idiC*, which may also contribute to the iron starvation induced modification of the electron transport chain, and which lies upstream of *idiB* in a common operon in *S. elongatus* PCC 7942 (Pietsch *et al.*, 2007). IdiC belongs to the thioredoxin-like ferredoxin family. Hence, a function of IdiC in the photosynthetic cyclic electron transport around PSI and/or in the respiratory electron transport is suggested based on its high similarity to NuoE, a peripheral subunit of the *E. coli* NDH–1 complex (Pietsch *et al.*, 2007). In contrast to *idiA* and *idiB*, *idiC* is essential and its expression is not dependent on IdiB. This suggests a rather general function of IdiC than in protection against iron starvation

It is worthwhile mentioning that in the same genomic region as *idiABC* one gene coding for DpsA and one coding for a Fur-like protein (see below) are found. *DpsA* is a starvation induced gene which is activated upon iron limitation (Sen *et al.*, 2000) in a Fur dependent manner (Hernández *et al.*, 2007). For the protein encoded by *dpsA* a function in protection of chromosomal DNA against oxidative damage is suggested (Latifi *et al.*, 2009). A *dpsA* mutant strain shows a reduced expression of *idiA* and *idiB* (Fig. 11.1, Michel *et al.*, 2003). Hence, this genomic region might be essential for the response of *S. elongatus* PCC 7942 to conditions in which iron is limiting.

On the chromosome of *S. elongatus* PCC 7942 four additional regions have been identified the transcripts of which accumulate due to iron-deficient growth conditions: *irpAB, fut, suf* and *ackA/pgam*. *ackA* and *pgam* are expressed in an IdiB-dependent manner (Nodop *et al.*, 2008) and are therefore part of the response to iron starvation. *AckA* encodes an acetate kinase and *pgam* phosphoglycerate mutase having a proposed function in acetate-phosphate metabolism. Nodop and coworkers (2008) discussed that the IdiB-dependent regulation of *ackA/pgam* is indicative for the existence of an additional substrate-level phosphorylation under iron-limiting conditions, namely via the conversion of acetyl-phosphate to acetate with concomitant synthesis of ATP. However, this relation is only proposed and has to be challenged in future, especially as their expression appears also to be diurnally regulated (Steunou *et al.*, 2006). The *suf* genes are responsible for the iron homeostasis and sensing redox stress (Balasubramanian *et al.*, 2006), and they are also involved in the biogenesis of the formation and repair of Fe/S clusters for PS I (Wang *et al.*, 2004). Hence, the function of the SUF system is not limited to the response to iron starvation. The iron-regulated protein A (IrpA), described in *Synechococcus* sp. PCC 7942, encodes a 36-kDa protein induced under iron deficiency, which is involved in iron acquisition (Durham *et al.*, 2003). It is a plasma membrane protein as determined by cell fractionation and immunogold detections in electron micrographs (Reddy *et al.*, 1988). Based on the analysis of *irpA* mutants a function in iron acquisition or storage was postulated but not yet experimentally confirmed. In *Synechococcus* sp. PCC 7942, *irpA* forms an operon with *irpB*, which, however, does not exist in all cyanobacterial genomes (Nodop *et al.*, 2008) and hence, its exact function remains to be established. It encodes a cytochrome family protein with two CXXC haem-binding sites. It should be noted that IrpB from *Synechococcus* sp. PCC 7942 does not share any sequence similarity with IrpB (also IRE-BP2 or IRP2) found in mammals, which is activated by iron starvation as well and which binds to iron response elements (IRE) and is involved in the assembly of iron–sulphur-clusters (Pantopoulos *et al.*, 1995).

As can be extracted from the above summary of components, many genes were identified to be regulated in response to iron limitation. However, with exception of the proteins coded by *isiAB* and *idiAB* the molecular function for the described components is only proposed. Therefore, efforts have to be directed toward the analysis of the function of proteins coded by these genes to understand the molecular mechanisms protecting the cyanobacteria. Of particular interest might be the analysis of factors not ubiquitously present in all cyanobacteria. They offer the possibility to learn more about the variety of evolved strategies to survive iron starvation.

Regulation of cyanobacterial gene expression by iron

The regulation of gene expression by iron is long known and was first explored in *E. coli* (Hantke, 1981; Chapter 5). Subsequently, Fur was identified as a Fe^{2+}-dependent repressor with central regulatory function, as inferred from early genetic studies. It was first demonstrated using an *in vitro* coupled-transcription translation system (Bagg and Neilands, 1987) which led to a model for Fur-mediated repression of target genes under iron-replete conditions. However, Fur protein binding to DNA can *in vivo* also be stimulated by other divalent cations such as Mn^{2+} (Hantke, 1987) and Fur was also found to be expressed in response to oxidative (Nunoshiba *et al.*, 1999) and acidic stress (Hall and Foster, 1996). Nevertheless, consistent with a role for Fur as sensor of intracellular iron levels its affinity for ferrous iron is comparable to the levels of free iron present in the cytosol (Keyer and Imlay, 1996).

The molecular mechanism how Fur interacts with DNA is still under debate. Most studies describe a model in which a protein dimer recognises a palindromic DNA sequence in an iron dependent manner (Bes *et al.*, 2001). One current model (e.g. Hantke, 2001) envisions the binding of Fe^{2+} loaded dimeric Fur protein to a specific DNA sequence, called the 'Fur box' (Fig. 11.1B). In *E. coli* the Fur box consensus sequence identified is GATAATGATAATCATTATCA, which was reinterpreted as three forward-forward-reverse tandem hexamer containing the recognition unit NAT(A/T)AT (Escolar *et al.*, 1998). The Fur box can be positioned in the non-transcribed or (and) transcribed regions of the gene regulated by the Fur system (Kunert *et al.*, 2003). In *E. coli* there is evidence for even more than one Fur box in a single promoter (Escolar *et al.*, 1997). At iron levels higher than required for normal growth,

Fur represses genes needed for iron uptake by blocking access of RNA polymerase (Faraldo-Gómez *et al.*, 2003). However, Fur does not only suppress, but also activate gene expression as observed while analysing *fur* mutants in *Vibrio cholerae* (Litwin and Calderwood, 1994; Chapter 7) or *katA* expression in *Staphylococcos aureus* (Horsburgh *et al.*, 2001). Recent work elegantly demonstrated that this regulation is indirect by repression of the sRNA RyhB, which in turn regulates the expression of non-essential iron containing proteins (Massé *et al.*, 2007). Hence, to date it is widely accepted that Fur, Fur-like proteins or proteins acting in a similar manner (e.g. Johnston *et al.*, 2007) are master regulators of iron-dependent expression.

The cyanobacterial Fur system

Fur proteins have also been found in several cyanobacteria, such as *Anabaena* sp. PCC 7119 (Bes *et al.*, 2001) or *Synechococcus* sp. PCC 7942 (Ghassemian and Straus, 1996), where they function as iron dependent regulators. Fur regulated cyanobacterial genes also contain a Fur box, but with variable arrangement as compared to *E. coli* (Kunert *et al.*, 2003; Hernández *et al.*, 2006b). Mutagenesis of this Fur box revealed that in cyanobacteria the exact position of A and T within the Fur binding sites is not critical (Hernández *et al.*, 2006b). As in *E. coli*, Mn^{2+} promotes the formation of FurA–DNA complexes in cyanobacteria (Hernández *et al.*, 2006b).

In *Anabaena* sp. PCC 7119, a filamentous cyanobacterium, which can form heterocysts and gas-vacuolated hormogonia, a Fur homologue was identified and two putative Fur target sequences have been found upstream of the *fur* start codon (Bes *et al.*, 2001). This supports the hypothesis that in cyanobacteria *fur* could be autoregulated as found in *E. coli*. *In vitro* studies with the Fur protein showed that it binds with even higher affinity to the *fur* than to the *isiB* promoter (Bes *et al.*, 2001), which is consistent with more than one Fur binding site in the *fur* promoter sequence.

Many cyanobacteria encode, however, three Fur homologues. The interplay between these three factors was intensively analysed in *Anabaena* sp. PCC 7120 where they are termed FurA (encoded by *all1691*; in *Synechocystis* sp. PCC 6803 encoded by *sll0567*), FurB (encoded by *all2473*; in *Synechocystis* sp. PCC 6803 encoded by *sll1937*) and FurC (encoded by *alr0957*; in *Synechocystis* sp. PCC 6803 not found) (Hernández *et al.*, 2004a, Kaneko *et al.*, 1996). FurA shows the highest similarity to Fur from heterotrophic bacteria and contains the conserved motif HHX HXX CXXC. By mutagenesis it was demonstrated that *furA* is indeed essential for cyanobacterial viability (Kunert *et al.*, 2003; Hernández *et al.*, 2004a, 2006a). At the transcriptional level FurA is autoregulated and probably modulated by FurB and FurC (Fig. 11.1). Post-transcriptionally, the *alr1690*-α-furA antisense RNA interferes with the *furA* transcript and directly controls the transcription of flavodoxin, the product of the *isiB* gene, and is moderately autoregulatory. FurB does not contain the conserved motif HHX HXX CXXC, but instead a cluster of three histidine residues was found (Kunert *et al.*, 2003). The analysis of knock-out mutants in *Synechocystis* sp. PCC 6803 showed that *furB* is not essential for growth under standard cultivation conditions. It might well be that FurB functions as an iron-response regulator (Irr) like protein since, like Irr proteins, FurB is more active in the absence of iron and shows a putative haem regulatory motif near its C-terminal sequence (CPVHNN). FurC only seems to modulate the binding activity of FurA and FurB towards the corresponding promoter regions, rather than to bind itself to any *fur*-type promoter (Hernández *et al.*, 2004). Thus, its molecular actions remains to be further established.

The communication between the cyanobacterial Fur and NtcA systems

In the absence of combined nitrogen, *Anabaena* sp. PCC 7120 fixes atmospheric nitrogen in a specialized cell type, the heterocysts, which differentiate from vegetative cells. To protect the oxygen-sensitive enzyme nitrogenase, their PSII is disassembled and the respiration rate is increased to create microoxic internal conditions (Nicolaisen *et al.*, 2009). NtcA is the principal transcription factor involved in the regulatory network of nitrogen metabolism. It regulates the initiation and progression of heterocyst development, activates numerous genes involved in nitrogen assimilation but can also work as repressor for example for the *rbcL* gene, which encodes a subunit of ribulose bisphosphate carboxylase/oxygenase

(Nicolaisen *et al.*, 2009). It is discussed that NtcA may coordinate iron acquisition and nitrogen metabolism by activating the expression of *pkn41* and *pkn42* (Cheng *et al.*, 2006). The *pkn22* operon, encoding for the protein kinase Pkn22 and peroxiredoxin PrxQ-A, was considered to be regulated by oxidative stress, but is also influenced by iron starvation and under the control of NtcA (Latifi *et al.*, 2009).

A direct crosstalk between heterocyst development and iron acquisition can be concluded from the strong activation of P_{furA}, but not P_{furB} and P_{furC}, in proheterocysts and heterocysts due to the binding of NtcA to the operators present in the upstream region of *furA* and *alr1690-α-furA* (Fig. 11.1, López-Gomollón *et al.*, 2007). Consistently, the promoter of *furA* shows several putative NtcA binding sites. The nitrogen status thereby influences *furA* expression (López-Gomollón *et al.*, 2007) and subsequently flavodoxin (*isiB*) synthesis (Hernández *et al.*, 2004b). However, in cyanobacteria many iron-responsive genes are regulated by NtcA, as for example *psaL, furA, furC, isiA* and *isiB* (López-Gomollón *et al.*, 2007b). Hence, the coordinated regulation of expression of 'photosynthetic genes' by FurA and NtcA offers a regulatory network able to respond to a range of environmental conditions.

The iron uptake system

The three major iron uptake systems are in general (i) the complexation with siderophores and the transport in a TonB-dependent manner; (ii) the uptake of ferrous iron via transporter like FeoB, which was originally described to use ATP hydrolysis as energy source but other reports showed a G protein-like functionality; and (iii) the uptake by ABC-transporters which have a specificity for iron but do not necessarily require outer membrane receptors or siderophores (Marlovits *et al.*, 2002; Andrews *et al.*, 2003). The molecular mechanisms of iron uptake are well studied in non-photosynthetic bacteria like *E. coli*, which are described in Chapter 5, but are rather unexplored in cyanobacteria. Only recently investigations on the TonB dependent outer membrane iron transporters in *Synechocystis* sp. PCC 6803 (Katoh *et al.*, 2001a) and *Anabaena* sp. PCC 7120 have been conducted (Mirus *et al.*, 2009). Additionally, SchT (Alr0397), a TonB-dependent transporter for schizokinen in *Anabaena* sp. PCC 7120 has been characterized in detail (Nicolaisen *et al.*, 2008). Furthermore, some information on periplasmic proteins involved in iron uptake and the plasma membrane localized machineries have been collected (Katoh *et al.*, 2001b, Badarau *et al.*, 2008; Koropatkin *et al.*, 2007).

Cyanobacterial siderophores

Nearly all soluble iron in oceans, coastal water and lakes, the natural habitat of many cyanobacteria, is probably complexed by organic ligands (Lewis *et al.*, 1995). Here, low-molecular-weight organic molecules termed siderophores secreted by microorganisms specifically chelate Fe^{3+} (Lewis *et al.*, 1995; Barbeau *et al.*, 2003). The conditional stability constant for the binding of Fe^{3+} by organic ligands is in order of 10^{20}, which indicates a very strong complexation of iron (Lewis *et al.*, 1995). Hydroxamate- and catecholate-type siderophores have been shown to be secreted by marine bacteria as well as iron-chelators with other functional groups (Lewis *et al.*, 1995).

Schizokinen, isolated from *Anabaena* sp. ATCC 27898, was the first structurally characterized siderophore (Table 11.2, Simpson and Neilands, 1976). It is a dihydroxamate derivative of citric acid and chelates iron via the two α-hydroxamate groups plus the α-hydroxycarboxylate group of the citrate moiety (Goldman *et al.*, 1983; Plowman *et al.*, 1984). Other tested siderophores were not able to mediate transport of iron into *Anabaena* (Lammers and Sanders-Loehr, 1982). From these experiments it was concluded that exclusively schizokinen is taken up by *Anabaena*. However, this conclusion has to be taken with care as only a very limited number of siderophores has been tested when compared to the vast amount of 500 such chelators known today. For example, Wilhelm and Trick (1994) reported two mechanisms for high-affinity iron acquisition used by cyanobacteria in aquatic systems after release of hydroxamate and iron chelation, (i) its uptake by a specific receptor, or alternatively (ii) the deferration of the ferrihydroxamate complex at the cell surface by a catechol-type siderophore and assimilation via the newly formed catechol–iron complex.

Table 11.2 Siderophores produced and taken up by cyanobacteria

Species	Siderophore	Reference
Synechococcus sp. PCC 7002 (*Agenellum quadruplicatum*)	*Catecholate*	Armstrong and van Baalen (1979)
	Hydroxamate	Barbeau *et al.* (2003)
	Schizokinen	
	Synechobactin A, B and C	Ito and Butler (2005)
Anabaena flos-aquae	*Schizokinen*	Lammers and Sanders-Loehr (1982)
	Hydroxamate	Goldmann *et al.* (1983)
Anabaena cylindrica	Schizokinen	Lammers and Sanders-Loehr (1982)
Anabaena sp. ATCC 27898	Schizokinen	Lammers and Sanders-Loehr (1982)
Anabaena sp. PCC 6411	Schizokinen *Aerobactin*	Goldmann *et al.* (1983)
Anabaena sp. PCC 7120	Schizokinen *Ferric aerobactin* *Ferrioxamin*	Goldmann *et al.* (1983)
Anabaena cylindrica Lemm 7122	*Schizokinen* Hydroxamate	Goldmann *et al.* (1983)
Anabaena cylindrica Lemm 1611	Hydroxamate	Goldmann *et al.* (1983)
Anabaena cylindrica	Anachelin	Beiderbeck *et al.* (2000), Ito *et al.* (2001)
Anacystis nidulans	Hydroxamate	McKnight *et al.* (1979)
Microcystis aeruginosa	Hydroxamate	McKnight *et al.* (1979)

Indicated are the siderophores which are produced by the different cyanobacteria. Siderophores not produced by the cyanobacteria but found to be transported are shown in italics.

The accumulation of schizokinen in the growth medium of *Anabaena* sp. PCC 7120 is proportional to cell density throughout the growth phase and continuous into stationary phase (Clarke *et al.*, 1987). Schizokinen can also bind copper but this complex cannot be incorporated. Based on this observation, a dual function of this siderophore was proposed. On the one hand, it is required for iron uptake and on the other hand it is required to prevent copper intoxication. A siderophore of similar structure has been observed in *Anabaena flos-aquae* and *A. nidulans* (Clarke *et al.*, 1987) and the isolation of a hydroxamate-type siderophore from an axenic culture of the cyanobacterium *Synechococcus* PCC 7002 (*A. quadruplicatum* PR6) was reported (Armstrong and van Baalen, 1979). Recently, the isolation and structural analysis of this siderophore was reported (Ito and Butler, 2005). It became evident that it indeed belongs to a family of siderophores with the same backbone structure as schizokinen. The three siderophores, named synechobactin A, B and C (Fig. 11.3) have a lauric acid, decanoic acid or a caprylic acid instead of the acetic acid found in schizokinen (Fig. 11.3). Structurally distinct siderophores are secreted by the fresh water cyanobacterium *Anabaena cylindrica,* namely the anachelins (Beiderbeck *et al.*, 2000; Ito *et al.*, 2001). These are catecholate-type siderophores with a central tripeptide consisting of L-Thr, D-Ser and L-Ser. Ito *et al.* (2001) described anachelin 1 and its esters as well as anachelin 2, whereas Beiderbeck *et al.* (2000) found anachelin H and anachelin 1. Gademann and Portmann (2008) discuss anachelin H as the native form. Based on the structure the authors suggested that this is the first siderophore exclusively produced by cyanobacteria (Beiderbeck *et al.*, 2000; Ito *et al.*, 2001).

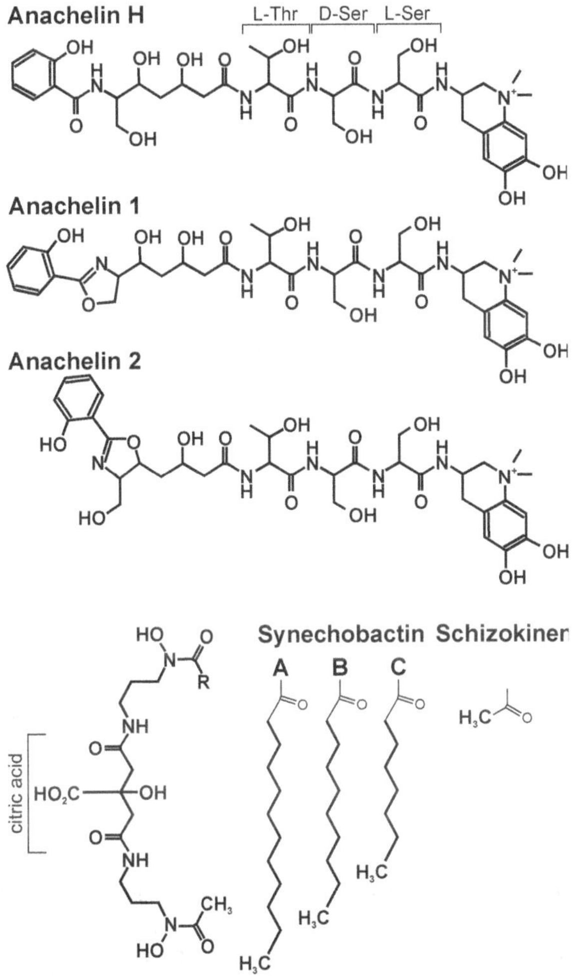

Figure 11.3 Structure of the siderophores secreted by cyanobacteria and identified so far. Shown is the structure of anachaline H, 1 and 2 (see text for description) secreted by *Anabaena cylindrica* (NIED-19) the backbone of synechobactin and schizokinen as well as the moiety differentiating between synechobactin A, B, C and schizokinen (the position is indicated by an R in the backbone).

In general, most siderophores are synthesized via a template-directed, nucleic acid-independent non-ribosomal mechanism involving two major classes of enzymes: non-ribosomal peptide synthetases (NRPSs) and polyketide synthetases (PKSs; e.g. Crosa and Walsh, 2002). The minimal core module of an NRPS consists of an adenylation (A) domain for the selection and the activation of amino acid monomers, a condensation (C) domain for catalysing the formation of peptide bonds and a thiolation or peptidyl carrier protein (P) domain with a swinging phosphopantetheine group for transferring the monomers or the growing peptide chain to various catalytic sites (e.g. Stachelhaus *et al.*, 1999; Welker and von Döhren, 2006). The PKSs catalyse the repetitive condensations of small carboxylic acid units, similar to fatty acid synthesis, leading to complex and diverse molecules. There are two main classes, class I consist of multifunctional proteins whereas class II are complexes of separate enzymes (Shen, 2003).

Jeanjean *et al.* (2008) described the first gene cluster involved in siderophore synthesis in *Anabaena* sp. PCC 7120, the only components known to be involved in the pathway to date. The cluster ranges from *all2635* encoding for a PKS to *all2658* coding for a sterol desaturase. Within the cluster there are seven genes coding for NRPS like proteins and two coding for PKS like proteins are present, which is indicative of a siderophore synthesis cluster. Interestingly, three genes coding for putative ABC transporter subunits of the HlyD family (Gentschev *et al.*, 2002) can also be found in the cluster (see below). A mutant in whom all NRPS encoding genes and one PKS encoding gene are deleted exhibited a reduced growth rate, a twofold reduction of intracellular iron and a tenfold reduction in secreted siderophores (Jeanjean *et al.*, 2008). The expression of the cluster is induced in response to iron starvation or oxidative stress, as exemplified for three selected genes. This again documents that in cyanobacteria the response to ROS and iron starvation is tightly interlinked. Unfortunately, the nature of the siderophore produced by the described cluster has remained elusive and Jeanjean and coworker even discuss the possibility of several different siderophores (Jeanjean *et al.*, 2008). Further, it is evident that the identified cluster is not responsible for the synthesis of the only structurally described siderophore schizokinen, which is structurally related to rhizobactin 1021. In *Sinorhizobium meliloti* rhizobactin 1021 is synthesized by proteins encoded by the *rhbABCDAEF* cluster regulated by RhrA (Lynch *et al.*, 2001) and a similar cluster can also be found in *Anabaena* sp. PCC 7120 (genes *all0390* to *all0396*; Nicolaisen *et al.*, 2008). Hence, the understanding of cyanobacterial siderophore content and synthesis is just emerging and remains to be extended in order to understand the response of cyanobacteria to iron limitation.

Photoreactivity of siderophores

Synechobactin secreted by the coastal marine cyanobacteria *Synechococcus* PCC 7002 was described to be a photoreactive amphiphilic siderophore (Ito and Butler, 2005). This conclusion was based on the observation that exposure of a Fe^{3+} complexed synechobactin to fluorescent light results in an iron-containing product with a molecular weight reduced by 46 Da. The same reduction of molecular weight by light exposure was reported for Fe^{3+} complexed petrobactin (Barbeau *et al.*, 2002). Photoreaction appears to be an important regulatory mechanism for iron acquisition in oceans. The molecular principles are best understood for aquachelins, which are similar to the siderophores produced by marine cyanobacteria (Barbeau *et al.*, 2001). Aquachelins contain one of a series of fatty-acid tails and a peptidic headgroup that forms a 1:1 Fe^{3+}:ligand complex via two hydroxamate groups and one β-hydroxyaspartate residue (Barbeau *et al.*, 2001). Remarkably, Fe^{3+}–aquachelin complexes readily undergo photolysis in natural sunlight. The nature of the Fe^{3+} binding groups is responsible for the photoreactivity of the siderophore, because hydroxamate groups are in general photochemically resistant regardless of Fe^{3+} complexation. Photolysis results in oxidative cleavage of the ligand and thereby in partial ligand breakdown and changes in iron speciation (Barbeau *et al.*, 2001). Two primary ligand photoproducts are formed, one more hydrophilic (the peptide portion) and one more hydrophobic (the fatty acid component) than the original Fe^{3+}–aquachelin complex. The photoreaction of the Fe^{3+}–aquachelin complexes is initiated by a β-hydroxyaspartate to Fe^{3+} (ligand-to-metal) charge-transfer reaction. During this process charges can photochemically be transferred from ligand to metal, which would generate a diurnal increase in Fe^{2+} and reactive Fe^{3+} (Barbeau *et al.*, 2003), affecting the bioavailability of siderophore-complexed Fe^{3+}. This appears to be an important regulatory mechanism and the fate of photolytically produced Fe^{2+} could include direct biological uptake by, for example, the FeoB system, oxidation to Fe^{3+}, or oxidation with subsequent complexation by other strong Fe^{3+} ligands, which is generally used by bacteria instead of the non-photolysed siderophore specific for only few species (Barbeau *et al.*, 2001).

Siderophore dependent iron uptake – the general mechanism

The general pathway of iron uptake was established mainly based on data existing for *E. coli* and *Pseudomonas aeruginosa* (for comparison read also Chapter 5). It starts with a high affinity association of a siderophore or protein (in case of lactoferin/transferin dependent systems) and a specific outer membrane transporter. Outer membrane transporters are also able to bind iron-free siderophores, but the biological function of this interaction is still unknown (Schalk *et al.*, 2004). Strikingly, organisms themselves often secrete only one siderophore, but many contain additional receptors for uptake of siderophores secreted by other species. Hence the diversity of iron uptake is not restricted to the pool of secreted siderophores. These outer membrane embedded receptors actively pump siderophores against a concentration gradient into the periplasm by utilizing an energy dependent transport mechanism

(Faraldo-Gómez *et al.*, 2003; Schalk *et al.*, 2004). They are supplied with energy from a cytoplasmic protein complex the so called Ton system (Faraldo-Gómez *et al.*, 2003). It consists of the integral membrane protein ExbB and the membrane anchored periplasmic proteins ExbD and TonB (Faraldo-Gómez *et al.*, 2003). The importance of TonB was first manifested in the 1970s by demonstrating that *tonB* mutants are unable to transport siderophores (Wang and Newton, 1971; Hantke and Braun, 1975). Based on this observation, the family of active outer membrane transport proteins has been termed TonB-dependent transporters.

TonB-dependent outer membrane receptors consist of a barrel-shaped membrane structure and contain a plug domain (Schalk *et al.*, 2004). This plug domain is located within the barrel of the outer membrane receptor blocking direct passage of siderphores across the outer membrane. Hence, siderophores bind first to an extracellular pocket (Schalk *et al.*, 2004). The energy from the interaction with TonB, which undergoes a conformational change, is then needed to cause a conformational change in the plug domain which either opens up a diffusion pore or directly drives the ligand into the periplasm (Faraldo-Gómez *et al.*, 2003). Further on, siderophore binding induces unwinding of the so called switch helix at the periplasmic site of the receptor, which represents the transmembrane signal that leads to the recruitment of TonB to the TonB box [TXXV(S/T)] of TonB-dependent transporters. The dislocation of the siderophore is energized by the chemiosmotic gradient of the plasma membrane transduced by TonB (Ferguson and Deisenhofer, 2002). The exact mechanism of this energy transfer, however, is still a matter of debate.

Once in the periplasm siderophores are further transported by an ABC transporter consisting of one or two proteins on the cytoplasmic membrane and an associated ATP-binding cassette protein (Ferguson and Deisenhofer, 2002). The transport of siderophores across the cytoplasmic membrane is less specific than over the outer membrane. It can be assumed that ABC permeases transport several different but structurally related substrates. FhuD is the best characterized siderophore binding protein. It contributes to a large extent to the substrate specificity of transport through the cytoplasmic membrane (Köster, 2001). FhuB is an extremely hydrophobic polytopic integral membrane protein. It interacts with FhuD, FhuC and transports different ferric hydroxamates (e.g. Köster and Braun, 1990; Köster and Böhm, 1992). FhuB folds autonomously into the lipid bilayer, its N- and C-terminus is located to the cytosol. FhuC in turn is an ATPase, which energizes the translocation process (Köster, 2001).

The expression of TonB-dependent iron transporters is regulated by the ferric-uptake regulator Fur (see above). However, one class of TonB-dependent iron transporter exhibits an additional possibility of gene regulation. They contain an N-terminal extension of around 70 amino acids, a so called transducer domain (Schalk *et al.*, 2004). With the help of this domain the association of the outer membrane transporter (abbreviated as OMT_N) with their iron loaded siderophore is sensed and the information transduced into the cell (Schalk *et al.*, 2004). The signal cascade initiated at the cell surface is transmitted to the cytoplasm by two downstream acting components, an inner membrane regulator protein and a cytoplasmic sigma factor belonging to the extracytoplasmic function family (ECF). Most outer membrane transporters have an empty binding site where the siderophore can bind, OMT_Ns instead are siderophore loaded and must either replace their iron-free siderophores or extract iron from the environment (Schalk *et al.*, 2004). For triggering the signal cascade the binding of iron-loaded siderophores to the OMT_N is sufficient, while transport itself is not required suggesting that both processes are uncoupled (Schalk *et al.*, 2004).

Siderophore-dependent iron uptake systems in cyanobacteria

As outlined above, iron uptake is induced upon iron starvation and repressed by iron in the medium at a concentration of about 1 μM. Cyanobacteria such as *Anabaena* sp. can accumulate iron against a concentration gradient (Lammers and Sanders-Loehr, 1982). This is indicative for the existence of an active transport/uptake system. The apparent K_m for ferric schizokinen uptake in *Anabaena* sp. is 0.04 μM (Lammers and Sanders-Loehr, 1982). Furthermore, iron uptake by cyanobacteria is sensitive to DCCD, an inhibitor of H^+-translocating ATPases in bacteria, mitochondria and chloroplasts (Lammers and Sanders-Loehr, 1982). This implies that ATP is required as energy source for the incorporation of iron into the cell, which is consistent with a model for siderophore dependent iron uptake based on the proteobacterial system. However, in cyanobacteria ATP is synthesized during endogenous respiration or in the dark by cyclic phosphorylation (Stanier and Cohen-Bazire, 1977). Therefore, it cannot be ruled out that the observed inhibition of iron uptake by DCCD is caused by interference with ATP formation or disturbance of the membrane potential, as both are required for iron transport as well.

The molecular nature of the iron uptake systems has been mainly investigated in three different cyanobacteria, *S. elongatus* PCC 7942, *Synechocystis* sp. PCC 6803 and *Anabaena* sp. PCC 7120. *S. elongatus* PCC 7942 and *Synechocystis* sp. PCC 6803 are unicellular cyanobacteria. In contrast, *Anabaena* sp. PCC 7120 is a filamentous cyanobacterium, which is able to fix atmospheric nitrogen (Nicolaisen *et al.*, 2009). Hence, as both types of cyanobacteria are investigated, the results observed may be extrapolated to other cyanobacteria.

As described above one of the genomic regions of *S. elongatus* PCC 7942 of which transcripts accumulated in iron-deficient growth conditions was *irpAB*. Downstream, but in opposite orientation to *irpAB* two genes have been found that code for proteins with similarity to SomA [SomA(1), gene *Synpcc7942_1463*] and SomB [SomB(1), gene *Synpcc7942_1464*; Nodop *et al.*, 2008]. SomA and SomB are outer-membrane proteins, which form porin-like beta-barrel structures (Hansel *et al.*, 1998). Interestingly, the genome of *S. elongatus* PCC 7942 contains an additional gene (*Synpcc7942_1607*) coding for a SomA like protein (SomA(2)). It was shown by microarray analysis that the transcript level of SomA(2) and SomB(1) is substantially increased upon iron starvation (Nodop *et al.*, 2008), but not the transcript of SomA(1). Furthermore, a Ftr1-like protein encoded by the gene *Synpcc7942_2421* was identified. Ftr1 is a permease of a high-affinity iron uptake system first identified in the plasma membrane of yeast (Stearman *et al.*, 1996; Larrondo *et al.*, 2007). Based on their microarray data Nodop *et al.* (2008) suggest that IrpA, IrpB, SomB(1) and SomA(2) and Ftr1 represent a novel iron acquisition system in *S. elongatus* PCC 7942, whose expression is regulated by IdiB. Despite this proposed pathway for iron uptake evidence for a second system has been found in *S. elongatus* PCC 7942. Genome analysis demonstrated the existence of three *fut* genes (*futA2, futB and futC;* Nodop *et al.*, 2008) which are expressed in iron dependent manner, as determined by microarray analysis (Nodop *et al.*, 2008), and a similar function as described for *Synechocystis* sp. PCC 6803 is proposed.

In *Synechocystis* sp. PCC 6803 the *fut* genes encode for subunits of an iron-transport system (Fig. 11.4; Katoh *et al.*, 2001a). This is composed of four subunits encoded by *futA1* (*slr1295*), *futA2* (*slr0513*), *futB* (*slr0327*) and *futC* (*sll1878*) (Katoh *et al.*, 2000, 2001a). All *fut* genes are constitutively expressed under normal conditions and show enhanced expression during iron depletion (Katoh *et al.*, 2001a). The simultaneous inactivation of these four genes did not result in a lethal phenotype even when *Synechocystis* sp. PCC 6803 was grown in media with low iron content. *FutA1 (slr1295)* and *futA2 (slr0513)* encode periplasmic binding proteins of an ATP-binding cassette (ABC) type iron transporter. It was shown that FutA1 can bind iron *in vitro* without an organic iron chelator such as a siderophore or citrate, similar to what is known for the Hit system in *Haemophilus* or the Fbp system in *Neisseria*. The natural iron source for the Fut system is still unclear. There are hints that the ligand for FutA2 is ferric ion as well (Katoh *et al.*, 2001b). In deletion mutants of *futA1* or *futA2* the level of iron uptake is only 37% and 84% that of the wildtype, respectively. In the double-mutant the uptake rate is even less than 5%, which suggests a redundant role of both gene products. Furthermore, the double mutant of *futA1* and *futA2* shows the same phenotype as a *futB* or *futC* single mutant. *futB* and *futC* encode for inner membrane bound or membrane-associated subunits. Both contain nucleotide-binding motifs and belong to the ABC transporter family. FutC was first identified to be involved in iron acquisition during analysis of potential nucleotide-binding components of ATP-binding cassette transporters (Katoh *et al.*, 2000). FutC seems to be a peripheral membrane protein (Katoh *et al.*, 2000). In line with a function in iron uptake, the knock-out mutant of *futC* shows a defect in iron acquisition. This transport system does not utilize citrate-iron chelates (Katoh *et al.*, 2000), so it is probably a complex between iron and a

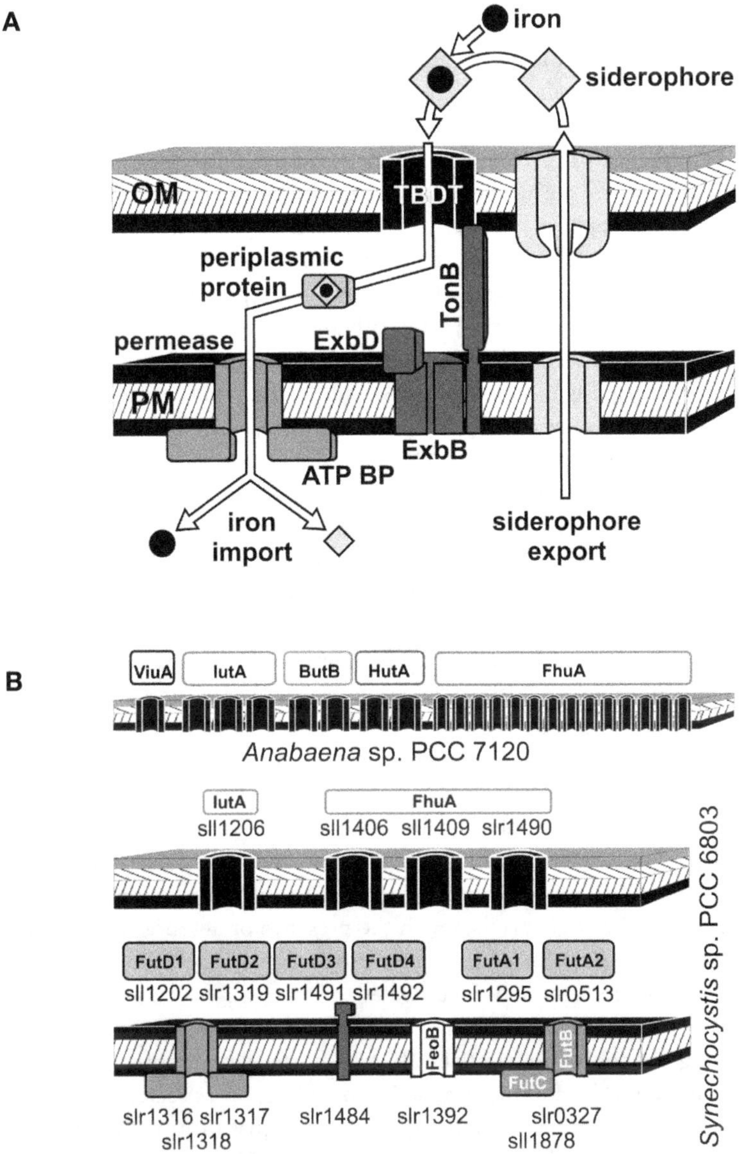

Figure 11.4 The iron transport system. In (A) the components generally involved in the siderophore mediated transport of iron across outer membrane, periplasm and plasma membrane are indicated. The same colour code is used in (B) where components of *Synechocystis* sp. PCC 6803 are shown. On top, the diversity of the TonB dependent transporter (TBDT) of the outer membrane of *Anabaena* sp. PCC 7120 is shown for comparison. Here, the number of the TonB dependent transporters with homology to the indicated transporter (according to Nicolaisen *et al.*, 2009; Mirus *et al.*, 2009) is indicated by the number of channels shown.

siderophore produced upon iron deprivation. However, at present there is no report on the production of siderophores by *Synechocystis* sp. PCC 6803.

Further components of putative iron transport systems were identified by sequence similarity in the genome of *Synechocystis* sp. PCC 6803. Sll1406, Sll1409, Slr1490, and Sll1206 are homologous to the TonB dependent outer membrane transporter, Sll1202, Slr1491, Slr1492, and Slr1319 share similarity with periplasmic components like FhuD (Fig. 11.4; Braun and Braun, 2002). Slr1316 and Slr1317 are two factors that are associated with or inserted in the plasma membrane like FhuB/C and Slr1392 was suggested to be a Feo-like plasma membrane transporter (Katoh *et al.*, 2001a). Here, only *sll1409, slr1319* and *slr1492* are constitutively expressed, all others being induced upon iron starvation (Katoh *et al.*, 2001a). Inactivation of these components did not result in a growth phenotype or an alteration of iron uptake in the presence of Ferrozine and only the knock-out of *slr1392* resulted in a reduced uptake in the presence of ascorbate when cells were grown in iron depleted medium (Katoh *et al.*, 2001a). However, as the siderophore specificity of the TonB dependent outer membrane transporter Sll1406, Sll1409, Slr1490, and Sll1206 is not yet established it remains elusive whether they recognize a siderophore produced by *Synechocystis* sp. PCC 6803 or a siderophore which might have been present in the natural habitat of these cyanobacteria. Particularly the latter model would explain why knock-out mutants of these genes have no effect under laboratory conditions.

A further model investigated is iron uptake by *Anabaena.* The uptake of iron utilising schizokinen was analysed for *Anabaena* sp. strains 6411 (ATCC27898), 7120, 7122 (ATCC27899) and 1611. Both, *Anabaena* sp. PCC 6411 and *Anabaena* sp. PCC 7120 are also able to take up aerobactin when supplied in high concentrations (Goldman *et al.*, 1983). *Anabaena cylindrica* sp. 7122 and 1611, possibly reflecting the same species, were not able to use schizokinen and are therefore considered to produce a different siderophore (Goldman *et al.*, 1983). However, the transport system for schizokinen was only recently identified. During a proteomic analysis of the outer membrane of *Anabaena* sp. PCC 7120 the TonB-dependent iron transporter Alr0397 was reported (Moslavac *et al.*, 2005, 2007). The expression of the corresponding gene is only moderately affected by iron limitation and it is not essential for viability. A generated insertion mutant exhibited a mild response to iron starvation and a diminished schizokinen-mediated iron uptake when compared with wild type (Nicolaisen *et al.*, 2008). This mild phenotype can be explained by the vast variety of TonB dependent transporters encoded by the genome of *Anabaena* sp. PCC 7120. In total, 22 genes have been assigned as putative TonB-dependent iron transporters (Mirus *et al.*, 2009). Remarkable in this context, only a single TonB exists (Mirus *et al.*, 2009). The identified TonB dependent transporters represent almost all known subclasses defined by their substrate molecules with the exception of the diferric dicitrate transporter of the FecA-type. Their expression pattern during iron and/or nitrogen starvation does not correlate with the presence or absence of Fur- and NtcA binding motifs (Mirus *et al.*, 2009). Hence, the mode of regulation and may be the link to other transcriptional activators such as IsiB or IdiB remains to be investigated.

The knowledge on the periplasmic plasma membrane localized proteins is rather limited. As described above, three genes coding for putative ABC transporter subunits of the HlyD family are within the cluster proposed to be involved in the production of siderophores secreted by *Anabaena* sp. PCC 7120 (Jeanjean *et al.*, 2008). Hence, it is tempting to speculate that these three genes encode a system possibly involved in the secretion of the siderophore produced. However, this function, as well as the identification of the other plasma membrane components involved in iron uptake, remains to be further approached.

Iron storage

Iron can be stored in cells complexed to bacterioferritins, proteins that are involved in the normal iron metabolism but at the same time in protection against iron intoxication. This iron pool can count up to 50% of the cellular iron quota (Keren *et al.*, 2004). The *Synechocystis* sp. PCC 6803 genome encodes two iron storage proteins, which are both members of the bacterioferritin (Bfr) family (Bertani *et al.*, 1997). Bfrs contain a ferrioxidase centre and a haem moiety is bound at the interface between the two subunits (Carrondo, 2003). However, the role of the haem group is poorly understood. Bfrs store iron in the cavity at the centre of their 24-mer ultrastructure and iron enters the Bfr complex as Fe^{2+} and is oxidized on its way to the central cavity.

Synechocystis Bfr genes belong to a subfamily in which one gene encodes for a protein with a conserved haem ligand and the other codes for a protein with conserved di-iron centre ligands (Bertani *et al.*, 1997). Mass spectrometric analysis of isolated Bfr complexes showed that they consist of the two proteins Sll1341 (BfrA) and Slr1890 (BfrB). Knock-out mutants of each gene or both genes together have been generated. In iron-depleting media the single knock-out mutants show a growth rate half to that of the wild-type. Analysing the fluorescence emission spectra at 77 K to detect the presence of CP43′ it became clear that also the knock-out mutants grown in normal growth media sense iron starvation. Hence, the two Bfr proteins are necessary for iron accumulation because a single one could not build a functional dimer with a haem or di-iron cluster. Atomic absorption spectroscopy revealed that in the early exponential growth phase of a culture the iron content of the mutant strains is reduced to 70% in comparison to wildtype ($9.3 \times 10^6 \pm 7.5 \times 10^5$ atoms/cell). This is consistent with the observation that iron is stored in the cells quite early in the exponential phase and then used in the stationary phase when the culture becomes dense and iron a limiting factor (Keren *et al.*, 2004). Beside the proteins in *Synechocystis*, ferritin homologues either with haem-binding or di-iron binding function have been found in *Synechococcus* sp. PCC 7002, *P. marinus* MED4, *T. elongatus* Bp-1, *G. violaceus* PCC 7421 and *Anabaena* sp. PCC 7120 (Keren *et al.*, 2004).

Another class of proteins involved in intracellular iron storage are members of the DNA-binding protein from stationary phase cells (Dps) family, which play an important role in protecting cells from oxidative and nutritional stress (Hernández *et al.*, 2007). Dps proteins are divergent members of the bacterioferritin/ferritin superfamily (Pena and Bullerjahn, 1995), and several homologues have been identified in different bacteria, playing roles in iron storage and DNA protection (Gupta and Chaterji, 2003). Depending on the nutritional status of the cells and the pressure of environmental conditions, Dps proteins can act as ferritins and as a defence against oxidative stress via either the ability in non-specific DNA-binding or the function in protection of DNA from Fenton-mediated damage (Hernández *et al.*, 2007). In *Anabaena* sp. PCC 7120 the ORF *alr3808* has high similarity to *dpsA* from *Synechococcus* sp. PCC 7942 and the characteristics of these two proteins are very similar. In *Synechococcus* DpsA was described as a DNA-binding haem-protein possessing a weak catalytic activity, with no ferrioxidase activities (Pena and Bullerjahn, 1995). It confers resistance to peroxide damage during periods of oxidative stress and long term nutrient starvation (Sen *et al.*, 2000; Latifi *et al.*, 2009). Electromobility shift assays showed that FurA binds specifically to the *dpsA* promoter and that transcription and translation are down-regulated in FurA overexpressing mutants. Therefore the DpsA in its role as iron-storage protein is regulated by FurA (Hernández *et al.*, 2007).

Iron dependency and transport by cyanobacteria on the day after tomorrow

More than 30 years ago the first reports on siderophore secretion and iron uptake by cyanobacteria were presented (Simpson and Neilands, 1976; Armstrong and van Baalen, 1979; Lammers and Sanders-Loehr, 1982; Goldmann *et al.*, 1983). Since then the major focus was placed towards the understanding of the response to iron starvation by cyanobacteria, particularly of photosystems (Fig. 11.1) and general morphology changes, as documented by a large number of publications. Only recent studies reflected on the molecular mechanisms of iron uptake systems including siderophore synthesis, iron acquisition systems and iron storage (Figures 11.3 and 11.4). One of the best explored systems is *Synechocystis* sp. PCC 6803 (e.g. Katoh *et al.*, 2001b). However, initial bioinformatic analysis strongly suggests that both, the response to iron starvation and the incorporation of iron into the cell are not uniform in cyanobacteria (Figures 11.2 and 11.4; Mirus *et al.*, 2009). Hence, in the future it will be important to understand which systems are generally present and which are unique for individual cyanobacteria. For the latter it will be interesting to understand whether the habitat or the genus defines the system(s) present. Furthermore, even in the best understood system *Synechocystis* sp. PCC 6803 direct evidence as well as *in vitro* and *in vivo* characterization of the functional modules are largely lacking, as for example the substrates for the four TonB dependent transporters are still elusive. Beside the molecular mechanism of iron transport across the cell envelope, the understanding of siderophore synthesis is just at its beginning and the recently published characterization of the synthesis operon in *Anabaena* sp. PCC 7120 (Jeanjean *et al.*, 2008) marks just a first step in understanding this process.

A very exciting observation is the crossover between different regulatory networks, although our understanding of the individual networks induced by nitrogen starvation (e.g. Schwarz and Forchhammer, 2005; Zhang *et al.*, 2006; Nicolaisen *et al.*, 2009) or iron starvation (Fig. 11.2) and on the links between them needs further refinement. Therefore, in the future it will be important to explore the molecular components, the mechanistic details and the regulatory network involved to obtain a larger understanding of iron metabolism of cyanobacteria.

Acknowledgements

Special thanks to Cheng-Cai Zhang and Enrique Flores for critical reading and helpful comments, to M.T. Bohnsack and A. Hahn for helpful discussions and careful reading of the manuscript. Financial support from EMBO (short term fellowship) to K.N., from the Deutsche Forschungsgemeinschaft (DFG, SFBTR1-C7) to E.S. and the Volkswagenstiftung to E.S. is acknowledged.

References

Andrews, S.C., Robinson, A.K., and Rodriguez-Quinones, F. (2003). Bacterial iron homeostasis. FEMS Microbiol. Rev. *27*, 215–237.

Andrizhiyevskaya, E.G. Schwabe, T.M., Germano, M., D'Haene, S., Kruip, J., and van Grondelle, R. (2002) Spectroscopic properties of PSI–IsiA supercomplexes from the cyanobacterium *Synechococcus* PCC 7942. Biochim. Biophys. Acta. *1556*, 265–272.

Armstrong. J.E., and van Baalen, C. (1979). Iron transport in microalgae: The isolation and biological activity of a hydroxamate siderophore from the blue-green algae *Agmenellum quadruplicatum*. J. Gen. Microbiol. *111*, 253–262.

Badarau, A., Firbank, S.J., Waldron, K.J., Yanagisawa, S., Robinson, N.J., Banfield, M.J., and Dennison, C. (2008). FutA2 is a ferric binding protein from *Synechocystis* PCC 6803. J. Biol. Chem. *283*, 12520–12527.

Bagchi, S.N., Pistorius, E.K., and Michel, K.P. (2003). A *Synechococcus* sp. PCC 7942 mutant with a higher tolerance towards bentazone. Photosynth. Res. *75*, 171–182.

Bagg, A., and Neilands, J.B. (1987). Ferric uptake regulation protein acts as a repressor, employing iron (II) as a cofactor to bind the operator of an iron transport operon in *Escherichia coli*. J. Biochem. *26*, 5471–5477.

Balasubramanian, R., Shen, G., Bryant, D.A., and Golbeck, J.H. (2006). Regulatory roles for IscA and SufA in iron homeostasis and redox stress responses in the cyanobacterium *Synechococcus* sp. strain PCC 7002. J. Bacteriol. *188*, 3182–3191.

Barbeau, K., Rue, E.L., Bruland, K.W., and Butler, A. (2001). Photochemical cycling of iron in the surface ocean mediated by microbial iron(III)-binding ligands. Nature *413*, 409–413.

Barbeau, K., Zhang, G., Live, D.H., and Butler, A. (2002). Petrobactin, a photoreactive siderophore produced by the oil-degrading marine bacterium *Marinobacter hydrocarbonoclasticus*. J. Am. Chem. Soc. *124*, 378–379.

Barbeau, K., Rue, E.L., Trick, C.G., Bruland, K.W., and Butler, A. (2003). Photochemical reactivity of siderophores produced by marine heterotrophic bacteria and cyanobacteria based on characteristic Iron(III) binding groups. Limnol. Oceanogr. *48*, 1069–1078.

Beiderbeck, H., Taraz, K., Budzikiewicz, H., and Walsby, A.E. (2000). Anachelin, the siderophore of the cyanobacterium *Anabaena cylindrica* CCAP 1403/2A. Z. Naturforschung. *55*, 681–687.

Bejerrum, C.J., and Canfield, D.E. (2002). Ocean productivity before about 1.9 Gyr ago limited by phosphorus assorption onto iron oxides. Nature *417*, 159–162.

Berman-Frank, I., Quigg, A., Finkel, Z.V., and Irwin, A.L. (2007). Nitrogen-fixation strategies and Fe requirements in cyanobacteria. Limnol. Oceanogr. *52*, 2260–2269.

Bertani, L.E., Huang, J.S., Weir, B.A., and Kirschvink, J.L. (1997). Evidence for two types of subunits in the bacterioferritin of *Magnetospirillum magnetotacticum*. Gene. *424*, 1051–1054.

Bes, M.T., Hernandez, J.A., Peleato, M.L., and Fillat, F.F. (2001). Cloning, overexpression and interaction of recombinant Fur from the cyanobacterium *Anabaena* sp. PCC 7119 with *isiB* and its own promoter. FEMS Microbiol. Lett. *194*, 187–192.

Bibby, T.S., Nield, J., and Barber, J. (2001a). Iron deficiency induces the formation of an antenna ring around trimeric photosystem I in cyanobacteria. Nature *412*, 743–745.

Bibby, T.S., Nield, J., and Barber, J. (2001b). Three-dimensional model and characterization of the iron-stress induced CP43′-photosystem I supercomplex isolated from the cyanobacterium *Synechocystis* PCC 6803. J. Biol. Chem. *22*, 22–30.

Boekema, E.J., Hifney, A., Yakushevska, A.E., Piotrowski, M., Keegstra, W., Berry, S., Michel, K.P., Pistorius, E.K., and Kruip, J. (2001). A giant chlorophyll–protein complex induced by iron deficiency in cyanobacteria. Nature *412*, 745–748.

Boussiba, S., and Richmond, A.E. (1980), C-phycocyanin as a storage protein in the blue-green alga *Spirulina platensis*. Arch. Microbiol. *125*, 143–147.

Boyer, G.L., Sullivan, J.J., Andersen, R.J., Harrison, P.J., and Taylor, F.J.R. (1987) Effects of nutrient limitation on toxin production and composition in the marine dinoflagellate *Protogonyaulax tamarensis*. Mar. Biol. *96*, 123–128.

Braun, V., Schäffer, S., Hantke, K., and Tröger, W. (1990). Regulation of gene expression by iron. In The Molecular Basis of Bacterial Metabolism, Hauska, G., and Thauer, R. eds. (New York, USA: Springer), pp. 164–179.

Braun, V., and Braun, M. (2002). Active transport of iron and siderophore antibiotics. Curr. Opin. Microbiol. *5*, 194–201.

Brown, C.M., and Trick, C.G. (1992). Response of the cyanobacterium, *Oscillatoria tenuis*, to low iron environments: the effect on growth rate and evidence for siderophore production. Arch. Microbiol. *157*, 349–354.

Brown, I.I., Mummey, D., and Cooksey, K.E. (2005). A novel cyanobacterium exhibiting an elevated tolerance for iron. FEMS Microbiol. Lett. *52*, 307–314.

Carrondo, M.A. (2003). Ferrritins, iron uptake and storage from the bacterioferritin point of view. EMBO J. *22*, 1959–1968.

Chang, C.F., Shuman, H., and Somlyo, A.P. (1986) Electron probe analysis, X-ray mapping, and electron energy-loss spectroscopy of calcium, magnesium, and monovalent ions in log-phase and in dividing Escherichia coli B cells. *J Bacteriol. 167*, 935–939.

Cheng, L.J., Shi, L., Latifi, A., and Zhang, C.C. (2006). A pair of iron-responsive genes encoding protein kinases with a Ser/thr-kinase domain and a His-kinase domain are regulated by NtcA in the cyanobacterium *Anabaena sp. PCC 7120*. J. Bacteriol. *188*, 4822–4829.

Clarke, S.E., Stuart, J., and Sanders-Loehr, J. (1987). Induction of siderophore activity in *Anabaena* spp. and moderation of copper toxicity. Appl. Environ. Microbiol. *53*, 917–922.

Cody, G.D., Boctor, N.Z., Filley, T.R., Hazen, R.M., Scott,J.H., Sharma, A., and Yoder, H.S., Jr. (2000). Primordial carbonylated iron-sulfur compounds and the synthesis of pyruvate. Science. *289*, 1337–1340.

Crosa, J.H., and Walsh, C.T. (2002). Genetics and assembly line enzymology of siderophore biosynthesis in bacteria. Microbiol. Mol. Biol. Rev. *66*, 223–249.

Doucette, G.J., Erdner, D.L., Peleato, M.L., Hartmann, J.J., and Anderson, D.M. (1996). Quantitative analysis of iron stress related proteins in *Thalassiosira weissflogii*: measurement of flavodoxin and ferredoxin using HPLC. Mar. Ecol. Prog. Ser. *130*, 269–276.

Durham, K.A., Porta, D., McKay, R.M.L., and Bullerjahn, G.S. (2003). Expression of the iron-responsive *irpA* gene from the cyanobacterium *Synechococcus* sp. strain PCC 7942. Arch. Microbiol. *179*, 131–134.

Escolar, L. de Lorenzo, V., and Perez-Martin, J. (1997). Metalloregulation *in vitro* of the aerobaction promoter in *Escherichia coli* by the FUR (ferric uptake regulator) protein. Mol. Microbiol. *26*, 799–808.

Escolar, L., Perez-Martin, J., and de Lorenzo, V. (1998). Binding of the Fur (ferric uptake regulator) repressor of *Escherichia coli* to arrays of the GATAAT sequence. J. Mol. Biol. *283*, 537–547.

Exss-Sonne, P., Tölle, J., Bader, K.P., Pistorius, E.K., and Michel, K.P. (2000). The IdiA protein of *Synechococcus* sp. PCC 7942 functions in protecting photosystem II under oxidative stress. Photosynth. Res. *63*, 145–157.

Faraldo-Gómez, J.D., and Sansom, M.S.P. (2003). Acquisition of siderophores in gram-negative bacteria. Nat. Rev. Mol. Cell Biol. *4*, 105–116.

Ferguson, A.D., and Deisenhofer, J. (2002). TonB-dependent receptors – structural perspectives. Biochim. Biophys. Acta. *1565*, 318–332.

Ferreira, F., and Strauss, N.A. (1994). Iron deprivation in cyanobacteria. J. Appl. Phycol. *6*, 199–210.

Fillat, M.F., Sandmann, G., and Gomez-Moreno, C. (1988). Flavodoxin from the nitrogen-fixing cyanobacterium *Anabaena* PCC 7119. Arch. Microbiol. *150*, 160–164.

Fillat, M.F., Edmondson, D.E., and Gomez-Moreno, C. (1990). Structural and chemical properties of a flavodoxin from *Anabaena* PCC 7119. Biochim. Biophys. Acta. *1040*, 301–307.

Gademann, K., and Portmann, C. (2008). Secondary metabolites from cyanobacteria: complex structures and powerful bioactivities. Curr. Org. Chem. *12*, 326–341.

Gentschev, I., Dietrich, G., Goebel, W.. (2002) The *Eschericha coli* alpha-hemolysin secretion system and its use in vaccine development. Trends Microbiol. *10*, 39–45.

Ghassemian, M., and Straus, N.A. (1996). Fur regulates the expression of iron-stress genes in the cyanobacterium *Synechococcus* sp. strain PCC 7942. J. Microbiol. *142*, 1469–1476.

Goldman, S.J., Lammers, P.J., Berman, M.S., and Sanders-Loehr, J. (1983). Siderophore-mediated uptake in different strains of *Anabaena* sp. J. Bacteriol. *156*, 1144–1150.

Guikema, J.A., and Sherman, L.A. (1983). Influence of iron deprivation on the membrane composition of anacystis nidulans 1. Plant Physiol. *73*, 250–256.

Gupta, S., and Chaterji, D. (2003). Bimodal protection of DNA by *Mycobacterium smegmatis* DNA-binding protein from stationary phase cells. J. Biol. Chem. *278*, 5235–5241.

Hall, H.K., and Foster, J.W. (1996). The role of FUR on the acidic tolerance response of *Salmonella typhimurium* is physiologically and genetically separable from its role in iron acquisition. J. Bacteriol. *178*, 5863–5691.

Hansel, A., Pattus, F., Jürgens, U.J., and Tadros, M.H. (1998). Cloning and characterization of the genes encoding for two porins in the unicellular cyanobacterium *Synechococcus* sp. PCC 6803. Biochim. Biophys. Acta. *1399*, 31–39.

Hantke, K., and Braun, V. (1975). Membrane receptor dependent iron transport in *Escherichia coli*. FEBS Lett. *49*, 301–305.

Hantke, K. (1981). Regulation of ferric iron transport in *Escherichia coli* K12: isolation of a constitutive mutant. Mol. Gen. Genet. *182*, 288–292.

Hantke, K. (1987). Selection procedure for deregulated iron transport mutants (*fur*) in *Escherichia coli* K 12: fur not only affects iron metabolism. Mol. Gen. Genet. *210*, 135–139.

Hantke, K. (2001). Iron and metal regulation in bacteria. Curr. Opin. Microbiol. *4*, 172–177.

Hardie, L.P., Balkwill, D.L., and Stevens, S.E. (1983). Effects of iron starvation on the ultrastructure of cyanobacterium *Agmenellum quadruplicatum*. Appl. Environ. Microbiol. *45*, 1007–1017.

Hernández, J.A., López-Gomollon, S., Bes, M.T., Fillat, M.F., and Peleato, M.L. (2004a). Three *fur* homologues from *Anabaena* sp. PCC 7120: exploring reciprocal protein-protein promoter recognition. FEMS Microbiol. Lett. *236*, 275–282.

Hernández, J.A., Peleato, M.L., Fillat, M.F., and Bes, M.T. (2004b). Heme binds to and inhibits the DNA-binding activity of the global regulator FurA from *Anabaena* sp. PCC 7120. FEBS Lett. *577*, 35–41.

Hernández, J.A., Muro-Pastor, A.M., Flores, E., Bes, M.T., Peleato, M.L., and Fillat, M.F. (2006a). Identification of a *furA* cis antisense RNA in the cyanobacterium *Anabaena* sp. PCC 7120. J. Mol. Biol. *355*, 325–334.

Hernández, J.A., López-Gomollón, S., Muro-Pastor, A., Valladares, A., Bes, M.T., Pelleato, M.L., and Fillat, M.F. (2006b). Interaction of FurA from *Anabaena* sp. PCC 7120 with DNA: A reducing environment and the presence of $Mn^{2+}$are positive effectors in the binding to *isiB* and *furA* promoters. Biometals *19*, 259–268.

Hernández, J.A., Pellicer, S., Huang, L., Peleato, M.L., and Fillat, M.F. (2007). FurA modulates gene expression of *alr3808*, a DpsA homologue in *Nostoc* (*Anabaena*) sp. PCC 7120. FEBS Lett. *581*, 1351–1356.

Horsburgh, M.J., Ingham, E., and Forster, S.J. (2001). In *Staphylococcus aureus*, Fur is an interactive regultor with PerP, contributes to virulence, and is necessary for oxidative stress resistance through positive regulation of catalase and iron homeostasis. J. Bacteriol. *183*, 468–475.

Ito, Y., Okada, S., and Murakami, M. (2001). Two structural isomeric siderophores from the freshwater cyanobacterium *Anabaena cylindrica* (NIES-19). Tetrahedron. *57*, 9093–9099.

Ito, Y., and Butler, A. (2005). Structure of synechobactins, new siderophores of the marine cyanobacterium *Synechococcus* sp. PCC 7002. J. Limnol. Oceanogr. *50*, 1918–1923.

Ivanov, A.G., Park, Y.-I., Miskiewicz, E., Ravem, J.A., Huner, N.P. A., and Öqusit. G. (2001). Iron stress restricts photosynthetic intersystem electron transport in *Synechoccus* sp. PCC 7942. FEBS Lett. *485*, 173–177.

Ivanov, A.G., Krol, M., Sveshnikov, D., Selstam, E., Sandström, S., Koochek, M., Park, Y.-I., Vasilev, S., Bruce, D. Öquist, G., and Huner, N.P.A. (2006). Iron deficiency in cyanobacteria causes monomerization of photosystem I trimes and reduces the capacity for state transitions and the effective absorption cross section of photosystem I *in vivo*. Plant Physiol. *141*, 1436–1445.

Jeanjean, R., Talla, E., Latifi, A., Havaux, M., Janicki, A., and Zhang, C.C. (2008). A large gene cluster encoding peptide synthetases and polyketide synthases is involved in production of siderophores and oxidative stress response in the cyanobacterium *Anabaena* sp. PCC 7120. Environ. Microbiol. *10*, 2574–2585.

Johnston, A.W., Todd, J.D., Curson, A.R., Lei, S., Nikolaidou-Katsaridou, N., Gelfand, M.S., and Rodionov, D.A. (2007). Living without Fur: the subtlety and complexity of iron-responsive gene regulation in the symbiotic bacterium *Rhizobium* and other alpha-proteobacteria. Biometals *20*, 501–511.

Jordan, P., Fromme, P., Witt, H.T., Klukas, O., Saenger, W., and Krauss, N. (2001). Three-dimensional structure of cyanobacterial photosystem I at 2.5 A resolution. Nature *411*, 909–917.

Kaneko, T., Sato, S., Kotani, H., Tanaka, A., Asamizu, E., Nakamura, Y., Miyajima, N., Hirosawa, M., Sugiura, M., Sasamoto, S., Kimura, T., hosuchi, T., Matsuno, A., Muraki, A., Nakazaki, N., Naruo, K., Okumura, S., Shimpo, S. Taheuchi, C., Wasa, T., Watanabe, A., Yamada, M., Yasuda, M., and Tabata, S. (1996). Sequence analysis of the unicellular cyanobacterium *Synechocystis* sp. strain PCC 6803. II. Sequence determination of the entire genome and assignment of potential protein-coding regions. DNA Res. *3*, 109–136.

Katoh, H., Grossmann, A.R., Hagino, N., and Ogawa, T. (2000). A gene of *Synechocystis* sp. strain PCC 6803 encoding a novel iron transporter. J. Bacteriol. *182*, 6523–6524.

Katoh, H., Haigon, N., Grossmann, A.R., and Ogawa, T. (2001a). Genes essential to iron transport in the cyanobacterium *Synechocystis* sp. strain PCC 6803. J. Bacteriol. *183*, 2779–2784.

Katoh, H., Hagino, N., and Ogawa, T. (2001b). Iron-binding activity of FutA1 subunit of an ABC-type iron transporter in the cyanobacterium *Synechocystis* sp. strain PCC 6803. Plant Cell Physiol. *42*, 823–827.

Keren, N., Aurora, R., and Pakrasi, H.B. (2004). Critical roles of bacterioferritins and iron storage and proliferation in cyanobacteria. Plant Physiol. *135*, 1666–1673.

Kerry, A., Laudenbach, D.L., and Trick, C.G. (1988). Influence of iron limitation and nitrogen source on growth and siderophore production by cyanobacteria. J. Phycol. *24*, 566–571.

Keyer, K., and Imlay, J.A. (1996). Superoxide accelerates DNA damage by elevating free-iron levels. Proc. Natl. Acad. Sci. U.S.A. *93*, 13635–13640.

Koropatkin, N., Randich, A.M., Bhattacharyya-Pakrasi, M., Pakrasi, H.B., and Smith, T.J. (2007). The structure of the iron-binding protein, FutA1, from *Synechocystis* 6803. J. Biol. Chem. *282*, 27468–27477.

Köster, W., and Braun, V. (1990). Iron(III) hydroxamate transport of *Escherichia coli*: restoration of iron supply by coexpression of the N- and C-terminal halves of the cytoplasmic membrane protein FhuB cloned on separate plasmids. Mol. Gen. Genet. *223*, 379–384.

Köster, W., and Böhm, B. (1992). Point mutations in two conserved glycine residues within the integral membrane protein FhuB affect iron(III) hydroxamate transport. Mol. Gen. Genet. *232,* 399–407.

Köster, W. (2001). ABC transporter-mediated uptake of iron, siderophores, heme and vitamin B12. Res. Microbiol. *152,* 291–301.

Kouril, R., Arteni, A.A., Lax, J., Yeremenko, N., D'Haene, S., Rögner, M., Matthijs, H.C.P., Dekker, J.P., and Boekema, E.J. (2005). Structure and functional role of supercomplexes of IsiA and photosystem I in cyanobacterial phototsynthesis. FEBS Lett. *579,* 3253–3257.

Kunert, A., Vinnemeier, J., Erdmann, N., and Hagemann, M. (2003). Repression by Fur is not the main mechansim controlling the iron-inducible *isiAB* operon in the cyanobacterium *Synechocystis* sp. PCC 6803. FEMS Microbiol. Lett. *227,* 255–262.

Küpper, H., Setlík, I., Seibert, S., Prásil, O., Setlikova, E., Strittmatter, M., Levitan, O., Lohscheider, J., Adamska, I., and Berman-Frank, I. (2008). Iron limitation in the marine cyanobacterium *Trichodesmium* reveals new insights into regulation of photosynthesis and nitrogen fixation. New Phytol. *179,* 784–798.

Kustka, A., Carpenter, E.J., and Sanudo-Wilhelmy, S.A. (2002). Iron and marine nitrogen fixation: Progress and future directions. Res. Microbiol. *153,* 255–262.

Lammers, P.J., and Sanders-Loehr, J. (1982). Active transport for ferric schizokinen in *Anabaena* sp. J. Bacteriol. *151,* 288–294.

Larrondo, L.F., Canessa, P., Meol, F., Polanco, R., and Vicuna, R. (2007). Cloning and characterization of the genes encoding the high-affinity iron-uptake protein complex Fet3/Ftr1 in the basidiomycete *Phanerochaete chryosporium.* J. Microbiol. *153,* 1772–1780.

Latifi, A., Jeanjean, R., Lemeille, S. Havaux, M., and Zhang, C.C. (2005). Iron starvation leads to oxidative stress in *Anabaena* sp. strain PCC 7120. J. Bacteriol. *187,* 6596–6598.

Latifi, A., Ruiz, M., and Zhang, C.C. (2009). Oxidative stress in cyanobacteria. FEMS Microbiol. Rev. 33, 258–78.

Laudenbach, D.E., Reith, M.E., and Straus, N.A. (1988). Isolation, sequence analysis, and transcriptional studies of the flavodoxin gene from *Anacystis nidulans* R2. J. Bacteriol. *170,* 258–265.

Lax, J.E., Arteni, A.A., Boekema, E.J., Pistorius, E.K., Michel, K.P., and Rögner, M. (2007). Structural response of Photosystem 2 to iron deficiency: characterization of a new photosystem 2–IdiA complex from the cyanobacterium *Thermosynechococcus elongatus* BP-1. Biochim. Biophys. Acta. *1767,* 528–534.

Lewis, B.L.O., Holt, P.D., Taylor, S.W., Wilhelm, S.W., Trick, C.G., Butler, A., and Luther III, G.W. (1995). Voltammetric estimation of iron(III) thermodynamic stability constants for catecholate siderophores isolated from marine bacteria and cyanobacteria. Mar. Chem. *50,* 179–188.

Litwin, C.M., and Calderwood, S.B. (1994). Analysis of the complexity of gene regulation by Fur in *Vibrio choleraee.* J. Bacteriol. *176,* 240–248.

López-Gomollon, S., Hernández, J.A., Wolk, C.P., Peleato, M.L., and Fillat, M.F. (2007a). Expression of *furA* is modulated by NtcA and strongly enhanced in heterocysts of *Anabaena* sp. PCC 7120. J. Microbiol. *153,* 42–50.

López-Gomollon, S., Hernández, J.A., Pellicer, S., Angarica, V.E., Peleato, M.L., and Fillat, M.F. (2007b). Cross-talk between iron and nitrogen regulatory networks in *Anabaena* (*Nostoc*) sp. PCC 7120: Identification of overlapping genes in FurA and NtcA regulons., J. Mol. Biol. *374,* 267–281.

Lynch, D., O'Brien, J., Welch, T., Clarke, P., Cuív, P.O., Crosa, J.H., and O'Connell, M. (2001). Genetic organization of the region encoding regulation, biosynthesis, and transport of rhizobactin 1*021,* a siderophore produced by *Sinorhizobium meliloti.* J. Bacteriol. *183,* 2576–2585.

Marlovits, T.C., Haase, W., Herrmann, C., Aller, S.G., and Unger, V.M. (2002). The membrane protein FeoB contains an intramolecular G protein essential for Iron(II) uptake in bacteria. Proc. Natl. Acad. Sci. U.S.A. *99,* 16243–16248.

Massé, E., Salvail, H., Desnoyers, G., and Arguin, M. (2007). Small RNAs controlling iron metabolism. Curr. Opin. Microbiol. *10,* 140–145.

McKnight, D.M., and Morel, F.M.M. (1979). Release of weak and strong copper-complexing agents by algae. J. Limnol. Oceanogr. *24,* 823–837.

Michel, K.P., Thole, H.H., and Pistorius, E.K. (1996). IdiA, a 34 kDa protein in the cyanobacteria *Synechococcus* sp. strains PCC 6301 and PCC 7942, is required for growth under iron and manganese limitations. J. Microbiol. *142,* 2635–2645.

Michel, K.P., Exss-Sonne, P., Scholten-Beck, G., Kahmann, U., Ruppel, H.G., and Pistorius, E.K. (1998). Immunocytochemical localization of IdiA, a protein expressed under iron or manganese limitation in the mesophilic cyanobacterium *Synechococcus* PCC 6301 and the thermophilic cyanobacterium *Synechococcus elongatus.* Planta. *205,* 73–81.

Michel, K.P., Kruger, F., Puhler, A., and Pistorius, E.K. (1999). Molecular characterization of *idiA* and adjacent genes in the cyanobacteria *Synechococcus* sp. strains PCC 6301 and PCC 7942. J. Microbiol. *145,* 1473–1484.

Michel, K.P., Pistorius, E.K., and Golden, S.S. (2001). Unusual regulatory elements for iorn deficiency induction of the *idiA* gene of *Synechococcus elongatus* PCC 7942. J. Bacteriol. *183,* 5015–5024.

Michel, K.P., Berry, S., Hifney, A., Kruip, J., and Pistorius, E.K. (2003). Adaptation to iron deficiency: a comparison between the cyanobacterium *Synechococcus elongatus* PCC 7942 wild-type and a DpsA-free mutant. Photosynth Res. *75,* 71–84.

Michel, K.P., and Pistorius, E.K. (2004). Adaptation of the photosynthetic electron transport chain in cyanobacteria to iron deficiency: the function of IdiA and IsiA. Physiol. Plant. *120,* 36–50.

Mirus, O., Strauss, S., Nicolaisen, K., von Haesseler, A., and Schleiff, E. (2009). Classification of TonB-dependent transporters and their eoccurence in cyanobacteria. BMC Biol. *12*, 7–86.

Morel, F.M., and Price, N.M. (2003). The biogeochemical cycles of trace metals in the oceans. Science. *300*, 944–947.

Moslavac, S., Bredemeier, R., Mirus, O., Granvogl, B., Eichacker, L.A., and Schleiff, E. (2005). Proteomic analysis of the outer membrane of *Anabaena* sp. strain PCC 7120. J. Proteome Res. *4*, 1330–1338.

Moslavac, S., Reisinger, V., Berg, M., Mirus, O., Vosyka, O., Plöscher, M., Flores, E., Eichacker, L.A., and Schleiff, E. (2007). The proteome of the heterocyst cell wall in *Anabaena* sp. PCC 7120. J. Biol. Chem. *388*, 823–829.

Nakamura, Y., Kaneko, T., Hirosawa, M., Miyajima, N., and Tabata, S. (1998). CyanoBase, a www database containing the complete nucleotide sequence of the genome of *Synechocystis* sp. strain PCC6803. Nucleic Acids Res. *26*, 63–67.

Nakamura, Y., Kaneko, T., and Tabata, S. (2000). CyanoBase, the genome database for *Synechocystis* sp. strain PCC6803: status for the year 2000. Nucleic Acids Res. *28*, 72.

Neilands, J.B. (1995). Siderophores: structure and function of microbial iron transport compounds. J. Biol. Chem. *270*, 26723–26726.

Nicolaisen, K., Moslavac, S., Samborski, A., Valdebenito, M., Hantke, K., Maldener, I., Muro-Pastor, A.M., Flores, E., and Schleiff, E. (2008). Alr0397 is an outer membrane transporter for the siderophore schizokinen in *Anabaena* sp. strain PCC 7120. J. Bacteriol. *190*, 7500–7507.

Nicolaisen, K., Hahn, A., and Schleiff, E. (2009). The cell wall in heterocyst formation by *Anabaena* sp. PCC 7120. J. Basic Microbiol. *49*, 5–24.

Nodop, A., Pietsch, D., Höcker, R., Becker, A. Pistorius, E.K., Forchhammer, K., and Michel, K.P. (2008). Transcript profiling reveals new insights into the acclimation of the mesophilic fresh-water cyanobacterium *Synechococcus elongatus* PCC 7942 to iron starvation. Plant Physiol. *147*, 747–763.

Nunoshiba, T., Obata, F., Boss, A.C., Oikawa, S., Mori, T., Kawanishi, S., and Yamamoto, K. (1999). Role of iron and superoxide for generation of hydroxyl radical, oxidative DNA lesions and mutagenesis. J. Biol. Chem. *274*, 34832–34837.

Okmen, G., Donmez, G., and Donmez, S. (2007). Influence of osmotic and metal stresses on nitrogenase activity of cyanobacteria isolated from paddy fields. African J. Biotech. *6*, 1828–1832.

Outten, C.E., and O'Halloran, T.V. (2001). Femtomolar sensitivity of metalloregulatory proteins controlling zinc homeostasis. Science. *292*, 2488–2492.

Pantopoulos, K., Gray, N.K., and Hentze, M.W. (1995). Differential regulation of two related RNA-binding proteins, iron regulatory protein (IRP) and IRPB. RNA *1*, 155–163.

Pena, M.M., and Bullerjahn, G.S. (1995). The DpsA protein of *Synechococcus* sp. strain PCC 7942 is a DNA-binding hemoprotein. Linkage of the Dps and bacterioferritin families. J. Biol. Chem. *271*, 22478–22482.

Pietsch, D., Staiger, D., Pistorius, E.K., and Michel, K.P. (2007). Characterization of the putative iron sulphur protein IdiC (ORF5) in *Synechococcus elongatus* PCC 7942. Photosynth. Res. *94*, 91–108.

Plowman, J.E., Loehr, T.M., Goldman, S.J., and Sanders-Loehr, J. (1984). Structure and siderophore activity of ferric schizokinen. J. Inorg. Biochem. *20*, 183–197.

Raven, J.A. (1988). The iron and molybdenum use efficiencies of plant growth with different energy, carbon and nitrogen. New Phytol. *109*, 279–287.

Reddy, K.J., Bullerjahn, G.S., Sherman, D.M., and Sherman, L.A. (1988). Cloning, nucleotide sequence, and mutagenesis of a gene (*irpA*) involved in iron-deficient growth of the cyanobacterium *Synechococcus* sp. strain PCC7942. J. Bacteriol. *170*, 4466–4476.

Roger, P.A., Tirol, A., Ardales, S., and Watanabe, I. (1986). Chemical composition of cultures and natural samples of N2-fixing blue-green algae from rice fields. Biol. Fertil. Soils. *2*, 131–146.

Russell, M.J., and Martin, W. (2004). The rocky roots of the acetyl-CoA pathway. Trends Biochem. Sci. *29*, 358–63.

Sandmann, G., Peleato, M.L., Fillat, M.F., Lazaro, M.C., and Gomez-Moreno, C. (1990). Consequences of the iron-dependent formation of ferredoxin and flavodoxin on photosynthesis and nitrogen fixation an *Anabaena* strains. Photosynth. Res. *26*, 119–125.

Sandström, S., Ivanov, A.G., Park, Y.I., Oquist, G., and Gustafsson, P. (2002). Iron stress responses in the cyanobacterium *Synechococcus* sp. PCC 7942. Physiol. Plant. *116*, 255–263.

Schalk, I.J., Yue, W.W., and Buchanan, S.K. (2004). Recognition of iron-free siderophores by TonB-dependent iron transporters. Mol. Microbiol. *54*, 14–22.

Schwarz, R., and Forchhammer, K. (2005). Acclimation of unicellular cyanobacteria to macronutrient deficiency: emergence of a complex network of cellular responses. Microbiology *151*, 2503–2514.

Sen, A., Dwivedi, K., Rice, K.A., and Bullerjahn, G.S. (2000). Growth phase and metal-dependent regulation of the *dpsA* gene in *Synechococcus* sp. strain PCC 7942. Arch. Microbiol. *173*, 352–357.

Shcolnick, S., and Keren, N. (2006). Metal homeostasis in cyanobacteria and chloroplasts. Balancing benefits and risks to the photosynthetic apparatus. Plant Physiol. *141*, 805–810.

Shen, B. (2003). Polyketide biosynthesis beyond the type I, II and III polyketide synthase paradigms. Curr. Opin. Chem. Biol. *7*, 285–295.

Sherman, D.M., and Sherman, L.A. (1983). Effect of iron deficiency and iron restoration on ultrastructure of *Anacystis nidulans*. J. Bacteriol. *156*, 393–401.

Shi, T., Sun, Y., and Falkowski, P.G. (2007). Effects of iron on the limitation on the expression of metabolic genes in the marine cyanobacterium *Tricodesmium erythraeum* IMS101. Environ. Microbiol. *9*, 2945–2956.

Simpson, F.B., and Neilands, J.B. (1976). Siderochromesin cyanophyceae: Isolation and characterization of schizokinen from *Anabaena* sp. J. Phycol. *12,* 44–48.

Singh, A.K., McIntyre, L.M., and Sherman, L.A. (2003). Microarray analysis of the genome-wide response to iron deficiency and iron reconstitution in the cyanobacterium *Synechocystis* sp. PCC 6803. Plant Physiol. *132,* 1825–1839.

Sleep, N.H., and Bird, D.K. (2008). Evolutionary ecology during the rise of dioxygen in the Earth's atmosphere. Philos. Trans. R. Soc. Lond. B. Biol. Sci. *363,* 2651–2664.

Stachelhaus, T., Mootz, H.D., and Marahiel, M.A. (1999). The specificity-conferring code of adenylation domains in nonribosomal peptide synthetases. J. Chem. Biol. *6,* 493–505.

Stanier, R.Y., and Cohen-Bazire, G. (1977). Phototrophic prokaryotes: the cyanobacteria. Annu. Rev. Microbiol. *31,* 225–274.

Stearman, R., Yuan, D.S., Yamaguchi-Iwai, Y., Klausner, R.D., and Dancis, A. (1996). A permease–oxidase complex involved in high-affinity iron uptake in yeast. Science. *271,* 1552–1557.

Steunou, A.S., Bhaya, D., Bateson, M.M., Melendrez, M.C., Ward, D.M., Brecht, E., Peters, J.W., Kühl, M., and Grossman, A.R. (2006). In situ analysis of nitrogen fixation and metabolic switching in unicellular thermophilic cyanobacteria inhabiting hot spring microbial mats.Proc. Natl. Acad. Sci. U.S.A. *103,* 2398–2403.

Straus, N.A. (1994) Iron deprivation: physiology and gene regulation. In The Molecular Biology of Cyanobacteria. D.A. Bryant, ed. (Dordrecht, the Netherlands: Kluwer Academic Publishers), pp. 731–750.

Swingley, W.D., Hohmann-Marriott, M.F., Le Olson, T., and Blankenship, R.E. (2005). Effect of iron on growth and ultrastructure of *Acaryochloris marina*. Appl. Environ. Microbiol. *71,* 8606–8610.

Ting, C.S., Rocap, G., King, J., and Chisholm, S.W. (2002). Cyanobacterial photosynthesis in the oceans: the origins and significance of divergent light-harvesting strategies. Trends Microbiol. *10,* 134–142.

Tölle, J., Michel, K.P., Kruip, J., Kahmann, U., Preisfeld, A., and Pistorius, E.K. (2002). Localization and function of the IdiA homologue Slr1295 in the cyanobacterium *Synechocystis* sp. strain PCC 6803. J. Microbiol. *148,* 3293–3305.

Wang, C.C., and Newton, A. (1971) An additional step in the transport of iron defined by the *tonB* locus of *Escherichia coli.* J. Biol. Chem. *246,* 2147–2151.

Wang, T., Shen, G., Balasubramanian, R., McIntosh, L., Bryant, D.A., and Golbeck, J.H. (2004). The *sufR* gene (*sll0088* in *Synechocystis* sp. strain PCC 6803) functions as a repressor of the *sufBCDS* operon in iron-sulfur cluster biogenesis in cyanobacteria. J. Bacteriol. *186,* 956–967.

Welker, M., and von Döhren, H. (2006). Cyanobacterial peptides – nature's own combinatorial biosynthesis. FEMS Microbiol. Rev. *30,* 530–563.

Wheeler, D.L., Chappey, C., Lash, A.E., Leipe, D.D., Madden, T.L., Schuler, G.D., Tatusova, T.A., and Rapp, B.A. (2000). Database resources of the national center for biotechnology information. Nucleic Acids Res. *28,* 10–14.

Wilhelm, S.W., and Trick, C.G. (1994). Iron-limited growth of cyanobacteria: Multiple siderophore production is a common response. J. Limnol. Oceanograph. *39,* 1979–1984.

Wilhelm, S.W. (1995). Ecology of iron-limited cyanobacteria: a review of physiological responses and implications for aquatic systems. Aquatic Microbial Ecology. *9,* 295–303.

Wilhelm, S.W., Maxwell, D.P., and Trick, C.G. (1996). Growth, iron requirements, and siderophore production in iron-limited *Synechococcus* PCC 7002. J. Limnol. Oceanogr. *41,* 89–97.

Winkelmann, G. (2001). Microbial transport systems. Weihnheim: Wiley-VCH.

Yousef, N., Pistorius, E.K., and Michel, K.P. (2003). Comparative analysis of *idiA* and *isiA* transcription under iron starvation and oxidative stress in *Synechococcus elongatus* PCC 7942 wild-type and selected mutants. Arch. Microbiol. *180,* 471–483.

Zhang, C.C., Laurent, S., Sakr, S., Peng, L., and Bédu, S. (2006). Heterocyst differentiation and pattern formation in cyanobacteria: a chorus of signals. J. Mol. Microbiol. *59,* 367–375.

Iron Uptake in Gram-positive (*Bacillus*) 12

Ahmed Gaballa and John D. Helmann

Abstract

Bacillus subtilis is a metabolically versatile soil microbe and Gram-positive model organism that displays a sophisticated adaptive response to conditions of iron limitation. The endogenous siderophore of *B. subtilis* is bacillibactin, a trimeric catecholate siderophore similar in structure to enterobactin. In addition to bacillibactin, *B. subtilis* can obtain iron from several xenosiderophores, ferric citrate, haem, and through a newly discovered elemental iron permease. The regulation of iron homeostasis in *B. subtilis* is complex and involves a ferric uptake regulator (Fur) protein as master regulator and at least two subsidiary regulatory systems. The most significant of these is an iron-sparing/prioritization response controlled by the small RNA FsrA and three auxiliary proteins (FbpABC). In addition, the bacillibactin uptake system is transcriptionally activated by an AraC family activator, Btr that directly senses bacillibactin. Iron uptake and homeostasis systems in *B. anthracis* and related organisms are largely similar to those in *B. subtilis* with some additional components. These include a second siderophore synthesis operon for petrobactin, which is important for virulence, and a more elaborate (or at least better understood) haem uptake system.

Introduction and historical background

Siderophores and their uptake systems are extremely widespread within the microbial world with the exception of a few specialist bacteria that have either greatly reduced (Elli *et al.*, 2000) or completely dispensed with an iron requirement (Posey and Gherardini, 2000). Even acidophilic bacteria that grow in iron-rich environments express complex adaptive responses for maintaining iron homeostasis (Osorio *et al.*, 2008; Quatrini *et al.*, 2007).

Siderophores and their role in bacterial physiology have been a subject of investigation for over 50 years (Neilands, 1995; Wandersman and Delepelaire, 2004) and much of the work during this time can be divided into three phases. Most early studies focused on the chemical identification and characterization of siderophores: studies greatly aided by the observation that their secretion into the growth medium could be strongly stimulated by growth of cells under iron-limiting conditions (Garibaldi and Neilands, 1956). In the second phase, physiological and early genetic studies established the broad outlines of the regulatory processes controlling siderophore biosynthesis and transport. Finally, in the third and still ongoing phase, genome-enabled technologies led to a greatly increased appreciation of the complexity of the iron homeostasis responses in a wide variety of bacterial systems (Andrews *et al.*, 2003). In each of these three phases, *Bacillus subtilis* and its relatives have figured prominently.

Throughout the decades of the 1950s through the 1970s most research on bacterial siderophores was focused on elucidation of their chemical structures and iron-binding properties. The first characterized siderophore was ferrichrome from the fungus *Ustilago sphaerogena* (Garibaldi and Neilands, 1955). When it became appreciated that many different organisms produced high-affinity iron-chelating compounds in response to iron deprivation, efforts to identify and chemically characterize the corresponding molecules intensified (Garibaldi and Neilands, 1956). In parallel with these efforts, other molecules identified initially as growth-promoting factors were subsequently found to function as siderophores (Lankford *et al.*, 1966; Snow, 1970).

Early structural studies in the Neilands' laboratory characterized several siderophores from culture supernatants of iron-starved bacilli. Indeed, the first bacterial siderophore to be structurally characterized was itoic acid (2,3-dihydroxybenzoylglycine) from *B. subtilis* (Ito and Neilands, 1958). In retrospect, itoic acid is actually a siderophore precursor and its production by *B. subtilis* is due to a mutation (*sfp*^{0}) present in most laboratory strains that prevents the production of the mature catecholate siderophore, bacillibactin (see below).

The hydroxamic acid siderophore schizokinin was characterized (originally as a growth-promoting factor) from *B. stearothermophilus* (Lankford *et al.*, 1966). Studies to define the structure of schizokinin (Mullis *et al.*, 1971) were a significant part of the doctoral thesis of Kary Mullis (while working in the Neilands' laboratory), who went on to later fame (and a Nobel Prize) for his discovery of the polymerase chain reaction (PCR). As more and more iron-chelating compounds were characterized it became clear that their synthesis is a widespread adaptation amongst microbes to conditions of iron limitation. This led to the introduction, by Lankford, of the term 'siderophore' to describe this class of compounds in 1973 (Neilands, 1981). For an overview of siderophores in microorganisms the reader is referred to Chapter 1.

The first efforts to explore the regulation of siderophore production in the bacilli were pursued by R.A.J. Warren and resulted in a series of publications spanning from 1968 to 1971 (Peters and Warren, 1968a,b; 1970; Walsh *et al.*, 1971; Walsh and Warren, 1971). This work established that production of itoic acid is regulated by iron availability and that accumulation commences upon cessation of growth under iron-limiting conditions (Peters and Warren, 1968a). Both 2,3-dihydroxybenzoic acid (DHB) and 2,3-dihydroxybenzoylglycine (DHBG) were found to accumulate in the culture supernatant (a mixture referred to as DHB(G)). The ability of iron to repress DHB(G) synthesis was enhanced if DHB(G) was also present, consistent with a model in which DHB(G) enabled iron transport into the cell and that iron levels were sensed by a cytosolic regulator. Further studies demonstrated that the iron uptake capacity of the starved cells was much greater than for cells grown in iron replete conditions and transport had the properties expected for an active system (Peters and Warren, 1968b). As expected for a siderophore-dependent process, uptake is highly specific for iron with little or no inhibition noted for divalent metal ions such as cobalt, nickel, zinc, or manganese (Walsh and Warren, 1971). Peters and Warren also observed that the presence of citrate in the medium could reduce DHB(G)-dependent iron uptake, but the mechanism of this effect was not defined (Peters and Warren, 1968b). Collectively, these early studies hinted at the breadth of the iron uptake potential in *B. subtilis*. For example, ferrichrome was shown to effectively inhibit the accumulation of DHB(G), consistent with the hypothesis that this might be an effective siderophore for *B. subtilis* (Peters and Warren, 1970). Although these studies clearly indicated that siderophore production was regulated by iron availability, the nature of the regulator(s) was not clear.

Genetic studies of iron uptake and its regulation were not vigorously pursued until the 1980s, although a few sporadic reports appeared earlier. One of the first was the identification by Bruce Ames of mutants of *Salmonella* that secreted catechols into the medium (Pollack *et al.*, 1970). Subsequently, mutants altered in the regulation of iron-uptake functions were characterized in *Salmonella* in 1978 in Rothfield's laboratory (Ernst *et al.*, 1978) and in *E. coli* in 1981 by Klaus Hantke (Hantke, 1981). These mutants, designated *fur* (ferric uptake regulation) were constitutive for the expression of outer membrane proteins normally only made under conditions of iron limitation. Subsequent cloning of the *fur* gene in Hantke's laboratory and biochemical studies by Bagg and Neilands established the broad outlines of Fur-mediated transcriptional control by demonstrating that the Fur repressor protein bound to its specific operator DNA in response to iron availability. By the end of the 1980s the broad outlines of iron homeostasis in the enteric bacteria had been established, but the full extent and complexity of the

Fur regulon, and the extent to which this mechanism might be applicable to other organisms, could not yet be imagined.

Based on the early studies from Hantke and Neilands, it was clear that the ferric-uptake regulator (Fur) protein functioned as an iron-dependent repressor of genes involved in siderophore biosynthesis and uptake in the enteric bacteria. In contrast, the first characterized iron-dependent repressor from a Gram-positive bacterium was the diphtheria toxin repressor, DtxR, a structurally distinct, albeit functionally analogous protein. This led to the simplistic, if naive, hypothesis that Gram-negative bacteria typically regulated iron homeostasis using Fur-type regulators whereas Gram-positive bacteria would instead use a DtxR-like protein. Subsequently, work from both our group (Chen *et al.*, 1993) and Hantke's group (Schneider and Hantke, 1993) provided evidence that iron-regulated genes in *B. subtilis* are associated with Fur box elements and therefore likely to be under control of a Fur-type regulatory protein.

The molecular genetic analysis of iron uptake in the Bacilli took a large leap forward with the study of Schneider and Hantke (Schneider and Hantke, 1993). This work established that *B. subtilis* uses at least three different types of hydroxamate siderophore (ferrichromes, ferrioxamine, and schizokinen) and that each is transported by a genetically distinct pathway. Using a strategy of selecting for resistance to antibiotics internalized through iron uptake pathways, they recovered mutants defective for growth in the presence of ferrichrome and/or ferrioxamine. The cloning and sequencing of the *fhuD* gene suggested the presence of a membrane-associated lipoprotein predicted to be part of an iron-regulated ferrichrome uptake system. The presence of candidate operator sites preceding the *fhuD* gene predicted the presence of a corresponding regulatory protein with DNA-binding selectivity similar to *E. coli* Fur.

Concurrent work in our laboratory was focused more specifically on the identification and characterization of metal-dependent regulatory proteins (Bsat *et al.*, 1998; Chen and Helmann, 1995; Chen *et al.*, 1993). To begin to delineate the mechanisms of metal-dependent regulation in *B. subtilis* we screened for metal-regulated genes using random *lacZ* fusions. Several metal-regulated genes that were induced by the presence of the iron-chelator EDDHA were subsequently characterized including *mrgA*, *mrgB*, and *mrgC*. Inspection of the promoter regions for these genes revealed the presence of Fur box-like sequences and mutational studies implicated this site in the *mrgC* gene in iron-dependent repression (Chen *et al.*, 1993). Together, these early studies (prior to the release in 1997 of the *B. subtilis* genome sequence) provided the first hints that *B. subtilis* (and, as we now know, many Gram-positive bacteria) also regulate iron homeostasis with a classic Fur-type repressor. Subsequent work in our group, facilitated by the release of the *B. subtilis* genome sequence in 1997, established that *B. subtilis* in fact encodes three members of the Fur family of regulatory proteins (Bsat *et al.*, 1998; Gaballa and Helmann, 1998). These include a *bona fide* Fur that controls iron homeostasis, a zinc homeostasis regulatory (Zur), and the peroxide stress response regulator (PerR) (reviewed in Lee and Helmann, 2007).

In the sections that follow, we will review our current knowledge of the iron homeostasis systems in *B. subtilis* and members of the *B. cereus* group. This group includes *B. anthracis*, *B. cereus*, and *B. thuringiensis* (therefore sometimes called the ACT group). *B. subtilis* encodes the biosynthetic machinery for a single catecholate siderophore (bacillibactin), whereas *B. anthracis* synthesizes two distinct siderophores (bacillibactin and petrobactin) (Fig. 12.1). However, siderophore biosynthesis represents only a small part of the overall genetic program dedicated to adaptation to iron limiting growth conditions. Bacilli also encode numerous uptake systems for other forms of iron (e.g. xenosiderophores, haem, citrate, and elemental iron) and a complex iron-sparing response that helps remodel the proteome to more efficiently utilize this limiting resource (Table 12.1).

Iron uptake in *Bacillus subtilis*

Bacillibactin biosynthesis, export and uptake

In the last few years, a relatively complete picture has emerged for siderophore biosynthesis and transport in *B. subtilis*. Under conditions of iron limitation, wild-type strains of *B. subtilis* produce bacillibactin (BB), a cyclic catecholate siderophore containing three iron-chelating 2,3-dihydroxybenzoate (DHB)

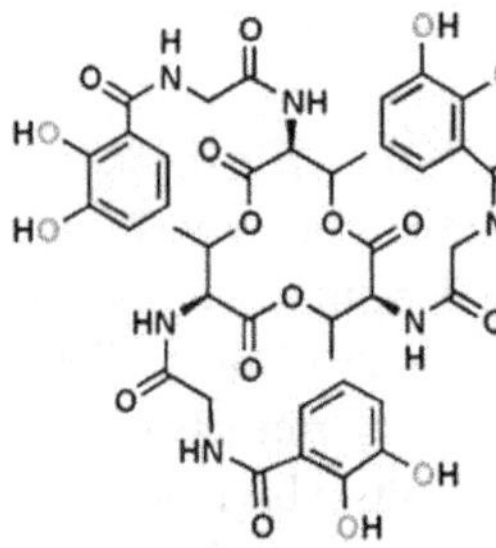

Bacillibactin (BB)

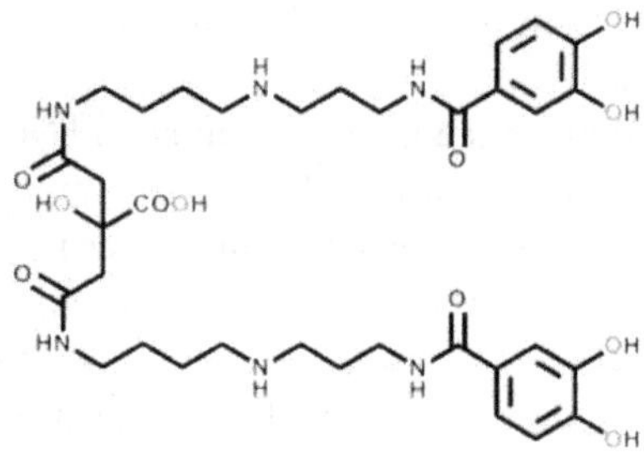

Petrobactin (PB)

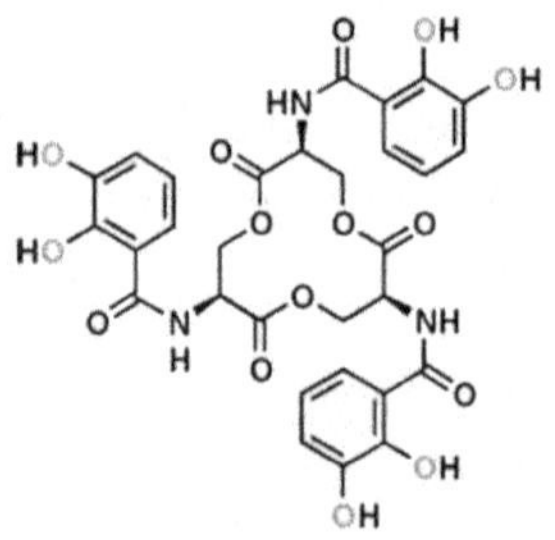

Enterobactin (Ent)

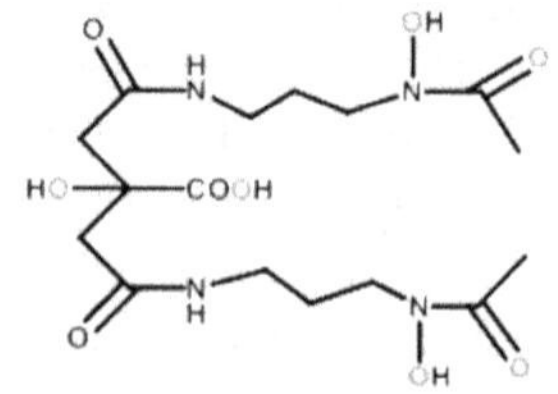

Schizokinen (Sch)

Figure 12.1 Representative siderophores made or utilized by *Bacilli*. *B. subtilis* produces bacillibactin (BB), which is structurally similar to (and often shares uptake systems with) enterobactin (Ent). *B. anthracis* produces both BB and petrobactin (PB). *B. megaterium* produces schizokinin. *B. subtilis* can use all the iron complexes of all four of these siderophores as an iron source.

moieties. BB is structurally similar to enterobactin (Ent), a prototypical catecholate siderophore from the enteric bacteria (Fig. 12.1). Together, these two molecules are the siderophores with the highest known formation constants for their cognate ferric complexes (Abergel *et al.*, 2009).

Synthesis of BB is catalysed by the products of the *dhbACEBF* operon (May *et al.*, 2001). Synthesis of the DHB precursor from chorismate is catalysed by isochorismate synthase (DhbC), the isochorismate lyase activity of the bifunctional DhbB protein, and 2,3-dihydo-2,3-dihydroxybenzoate dehydrogenase (DhbA). Three DHB moieties are linked with glycine and threonine and then cyclized into a macrolactone to yield the functional BB siderophore using a non-ribosomal peptide synthetase (NRPS) assembly-line mechanism catalysed by the DhbEBF enzyme complex (Fig. 12.2). In this complex, DhbE activates DHB by adenylation of the carboxylate (generating DHB-O-AMP). The DHB group is then transferred to a phosphopantetheinyl (Ppant) carrier group on the DhbB protein to from a reactive thioester. The DhbF subunit, in turn, carries two adenylation (A) and two Ppant-containing peptidyl carrier protein (PCP) domains. The DhbF protein activates (via adenylation) both Gly (in the amino-terminal domain) and Thr (in the carboxyl-terminal domain) to generate the corresponding Ppant thioesters. The condensation domains (C) then catalyse the transfer of the activated DHB group to Gly, and subsequent peptide bond formation between Gly and Thr generates the DHB-Gly-Thr precursor unit. This unit is 'stored' on a conserved Ser residue of the final termination (Te) domain at the carboxy-terminus of the DhbF polypeptide. A second unit is then synthesized by the same process, and the two DHB-Gly-Thr units are joined by esterification between the Thr side-chain (of the first) and the Thr carboxylate (of the second), leading to a dimeric precursor still attached to the Te domain. A third repetition of this process

Table 12.1 Selected genetic determinants of iron homeostasis in *Bacillus subtilis*

Gene/operon	Regulator(s)	Function
dhbACEBF	Fur	BB synthesis
btr	Fur	Substrate induction of BB uptake
feuABCybbA	Fur, Btr	BB uptake
yxeB	Fur	SBP:ferrioxamine (with FhuBCG)
fhuD	Fur	SBP: ferrichrome (with FhuBCG)
fhuBCG	Fur	Hydroxamate uptake
yfiY	Fur	SBP: schizokinen, arthrobactin
yfiZyfhA	Fur	schizokinen, arthrobactin uptake
yusV	Fur	ATPase for YfiYZYfhA and FeuABC
yfhQ	Fur	SBP
yclNOPQ	Fur	ABC transporter; substrate unknown
ywbLMN	Fur	Elemental iron permease
besA	Fur	Fe-BB esterase
yfmCDEF	Fur	Fe-citrate uptake
ycgT	Fur	Thioredoxin reductase homologue
ykuNOP	Fur	Flavodoxins
fsrA	Fur	sRNA; iron-sparing response
fbpAB	Fur	Iron-sparing response
fbpC	Fur	Iron-sparing response
hmoA	Fur	Haem oxidase
hmoB	haem	Haem oxidase
mrgA	PerR	Fe-binding mini-ferritin
dps	σ^B	Fe-binding mini-ferritin

leads to the linear BB precursor [(DHB-Gly-Thr)$_3$] which is cyclized and released from the Te domain to generate the complete cyclic lactone core (built from Thr) and the Gly-DHB iron-binding 'arms'.

The production of Itoic acid (i.e. DHB(G)) is due, in laboratory strains of *B. subtilis* W168, to a mutation in the *sfp* gene encoding the phosphopantetheinyl (Ppant) transferase required for modification of this large biosynthetic complex (Walsh *et al.*, 1997). Biochemical experiments established that this gene (originally named for its role in surfactin production) encodes a Ppant-transferase (acyl-carrier protein synthase) apparently dedicated to modification of enzymes needed for production of secondary metabolites (e.g. BB and surfactin). Without its appended Ppant carrier arm, the NRPS enzyme is unable to assemble BB. However, the presence of DHBG in iron-starved cultures of *B. subtilis* suggests that there is some enzyme modification and that the BB synthetase complex is at least partially functional. Since *B. subtilis* also carries an essential acyl-carrier protein synthase (AcpS) required for fatty acid synthesis, it is likely that there is at least a small level of cross-activation of the BB NRPS by this 'housekeeping' enzyme. A comprehensive review on the enzymatic properties of these enzymes and BB biosynthesis has been recently published (Miethke and Marahiel, 2007).

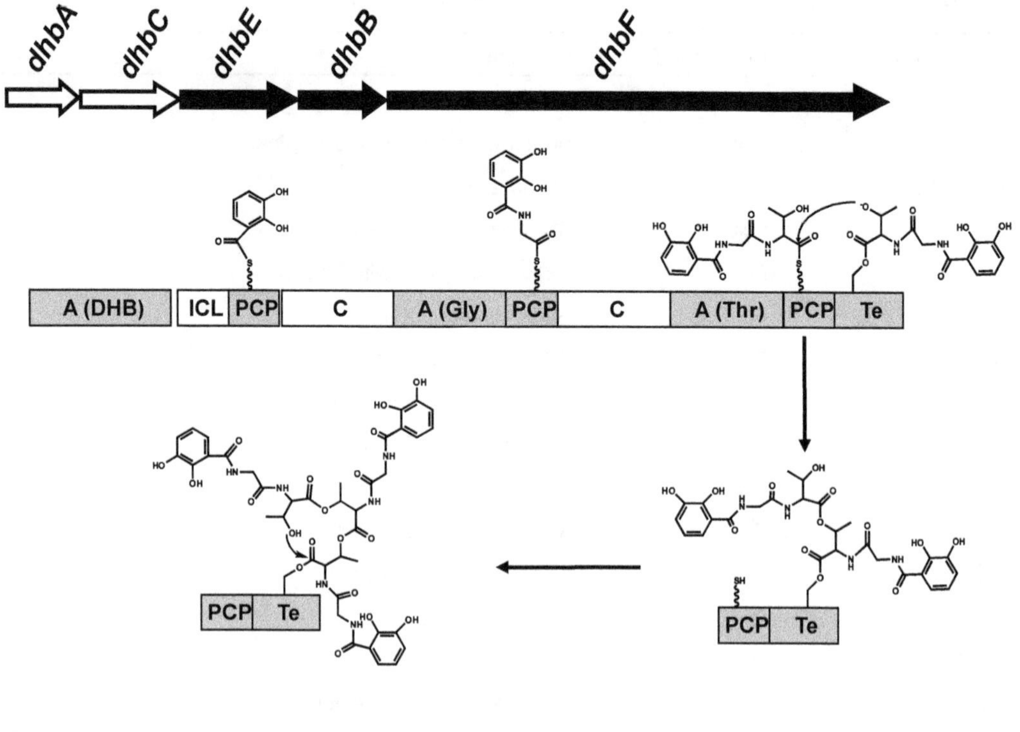

Figure 12.2 Biosynthesis of BB in *B. subtilis*. The Fur-regulated *dhbACEBF* operon (top) encodes the biosynthetic machinery for BB (bottom). The DhbEBF assembly functions as a non-ribosomal peptide synthase (NRPS) to assemble BB from 2,3-dihydroxybenzoic acid, glycine, and threonine (see text for details). The relevant protein domains are indicated as: adenylation (A), peptidyl-carrier protein (PCP) carrying the 4′-phosphopantetheinyl carrier arm, condensation (C), isochorismate lyase (ICL) and termination (Te).

Once synthesized, BB must be exported from the cell. In general, siderophore export systems have been hard to identify and there are only a few well-characterized examples in bacteria (Allard *et al.*, 2006; Franza *et al.*, 2005; Furrer *et al.*, 2002; Page *et al.*, 2003; Tanabe *et al.*, 2006). In most cases, siderophore export involves a major facilitator superfamily type transport system. Recently, the involvement of YmfE in bacillibactin export was reported (Miethke *et al.*, 2008). Surprisingly, *ymfE* is not regulated by iron but by the global MerR-type Mta (Miethke *et al.*, 2008). An *ymfE* mutant has a pronounced defect in secretion of the fully synthesized BB while it secretes high concentration of the DHB precursor.

Once exported from the cell, BB binds ferric ion with high avidity comparable to that measured for enterobactin (Dertz *et al.*, 2006). The resulting Fe–BB complex is transported through an ABC transporter, FeuABC-YusV (Ollinger *et al.*, 2006). This system is required for Fe-BB uptake and mutants are unable to grow in the presence of strong iron chelators unless a different class of siderophore is added (Miethke *et al.*, 2006; Ollinger *et al.*, 2006). The uptake system is constituted from a substrate binding lipoprotein (SBP), FeuA, with high affinity for Fe-BB (K_d~15 nM; Abergel *et al.*, 2009), two transmembrane proteins, FeuB and FeuC, that form the channel across the membrane, and the YusV ATPase which provides the system with the energy required for uptake (Miethke *et al.*, 2006; Ollinger *et al.*, 2006). The *feuABC* genes are co-transcribed as an operon together with the *ybbA* gene encoding a putative siderophore esterase. In contrast, the *yusV* gene is unlinked and encodes an ATPase that powers several different ABC uptake system (Ollinger *et al.*, 2006). In contrast with *B. subtilis*, in the ACT group the corresponding ATPase is FeuD and is co-transcribed with the *feuABC* genes.

In addition to BB, the FeuABC-YusV system is able to transport the structurally similar enterobactin siderophore produced by enteric bacteria and also contributes to the uptake of DHB(G) (Dertz *et al.*, 2006; Ollinger *et al.*, 2006). Enterobactin (Ent) is a cyclic lactone of the simpler DHB-Ser precursor. Most significantly, the Fe–Ent complex has the opposite chirality of the Fe–BB complex (Bluhm *et al.*, 2002). Regardless, both Fe–siderophore complexes are recognized by FeuA with similar affinity and are subsequently internalized in *B. subtilis* (Dertz et *al.*, 2006).

Upon internalization of the Fe–BB complex the iron is released through the cleavage of the trilactone backbone of the ferri-siderophore by the esterase BesA (Miethke *et al.*, 2006). The *besA* (formerly *yuiI*) gene is immediately upstream of, and co-directional with, the *dhb* operon for BB synthesis. Although *besA* is independently transcribed, both the *besA* and *dhb* operons are regulated by Fur (Baichoo *et al.*, 2002). While BesA is able to cleave both BB and Fe-BB *in vitro*, it shows higher catalytic efficiency towards Fe–BB complex (Miethke *et al.*, 2006). Intriguingly, the *B. subtilis* genome encodes another iron-regulated esterase, YbbA, which is expressed as part of the operon that encodes the *feuABC* uptake system. Despite this co-localization, YbbA cannot compensate for a *besA* mutation and is still of unknown function (Miethke *et al.*, 2006). BesA can cleave both Fe-BB and Fe–Ent complexes to allow iron utilization (Abergel *et al.*, 2009).

Uptake of exogenous siderophores

In addition to BB, *B. subtilis* is able to utilize exogenous siderophores produced by other organisms (referred to as xenosiderophores). *B. subtilis* is able to transport siderophores from all sub-groups including catechols (e.g. BB, enterobactin), hydroxamates (e.g. ferrioxamine and ferrichrome), carboxylates (e.g. citric acid) and mixed type siderophores (e.g. schizokinen and arthrobactin) (Dertz *et al.*, 2006; Ollinger *et al.*, 2006; Schneider and Hantke, 1993). Identification of the responsible transport systems has been facilitated by efforts to define the Fur regulon (Baichoo *et al.*, 2002). In response to iron deprivation, *B. subtilis* de-represses the expression of more than 20 operons, including several encoding ABC transporters for ferri-siderophore uptake (Table 12.1 and Fig. 12.3).

To identify the specificity of these different transport systems we generated null mutants in 14 different Fur-regulated transcription units to assess their possible role in iron transport (Ollinger *et al.*, 2006). These studies identified the FeuABC transporter as required for growth stimulation by both BB and Ent when cells are grown on plates containing the strong iron chelator ethylenediamine-di(*o*-hydroxyphenylacetic acid) (EDDA). In addition, growth required the YusV ATPase, as noted above. *B. subtilis* strains are unable to grow on this medium due to the presence of the *sfp*^{0} mutation. However, if this mutation is corrected, thereby restoring production of BB, growth is also restored. Previously, it was suggested that there was perhaps some Ent transport through a secondary system since Fe-BB was unable to completely block the transport of Fe-Ent (Dertz *et al.*, 2006). However, this system does not appear to be sufficient to support growth in the absence of the FeuABC system as originally reported by Ollinger *et al.* (2006) and recently confirmed by Abergel *et al.* (2009).

Mutational studies defined the FhuBCG system as required for the uptake of the hydroxamate-based siderophores ferrichrome and ferrioxamine, and ferrichrome utilization specifically requires the FhuD SBP (Schneider and Hantke, 1993). More recently, YxeB was identified as the SBP that facilitates ferrioxamine uptake through this same ABC transporter (Ollinger *et al.*, 2006). Growth stimulation by two other siderophores, arthrobactin and schizokinen, required the YfiYZ-YfhA ABC transporter. Finally, the YfmCDEF ABC transporter was identified as a ferric-citrate uptake system (Ollinger *et al.*, 2006). Ferric citrate is unable to support growth on medium containing EDDA, as predicted from its weaker iron-binding affinity under physiological conditions. However, in medium lacking EDDA and containing citrate, this is the major pathway for iron uptake (citrate is a better iron chelator than DHB(G)).

Elemental iron uptake

DHB(G) itself makes a relatively modest contribution to iron utilization by *B. subtilis sfp*^{0} strains, and then only in medium lacking citrate (Ollinger *et al.*, 2006). Indeed, in low iron medium lacking citrate

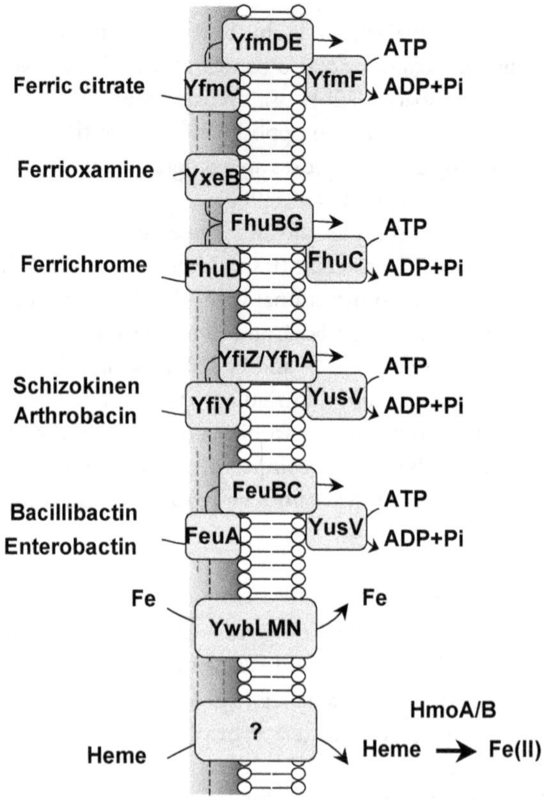

Figure 12.3 Major iron uptake pathways in *B. subtilis*. The Fur regulon encodes four distinct ABC transporters of known specificity (as illustrated), an elemental iron permease of the OFeT family, and two haem monooxygenases that may contribute to growth on haem as sole iron source.

B. subtilis growth is largely dependent on an elemental iron uptake system encoded by the *ywbLMN* operon. The YwbLMN proteins share similarity with the copper oxidase-dependent elemental iron uptake system of *Saccharomyces cerevisiae* (Fet3p/Ftr1p), as originally noted (Stearman *et al.*, 1996). *E. coli* also encodes a similar elemental iron uptake system which has been renamed as the EfeUOB system (Cao *et al.*, 2007).

The functional model of elemental iron uptake, as exemplified by Fet3p/Ftr1p, is rather complex. First, ferric iron Fe(III) is reduced to ferrous iron Fe(II) by the extracellular iron reductase Fre. Fe(II) is bound by the copper-dependent oxidase Fet3p and the re-oxidized Fe(III) is finally transferred to Ftr1p for uptake (Kosman, 2003). Initially, it was suggested that the *B. subtilis* YwbLMN system might work similarly (Ollinger *et al.*, 2006). Indeed, YwbL is homologous to the permease component and YwbN is a secreted peroxidase and is the only known substrate for a dedicated twin-arginine translocase (Tat) system (TatAyCy) in *B. subtilis* (Jongbloed *et al.*, 2004).

Recent genetic and biochemical studies on the orthologous system from *E. coli* support a model for EfeUOB function in which ferrous iron is transported into the cell. Although EfeU is a member of the oxidase-dependent iron transporter (OFeT) family of proteins (Grosse *et al.*, 2006), the role of oxidation/reduction in this system is not yet clear.

Subsequent genetic analyses have confirmed that all three components, EfeUOB, are required for iron utilization and are preferentially expressed at low pH (Cao *et al.*, 2007). Whereas *E. coli* can transport Fe(II) anaerobically via the FeoAB uptake systems, under aerobic conditions and at low pH the EfeUOB system is more important. As for *B. subtilis*, the EfeB protein is a haem peroxidase secreted by the Tat pathway. The roles of the two extracytoplasmic components in this system (EfeO and EfeB) are largely unknown. However, recent evidence suggests that EfeB may function as a haem deferrochelatase to remove iron from the haem tetrapyrrole thereby enabling uptake of haem iron without uptake of haem per se. Since haem can be toxic in cells, this may provide an alternative to the more familiar haem uptake and haem-monooxygenase-dependent pathways for the utilization of haem iron (Létoffé *et al.*, 2009). Indeed, expression of the intact *efeUOB* operon in *E. coli* strains amended to express the outer membrane haem receptor HasR allows the utilization of haem iron apparently without transport of haem into the cytosol (Létoffé *et al.*, 2009).

Haem uptake

Haem is widely recognized as an iron source by pathogenic bacteria and, in a few cases, soil bacteria (Wandersman and Delepelaire, 2004). Recently, *B. subtilis* was also found to use haem as an iron source (A. Gaballa and JDH, unpublished results). Growth on haem as sole iron source may be facilitated by two haem monooxygenases, HmoA (YetG) and HmoB (HhgC). These two proteins are differentially expressed with *hmoA* induced under iron-limiting growth conditions (under Fur control) and *hmoB* induced in the presence of haemin (by an unknown regulatory mechanism). Both purified proteins bind and apparently oxidize haemin *in vitro*, and either can support the use of haem as iron source (A. Gaballa and JDH, unpublished results).

Multiple sequence alignments and phylogenetic analysis demonstrates that HmoB belongs to the well-characterized *B. anthracis* and *S. aureus* IsdG family (Wu *et al.*, 2005). HmoA encodes a similar protein but belongs to a distinct, and still uncharacterized, subgroup that is characterized by replacement of a highly conserved Asn residue within the enzyme active site by a basic amino acid. The implications of this substitution on the specificity and enzymology of these proteins are not yet known, nor is it yet clear how haem is transported into the cell.

Regulation of iron homeostasis in *B. subtilis*

Iron homeostasis in *B. subtilis* is controlled primarily through the global regulator Fur (Bsat *et al.*, 1998) which works in concert with several subsidiary regulators. These include a small RNA-mediated iron-sparing response (Gaballa *et al.*, 2008), operon specific activation of BB uptake by the AraC-family Btr protein (Gaballa and Helmann, 2007), and possible post-transcriptional regulation mediated by aconitase in its role as an iron-responsive element (IRE)-binding protein (Alen and Sonenshein, 1999).

Fur is the global regulator for iron uptake system in *B. subtilis*

Fur is the major regulatory of iron homeostasis in *B. subtilis*, as it is in many other organisms (Bagg and Neilands, 1987; Carpenter *et al.*, 2009; Lee and Helmann, 2007). Under iron-replete growth condition, Fur represses expression of siderophore biosynthesis and iron uptake systems. Altogether, Fur represses more than 40 genes in ~23 operons by direct interaction with the corresponding regulatory region (Baichoo *et al.*, 2002). In addition, Fur indirectly activates the expression of iron-containing enzymes (Bagg and Neilands, 1987; Gaballa *et al.*, 2008).

We initially defined the Fur regulon through comprehensive transcriptome analysis using DNA microarray analysis to compare RNA populations isolated from *fur* mutant with wild type cells when grown under iron replete conditions (when Fur is active as a repressor) (Baichoo *et al.*, 2002). In addition, we determined the transcriptome changes that accompany iron starvation by comparing cells grown in medium that was either iron replete or iron-deficient (2,2′-dipyridyl treated). Nearly all genes that were de-repressed by at least 3-fold by 2,2′-dipyridyl were equivalently de-repressed in the *fur* mutant strain. Thus, Fur is required for the regulation of the vast majority of genes that are strongly induced by iron starvation. In essentially all cases, these Fur-regulated, iron-repressed operons are regulated by Fur binding directly to the corresponding regulatory region. This was initially inferred by scanning the genome sequence for the 19-bp Fur binding site consensus (Fur box) as defined for *E. coli*. In addition, purified Fur protein was found to bind to these sites using DNAse I footprinting and electrophoretic mobility shift assays. Subsequent to this initial global characterization (Baichoo *et al.*, 2002), several additional direct targets for Fur regulation have been defined including the HmoA haem monooxygenase mentioned above, a thioredoxin reductase homologue YcgT (Ollinger *et al.*, 2006), and the small regulatory RNA FsrA (Gaballa *et al.*, 2008). While the majority of Fur regulon members with assigned functions are important for various iron uptake pathways (Ollinger *et al.*, 2006), several members of the Fur regulon are still of unknown function. These include a predicted ferri-siderophore esterase, (YbbA), two flavodoxins (YkuN and YkuP), and the thioredoxin reductase homologue, YcgT. Although their physiological role has not been definitively established, these two *B. subtilis* flavodoxins can transfer electrons to both BioI (Lawson *et al.*, 2004) and nitric oxide synthase (Wang *et al.*, 2007).

Studies with *B. subtilis* Fur contributed to a reassessment of the mode of DNA-recognition by Fur proteins in general (Baichoo and Helmann, 2002). Classically, the Fur box is a 19 bp inverted repeat (GATAATGATwATCATTATC) but, due to internal symmetries, this sequence could potentially interact with the dimeric Fur repressor in a number of ways. In principle, this 19 bp sequence could represent an extended binding site for a single Fur dimer. Alternatively, this sequence can be viewed as repeated arrays of the GATAAT subsequence in head-head-tail orientation (Escolar *et al.*, 1998, 1999). According to this latter model, Fur recognizes repeated arrays of this hexanucleotide motif. Using a series of oligonucleotide fragments, we demonstrated that *B. subtilis* Fur binds, with high affinity, to a minimal 7–1–7 (15 bp) sequence (tGATAATnATTATCa). This represents the minimal recognition site for one dimer of Fur (Baichoo and Helmann, 2002) and, in fact, there are several Fur-binding sites that match this shorter consensus and do not contain evidence for conservation beyond this region above that expected from chance alone. In most cases, however, Fur-binding sites do match the longer 19 bp consensus. At these sites, Fur is proposed to bind to two, overlapping 7–1–7 motifs which, together, generate that classically defined 19 bp consensus. In some cases, a functional Fur operator will probably bind more than two dimers and even more extended regions of sequence conservation can be seen. The precise architecture of these complexes has not yet been defined.

The Fur box consensus sites in *B. subtilis* are often quite well conserved and a significant fraction (over half) of the Fur regulon can be identified by simply searching the genome for intergenic sites that closely match the 19 bp Fur box consensus (Baichoo *et al.*, 2002). In general, nearly all sites that match consensus at 16/19 positions are functional as Fur boxes, consistent with the expectation that such sites would not occur by chance. Indeed, one site originally suggested to be a 'false-positive', despite being a 17/19 match to consensus, was subsequently found to regulate the FsrA sRNA (Gaballa *et al.*, 2008). On the other hand, there are functional Fur boxes (particularly those that match only or primarily to the shorter 7–1–7 consensus) that would not be easily detected by simply pattern searching algorithms. A further complexity is introduced by the presence, in many genomes, of multiple Fur paralogues. In the case of *B. subtilis*, the three Fur family paralogues (Fur, PerR, and Zur) recognize operator sites that differ in only one or two key positions per half site (Fuangthong and Helmann, 2003). Despite these limitations, by simple searching newly sequenced genomes for Fur-box like elements it is often possible to obtain at least an initial glimpse into the nature of the Fur regulon.

BB-dependent activation of the feuABCybbA operon

In addition to its direct repression by Fur, the BB uptake system is also under positive regulation by an AraC-family activator that functions as a Bacillibactin transport regulator and hence designated Btr (Gaballa and Helmann, 2007). Btr is unusual amongst AraC family regulators which typically contain a non-conserved amino-terminal domain and conserved carboxyl-terminal DNA-binding motif. In the case of Btr, the AraC-like transcription factor domain is fused to a carboxyl-terminal domain homologous to FeuA, the SBP that binds Fe-BB. As a result, Btr, at 529 aa in length, is double the size of typical AraC family activator proteins.

Interestingly, the *btr* gene is encoded immediately upstream of and co-directional with the *feuABCybbA* operon which it regulates. It is therefore easy to imagine that the *feuA* gene was, at some point, duplicated with the upstream copy then fused to a DNA-binding domain. In this way, a DNA-binding regulator has acquired the ability to sense its ligand by acquisition of a protein domain normally functional outside the cell. As a result, Btr is a novel one-component regulator that can mediate the induction of the *feuABCybbA* operon in response to the presence (intracellularly) of substrate for the transporter. Although Btr activates *feuABC* expression even in the absence of BB, binding of BB (or Ent) to Btr enhances expression a least 3-fold. Btr activates the *feuA* promoter by binding to a 9-N_{13}-9 direct repeat region overlapping the −35 region (Gaballa and Helmann, 2007). We have shown that Btr, in contrast to known bacterial activators, works by enhancing RNA-polymerase promoter clearance only in the presence of the co-activator BB (our unpublished results).

The ability of ferri–siderophore complexes to induce their own uptake systems is well established in other systems, mostly described in Gram-negative bacteria. In some cases, the mechanism involves signal transduction from the outer-membrane receptor to an inner membrane sigma-regulator/anti-sigma protein which then allows function of a specific sigma factor to activate transcription of the transporter. This type of signalling pathway is best characterized for ferric citrate uptake in *E. coli* where the FecR membrane protein regulates the activity of the FecI sigma factor (Braun and Mahren, 2005). An alternative mechanism involves AraC family regulators such as *Yersinia pestis* YbtA (Fetherston *et al.*, 1996), *Pseudomonas aeruginosa* PchR (Michel *et al.*, 2005), and *Bordetella* spp. AlcR and BfeR (Anderson and Armstrong, 2004; Brickman *et al.*, 2001). Unlike the situation with Btr, the nature of the ligand binding domain in these AraC family regulators is unclear.

The iron-sparing response in *B. subtilis*

An additional mechanism of iron homeostasis is the re-allocation of limited iron from non-essential to essential functions. This is generally mediated by the repression, under conditions of iron-limitation, of non-essential, iron-containing proteins (Masséet *al.*, 2007). This general adaptation mechanism can be considered an 'iron-sparing response' since it spares the iron that would otherwise be incorporated into non-essential proteins. Early studies had identified roles of *E. coli* Fur in the positive regulation of some iron-containing proteins such as Fe–SOD (Fee, 1991). However, the true impact of this type of regulation was not immediately appreciated. One key result was the unexpected finding that *E. coli fur* mutant cells actually contain significantly less iron than their isogenic wild-type parents (Abdul-Tehrani *et al.*, 1999). Since *fur* mutants are derepressed for iron uptake functions, this result was counterintuitive. Subsequently, the studies of Massé and Gottesman identified the Fur-regulated small regulatory RNA, RyhB, as key mediator of positive gene regulation in *E. coli* that serves to help prioritize iron usage (Massé and Gottesman, 2002). This sRNA, together with the RNA chaperone Hfq, mediates the repression of numerous proteins under iron-deficient conditions and can thereby account for the overall reduced iron contents of *fur* mutant strains (Kadner, 2005; Masséet *al.*, 2005).

B. subtilis also has a robust iron-sparing response mediated by a regulatory small RNA designated FsrA (Fur-regulated small RNA A) (Gaballa *et al.*, 2008). Although functionally analogous to RyhB, there is no obvious sequence similarity between FsrA and RyhB and it is likely that these systems have evolved independently. Functionally analogous, but mechanistically distinct, iron-sparing systems (based on regulatory proteins rather than sRNAs) are found in *Saccharomyces cerevisiae* (Vergara and Thiele, 2008) and *Corynebacterium glutamicum* (Wennerhold *et al.*, 2005). The independent evolution of these various systems indicates that the ability to prioritize the usage of iron, a limiting nutrient in a variety of environments, is an adaptation of considerable significance.

The *B. subtilis* iron-sparing response is coordinated by FsrA working in concert with three Fur-regulated small, basic proteins designated FbpABC. Because these proteins are quite small, only one (*fbpB*; previously *ydbN*) was previously annotated as a protein-coding region. Reannotation of the *ydbN* region revealed the presence of an operon encoding two small proteins, now designated FbpA (53 aa) and FbpB (59 aa). The 29 aa FbpC peptide is the product of one of the earliest identified Fur regulon members, previously named *mrgC* (Chen *et al.*, 1993). Genetic and physiological studies suggest that FsrA acts to down-regulate the expression of numerous iron-containing proteins and this effect is variably dependent on the FbpABC proteins. It is also possible that the *fbpAB* and *fbpC* transcripts may, in addition to their protein-coding potential, function directly as regulatory RNA.

The effects of the FsrA/FbpABC-mediated iron-sparing response on *B. subtilis* physiology are profound (Gaballa *et al.*, 2008). These mediators down-regulate the expression of iron-containing proteins including succinate dehydrogenase, aconitase, cytochromes, haem biosynthetic enzymes, and glutamate synthase. FsrA represses translation of the *sdhCAB* and *gltAB* operons through direct RNA–RNA interactions that presumably inhibit ribosome binding. This regulation is largely independent of FbpABC. On the other hand, FsrA and FbpAB are together involved together in the repression of the lactate utilization system, LutABC.

As a result of these effects, *fur* mutants are impaired in growth relative to wild-type under a variety of conditions. For example, as originally noted in *E. coli* (Hantke, 1987), *B. subtilis fur* mutants are unable to grow on succinate as sole carbon source (Gaballa *et al.*, 2008). As a result of FsrA-mediated repression of the *gltAB* operon, *fur* mutants are also unable to grow on ammonium as nitrogen source. The poor growth of the *fur* mutant strain, as observed in rich medium such as LB, can be largely or completely suppressed by the additional mutation of *fsrA*. Thus, the poor growth of *fur* mutants is probably not due to constitutive expression of iron uptake functions or to iron overload, as once thought, but is a reflection of the inappropriate repression of numerous metabolic enzymes regulated as part of the iron-sparing response. In wild-type cells, this response would only be triggered under conditions of severe iron depletion: conditions likely to lead to a cessation of growth after, at most, a few additional rounds of cell division. Under these starvation conditions, there is little need for *de novo* synthesis of relatively abundant metabolic enzymes such as succinate dehydrogenase and glutamate synthase since these enzymes will only be diluted two-fold with each cell division and are not likely to become limiting for growth. Their repression therefore allows what little iron might be scavenged by the cell to be selectively incorporated into higher priority targets.

Under conditions of iron sufficiency, the converse situation exists: Fur-mediated repression of the FsrA/FbpABC gene products allows the expression of iron-rich proteins. In *S. enterica*, the expression of the iron storage proteins bacterioferritin and ferritin A is induced by high iron (Velayudhan *et al.*, 2007). *B. subtilis* encodes two proteins with the potential to store iron intracellularly and both are members of the Dps family of mini-ferritins (Hintze and Theil, 2006). The MrgA protein is controlled by PerR and induced under oxidative stress conditions (Chen and Helmann, 1995) whereas Dps is regulated by the general stress response regulator, σ^B (Antelmann *et al.*, 1997). Preliminary indications suggest that at least Dps can function as a reservoir of iron (R. Johnson and JDH, unpublished results). *B. anthracis* also encodes two Dps family proteins which have been characterized biochemically. Interestingly, these two proteins differ in the oxidants that optimally mediate iron mineralization (Liu *et al.*, 2006) but the roles of these proteins in iron storage, if any, are not yet known.

Iron uptake systems in *B. anthracis* and *B. cereus*

The iron uptake and homeostasis systems of the ACT group organisms are likely to be quite similar, in broad outline, to that described for *B. subtilis*. Two notable exceptions, which will be the focus of this section, are the presence of two endogenous siderophores (BB and petrobactin, PB; Fig. 12.1) and the considerably greater knowledge of haem transport mechanisms in *B. anthracis*. *B. anthracis* has an impressive arsenal of iron-uptake machinery including at least 15 ABC transporters known or likely to be involved in uptake of iron and iron-chelates, biosynthesis genes for both BB and PB, and a complex system for haem-scavenging and transport.

Siderophore synthesis and transport

Like *B. subtilis*, both *B. anthracis* and *B. cereus* synthesize and transport BB as iron source. Although, as noted above, BB has a remarkably high affinity for iron (Dertz *et al.*, 2006), it is ineffective during pathogenesis since it is specifically bound by the mammalian protein siderocalin (Abergel *et al.*, 2006). Bacterial pathogens have evolved a number of countermeasures to prevent siderophore sequestration by siderocalin. In the case of *Salmonella*, the glycosylation of Ent to generate derivatives (known as salmochelins) accomplishes this goal (Fischbach *et al.*, 2006). In the case of *B. anthracis*, the strategy involves the synthesis of a second siderophore, PB (Abergel *et al.*, 2006; Koppisch *et al.*, 2005). *B. anthracis* growth in macrophages and virulence in mice requires PB (Cendrowski *et al.*, 2004).

PB is an unusual 3,4-dihydroxybenzoic acid-containing siderophore assembled from spermidine, citrate, and 3,4-dihydroxybenzoic acid and was first isolated from the marine bacterium *Marinobacter hydrocarbonoclasticus* (Barbeau *et al.*, 2002). The Fe–PB complex is photoreactive which results in iron reduction, decarboxylation of the citrate moiety, and oxidation to form a 3-ketoglutarate derivative (Barbeau *et al.*, 2002). The resulting photoproduct is still able to bind iron with high affinity and can be utilized as siderophore (Abergel *et al.*, 2008). It has been suggested that this photoreactivity may serve to reduce the iron in marine systems (Barbeau, 2006), but the possible benefit in pathogenic and soil bacteria is unclear.

The biosynthetic pathway of PB has been elucidated in recent years and is a hybrid pathway involving both NRPS-type enzymes and enzymes classified as NRPS-independent siderophore (NIS) family proteins (Barry and Challis, 2009). The enzymes required for PB biosynthesis are encoded in a single operon, *asbABCDEF* (Fig. 12.4). The initial step in PB biosynthesis involves the synthesis of 3,4-dihydroxybenzoic acid from (–)–3-dehydroshikimate by the AsbF dehydratase (Fox *et al.*, 2008; Pfleger *et al.*, 2008). 3,4-dihydroxybenzoic acid is then activated by the AsbC adenylating enzyme and transferred to the acyl-carrier protein AsbD (Pfleger *et al.*, 2007). The PPTase (analogous to Sfp for BB biosynthesis) presumably necessary for the phosphopantetheinylation of AsbD is presently unknown. AsbE then condenses spermidine with 2,3-dihydroxybenzoic acid to form DHB-spermidine. In the final steps, AsbA condenses DHB-spermidine with citrate and AsbB condenses the resulting DHB-spermidine-citrate with a second molecule of DHB-spermidine to generate PB (Lee *et al.*, 2007; Oves-Costales *et al.*, 2009).

Despite the importance of PB for virulence of *B. anthracis* its uptake system is still uncharacterized. In recent work, however, several different *B. cereus* SBPs, presumed to be part of siderophore uptake

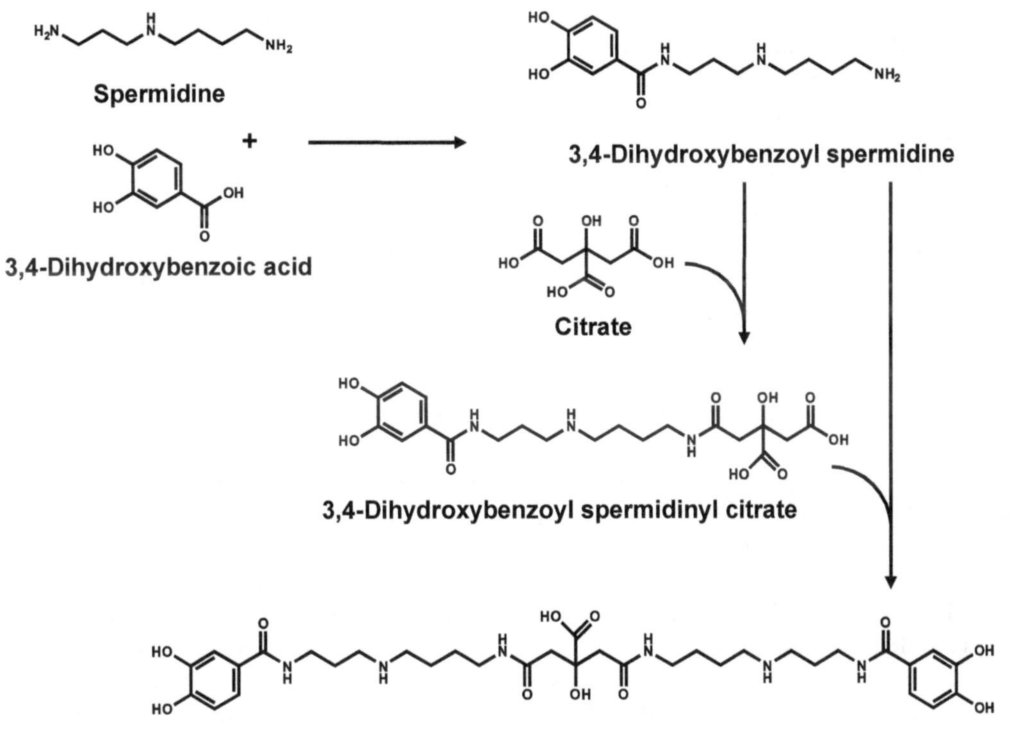

Figure 12.4 Biosynthesis of PB in *B. anthracis*. The Fur-regulated *asbABCDEF* operon encodes enzymes for the biosynthesis of PB from 3,4-dihydroxybenzoate (itself synthesized by AsbF), citrate, and spermidine.

systems, were over-expressed, purified and used for *in vitro* binding assays (Zawadzka *et al.*, 2009). It was shown that FatB (BC_5106) is able to bind both PB and its precursor 3,4-dihydroxybenzoic acid whereas FpuA (BC_4528) only binds PB (Zawadzka *et al.*, 2009). In addition, this study identified SBPs that interact with BB and Ent (FeuA), schizokinen (YfiY), and desferrioxamine and ferrichrome (YxeB). These latter assignments are largely consistent with the genetic analyses of the orthologous systems in *B. subtilis* (Ollinger *et al.*, 2006), as noted above.

Haem as an iron source in *B. anthracis*

The ability to acquire iron during infection is an essential attribute for many pathogenic bacteria. In fact, it has been suggested that iron is the most significant limiting nutrient during infection (Skaar *et al.*, 2006). *B. anthracis* excretes two siderophores to chelate free iron and these may also serve to mobilize iron bound, for example, to transferrins. However, during systemic infections *B. anthracis* accumulates in host tissues to enormous levels (up to 10^9 cells per ml in the bloodstream) and this requires the ability to access the single most abundant source of iron in this compartment: haemoglobin. For an overview of haem uptake systems in microorganisms the reader is referred to Chapter 2.

B. anthracis expresses a complete system that functions to extract haem from haemoproteins and subsequently deliver either the haem or the haem-derived iron to the cytosol. This system is orthologous to the well-characterized haem uptake system encoded by *S. aureus isd* (iron regulated surface determinants) operon described in Chapter 13 for *Staphylococcus aureus* (Tong and Guo, 2009). In addition, perhaps due to the presence of a thick capsule, *B. anthracis* has additionally evolved secreted haem-scavenging proteins, IsdX1 and IsdX2 (also called IsdJ and IsdK), with one or more conserved domains (referred to as NEAT for near transporter) (Gat *et al.*, 2008; Maresso *et al.*, 2008). In these systems, haem utilization initiates with the binding of haemoglobin to the secreted IsdX1 and IsdX2 proteins. Haem, removed from haemoglobin by IsdX1 and IsdX2, is then thought to be shuttled to the membrane localized transporters in a process probably involving other proteins encoded by the *isd* operon including IsdC (Maresso *et al.*, 2008; Fabian *et al.*, 2009), which is linked to the peptidoglycan (Maresso *et al.*, 2006). Three additional proteins, IsdDEF, form the transporter which delivers haem to the cytosol where iron is removed by the haem monooxygenase IsdG (Skaar *et al.*, 2006).

Concluding remarks

B. subtilis has long served as a model organism for the Gram-positive bacteria. As a metabolically versatile soil microorganism, this organism has also evolved a broad repertoire of iron uptake pathways. As presently understood, these pathways include a variety of ABC transporters for the uptake of both the endogenous siderophore (BB) as well as a variety of xenosiderophores. In addition, both ferric citrate and elemental iron uptake systems have been defined and recent studies hint at the presence of a functional haem utilization system. Organisms of the ACT group share many of these same uptake pathways, and probably additional components not represented in *B. subtilis*.

Studies in *B. subtilis* also provide a useful reference point for thinking about iron homeostasis in general (Moore and Helmann, 2005). The discovery of *B. subtilis* Fur was the first evidence that this class of regulatory protein, previously described in the enteric bacteria, was also present in Gram-positive microbes. More recently, additional regulatory proteins and systems have been described including the Btr protein responsible for the substrate induction of Fe-BB uptake, and the FsrA/FbpABC iron-sparing response. The availability of increasing numbers of bacterial genome sequences coupled with high-throughput approaches for transcriptome and proteome analysis promises rapid progress in defining the iron starvation responses in many bacterial systems. The thorough delineation of iron homeostasis systems in *B. subtilis* will provide an important resource for exploring the similarities and differences in both closely related organisms (e.g. other *Bacillus* species) as well as for more distantly related Gram-positive organisms (e.g. staphylococci, the Actinobacteria). The most significant contributions, perhaps, are in the genetic characterization of gene systems (functional genomics) and regulatory pathways enabled by the excellent genetic tools available in this organism. Evidence of this is already apparent from the rapid progress made in recent years in defining the orthologous pathways in *B. anthracis*.

Acknowledgements

Work on iron homeostasis in the Helmann laboratory is supported by a research grant from the National Institutes of Health (GM059323).

References

Abdul-Tehrani, H., Hudson, A.J., Chang, Y.S., Timms, A.R., Hawkins, C., Williams, J.M., Harrison, P.M., Guest, J.R., and Andrews, S.C. (1999). Ferritin mutants of *Escherichia coli* are iron deficient and growth impaired, and *fur* mutants are iron deficient. J. Bacteriol. *181*, 1415–1428.

Abergel, R.J., Wilson, M.K., Arceneaux, J.E., Hoette, T.M., Strong, R.K., Byers, B.R., and Raymond, K.N. (2006). Anthrax pathogen evades the mammalian immune system through stealth siderophore production. Proc. Natl. Acad. Sci. U.S.A. *103*, 18499–18503.

Abergel, R.J., Zawadzka, A.M., Hoette, T.M., and Raymond, K.N. (2009). Enzymatic hydrolysis of trilactone siderophores: where chiral recognition occurs in enterobactin and bacillibactin iron transport (1). J. Am. Chem. Soc. *131*, 12682–12692.

Abergel, R.J., Zawadzka, A.M., and Raymond, K.N. (2008). Petrobactin-mediated iron transport in pathogenic bacteria: coordination chemistry of an unusual 3,4-catecholate/citrate siderophore. J. Am. Chem. Soc. *130*, 2124–2125.

Alen, C., and Sonenshein, A.L. (1999). *Bacillus subtilis* aconitase is an RNA-binding protein. Proc. Natl. Acad. Sci. U.S.A. *96*, 10412–10417.

Allard, K.A., Viswanathan, V.K., and Cianciotto, N.P. (2006). lbtA and lbtB are required for production of the *Legionella pneumophila* siderophore legiobactin. J. Bacteriol. *188*, 1351–1363.

Anderson, M.T., and Armstrong, S.K. (2004). The BfeR regulator mediates enterobactin-inducible expression of *Bordetella* enterobactin utilization genes. J. Bacteriol. *186*, 7302–7311.

Andrews, S.C., Robinson, A.K., and Rodriguez-Quinones, F. (2003). Bacterial iron homeostasis. FEMS Microbiol. Rev. *27*, 215–237.

Antelmann, H., Engelmann, S., Schmid, R., Sorokin, A., Lapidus, A., and Hecker, M. (1997). Expression of a stress- and starvation-induced *dps/pexB*-homologous gene is controlled by the alternative sigma factor sigmaB in *Bacillus subtilis*. J. Bacteriol. *179*, 7251–7256.

Bagg, A., and Neilands, J.B. (1987). Molecular mechanism of regulation of siderophore-mediated iron assimilation. Microbiol. Rev. *51*, 509–518.

Baichoo, N., and Helmann, J.D. (2002). Recognition of DNA by Fur: a reinterpretation of the Fur box consensus sequence. J. Bacteriol. *184*, 5826–5832.

Baichoo, N., Wang, T., Ye, R., and Helmann, J.D. (2002). Global analysis of the *Bacillus subtilis* Fur regulon and the iron starvation stimulon. Mol. Microbiol. *45*, 1613–1629.

Barbeau, K. (2006). Photochemistry of organic iron(III) complexing ligands in oceanic systems. Photochem. Photobiol. *82*, 1505–1516.

Barbeau, K., Zhang, G., Live, D.H., and Butler, A. (2002). Petrobactin, a photoreactive siderophore produced by the oil-degrading marine bacterium *Marinobacter hydrocarbonoclasticus*. J. Am. Chem. Soc. *124*, 378–379.

Barry, S.M., and Challis, G.L. (2009). Recent advances in siderophore biosynthesis. Curr. Opin. Chem. Biol. *13*, 205–215.

Bluhm, M.E., Kim, S.S., Dertz, E.A., and Raymond, K.N. (2002). Corynebactin and enterobactin: related siderophores of opposite chirality. J. Am. Chem. Soc. *124*, 2436–2437.

Braun, V., and Mahren, S. (2005). Transmembrane transcriptional control (surface signalling) of the *Escherichia coli* Fec type. FEMS Microbiol. Rev. *29*, 673–684.

Brickman, T.J., Kang, H.Y., and Armstrong, S.K. (2001). Transcriptional activation of *Bordetella* alcaligin siderophore genes requires the AlcR regulator with alcaligin as inducer. J. Bacteriol. *183*, 483–489.

Bsat, N., Herbig, A., Casillas-Martinez, L., Setlow, P., and Helmann, J.D. (1998). *Bacillus subtilis* contains multiple Fur homologues: identification of the iron uptake (Fur) and peroxide regulon (PerR) repressors. Mol. Microbiol. *29*, 189–198.

Cao, J., Woodhall, M.R., Alvarez, J., Cartron, M.L., and Andrews, S.C. (2007). EfeUOB (YcdNOB) is a tripartite, acid-induced and CpxAR-regulated, low-pH Fe^{2+} transporter that is cryptic in *Escherichia coli* K-12 but functional in *E. coli* O157:H7. Mol. Microbiol. *65*, 857–875.

Carpenter, B.M., Whitmire, J.M., and Merrell, D.S. (2009). This is not your mother's repressor: the complex role of *fur* in pathogenesis. Infect. Immun. *77*, 2590–2601.

Cendrowski, S., MacArthur, W., and Hanna, P. (2004). *Bacillus anthracis* requires siderophore biosynthesis for growth in macrophages and mouse virulence. Mol. Microbiol. *51*, 407–417.

Chen, L., and Helmann, J.D. (1995). *Bacillus subtilis* MrgA is a Dps(PexB) homologue: evidence for metalloregulation of an oxidative-stress gene. Mol. Microbiol. *18*, 295–300.

Chen, L., James, L.P., and Helmann, J.D. (1993). Metalloregulation in *Bacillus subtilis*: isolation and characterization of two genes differentially repressed by metal ions. J. Bacteriol. *175*, 5428–5437.

Dertz, E.A., Xu, J., Stintzi, A., and Raymond, K.N. (2006). Bacillibactin-mediated iron transport in *Bacillus subtilis*. J. Am. Chem. Soc. *128*, 22–23.

Elli, M., Zink, R., Rytz, A., Reniero, R., and Morelli, L. (2000). Iron requirement of *Lactobacillus* spp. in completely chemically defined growth media. J. Appl. Microbiol. *88*, 695–703.

Ernst, J.F., Bennett, R.L., and Rothfield, L.I. (1978). Constitutive expression of the iron-enterochelin and ferrichrome uptake systems in a mutant strain of *Salmonella typhimurium*. J. Bacteriol. *135*, 928–934.

Escolar, L., Perez-Martin, J., and de Lorenzo, V. (1998). Binding of the Fur (ferric uptake regulator) repressor of *Escherichia coli* to arrays of the GATAAT sequence. J. Mol. Biol. *283*, 537–547.

Escolar, L., Perez-Martin, J., and de Lorenzo, V. (1999). Opening the iron box: transcriptional metalloregulation by the Fur protein. J. Bacteriol. *181*, 6223–6229.

Fee, J.A. (1991). Regulation of *sod* genes in *Escherichia coli*: relevance to superoxide dismutase function. Mol. Microbiol. *5*, 2599–2610.

Fabian, M., Solomaha, E., Olson, J.S., and Maresso, A.W. (2009). Heme transfer to the bacterial cell envelope occurs via a secreted hemophore in the gram-positive pathogen *Bacillus anthracis*. J. Biol. Chem. *284*, 32138–32146.

Fetherston, J.D., Bearden, S.W., and Perry, R.D. (1996). YbtA, an AraC-type regulator of the *Yersinia pestis* pesticin/yersiniabactin receptor. Mol. Microbiol. *22*, 315–325.

Fischbach, M.A., Lin, H., Liu, D.R., and Walsh, C.T. (2006). How pathogenic bacteria evade mammalian sabotage in the battle for iron. Nat. Chem. Biol. *2*, 132–138.

Fox, D.T., Hotta, K., Kim, C.Y., and Koppisch, A.T. (2008). The missing link in petrobactin biosynthesis: *asbF* encodes a (-)-3-dehydroshikimate dehydratase. Biochemistry. *47*, 12251–12253.

Franza, T., Mahe, B., and Expert, D. (2005). *Erwinia chrysanthemi* requires a second iron transport route dependent of the siderophore achromobactin for extracellular growth and plant infection. Mol. Microbiol. *55*, 261–275.

Fuangthong, M., and Helmann, J.D. (2003). Recognition of DNA by three ferric uptake regulator (Fur) homologs in *Bacillus subtilis*. J. Bacteriol. *185*, 6348–6357.

Furrer, J.L., Sanders, D.N., Hook-Barnard, I.G., and McIntosh, M.A. (2002). Export of the siderophore enterobactin in *Escherichia coli*: involvement of a 43 kDa membrane exporter. Mol. Microbiol. *44*, 1225–1234.

Gaballa, A., Antelmann, H., Aguilar, C., Khakh, S.K., Song, K.B., Smaldone, G.T., and Helmann, J.D. (2008). The *Bacillus subtilis* iron-sparing response is mediated by a Fur-regulated small RNA and three small, basic proteins. Proc. Natl. Acad. Sci. U.S.A. *105*, 11927–11932.

Gaballa, A., and Helmann, J.D. (1998). Identification of a zinc-specific metalloregulatory protein, Zur, controlling zinc transport operons in *Bacillus subtilis*. J. Bacteriol. *180*, 5815–5821.

Gaballa, A., and Helmann, J.D. (2007). Substrate induction of siderophore transport in *Bacillus subtilis* mediated by a novel one-component regulator. Mol. Microbiol. *66*, 164–173.

Garibaldi, J.A., and Neilands, J.B. (1955). Isolation and properties of ferrichrome A. J. Am. Chem. Soc. 77.

Garibaldi, J.A., and Neilands, J.B. (1956). Formation of iron-binding compounds by micro-organisms. Nature *177*, 526–527.

Gat, O., Zaide, G., Inbar, I., Grosfeld, H., Chitlaru, T., Levy, H., and Shafferman, A. (2008). Characterization of *Bacillus anthracis* iron-regulated surface determinant (Isd) proteins containing NEAT domains. Mol. Microbiol. *70*, 983–999.

Grosse, C., Scherer, J., Koch, D., Otto, M., Taudte, N., and Grass, G. (2006). A new ferrous iron-uptake transporter, EfeU (YcdN), from *Escherichia coli*. Mol. Microbiol. *62*, 120–131.

Hantke, K. (1981). Regulation of ferric iron transport in *Escherichia coli* K12: isolation of a constitutive mutant. Mol. Gen. Genet. *182*, 288–292.

Hantke, K. (1987). Selection procedure for deregulated iron transport mutants (*fur*) in *Escherichia coli* K 12: fur not only affects iron metabolism. Mol. Gen. Genet. *210*, 135–139.

Hintze, K.J., and Theil, E.C. (2006). Cellular regulation and molecular interactions of the ferritins. Cell. Mol. Life. Sci. *63*, 591–600.

Ito, T., and Neilands, J.B. (1958). Products of 'low-iron fermentation' with *Bacillus subtilis*: isolation, characterization and synthesis of 2,3-dihydroxybenzoylglycine. J. Am. Chem. Soc. *80*, 4645–4647.

Jongbloed, J.D., Grieger, U., Antelmann, H., Hecker, M., Nijland, R., Bron, S., and van Dijl, J.M. (2004). Two minimal Tat translocases in *Bacillus*. Mol. Microbiol. *54*, 1319–1325.

Kadner, R.J. (2005). Regulation by iron: RNA rules the rust. J. Bacteriol. *187*, 6870–6873.

Koppisch, A.T., Browder, C.C., Moe, A.L., Shelley, J.T., Kinkel, B.A., Hersman, L.E., Iyer, S., and Ruggiero, C.E. (2005). Petrobactin is the primary siderophore synthesized by *Bacillus anthracis* str. Sterne under conditions of iron starvation. Biometals *18*, 577–585.

Kosman, D.J. (2003). Molecular mechanisms of iron uptake in fungi. Mol. Microbiol. *47*, 1185–1197.

Lankford, C.E., Walker, J.R., Reeves, J.B., Nabbut, N.H., Byers, B.R., and Jones, R.J. (1966). Inoculum-dependent division lag of *Bacillus* cultures and its relation to an endogenous factor(s) ('schizokinen'). J. Bacteriol. *91*, 1070–1079.

Lawson, R.J., von Wachenfeldt, C., Haq, I., Perkins, J., and Munro, A.W. (2004). Expression and characterization of the two flavodoxin proteins of *Bacillus subtilis*, YkuN and YkuP: biophysical properties and interactions with cytochrome P450 BioI. Biochemistry. *43*, 12390–12409.

Lee, J.W., and Helmann, J.D. (2007). Functional specialization within the Fur family of metalloregulators. Biometals *20*, 485–499.

Lee, J.Y., Janes, B.K., Passalacqua, K.D., Pfleger, B.F., Bergman, N.H., Liu, H., Hakansson, K., Somu, R.V., Aldrich, C.C., Cendrowski, S., *et al.* (2007). Biosynthetic analysis of the petrobactin siderophore pathway from *Bacillus anthracis*. J. Bacteriol. *189*, 1698–1710.

Létoffé, S., Heuck, G., Delepelaire, P., Lange, N., and Wandersman, C. (2009). Bacteria capture iron from heme by keeping tetrapyrrol skeleton intact. Proc. Natl. Acad. Sci. U.S.A. *106*, 11719–11724.

Liu, X., Kim, K., Leighton, T., and Theil, E.C. (2006). Paired *Bacillus anthracis* Dps (mini-ferritin) have different reactivities with peroxide. J. Biol. Chem. *281*, 27827–27835.

Maresso, A.W., Chapa, T.J., and Schneewind, O. (2006). Surface protein IsdC and sortase B are required for heme-iron scavenging of *Bacillus anthracis*. J. Bacteriol. *188*, 8145–8152.

Maresso, A.W., Garufi, G., and Schneewind, O. (2008). *Bacillus anthracis* secretes proteins that mediate heme acquisition from hemoglobin. PLoS Pathog *4*, e1000132.

Massé, E., and Gottesman, S. (2002). A small RNA regulates the expression of genes involved in iron metabolism in *Escherichia coli*. Proc. Natl. Acad. Sci. U.S.A. *99*, 4620–4625.

Massé, E., Salvail, H., Desnoyers, G., and Arguin, M. (2007). Small RNAs controlling iron metabolism. Curr Opin Microbiol *10*, 140–145.

Massé, E., Vanderpool, C.K., and Gottesman, S. (2005). Effect of RyhB small RNA on global iron use in *Escherichia coli*. J. Bacteriol. *187*, 6962–6971.

May, J.J., Wendrich, T.M., and Marahiel, M.A. (2001). The *dhb* operon of *Bacillus subtilis* encodes the biosynthetic template for the catecholic siderophore 2,3-dihydroxybenzoate-glycine-threonine trimeric ester bacillibactin. J. Biol. Chem. *276*, 7209–7217.

Michel, L., Gonzalez, N., Jagdeep, S., Nguyen-Ngoc, T., and Reimmann, C. (2005). PchR-box recognition by the AraC-type regulator PchR of *Pseudomonas aeruginosa* requires the siderophore pyochelin as an effector. Mol. Microbiol. *58*, 495–509.

Miethke, M., Klotz, O., Linne, U., May, J.J., Beckering, C.L., and Marahiel, M.A. (2006). Ferri-bacillibactin uptake and hydrolysis in *Bacillus subtilis*. Mol. Microbiol. *61*, 1413–1427.

Miethke, M., and Marahiel, M.A. (2007). Siderophore-based iron acquisition and pathogen control. Microbiol. Mol. Biol. Rev. *71*, 413–451.

Miethke, M., Schmidt, S., and Marahiel, M.A. (2008). The major facilitator superfamily-type transporter YmfE and the multidrug-efflux activator Mta mediate bacillibactin secretion in *Bacillus subtilis*. J. Bacteriol. *190*, 5143–5152.

Moore, C.M., and Helmann, J.D. (2005). Metal ion homeostasis in *Bacillus subtilis*. Curr. Opin. Microbiol. *8*, 188–195.

Mullis, K.B., Pollack, J.R., and Neilands, J.B. (1971). Structure of schizokinen, an iron-transport compound from *Bacillus megaterium*. Biochemistry. *10*, 4894–4898.

Neilands, J.B. (1981). Microbial iron compounds. Annu. Rev. Biochem. *50*, 715–731.

Neilands, J.B. (1995). Siderophores: structure and function of microbial iron transport compounds. J. Biol. Chem. *270*, 26723–26726.

Ollinger, J., Song, K.B., Antelmann, H., Hecker, M., and Helmann, J.D. (2006). Role of the Fur regulon in iron transport in *Bacillus subtilis*. J. Bacteriol. *188*, 3664–3673.

Osorio, H., Martinez, V., Nieto, P.A., Holmes, D.S., and Quatrini, R. (2008). Microbial iron management mechanisms in extremely acidic environments: comparative genomics evidence for diversity and versatility. BMC Microbiol. *8*, 203.

Oves-Costales, D., Song, L., and Challis, G.L. (2009). Enantioselective desymmetrisation of citric acid catalysed by the substrate-tolerant petrobactin biosynthetic enzyme AsbA. Chem. Commun. (Camb), 1389–1391.

Page, W.J., Kwon, E., Cornish, A.S., and Tindale, A.E. (2003). The *csbX* gene of *Azotobacter vinelandii* encodes an MFS efflux pump required for catecholate siderophore export. FEMS Microbiol. Lett. *228*, 211–216.

Peters, W.J., and Warren, R.A.J. (1968a). Itoic Acid Synthesis in *Bacillus subtilis*. J. Bacteriol. *95*, 360–366.

Peters, W.J., and Warren, R.A.J. (1968b). Phenolic acids and iron transport in *Bacillus subtilis*. Biochim. Biophys. Acta. *165*, 225–232.

Peters, W.J., and Warren, R.A.J. (1970). The mechanism of iron uptake in *Bacillus subtilis*. Can. J. Microbiol. *16*, 1285–1291.

Pfleger, B.F., Kim, Y., Nusca, T.D., Maltseva, N., Lee, J.Y., Rath, C.M., Scaglione, J.B., Janes, B.K., Anderson, E.C., Bergman, N.H., *et al.* (2008). Structural and functional analysis of AsbF: origin of the stealth 3,4-dihydroxybenzoic acid subunit for petrobactin biosynthesis. Proc. Natl. Acad. Sci. U.S.A. *105*, 17133–17138.

Pfleger, B.F., Lee, J.Y., Somu, R.V., Aldrich, C.C., Hanna, P.C., and Sherman, D.H. (2007). Characterization and analysis of early enzymes for petrobactin biosynthesis in *Bacillus anthracis*. Biochemistry. *46*, 4147–4157.

Pollack, J.R., Ames, B.N., and Neilands, J.B. (1970). Iron transport in *Salmonella typhimurium*: mutants blocked in the biosynthesis of enterobactin. J. Bacteriol. *104*, 635–639.

Posey, J.E., and Gherardini, F.C. (2000). Lack of a role for iron in the Lyme disease pathogen. Science. *288*, 1651–1653.

Quatrini, R., Lefimil, C., Veloso, F.A., Pedroso, I., Holmes, D.S., and Jedlicki, E. (2007). Bioinformatic prediction and experimental verification of Fur-regulated genes in the extreme acidophile *Acidithiobacillus ferrooxidans*. Nucleic Acids Res. *35*, 2153–2166.

Rabsch, W., Methner, U., Voigt, W., Tschape, H., Reissbrodt, R., and Williams, P.H. (2003). Role of receptor proteins for enterobactin and 2,3-dihydroxybenzoylserine in virulence of *Salmonella enterica*. Infect. Immun. *71*, 6953–6961.

Rabsch, W., Voigt, W., Reissbrodt, R., Tsolis, R.M., and Baumler, A.J. (1999). *Salmonella typhimurium* IroN and FepA proteins mediate uptake of enterobactin but differ in their specificity for other siderophores. J. Bacteriol. *181*, 3610–3612.
Schneider, R., and Hantke, K. (1993). Iron-hydroxamate uptake systems in *Bacillus subtilis*: identification of a lipoprotein as part of a binding protein-dependent transport system. Mol. Microbiol. *8*, 111–121.
Skaar, E.P., Gaspar, A.H., and Schneewind, O. (2006). *Bacillus anthracis* IsdG, a heme-degrading monooxygenase. J. Bacteriol. *188*, 1071–1080.
Snow, G.A. (1970). Mycobactins: iron-chelating growth factors from mycobacteria. Bacteriol. Rev. *34*, 99–125.
Stearman, R., Yuan, D.S., Yamaguchi-Iwai, Y., Klausner, R.D., and Dancis, A. (1996). A permease–oxidase complex involved in high-affinity iron uptake in yeast. Science. *271*, 1552–1557.
Tanabe, T., Nakao, H., Kuroda, T., Tsuchiya, T., and Yamamoto, S. (2006). Involvement of the *Vibrio parahaemolyticus pvsC* gene in export of the siderophore vibrioferrin. Microbiol. Immunol. *50*, 871–876.
Tong, Y., and Guo, M. (2009). Bacterial heme-transport proteins and their heme-coordination modes. Arch. Biochem. Biophys. *481*, 1–15.
Velayudhan, J., Castor, M., Richardson, A., Main-Hester, K.L., and Fang, F.C. (2007). The role of ferritins in the physiology of *Salmonella enterica* sv. Typhimurium: a unique role for ferritin B in iron–sulphur cluster repair and virulence. Mol. Microbiol. *63*, 1495–1507.
Vergara, S.V., and Thiele, D.J. (2008). Post-transcriptional regulation of gene expression in response to iron deficiency: co-ordinated metabolic reprogramming by yeast mRNA-binding proteins. Biochem. Soc. Trans. *36*, 1088–1090.
Walsh, B.L., Peters, W.J., and Warren, R.A.J. (1971). The regulation of phenolic acid synthesis in *Bacillus subtilis*. Can. J. Microbiol. *17*, 53–59.
Walsh, B.L., and Warren, R.A.J. (1971). The iron-uptake system of *Bacillus subtilis*. Can. J. Microbiol. *17*.
Walsh, C.T., Gehring, A.M., Weinreb, P.H., Quadri, L.E., and Flugel, R.S. (1997). Post-translational modification of polyketide and nonribosomal peptide synthases. Curr. Opin. Chem. Biol. *1*, 309–315.
Wandersman, C., and Delepelaire, P. (2004). Bacterial iron sources: from siderophores to hemophores. Annu. Rev. Microbiol. *58*, 611–647.
Wang, Z.Q., Lawson, R.J., Buddha, M.R., Wei, C.C., Crane, B.R., Munro, A.W., and Stuehr, D.J. (2007). Bacterial flavodoxins support nitric oxide production by *Bacillus subtilis* nitric-oxide synthase. J. Biol. Chem. *282*, 2196–2202.
Wennerhold, J., Krug, A., and Bott, M. (2005). The AraC-type regulator RipA represses aconitase and other iron proteins from *Corynebacterium* under iron limitation and is itself repressed by DtxR. J. Biol. Chem. *280*, 40500–40508.
Wu, R., Skaar, E.P., Zhang, R., Joachimiak, G., Gornicki, P., Schneewind, O., and Joachimiak, A. (2005). *Staphylococcus aureus* IsdG and IsdI, heme-degrading enzymes with structural similarity to monooxygenases. J. Biol. Chem. *280*, 2840–2846.
Zawadzka, A.M., Abergel, R.J., Nichiporuk, R., Andersen, U.N., and Raymond, K.N. (2009). Siderophore-mediated iron acquisition systems in *Bacillus cereus*: Identification of receptors for anthrax virulence-associated petrobactin. Biochemistry. *48*, 3645–3657.

Iron Uptake in Staphylococci

13

Michelle L. Reniere, Gleb Pishchany and Eric P. Skaar

Abstract

Staphylococcus aureus causes a significant amount of human morbidity and mortality. The ability of *S. aureus* to cause disease is dependent upon its acquisition of iron from the host. *S. aureus* can obtain iron from various sources during infection, including haem and transferrin. The most abundant iron source in humans is haem iron bound by haemoglobin contained within erythrocytes. *S. aureus* is known to lyse erythrocytes through secretion of pore-forming toxins, providing access to host haemoglobin. Proteins of the iron-regulated surface determinant (Isd) system bind host haemoproteins, remove the haem cofactor, and shuttle haem into the cytoplasm for use as a nutrient iron source. Deletion of Isd system components decreases staphylococcal virulence, underscoring the importance of haem iron acquisition during infection. In addition to haem, *S. aureus* can utilize transferrin iron through the secretion of siderophores. Several staphylococcal siderophores have been described, some of which have defined roles during the pathogenesis of staphylococcal infections. A greater understanding of staphylococcal iron acquisition may lead to the development of novel therapeutic strategies that target nutrient uptake and decrease the threat of this increasingly drug-resistant bacterial pathogen.

Iron and *Staphylococcus aureus*

Staphylococci are non-motile Gram-positive cocci which occur in grape-like clusters due to alternating division planes and incomplete cell wall separation between daughter cells. Currently, more than thirty staphylococcal species have been identified and those which are clinically relevant are studied far more extensively than their less pathogenic relatives. In this regard, the pathogenic mechanisms of *Staphylococcus aureus* have received the most attention, with iron acquisition systems being some of the most well-studied virulence factors. *S. aureus* stands out among other members of the genus due to its ability to produce an arsenal of virulence factors that enable it to invade and multiply within virtually any host tissue. Consistent with this, *S. aureus* exceeds other staphylococci in its ability to acquire iron (Bergeron *et al.*, 1983; Schade, 1963; Trivier and Courcol, 1996). Owing to the significant clinical relevance of *S. aureus* and the fact that the iron uptake systems of *S. aureus* have been studied in considerable detail, this chapter will primarily focus on iron uptake in *S. aureus*.

S. aureus is a commensal organism colonizing the anterior nares of approximately thirty percent of the human population (Kuehnert *et al.*, 2006). Upon breakage of host protective barriers, *S. aureus* is capable of establishing infection by avoiding the immune system, adhering to cellular surfaces, and acquiring nutrients from host tissues. *S. aureus* infections are an increasing problem in developed countries, as the incidence of drug-resistant strains is steadily rising. In fact, invasive methicillin-resistant *S. aureus* (MRSA) infections are responsible for more deaths annually in the USA than AIDS (CDC, 2007; Klevens *et al.*, 2007). The most common invasive infections caused by *S. aureus* include bacteraemia, pneumonia, cellulitis, osteomyelitis, endocarditis and septic shock (Klevens *et al.*, 2007; Kuehnert *et al.*,

2005). The diverse ecological niches encountered by *S. aureus* necessitate a variety of systems devoted to the acquisition of nutrients during infection.

The success of *S. aureus* as an infectious agent relies on a diverse panel of virulence factors. It has become increasingly evident that the iron status of *S. aureus* and the production and function of many of these virulence factors are interconnected. These emerging findings suggest a paradigm whereby *S. aureus* senses iron availability and modifies its physiology to effectively colonize its host. One method by which *S. aureus* adapts to iron-starved environments is through iron-dependent alterations in gene expression mediated by the ferric uptake regulator (Fur). Fur is a ubiquitous transcription factor in bacteria that regulates gene expression in response to alterations in cellular iron status. Iron-bound Fur binds consensus DNA sequences located upstream of the transcriptional start sites of iron-regulated genes and represses transcription. Repression is alleviated when iron is scarce as Fur releases from the DNA, allowing access to RNA polymerase (Baichoo and Helmann, 2002). Over 40 cytoplasmic proteins are negatively regulated by Fur in *S. aureus*, highlighting the complexity of the bacterial response to iron deprivation (Friedman *et al.*, 2006). In addition to iron acquisition, many Fur-regulated proteins are involved in pathogenic functions such as biofilm formation, adhesion, evasion of the immune system and inhibition of wound healing, suggesting that iron-dependent regulation of these proteins has a considerable impact on the host–pathogen interaction (Athanasopoulos *et al.*, 2006; Baldassarri *et al.*, 2001; Chavakis *et al.*, 2002, 2005; Hussain *et al.*, 2001; Johnson *et al.*, 2005, 2008; Palma *et al.*, 1999). The importance of Fur-mediated iron-dependent gene regulation during pathogenesis is demonstrated by the significant attenuation of a *fur* mutant in a murine skin abscess model of *S. aureus* infection (Horsburgh *et al.*, 2001).

S. aureus is capable of acquiring iron from several different sources during infection. In the human body 75–80% of total iron is found as a component of haem (Drabkin, 1951). Haem, or iron-protoporphyrin IX, is a reactive molecule that catalyses peroxidase and oxidase reactions, which can result in damage to proteins and DNA. Within vertebrates, haem is sequestered by haemoproteins to prevent oxidative damage. Haemoglobin, the major oxygen transport protein contained within erythrocytes, is the most abundant haemoprotein in vertebrates (Crichton, 2001). Haemoglobin that is released into the serum upon erythrocyte lysis is quickly bound by circulating haptoglobin, which then traffics to the liver for degradation. *S. aureus* expresses specific transport systems dedicated to haem iron acquisition from host haemoglobin and haemoglobin–haptoglobin complexes, which will be discussed further in subsequent sections of this chapter. In addition to haemoglobin, haem iron is bound by myoglobin, the oxygen storage protein in vertebrates. *S. aureus* is able to utilize myoglobin as an iron source *in vitro*; however, the mechanism by which this occurs and its relevance to pathogenesis are unknown (Torres *et al.*, 2006).

The majority of the remaining non-haem iron within vertebrates is bound by ferritin, transferrin, or lactoferrin. Ferritin is an intracellular iron storage protein that is unlikely to be used as an iron source by extracellular bacteria such as *S. aureus*. In contrast, transferrin iron is found in the serum and accounts for up to seven percent of the total iron in the human body, making it an attractive iron source to extracellular pathogens (Drabkin, 1951). Iron acquisition from transferrin is mediated primarily by secreted siderophores in *S. aureus* (introduced in Chapter 1). Lactoferrin is found in neutrophil granules and many bodily fluids, including saliva, tears, and nasal secretions (Bullen and Griffiths, 1999). Rather than promoting growth by providing *S. aureus* with iron, lactoferrin exhibits a bacteriostatic effect on *S. aureus in vitro*. This growth inhibition is attributed to both its iron-chelating properties and serine protease activity (Aguila *et al.*, 2001; Clarke and Foster, 2008). Although it is not yet clear how lactoferrin functions *in vivo* during staphylococcal infection, it is not expected to be a viable iron source to *S. aureus*.

Recently studies have begun to investigate the expression of iron acquisition systems in specific tissues during infection, as the expression of iron-regulated genes within *S. aureus* may correlate with the levels of available iron within particular host tissues. During a systemic infection, Fur-regulated genes are expressed at high levels in *S. aureus* infected murine hearts and kidneys, whereas expression is not detected in whole livers (Pishchany *et al.*, 2009; Reniere and Skaar, 2008). These observations indicate that less iron is available to *S. aureus* during infection of the heart and kidneys as compared to infection of the liver, consistent with the significant contribution of the liver to iron storage and recycling (Drabkin, 1951). Considering the impact of iron concentration on staphylococcal virulence, organ specific variations in iron availability may contribute significantly to the types of *S. aureus* infections that present clinically.

A number of reports support the relevance of iron availability to the clinical disease course of *S. aureus* infections. For example, in patients with tunnelled dialysis catheters, increased iron dose is associated with increased risk of developing infections, including *S. aureus* bacteraemia (Jean *et al.*, 2002). Further, the administration of iron sucrose results in non-transferrin-bound iron in the serum of haemodialysis patients and this increased iron concentration enhances the growth of *S. aureus* and the risk of developing sepsis (Barton Pai *et al.*, 2006; Herron-Olson *et al.*, 2007). Finally, *S. aureus* has become a leading cause of bacterial infections in individuals with thalassaemia, a congenital defect that results in reduced haemoglobin synthesis and anaemia. The most common treatment for these patients is injection of the iron chelator desferrioxamine. Unfortunately, treatment with desferrioxamine further increases the rate of infections in these patients, presumably due to the ability of *S. aureus* to utilize desferrioxamine as an iron source (Brock and Ng, 1983; Rahav *et al.*, 2006).

In light of the severity of *S. aureus* infections and the unremitting antibiotic resistance exhibited by this organism, continuous efforts to describe iron acquisition strategies and their role in the pathogenesis of *S. aureus* may be crucial to the development of future therapeutic agents. In keeping with this, a description of the current understanding of staphylococcal iron acquisition systems is provided in the following section.

Staphylococcal siderophores

The study of siderophore production and utilization by *Staphylococcus* spp. was initiated upon the observation that growth of *S. epidermidis* is inhibited by transferrin-mediated iron chelation, whereas *S. aureus* growth is not altered in the same conditions. This inhibition can be reversed by the addition of *S. aureus* supernatants to *S. epidermidis* cultures, suggesting that the ability to acquire iron from transferrin is attributable to factors secreted by *S. aureus* (Marcelis *et al.*, 1978). It was subsequently shown that siderophores produced and secreted by *S. aureus* are capable of removing iron from transferrin for use as a nutrient source. However, coagulase-negative staphylococci (CoNS) such as *S. epidermidis*, are not capable of utilizing transferrin as an iron source, supporting these earlier observations (Lindsay *et al.*, 1995; Park *et al.*, 2005). Enhanced siderophore-mediated iron acquisition by *S. aureus* is also demonstrated by the fact that growth inhibition of *S. aureus* compared to CoNS requires addition of higher concentrations of iron chelators (Marcelis *et al.*, 1978; Valenti *et al.*, 1980). Together these observations suggest that *S. aureus* has evolved to more actively and effectively acquire siderophore-iron from its environment compared to other *Staphylococcus* spp., a hypothesis that correlates with the superior pathogenic capabilities of *S. aureus.*

S. aureus is known to sense and adapt to its local microenvironment, and in the case of iron depletion, it responds with robust protein expression changes, including an increase in the expression of proteins involved in siderophore production (Friedman *et al.*, 2006; Lindsay and Riley, 1994). Siderophore production is tightly regulated, as demonstrated by the reduction in siderophore synthesis upon addition of holo-transferrin to staphylococcal cultures. *S. aureus* is able to acquire transferrin-bound iron, and in doing so, decreases the production of siderophores (Lim *et al.*, 1998). Importantly, *S. aureus* has been shown to depend on siderophore-mediated iron capture in clinically relevant environments such as serum, peritoneal dialysate, and epithelial cells (Chung *et al.*, 2003; Garzoni *et al.*, 2007; Schade, 1963). Moreover, siderophore biosynthesis genes are induced *in vivo*, as demonstrated by microarray studies of *S. aureus* grown in the peritoneal cavity of a mouse (Allard *et al.*, 2006). Taken together, these data indicate a role for siderophore-mediated iron acquisition in *S. aureus* infection. Specifically, *S. aureus* has been reported to secrete four siderophores: staphyloferrin A, staphyloferrin B, staphylobactin, and aureochelin.

Carboxylate siderophores

The first staphylococcal siderophores to be isolated were named staphyloferrin A and staphyloferrin B (Meiwes *et al.*, 1990). Both of these small hydrophilic molecules were purified from *S. hyicus* and were found to be produced in iron-deplete conditions by many staphylococcal species, including *S. aureus*. Staphyloferrin A is produced in response to iron deprivation and is actively transported into staphylococcal cells, although the transport system responsible for staphyloferrin A import has yet to be determined (Konetschny-Rapp *et al.*, 1990). The structure of staphyloferrin A revealed that it is composed of D-ornithine linked to two citrate molecules through amide bonds. Utilization of D-ornithine in staphyloferrin A synthesis is reflected by an increase in staphyloferrin A production upon addition of D-ornithine to the medium (Meiwes *et al.*, 1990). Staphyloferrin A synthesis genes (*sfna*) were identified nearly two decades after the siderophore was first isolated (Fig. 13.1). The staphyloferrin A operon (*sfnaA-D*) is Fur-regulated, as evidenced by the iron-regulated secretion of staphyloferrin A and the presence of consensus Fur binding sites upstream from the transcriptional start site. This operon encodes for two non-ribosomal peptide synthetases (SfnaB and SfnaD), a putative transporter (SfnaA), and a racemase (SfnaC) (Cotton *et al.*, 2009). The mechanism of staphyloferrin A biosynthesis is the only staphylococcal siderophore biosynthesis pathway that has been elucidated to date.

Staphyloferrin B is also an iron-regulated carboxylate siderophore, however its synthesis is unaffected by the presence of ornithine (Meiwes *et al.*, 1990). Staphyloferrin B is asymmetrical and extremely hydrophilic. Although the pathway for staphyloferrin B synthesis has not yet been discerned, hydrolysis products of staphyloferrin B include L-diaminopropionic acid, citrate, ethylenediamine, and succinic semialdehyde. Involvement of L-diaminopropionic acid in staphyloferrin B synthesis is supported by increased production of staphyloferrin B when diaminopropionic acid is added to the culture (Drechsel *et al.*, 1993). Staphyloferrin B is produced by a variety of staphylococci, although studies with staphyloferrin B have typically been performed using product purified from *S. hyicus* (Drechsel *et al.*, 1993; Haag *et al.*, 1994). Staphyloferrin B stimulates bacterial growth in iron-deplete conditions through a mechanism that involves staphylococcal uptake of iron chelated by staphyloferrin B, as demonstrated by ^{55}Fe acquisition studies (Haag *et al.*, 1994). As with staphyloferrin A, a transport system responsible for staphyloferrin B uptake in staphylococci has yet to be identified.

Hydroxamate uptake

Early reports suggested that staphylococci are able to utilize foreign siderophores produced by Gram-negative enteric bacteria (Maskell, 1980). More specifically, all *S. aureus* strains are capable of utilizing ferrioxamine and other hydroxamate-type siderophores, although production of hydroxamate siderophores has not been demonstrated for staphylococci (Diarra *et al.*, 2002; Konetschny-Rapp *et al.*, 1990). Hydroxamate uptake is dependent upon proteins encoded by an operon originally named *fhuCBG* (*f*erric *h*ydroxamate *u*ptake operon). Based on homology with other bacterial transporters, *fhuB* and *fhuG* are predicted to encode for permeases, while *fhuC* (subsequently renamed *fhuA*) encodes an ATP-binding protein (Fig. 13.1). Inactivation of *fhuG* results in defective hydroxamate utilization, although these mutants are able to utilize other types of siderophores. This phenotype can be complemented by providing either the *fhuABG* operon or *fhuG* singly *in trans* (Sebulsky *et al.*, 2000). The lipoproteins FhuD1 and FhuD2 (encoded separately from the *fhuABG* operon, see Fig. 13.1) are also utilized for hydroxamate transport and were identified based on homology to the FhuD lipoproteins from *Bacillus subtilis* and *Escherichia coli*. In accordance with their role in iron acquisition, *fhuABG*, *fhuD1*, and *fhuD2* are regulated by iron in a Fur-dependent manner (Sebulsky and Heinrichs, 2001). The contribution of FhuD1 and FhuD2 to hydroxamate transport is supported by the fact that these proteins bind hydroxamate siderophores *in vitro* (Sebulsky *et al.*, 2003; Sebulsky *et al.*, 2004). FhuD1-deficient strains do not display an impairment in hydroxamate uptake; however, an *fhuD2* mutant is impaired in transport of all tested hydroxamates. Complementation studies show that *fhuD2* expressed *in trans* within the double *fhuD1fhuD2* mutant is sufficient for transport of all hydroxamates, while *fhuD1* rescues import of only ferrichrome and desferoxamine. Thus, it appears that FhuD2 is involved in transport of

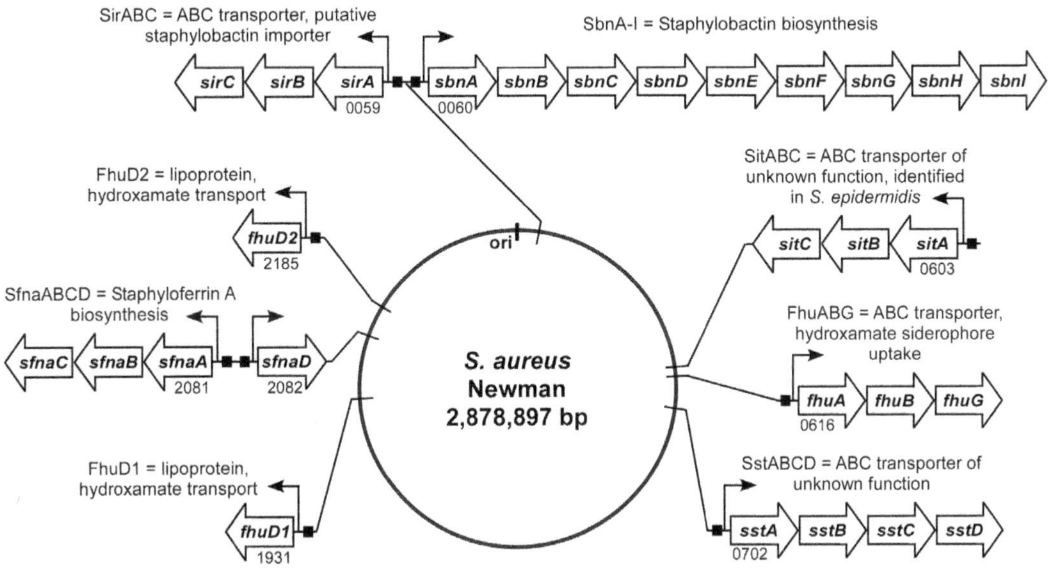

Figure 13.1 Genetic localization of genes related to siderophore-mediated iron uptake. The open arrows represent genes associated with siderophore-mediated iron acquisition, and their relative location is indicated on the circular representation of the *S. aureus* genome. The black boxes represent putative Fur binding sites, while the black arrows represent transcription start sites. Gene number in Newman (NWMN) notation is indicated under the first gene in each operon.

a wide range of hydroxamate siderophores, whereas FhuD1 transports a more limited subset (Sebulsky and Heinrichs, 2001). FhuD1 is not found in all *S. aureus* strains and its chromosomal location varies across strains, suggesting that the presence of *fhuD1* has arisen as the result of gene duplication (Sebulsky *et al.*, 2004). Thus, the advantage of expressing FhuD1 is unclear. It is possible that as-yet-unidentified siderophores are acquired through this lipoprotein. Notably, the ability of *S. aureus* to acquire hydroxamate siderophores has been suggested as a mechanism to distinguish this species from other members of the genus. The addition of desferrioxamine to an iron-poor antibiotic-selective medium selects for *S. aureus* against other staphylococci that are less capable of utilizing desferoxamine as an iron source. (Monsen *et al.*, 2003). Although these data suggest an important function for hydroxamate uptake in *S. aureus*, more experiments are necessary to demonstrate the role of FhuABG, FhuD1, and FhuD2 in hydroxamate siderophore transport.

Staphylobactin

Recently, a siderophore named staphylobactin has been isolated from the supernatants of *S. aureus*, however the structure of this putative siderophore has not been elucidated. The operon necessary for staphylobactin synthesis is regulated by iron in a Fur-dependent manner and is named *sbn* (*s*iderophore *b*iosy*n*thesis, Fig. 13.1). The *sbn* deletion strain is deficient for growth in iron-deplete conditions and displays decreased virulence in a murine model of infection (Dale *et al.*, 2004a). Adjacent to *sbn* is an operon encoding for a putative iron- and Fur-regulated transporter *sirABC* (*s*taphylococcal *i*ron *r*egulated) homologous to known bacterial siderophore transporters (Heinrichs *et al.*, 1999). *sirA* appears to encode for a lipoprotein and *sirBC* for permeases, while an apparent ATP-binding protein is missing from the operon. The *sir* operon does not play a role in staphylobactin synthesis, but is needed for its utilization, indicating a role for SirABC in staphylobactin import (Dale *et al.*, 2004b). Notably, *sbn* and *sir* are absent from CoNS.

The ATP-binding protein that is required for staphylobactin internalization appears to be FhuA. FhuA shares homology to known ATPases and the residues within FhuA which are thought to be necessary for

ATP–binding are also needed for siderophore uptake. Further, strains lacking *fhuA* are unable to utilize staphylobactin, indicating that it may function as an ATP-binding protein for both *fhu* and *sir* transport systems. Similar to *sbn* deletion strains, *S. aureus* strains lacking *fhuABG* exhibit reduced virulence in a murine model of infection (Speziali *et al.*, 2006).

Other systems

Information regarding additional siderophore-iron acquisition systems in *S. aureus* is limited. A catecholate siderophore named aurochelin was isolated from the supernatants of iron-starved staphylococci more than a decade ago but this molecule has not been characterized (Courcol *et al.*, 1997). The structure, biosynthesis, and transport of aureochelin remain to be discerned. Additional putative siderophore transporters have been identified solely based on homology to known transporters. The *sstABCD* and *sitABC* operons are iron-regulated and are predicted to encode for siderophore transporters (Fig. 13.1). SitABC was originally identified in *S. epidermidis*, while SstABCD was identified in *S. aureus* (Cockayne *et al.*, 1998; Morrissey *et al.*, 2000). Although it has been postulated that SstABCD transports siderophores, reduction of SstD levels does not decrease *S. aureus* growth in iron-limiting conditions *in vitro* or *in vivo* (Morrissey *et al.*, 2000). Thus, a complete description of the genes and gene products involved in siderophore biosynthesis and transport has not yet been achieved. Although siderophore-mediated iron acquisition is important, *S. aureus* preferentially utilizes haem iron *in vitro* when transferrin iron is simultaneously available, indicating that haem may be a more relevant iron source for *S. aureus* at certain points during infection (Skaar *et al.*, 2004b). In keeping with the importance of haem iron uptake to staphylococcal pathogenesis, the haem iron acquisition system of *S. aureus* has been extensively characterized and is described in the following section.

The iron regulated surface determinant (Isd) system

During infection of vertebrates bacteria encounter an environment in which over 99.9% of iron is intracellular, and therefore inaccessible to siderophores (Bullen and Griffiths, 1999). Haem iron bound by haemoglobin is by far the most abundant form of iron in the host, however, intracellular haemoglobin is inaccessible to extracellular pathogens such as *S. aureus*. In order to acquire haem iron from haemoglobin, *S. aureus* must first eliminate the protective barrier afforded by host erythrocytes. To this end, *S. aureus* secretes several haemolysins which form pores in host cell membranes, causing lysis and release of cellular contents (Menestrina *et al.*, 2003). Upon erythrocyte lysis, *S. aureus* binds the liberated haemoglobin and haemoglobin–haptoglobin complexes via proteins of the iron-regulated surface determinant (Isd) system (Skaar and Schneewind, 2004). The Isd system is comprised of nine iron-regulated proteins (IsdA-I) that act in concert to bind host haemoproteins, transport haem into the bacterial cytoplasm, and degrade haem to release nutrient iron (Fig. 13.2). To date, the Isd system is the most extensively characterized haem-uptake system in Gram-positive bacteria and homologues of individual components have been found in both Gram-positive and Gram-negative bacteria, including: *S. epidermidis, Bacillus anthracis, Bacillus halodurans, Bacillus cereus* (Chapter 12), *Bradyrhizobium japonicum* (Chapter 3), *Brucella melitensis, Clostridium tetani, Listeria monocytogenes*, and *Listeria innocua* (Puri and O'Brian, 2006; Skaar and Schneewind, 2004). A detailed description of the Isd system is provided in the following sections.

Haem/hemoprotein receptors (IsdA, IsdB, IsdC and IsdH)

S. aureus is known to take advantage of the abundance of haem iron in the host, as it preferentially imports haem iron when other iron sources are concurrently available (Skaar *et al.*, 2004b). The first obstacle in haem iron acquisition during infection is the capture of host haemoproteins and the removal of haem. To this end, the haemoprotein receptors IsdB and IsdH (also called HarA) bind haemoglobin and haemoglobin–haptoglobin complexes and remove the haem cofactor (Dryla *et al.*, 2007; Mazmanian *et al.*, 2003; Pilpa *et al.*, 2009; Zhu *et al.*, 2008). IsdH is the only staphylococcal haptoglobin-binding protein, while both IsdB and IsdH exhibit haemoglobin binding activities.

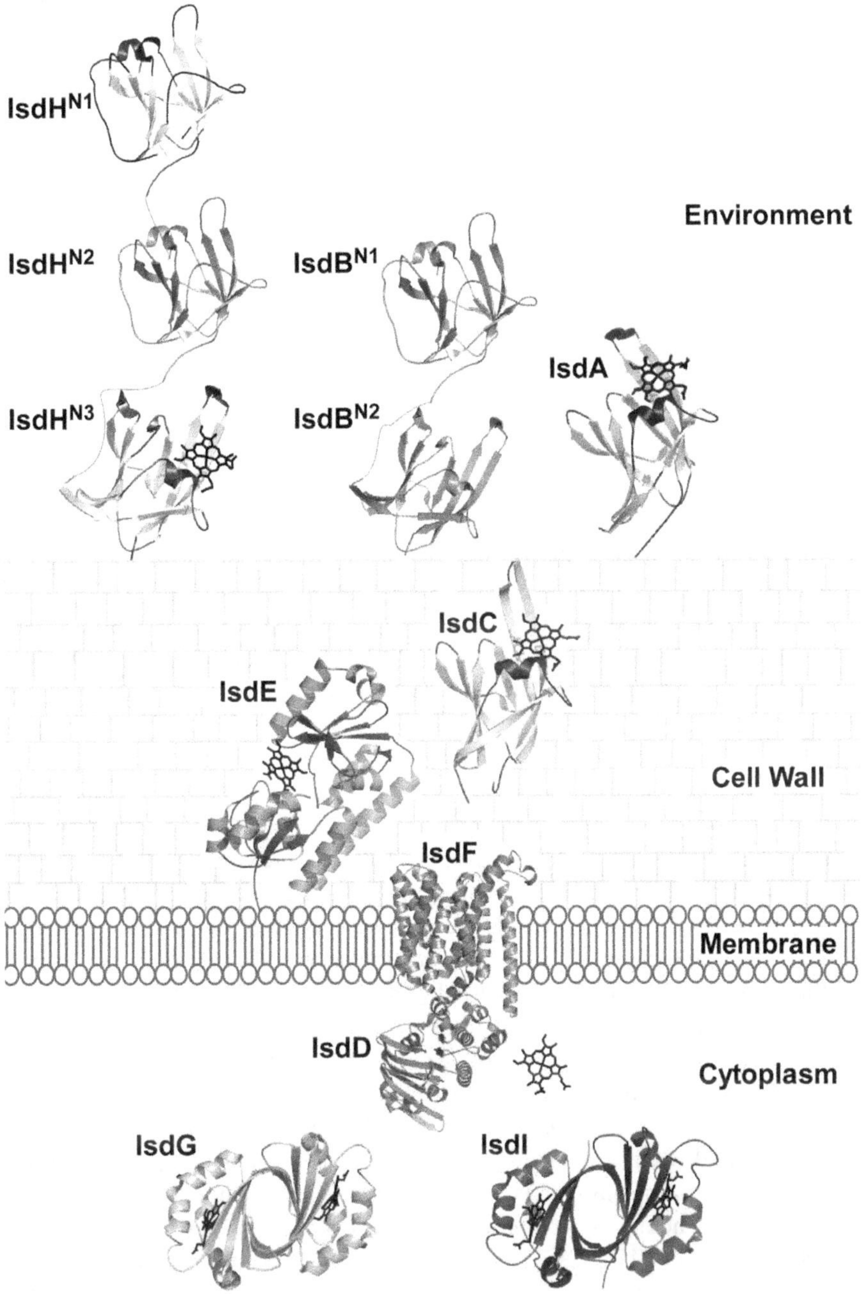

Figure 13.2 Model of Isd system in *S. aureus*. The model for haem transport predicts that once haem is removed from haemoglobin by IsdB and/or IsdH, it is transferred into the cell in the following order: IsdA → IsdC → IsdE → IsdDF →IsdGI. $IsdH^{N1}$, $IsdH^{N2}$, and $IsdB^{N1}$ domains bind host haemoproteins and transfer the haem cofactor to $IsdH^{N3}$ and $IsdB^{N2}$. IsdA then captures haem and transfers it to IsdC through formation of an IsdA–IsdC complex. Haem is then passed to IsdE and subsequently through the cytoplasmic membrane via IsdDF. Once haem enters the cytoplasm it can be degraded by IsdG and IsdI in order to liberate iron for use as a nutrient source. PDB codes are as follows: $IsdH^{N1}$ (2H3K), $IsdH^{N3}$–haem complex (2Z6F), IsdA NEAT domain complexed with haem (2ITF), IsdC–haem complex (2O6P), IsdE–haem complex (2Q8Q), IsdG-N7A complexed with haem (2ZDO), IsdI complexed with cobalt protoporphyrin IX (2ZDP). Protein domains without known structures are based on sequence similarity to proteins with a published structure. $IsdH^{N1}$ shares 59% sequence identity with $IsdH^{N2}$ and 46% sequence identity with $IsdB^{N1}$. $IsdH^{N3}$ shares 56% sequence identity with $IsdB^{N2}$. IsdD and IsdF belong to the vitamin B12 ABC transporter family, the prototype of which is BtuCD from *E. coli* (2QI9).

Once haem is acquired from host haemoproteins, it must next be transported across the thick Gram-positive cell wall, a process mediated by IsdA and IsdC. These cell-wall anchored proteins bind haem with modest affinity and traffic it across the dense layers of peptidoglycan (Sharp *et al.*, 2007; Vermeiren *et al.*, 2006). All haem and haemoprotein receptors are anchored to the cell wall through a transpeptidation reaction catalysed by sortase proteins. Sortase A anchors IsdA, IsdB, and IsdH through an LPXTG motif near the C-terminus of these proteins, whereas Sortase B anchors IsdC through a similarly located NPQTN motif (Mazmanian *et al.*, 2002, 2003). The mechanism of sortase-catalysed transpeptidation has been reviewed previously (Maresso and Schneewind, 2006; Mazmanian *et al.*, 2001). In brief, sortases catalyse the covalent linkage of protein substrates to immature murein subunits of the growing cell wall. Following cleavage between the threonine and asparagine/glycine of the anchoring motif, the threonine is connected to the pentaglycine cross-bridge of the *S. aureus* cell wall peptidoglycan via an amide-linkage (Mazmanian *et al.*, 1999; Ton-That *et al.*, 2000). In *S. aureus*, sortase A (*srtA*) is constitutively expressed and is responsible for anchoring over twenty distinct proteins to the cell wall (Mazmanian *et al.*, 1999). Inactivation of *srtA* severely impairs the virulence of *S. aureus* in an acute infection model, demonstrating the importance of cell wall-anchored proteins during infection (Mazmanian *et al.*, 2000). Interestingly, 15% of the LPXTG-containing proteins identified in *S. aureus* are members of the Isd system, indicating that the inability to bind host haemoproteins and acquire haem iron may be partially responsible for the dramatic virulence defect of the *srtA* mutant strain. To date, IsdC is the only protein known to be anchored by sortase B (Mazmanian *et al.*, 2002). In accordance with its dedicated role in IsdC anchoring, *srtB* is co-transcribed with *isdCDEFG* and is iron regulated (Mazmanian *et al.*, 2002). Inactivation of *srtB* decreases staphylococcal virulence in a murine renal abscess model of infection after 9 days of infection but no difference was observed after only five days of infection, a time point in which the *srtA* mutants were severely impaired (Mazmanian *et al.*, 2002). Thus, SrtA and SrtB perform vital and diverse roles during staphylococcal infection.

Common to all Isd receptor proteins is the presence of one or more NEAT (NEAr iron Transporter) domains. NEAT domains were identified based on their genomic localization within genes neighbouring putative iron transport systems, and these domains consist of a conserved region encoding 125 amino acids (Andrade *et al.*, 2002). IsdA and IsdC are each comprised of one NEAT domain, while IsdB contains two and IsdH contains three NEAT domains. The structure of the first NEAT domain to be elucidated was that of the first NEAT domain of IsdH ($IsdH^{N1}$) (Pilpa *et al.*, 2006). The structure revealed that although unrelated at the primary sequence, NEAT domains belong to the immunoglobulin (Ig) superfamily, adopting a β-sandwich fold consisting of two five-stranded antiparallel β-sheets (Fig. 13.2). $IsdH^{N1}$ binds both haemoglobin and haptoglobin through an aromatic motif formed by loops at the end of the β-barrel structure (Pilpa *et al.*, 2009). Haemoglobin binds $IsdH^{N1}$ in a 1:2 ratio, suggesting that the presence of two similar NEAT domains in IsdH may increase the binding affinity for haemoglobin. Alternatively, IsdH NEAT domains could each bind homodimers of haemoglobin and haptoglobin, as most Ig-like proteins bind homo- or heterodimeric ligands (Pilpa *et al.*, 2006).

$IsdH^{N1}$ and the second IsdH NEAT domain ($IsdH^{N2}$) are 53% identical and are both capable of binding haemoglobin with nanomolar affinity. The third NEAT domain of IsdH ($IsdH^{N3}$) is only 16% similar to $IsdH^{N1}$ and $IsdH^{N2}$ and is responsible for binding haem. $IsdH^{N3}$ can capture haem from haemoglobin and the rate of transfer is increased two-fold when haemoglobin is bound by $IsdH^{N1}$ (Pilpa *et al.*, 2008). Each $IsdH^{N3}$ domain binds several haem molecules, but only weakly (K_d 2.8 μM) (Pilpa *et al.*, 2008). Presumably this allows for efficient binding and release of haem as it is transferred into the cell (Watanabe *et al.*, 2008).

While IsdH is capable of binding haemoglobin *in vitro*, inactivation of *isdH* does not alter the ability of *S. aureus* to utilize haemoglobin as an iron source. However, *isdB* mutants are impaired in their ability to use haemoglobin as a sole iron source, indicating that *in vivo* IsdB is the primary staphylococcal haemoglobin receptor (Torres *et al.*, 2006). Full length IsdB binds haemoglobin with nanomolar affinity and this binding is attributed to the first NEAT domain ($IsdB^{N1}$), which is 46% identical to the haemoglobin-binding domain of IsdH ($IsdH^{N1}$) (Dryla *et al.*, 2007; Mazmanian *et al.*, 2003). The second IsdB NEAT domain ($IsdB^{N2}$) is 56% identical to $IsdH^{N3}$ and binds haem similarly (Muryoi *et al.*, 2008). In addition to using haemoglobin as an iron source, it is possible that binding haemoglobin with such strong affinity can also be used as an immune evasion technique. By expressing large amounts of IsdB and IsdH, *S. aureus* can coat its surface with haemoglobin and haemoglobin–haptoglobin complexes that may, in effect, make the bacteria invisible to host immune surveillance (Dryla *et al.*, 2007).

IsdA and IsdC are each composed of a single NEAT domain that binds haem (Sharp *et al.*, 2007; Vermeiren *et al.*, 2006). IsdC is thought to function as the central haem channel from the outer cell wall to the cytoplasmic membrane due to its unique SrtB-catalysed anchoring that positions it within the cell wall rather than exposed to the surface. IsdA is one of the most abundant proteins on the cell surface of iron-starved *S. aureus* and it has been shown to bind a diverse array of host ligands. In addition to haem, IsdA has been shown to bind the antimicrobial polypeptide lactoferrin, the extracellular matrix proteins fibrinogen and fibronection, and has been implicated in transferrin-binding, which will be discussed in more detail below (Clarke and Foster, 2008; Clarke *et al.*, 2004). The biological ramifications of this broad binding specificity have not been established, but presumably IsdA-mediated ligand binding promotes the ability of *S. aureus* to adapt to its environment and colonize its host.

Haem transport into the cytoplasm

The transfer of haem from host haemoproteins to components of the Isd system has recently been elucidated *in vitro* (Liu *et al.*, 2008; Muryoi *et al.*, 2008). As mentioned above, $IsdB^{N1}$, $IsdH^{N1}$, and $IsdH^{N2}$ bind host haemoproteins and facilitate removal of the haem cofactor. The mechanism by which haem is removed from haemoglobin has yet to be determined. $IsdB^{N2}$ and $IsdH^{N3}$ bind haem once it is removed from haemoglobin and subsequently transfer haem to IsdA (Zhu *et al.*, 2008). It is also possible for the haem-binding domains of IsdB and IsdH to bypass IsdA and transfer haem to IsdC. However, this transfer reaction is indirect and inefficient (Muryoi *et al.*, 2008; Zhu *et al.*, 2008). Considering the rapidity of haem uptake by *S. aureus* (Mazmanian *et al.*, 2003), it seems likely that the more efficient and direct transfer of haem to IsdA would be expected to occur *in vivo*. IsdA then transfers haem to apo-IsdC through formation of a holo–IsdA-apo-IsdC complex. Although this reaction is reversible, it favours formation of holo-IsdC due to the higher affinity of IsdC for haem as compared to IsdA (Liu *et al.*, 2008). IsdC then transfers haem to IsdE, the lipoprotein of the ATP-binding cassette (ABC) transporter (Grigg *et al.*, 2007; Muryoi *et al.*, 2008). Transfer of haem from IsdB/H through IsdA to IsdC and finally IsdE is unidirectional, rapid, and complete (Fig. 13.2).

Comparatively little is known about haem transport through the cytoplasmic membrane. IsdD and IsdF are putative ABC transporters, in which IsdD is a membrane-associated ATP-binding protein and IsdF is a membrane-spanning permease (Mazmanian *et al.*, 2003). It has been proposed that the IsdE–haem complex will transfer haem through IsdF, using energy generated by IsdD (Reniere *et al.*, 2007). In support of this hypothesis, IsdD and IsdE are capable of binding haem (Grigg *et al.*, 2007; Mazmanian *et al.*, 2003). Notably, inactivation of *isdDEF* impairs, but does not abolish, *S. aureus* growth on haem as a sole iron source, implying that an additional system exists for haem acquisition in *S. aureus*. To this end, the haem transport system (Hts) was identified based on homology to known ABC-type iron transporters (Skaar *et al.*, 2004b). The Hts system is comprised of three proteins: HtsA is a predicted membrane-associated lipoprotein, while HtsB and HtsC are putative ABC transporter permeases that closely resemble HmuU, the haem transport permease of *Corynebacterium diphtheriae* (Drazek *et al.*, 2000). Inactivation of *htsABC* significantly reduces haem iron acquisition *in vitro* and decreases virulence in a murine model of infection (Skaar *et al.*, 2004b). Little work has been done to continue exploring these initial observations so the specific roles that HtsABC and IsdDEF play in haem transfer across the cytoplasmic membrane are unclear.

The intracellular fate of haem

Once haem enters the cytoplasm, it has been proposed that *S. aureus* can utilize haem in one of two ways: as a nutrient iron source or as a cofactor for bacterial haemoproteins (Reniere *et al.*, 2007). In order to use haem as an iron source the porphyrin ring must be cleaved to release the coordinated iron atom. *S. aureus* expresses two haem oxygenases dedicated to this task, IsdG and IsdI (Skaar *et al.*, 2004a). *isdG* is the ultimate gene in the *isdCDEFsrtBisdG* operon, while *isdI* is co-transcribed with an uncharacterized open reading frame at a distinct genomic location (Skaar *et al.*, 2004a). IsdG and IsdI are the first members of a novel family of haem oxygenases (HOs) that do not resemble previously identified HOs. IsdG and IsdI are 64% identical at the amino acid level and their crystal structures can be superimposed with a root mean square deviation of 1.0 Å, indicating a high degree of structural similarity (Wu *et al.*, 2005). They form homodimeric β-barrels that adopt a ferredoxin-like fold. Each monomer contains an active site and the active site residues asparagine, histidine, and tryptophan are conserved between IsdG and IsdI (Lee *et al.*, 2008; Wu *et al.*, 2005). This is in stark contrast to all other known bacterial HOs, the first of which to be identified was HmuO from *C. diphtheriae* (Schmitt, 1997). The HmuO family of haem oxygenases are similar to vertebrate haem oxygenases in that they exhibit a monomeric α-helical fold with a single active site containing a GXXXG motif that is required for catalytic activity (Schuller *et al.*, 1999). The HmuO family of HOs degrade haem to release α-biliverdin, carbon monoxide, and iron, consistent with its homology to vertebrate HOs (Wilks and Schmitt, 1998). The mechanism of IsdG- and IsdI-catalysed haem degradation is not yet known, although preliminary data indicate that the mechanism is identical between the two proteins and different from the canonical haem oxygenase family.

Although IsdG and IsdI are structurally and functionally similar, the regulation and expression of these enzymes is distinct. Both *isdG* and *isdI* are regulated at the transcriptional level by iron in a Fur-dependent manner. This pattern of transcriptional regulation is similar to the regulation of other *isd* genes and allows *S. aureus* to express the haem uptake machinery only when in need of iron (Mazmanian *et al.*, 2003; Reniere and Skaar, 2008). In addition to transcriptional regulation by iron, IsdG is regulated in a Fur-independent manner at the post-transcriptional level by the availability of haem. In the presence of haem, IsdG protein levels are significantly increased compared to iron-deplete conditions lacking haem and this is due to the increased stability of IsdG when bound to its substrate (Reniere and Skaar, 2008). It is proposed that this dual regulation allows *S. aureus* to fine-tune the amount of haem oxygenase activity in the cell in order to degrade haem efficiently and prevent an accumulation of potentially toxic levels of haem.

As well as being a valuable source of iron, haem is potentially useful to *S. aureus* in its intact form. Intact haem is required as a cofactor in numerous enzymes, many of which are found in the cytoplasmic membrane, such as cytochromes that are necessary for aerobic respiration (Thony-Meyer, 1997). It is possible that when iron is not a limiting nutrient, *S. aureus* will acquire low levels of haem for use in bacterial haemoproteins, as scavenging haem requires less energy than synthesizing haem *de novo* (Perkins-Balding, 2004). This hypothesis is supported by evidence that exogenous haem is preferentially acquired and segregated intact to the membrane when non-haem iron sources are concurrently available (Skaar *et al.*, 2004b). More experiments are necessary to elucidate the fate of exogenously acquired intact haem and its role in bacterial haemoproteins.

The haem paradox – balancing haem uptake against haem toxicity

Haem is a useful molecule in biological reactions due to the redox potential of the coordinated iron. However, this reactivity also makes haem toxic when accumulated at high concentrations. Consequently, bacterial haem acquisition systems must be tightly regulated in order to allow for the import of haem to satisfy the nutrient iron requirements of the cell while avoiding the accumulation of surplus haem. When exposed to high levels of exogenous haem, such as may be encountered during infection, *S. aureus* avoids haem toxicity through elaboration of the haem-regulated transporter HrtAB (Friedman *et al.*, 2006). Pre-exposure of *S. aureus* to sub-inhibitory concentrations of haem renders it resistant to concentrations of haem that are otherwise toxic to naïve staphylococci. This adaptive resistance is mediated entirely by HrtAB which was identified based on its dramatic up-regulation (45-fold) when *S. aureus* is exposed to haem. Consistent with its vital contribution to haem detoxification, inactivation of *hrtA* renders *S. aureus* acutely sensitive to haem toxicity (Friedman *et al.*, 2006).

HrtAB is an ABC-type transport system that is predicted to function as an efflux pump, although the exact mechanism by which it confers haem resistance is unknown. In contrast, the mechanism by which *hrtAB* is activated upon haem exposure has been described in detail. *hrtAB* is divergently transcribed immediately adjacent to a two-component system named the haem-sensor system (HssRS) (Torres *et al.*, 2007). Bacterial two-component systems are comprised of a histidine kinase sensor and a cytoplasmic response regulator that act in concert to modulate gene expression in response to environmental signals. The activation of HssRS occurs upon haem exposure, whereupon HssS undergoes autophosphorylation followed by transphosphorylation of HssR. Phosphorylated HssR then binds to a direct repeat sequence within the *hrtAB* promoter and drives transcription (Stauff *et al.*, 2007). The combined activities of HssRS and HrtAB result in the alleviation of haem toxicity and the increased ability of *S. aureus* to grow in the presence of haem. Interestingly, orthologues of *hssRS* and *hrtAB* are found in many Gram-positive pathogens, including *S. epidermidis, S. saprophyticus, L. monocytogenes, Bacillus anthracis,* and *B. cereus.* However, orthologues of *hssRS* and *hrtAB* are not found in organisms from these same genera that are not expected to encounter blood or haem in their life cycle such as *B. subtilis.* Taken together, these data suggest a role for haem detoxification by HssRS and HrtAB, a process that appears to be important for survival of pathogenic and saprophytic organisms that come into contact with vertebrate blood.

Haem acquisition and infection

Haem iron acquisition is a vital process during *S. aureus* pathogenesis. Inactivation of *isdB* alone or in combination with *isdH* results in a 1-log reduction in bacterial colonization in the spleens and kidneys of infected mice as compared to wild type. In contrast, inactivation of *isdH* singly does not alter virulence, further indicating that IsdB is the primary staphylococcal haemoglobin receptor *in vivo* (Torres *et al.*, 2006). The intracellular components of the Isd system have also been implicated in *S. aureus* pathogenesis. Inactivation of *isdG* alone or in combination with *isdI* decreases bacterial load in the hearts and kidneys of infected animals (Reniere and Skaar, 2008). Inactivation of *isdI* alone diminishes the pathogenesis in the hearts of infected animals, but not the kidneys. The organ-specific difference in virulence between the *isdG* and *isdI* mutants suggests that these haem oxygenases may be utilized at different points during infection, consistent with their distinct regulation described previously. Considering that *S. aureus* encodes multiple haem and haemoprotein receptors (IsdA, IsdB, IsdC, IsdH), two membrane-associated haem transporters (IsdDEF, HtsABC), and two cytoplasmic haem degrading enzymes (IsdG, IsdI), it is not surprising that deletion of only one or two haem transport components does not completely attenuate *S. aureus* virulence. It is also important to consider that the Isd mutant strains used in these studies express wild type levels of siderophores, and are therefore able to obtain nutrient iron independently of haem. In this regard, it is difficult to accurately assess the true significance of haem acquisition during infection and the phenotypes that are observed are most likely under-representations of the virulence contribution provided by these proteins.

Differences in the iron content of host tissues may contribute to the organ-specific requirements of the various iron transport systems. For example, inactivation of *isdB, isdG,* or *isdI* has no effect on colonization of murine livers, indicating that iron may be less restricted in this organ (Pishchany *et al.*, 2009; Reniere and Skaar, 2008; Torres *et al.*, 2006). Indeed, iron in the liver is the primary source of iron in the human body (Drabkin, 1951). Accordingly, *S. aureus* infection of patients following liver transplantation is used in the clinical setting as a predictor of poor prognosis. These individuals suffer from iron overload and therefore exhibit increased susceptibility to *S. aureus* bacteraemia, resulting in mortality rates of up to eighty percent (Singh and Sun, 2008). Although *S. aureus* does not appear to be iron starved during colonization of the liver, this situation changes upon the formation of hepatic abscesses. Specifically, it has been shown that *isdI* is expressed in kidney and liver abscesses in animal models of infection (Reniere and Skaar, 2008). These findings suggest that abscess formation creates an iron-starved microenvironment within an otherwise iron rich organ. Abscess formation, typical of *S. aureus* infection, results from an influx of neutrophils to the site of infection (Lowy, 1998). In addition to antimicrobial peptides and reactive oxygen species, the neutrophil arsenal also includes the metal-chelating protein calprotectin

(Corbin et al., 2008). Calprotectin sequesters zinc and manganese away from bacteria in an abscess, indicating that metal chelation is a viable antibacterial strategy employed by the host in abscesses, a strategy that may also be used to chelate iron.

Although inactivation of the haem uptake machinery decreases the virulence of *S. aureus*, the opposite scenario is encountered upon deletion of the system responsible for haem detoxification. Inactivation of *hrtA* or *hssR* increases bacterial colonization in the livers of infected animals by greater than two logs (Torres *et al.*, 2007). This hepatic hyper-virulence is due to the reduced recruitment of neutrophils to the site of infection. A mechanistic explanation for this finding was provided by the observation that *S. aureus* lacking *hrtA* exhibit increased expression and/or secretion of a number of immunomodulatory proteins, resulting in decreased numbers of phagocytes in the livers of infected animals (Torres *et al.*, 2007). The mechanism responsible for the increased secretion of these immunomodulatory proteins is unknown. It is possible that in the absence of HrtAB, excess haem or toxic haem byproducts accumulate in the cytoplasm, increasing cellular stress. This stress may lead to increased expression of immunomodulatory proteins aimed at making the local environment more favourable to *S. aureus* colonization. Additional experimentation is needed to identify the mechanism of increased virulence production in the *hrtA* mutant in order to better understand the effects of haem homeostasis on *S. aureus* virulence factor production and pathogenesis.

In light of the importance of IsdB during infection and its surface-exposed localization, it has been proposed that IsdB represents a promising vaccine candidate. Indeed, mice immunized with purified IsdB exhibited greater survival after challenge with *S. aureus* as compared to mock-immunized mice and survival rates correlated with αIsdB antibody titres (Kuklin *et al.*, 2006). However, survival rates did not reach 100% when animals were vaccinated with IsdB alone. A subsequent study found that the four staphylococcal proteins that generate the highest amount of protective immunity are IsdA, IsdB, SdrD, and SdrE. The functions of SdrD and SdrE are unknown, however vaccination with a combination of all four antigenic proteins afforded complete protection against a lethal *S. aureus* challenge in a murine infection model (Stranger-Jones *et al.*, 2006). These results suggest that surface exposed proteins of the Isd system have significant potential as vaccine candidates although additional work is needed to determine if this protection will extend to human subjects.

In addition to the multiple functions described above, IsdA and IsdB promote resistance to neutrophil killing *in vitro* (Palazzolo-Ballance *et al.*, 2008). Neutrophils are the host's first line of defence against invading pathogens. After exposure to hydrogen peroxide, a key component of a neutrophil's antimicrobial arsenal, transcript levels of *isdA-G* are up-regulated, suggesting that haem iron acquisition is important for resistance to peroxides. Moreover, inactivation of *isdAB* decreases *S. aureus* survival when incubated in the presence of neutrophils (Palazzolo-Ballance *et al.*, 2008). The mechanism by which Isd proteins mediate resistance to hydrogen peroxide is not yet clear. Haem is a cofactor of bacterial enzymes that are involved in protection against reactive oxygen species, such as catalase (Mandell, 1975). In the absence of the Isd system it is possible that there is not sufficient haem available in the cell for these enzymes to function, rendering *S. aureus* more susceptible to reactive oxygen-mediated killing. Taken together, these studies underscore the importance of the Isd system in *S. aureus* colonization, survival, and disease, demonstrating that haem iron acquisition is vital to *S. aureus* pathogenesis.

Other iron sources

Transferrin iron, although not as abundant as haem iron in the human body, is found in the serum, making it an attractive iron source to extracellular pathogens (Drabkin, 1951). While siderophores secreted by *S. aureus* can obtain iron from transferrin, several studies have investigated direct transferrin-binding by *S. aureus* as an alternative mechanism for acquiring iron from this host protein. The staphylococcal transferrin receptor protein (Tpn) was identified by Modun *et al.* to be a transferrin-binding protein related to glyceraldehyde-3-phosphate dehydrogenase (GAPDH) (Modun *et al.*, 2000). This was surprising, as GAPDH is normally a glycolytic protein localized in the cytoplasm. It was subsequently shown that Tpn (annotated as Gap) does not bind transferrin and is not a *bona fide* transferrin receptor. Transferrin binding was instead attributed to the staphylococcal transferrin binding protein A (StbA) based on its iron-regulated expression and ability to confer transferrin-binding on *E. coli* when exogenously expressed (Taylor and Heinrichs, 2002). Coincident with this report, the same protein was identified as IsdA, the haem binding receptor of the Isd haem-uptake machinery (Mazmanian *et al.*, 2003). A more recent study concluded that StbA/IsdA is not a transferrin receptor, and that transferrin iron acquisition is primarily mediated by secretion of siderophores rather than transferrin-binding to a staphylococcal receptor (Park *et al.*, 2005). Taking these facts together, it appears that the transferrin binding activity attributed to IsdA is not an iron acquisition strategy but is instead consistent with the broad ligand binding activity of this protein (Clarke *et al.*, 2004). It is still unclear what role transferrin-binding may play in staphylococcal iron acquisition during infection, although it is formally possible that *S. aureus* does not produce a transferrin receptor involved in iron acquisition.

In addition to binding host iron-containing proteins as a mechanism of iron acquisition, *S. aureus* can modulate its environment in order to make iron more bioavailable. Specifically, *S. aureus* is known to alter the expression of genes that control central metabolic pathways in order to increase free iron concentrations (Friedman *et al.*, 2006). Upon iron-deprivation, glycolytic enzymes are increased while TCA cycle enzymes are down-regulated, resulting in pyruvate being shuttled into fermentative pathways. This leads to an increase in the production of acidic products, such as lactate. Lactate that is released into the medium decreases the pH of the local microenvironment and decreases the oxidation reduction potential (Eh). The combination of these effects leads to a conversion of iron to its more soluble and bioavailable ferrous form as well as an increased release of iron from host proteins (Friedman *et al.*, 2006). In this way, *S. aureus* may be able to manipulate the local environment at the site of infection in order to obtain the nutrients it requires for survival.

Concluding remarks

There is still much to be learned regarding the mechanism and importance of *S. aureus* iron acquisition during infection. Many of the haem-binding properties of the Isd system have been studied in the laboratory setting, but the impact of these interactions during staphylococcal infection is less clear. Haemoglobin-binding has been demonstrated for IsdB and IsdH, although it is not yet known how IsdB and IsdH remove the haem cofactor from haemoglobin. The aforementioned studies suggest that IsdB is a viable vaccine candidate in a murine model of infection. Therefore, understanding the mechanism by which IsdB functions during infection is particularly important for the further development of this vaccine for use in prevention of human infections.

Once haem is transported into the cell, little is known about its fate. It has been proposed that non-degraded haem can be used as a cofactor for bacterial haemoproteins, although this has not yet been proven experimentally. Haem can also be degraded for use as an iron source, but the mechanism of IsdG- and IsdI-mediated haem degradation has not yet been determined. In eukaryotes haem is degraded to iron, carbon monoxide, and biliverdin. Carbon monoxide is an effective anti-inflammatory molecule and biliverdin is reduced to bilirubin, a potent antioxidant (Chung *et al.*, 2008; Stocker *et al.*, 1987; Unno *et al.*, 2007). In a similar manner, the products of IsdG- and IsdI-mediated degradation may have important functions inside the bacterial cell. Elucidating the mechanism of haem degradation may also provide insight for the design of small molecule inhibitors of these enzymes. Moreover, all proteins involved in iron acquisition are potential targets for small molecule inhibitors that can be utilized as therapeutics, a supposition supported by the contribution of iron acquisition systems to staphylococcal pathogenesis in murine infection models.

The staphylococcal siderophore field also has many unanswered questions. Although *S. aureus* is known to produce four siderophores, only staphyloferrin A and staphyloferrin B have been structurally characterized. Additionally, *S. aureus* strains deficient in staphyloferrin A and/or staphyloferrin B production have not been created so the contribution of each of these siderophores to staphylococcal iron acquisition is unknown. There have also been several genes identified that encode for putative siderophore transporters, but the specific siderophores they transport are unknown. Given that mutations in haem iron acquisition systems do not completely abolish *S. aureus* virulence, non-haem iron acquisition probably plays an important role during infection. In this regard, it is important to fully characterize siderophore production and utilization, and to determine when during infection specific iron acquisition systems are utilized.

Considering the importance of *S. aureus* as a human pathogen and the likelihood that this organism will eventually exhibit complete drug resistance, it is imperative that we gain a better understanding of *S. aureus* pathogenesis. As iron is necessary for the survival and growth of *S. aureus*, inhibiting metal acquisition systems represents an attractive therapeutic strategy against drug-resistant bacteria.

Acknowledgements

We would like to thank the members of the Skaar laboratory for critical reading of this chapter. Research in the Skaar laboratory is supported by the Searle Scholars Program, and United States Public Health Service Grants AI69233, AI073843, and U54 AI057157 from the National Institute of Allergy and Infectious Diseases and The Southeast Regional Center for Excellence for Emerging Infections and Biodefense. Eric Skaar, Ph.D. holds an Investigator in the Pathogenesis of Infectious Disease Award from the Burroughs Welcome Fund. Michelle Reniere is funded by NIH Training Grant in Mechanisms of Vascular Disease, 5 T32 HL07751.

References

Aguila, A., Herrera, A.G., Morrison, D., Cosgrove, B., Perojo, A., Montesinos, I., Perez, J., Sierra, G., Gemmell, C.G., and Brock, J.H. (2001). Bacteriostatic activity of human lactoferrin against *Staphylococcus aureus* is a function of its iron-binding properties and is not influenced by antibiotic resistance. FEMS Immunol. Med. Microbiol. *31*, 145–152.

Allard, M., Moisan, H., Brouillette, E., Gervais, A.L., Jacques, M., Lacasse, P., Diarra, M.S., and Malouin, F. (2006). Transcriptional modulation of some *Staphylococcus aureus* iron-regulated genes during growth *in vitro* and in a tissue cage model *in vivo*. Microbes Infect. *8*, 1679–1690.

Andrade, M.A., Ciccarelli, F.D., Perez-Iratxeta, C., and Bork, P. (2002). NEAT: a domain duplicated in genes near the components of a putative Fe3+ siderophore transporter from Gram-positive pathogenic bacteria. Genome Biol. 3, RESEARCH0047.

Athanasopoulos, A.N., Economopoulou, M., Orlova, V.V., Sobke, A., Schneider, D., Weber, H., Augustin, H.G., Eming, S.A., Schubert, U., Linn, T., *et al.* (2006). The extracellular adherence protein (Eap) of *Staphylococcus aureus* inhibits wound healing by interfering with host defense and repair mechanisms. Blood. *107*, 2720–2727.

Baichoo, N., and Helmann, J.D. (2002). Recognition of DNA by Fur: a reinterpretation of the Fur box consensus sequence. J. Bacteriol. *184*, 5826–5832.

Baldassarri, L., Bertuccini, L., Ammendolia, M.G., Arciola, C.R., and Montanaro, L. (2001). Effect of iron limitation on slime production by *Staphylococcus aureus*. Eur. J. Clin. Microbiol. Infect. Dis. *20*, 343–345.

Barton Pai, A., Pai, M.P., Depczynski, J., McQuade, C.R., and Mercier, R.C. (2006). Non-transferrin-bound iron is associated with enhanced *Staphylococcus aureus* growth in hemodialysis patients receiving intravenous iron sucrose. Am. J. Nephrol. *26*, 304–309.

Bergeron, R.J., Elliott, G.T., Kline, S.J., Ramphal, R., and St James, L., 3rd (1983). Bacteriostatic and fungostatic action of catecholamide iron chelators. Antimicrob. Agents Chemother. *24*, 725–730.

Brock, J.H., and Ng, J. (1983). The effect of desferrioxamine on the growth of *Staphylococcus aureus*, *Yersinia enterocolitica* and *Streptococcus faecalis* in human-serum – uptake of desferrioxamine-bound iron. FEMS Microbiol. Lett. *20*, 439–442.

Bullen, J.J., and Griffiths, E. (1999). Iron and Infection: Molecular, Physiological and Clinical Aspects (New York: John Wiley and Sons).

CDC (2007). HIV/AIDS Surveillance Report. *17*.

Chavakis, T., Hussain, M., Kanse, S.M., Peters, G., Bretzel, R.G., Flock, J.I., Herrmann, M., and Preissner, K.T. (2002). *Staphylococcus aureus* extracellular adherence protein serves as anti-inflammatory factor by inhibiting the recruitment of host leukocytes. Nat. Med. *8*, 687–693.

Chavakis, T., Wiechmann, K., Preissner, K.T., and Herrmann, M. (2005). *Staphylococcus aureus* interactions with the endothelium: the role of bacterial 'secretable expanded repertoire adhesive molecules' (SERAM) in disturbing host defense systems. Thromb. Haemost. *94*, 278–285.

Chung, J.H., Park, M.H., Kim, J.H., Lim, Y., and Shin, S.H. (2003). Growth and siderophore production of staphylococci in human peritoneal dialysate. J. Korean Med. Sci. *18*, 158–162.

Chung, S.W., Liu, X., Macias, A.A., Baron, R.M., and Perrella, M.A. (2008). Heme oxygenase-1-derived carbon monoxide enhances the host defense response to microbial sepsis in mice. J. Clin. Invest. *118*, 239–247.

Clarke, S.R., and Foster, S.J. (2008). IsdA protects *Staphylococcus aureus* against the bactericidal protease activity of apolactoferrin. Infect. Immun. *76*, 1518–1526.

Clarke, S.R., Wiltshire, M.D., and Foster, S.J. (2004). IsdA of *Staphylococcus aureus* is a broad spectrum, iron-regulated adhesin. Mol. Microbiol. *51*, 1509–1519.

Cockayne, A., Hill, P.J., Powell, N.B., Bishop, K., Sims, C., and Williams, P. (1998). Molecular cloning of a 32-kilodalton lipoprotein component of a novel iron-regulated *Staphylococcus epidermidis* ABC transporter. Infect. Immun. *66*, 3767–3774.

Corbin, B.D., Seeley, E.H., Raab, A., Feldmann, J., Miller, M.R., Torres, V.J., Anderson, K.L., Dattilo, B.M., Dunman, P.M., Gerads, R., *et al.* (2008). Metal chelation and inhibition of bacterial growth in tissue abscesses. Science. *319*, 962–965.

Cotton, J.L., Tao, J., and Balibar, C.J. (2009). Identification and characterization of the *Staphylococcus aureus* gene cluster coding for staphyloferrin A. Biochemistry. *48*, 1025–1035.

Courcol, R.J., Trivier, D., Bissinger, M.C., Martin, G.R., and Brown, M.R. (1997). Siderophore production by *Staphylococcus aureus* and identification of iron-regulated proteins. Infect. Immun. *65*, 1944–1948.

Crichton, R. (2001). Inorganic Biochemistry of Iron Metabolism: From Molecular Mechanisms to Clinical Consequences, Vol 2 (West Sussex, England: John Wiley & Sons, Ltd.).

Dale, S.E., Doherty-Kirby, A., Lajoie, G., and Heinrichs, D.E. (2004a). Role of siderophore biosynthesis in virulence of *Staphylococcus aureus*: identification and characterization of genes involved in production of a siderophore. Infect. Immun. *72*, 29–37.

Dale, S.E., Sebulsky, M.T., and Heinrichs, D.E. (2004b). Involvement of SirABC in iron-siderophore import in *Staphylococcus aureus*. J. Bacteriol. *186*, 8356–8362.

Diarra, M.S., Petitclerc, D., and Lacasse, P. (2002). Response of *Staphylococcus aureus* isolates from bovine mastitis to exogenous iron sources. J. Dairy Sci. *85*, 2141–2148.

Drabkin, D. (1951). Metabolism of the hemin chromoproteins. Physiological Reviews. *31*, 345–431.

Drazek, E.S., Hammack, C.A., and Schmitt, M.P. (2000). *Corynebacterium diphtheriae* genes required for acquisition of iron from haemin and haemoglobin are homologous to ABC haemin transporters. Mol. Microbiol. *36*, 68–84.

Drechsel, H., Freund, S., Nicholson, G., Haag, H., Jung, O., Zahner, H., and Jung, G. (1993). Purification and chemical characterization of staphyloferrin B, a hydrophilic siderophore from staphylococci. Biometals *6*, 185–192.

Dryla, A., Hoffmann, B., Gelbmann, D., Giefing, C., Hanner, M., Meinke, A., Anderson, A.S., Koppensteiner, W., Konrat, R., von Gabain, A., and Nagy, E. (2007). High-affinity binding of the staphylococcal HarA protein to haptoglobin and hemoglobin involves a domain with an antiparallel eight-stranded beta-barrel fold. J. Bacteriol. *189*, 254–264.

Friedman, D.B., Stauff, D.L., Pishchany, G., Whitwell, C.W., Torres, V.J., and Skaar, E.P. (2006). *Staphylococcus aureus* redirects central metabolism to increase iron availability. PLoS Pathog. *2*, e87.

Garzoni, C., Francois, P., Huyghe, A., Couzinet, S., Tapparel, C., Charbonnier, Y., Renzoni, A., Lucchini, S., Lew, D.P., Vaudaux, P., *et al.* (2007). A global view of *Staphylococcus aureus* whole genome expression upon internalization in human epithelial cells. BMC Genomics. *8*, 171.

Grigg, J.C., Vermeiren, C.L., Heinrichs, D.E., and Murphy, M.E. (2007). Heme coordination by *Staphylococcus aureus* IsdE. J. Biol. Chem. *282*, 28815–28822.

Haag, H., Fiedler, H.P., Meiwes, J., Drechsel, H., Jung, G., and Zahner, H. (1994). Isolation and biological characterization of staphyloferrin B, a compound with siderophore activity from staphylococci. FEMS Microbiol. Lett. *115*, 125–130.

Heinrichs, J.H., Gatlin, L.E., Kunsch, C., Choi, G.H., and Hanson, M.S. (1999). Identification and characterization of SirA, an iron-regulated protein from *Staphylococcus aureus*. J. Bacteriol. *181*, 1436–1443.

Herron-Olson, L., Fitzgerald, J.R., Musser, J.M., and Kapur, V. (2007). Molecular correlates of host specialization in *Staphylococcus aureus*. PLoS ONE. *2*, e1120.

Horsburgh, M.J., Ingham, E., and Foster, S.J. (2001). In *Staphylococcus aureus*, fur is an interactive regulator with PerR, contributes to virulence, and is necessary for oxidative stress resistance through positive regulation of catalase and iron homeostasis. J. Bacteriol. *183*, 468–475.

Hussain, M., Becker, K., von Eiff, C., Schrenzel, J., Peters, G., and Herrmann, M. (2001). Identification and characterization of a novel 38.5-kilodalton cell surface protein of *Staphylococcus aureus* with extended-spectrum binding activity for extracellular matrix and plasma proteins. J. Bacteriol. *183*, 6778–6786.

Jean, G., Charra, B., Chazot, C., Vanel, T., Terrat, J.C., Hurot, J.M., and Laurent, G. (2002). Risk factor analysis for long-term tunneled dialysis catheter-related bacteremias. Nephron. *91*, 399–405.

Johnson, M., Cockayne, A., and Morrissey, J.A. (2008). Iron-regulated biofilm formation in *Staphylococcus aureus* Newman requires *ica* and the secreted protein Emp. Infect. Immun. *76*, 1756–1765.

Johnson, M., Cockayne, A., Williams, P.H., and Morrissey, J.A. (2005). Iron-responsive regulation of biofilm formation in *Staphylococcus aureus* involves *fur*-dependent and *fur*-independent mechanisms. J. Bacteriol. *187*, 8211–8215.

Klevens, R.M., Morrison, M.A., Nadle, J., Petit, S., Gershman, K., Ray, S., Harrison, L.H., Lynfield, R., Dumyati, G., Townes, J.M., *et al.* (2007). Invasive methicillin-resistant *Staphylococcus aureus* infections in the United States. JAMA. *298*, 1763–1771.

Konetschny-Rapp, S., Jung, G., Meiwes, J., and Zahner, H. (1990). Staphyloferrin A: a structurally new siderophore from staphylococci. Eur. J. Biochem. *191*, 65–74.

Kuehnert, M.J., Hill, H.A., Kupronis, B.A., Tokars, J.I., Solomon, S.L., and Jernigan, D.B. (2005). Methicillin-resistant-*Staphylococcus aureus* hospitalizations, United States. Emerg. Infect. Dis. *11*, 868–872.

Kuehnert, M.J., Kruszon-Moran, D., Hill, H.A., McQuillan, G., McAllister, S.K., Fosheim, G., McDougal, L.K., Chaitram, J., Jensen, B., Fridkin, S.K., *et al.* (2006). Prevalence of *Staphylococcus aureus* nasal colonization in the United States, 2001–2002. J. Infect. Dis. *193*, 172–179.

Kuklin, N.A., Clark, D.J., Secore, S., Cook, J., Cope, L.D., McNeely, T., Noble, L., Brown, M.J., Zorman, J.K., Wang, X.M., *et al.* (2006). A novel *Staphylococcus aureus* vaccine: iron surface determinant B induces rapid antibody responses in rhesus macaques and specific increased survival in a murine *S. aureus* sepsis model. Infect. Immun. *74*, 2215–2223.

Lee, W.C., Reniere, M.L., Skaar, E.P., and Murphy, M.E. (2008). Ruffling of metalloporphyrins bound to IsdG and IsdI, two heme-degrading enzymes in *Staphylococcus aureus*. J. Biol. Chem. *283*, 30957–30963.

Lim, Y., Shin, S.H., Lee, S.I., Kim, I.S., and Rhee, J.H. (1998). Iron repressibility of siderophore and transferrin-binding protein in *Staphylococcus aureus*. FEMS Microbiol. Lett. *163*, 19–24.

Lindsay, J.A., and Riley, T.V. (1994). Staphylococcal iron requirements, siderophore production, and iron-regulated protein expression. Infect. Immun. *62*, 2309–2314.

Lindsay, J.A., Riley, T.V., and Mee, B.J. (1995). *Staphylococcus aureus* but not *Staphylococcus epidermidis* can acquire iron from transferrin. Microbiology *141*, 197–203.

Liu, M., Tanaka, W.N., Zhu, H., Xie, G., Dooley, D.M., and Lei, B. (2008). Direct hemin transfer from IsdA to IsdC in the iron-regulated surface determinant (Isd) heme acquisition system of *Staphylococcus aureus*. J. Biol. Chem. *283*, 6668–6676.

Lowy, F.D. (1998). *Staphylococcus aureus* infections. N. Engl. J. Med. *339*, 520–532.

Mandell, G.L. (1975). Catalase, superoxide dismutase, and virulence of *Staphylococcus aureus*. *In vitro* and *in vivo* studies with emphasis on staphylococcal--leukocyte interaction. J. Clin. Invest. *55*, 561–566.

Marcelis, J.H., den Daas-Slagt, H.J., and Hoogkamp-Korstanje, J.A. (1978). Iron requirement and chelator production of staphylococci, *Streptococcus faecalis* and enterobacteriaceae. Antonie Van Leeuwenhoek; *44*, 257–267.

Maresso, A.W., and Schneewind, O. (2006). Iron acquisition and transport in *Staphylococcus aureus*. Biometals *19*, 193–203.

Maskell, J.P. (1980). The functional interchangeability of enterobacterial and staphylococcal iron chelators. Antonie Van Leeuwenhoek. *46*, 343–351.

Mazmanian, S.K., Liu, G., Jensen, E.R., Lenoy, E., and Schneewind, O. (2000). *Staphylococcus aureus* sortase mutants defective in the display of surface proteins and in the pathogenesis of animal infections. Proc. Natl. Acad. Sci. U.S.A. *97*, 5510–5515.

Mazmanian, S.K., Liu, G., Ton-That, H., and Schneewind, O. (1999). *Staphylococcus aureus* sortase, an enzyme that anchors surface proteins to the cell wall. Science. *285*, 760–763.

Mazmanian, S.K., Skaar, E.P., Gaspar, A.H., Humayun, M., Gornicki, P., Jelenska, J., Joachmiak, A., Missiakas, D.M., and Schneewind, O. (2003). Passage of heme-iron across the envelope of *Staphylococcus aureus*. Science. *299*, 906–909.

Mazmanian, S.K., Ton-That, H., and Schneewind, O. (2001). Sortase-catalysed anchoring of surface proteins to the cell wall of *Staphylococcus aureus*. Mol. Microbiol. *40*, 1049–1057.

Mazmanian, S.K., Ton-That, H., Su, K., and Schneewind, O. (2002). An iron-regulated sortase anchors a class of surface protein during *Staphylococcus aureus* pathogenesis. Proc. Natl. Acad. Sci. U.S.A. *99*, 2293–2298.

Meiwes, J., Fiedler, H.P., Haag, H., Zahner, H., Konetschny-Rapp, S., and Jung, G. (1990). Isolation and characterization of staphyloferrin A, a compound with siderophore activity from *Staphylococcus hyicus*. DSM 20459. FEMS Microbiol. Lett. *55*, 201–205.

Menestrina, G., Dalla Serra, M., Comai, M., Coraiola, M., Viero, G., Werner, S., Colin, D.A., Monteil, H., and Prevost, G. (2003). Ion channels and bacterial infection: the case of beta-barrel pore-forming protein toxins of *Staphylococcus aureus*. FEBS Lett. *552*, 54–60.

Modun, B., Morrissey, J., and Williams, P. (2000). The staphylococcal transferrin receptor: a glycolytic enzyme with novel functions. Trends Microbiol. *8*, 231–237.

Monsen, T., Olofsson, C., Granstrom, S., and Wistrom, J. (2003). A preliminary evaluation of a new selective agar supplemented with desferrioxamine for detection of methicillin-resistant *Staphylococcus aureus*. J. Hosp. Infect. *54*, 216–221.

Morrissey, J.A., Cockayne, A., Hill, P.J., and Williams, P. (2000). Molecular cloning and analysis of a putative siderophore ABC transporter from *Staphylococcus aureus*. Infect. Immun. *68*, 6281–6288.

Muryoi, N., Tiedemann, M.T., Pluym, M., Cheung, J., Heinrichs, D.E., and Stillman, M.J. (2008). Demonstration of the iron-regulated surface determinant (Isd) heme transfer pathway in *Staphylococcus aureus*. J. Biol. Chem. *283*, 28125–28136.

Palazzolo-Ballance, A.M., Reniere, M.L., Braughton, K.R., Sturdevant, D.E., Otto, M., Kreiswirth, B.N., Skaar, E.P., and DeLeo, F.R. (2008). Neutrophil microbicides induce a pathogen survival response in community-associated methicillin-resistant *Staphylococcus aureus*. J. Immunol. *180*, 500–509.

Palma, M., Haggar, A., and Flock, J.I. (1999). Adherence of *Staphylococcus aureus* is enhanced by an endogenous secreted protein with broad binding activity. J. Bacteriol. *181*, 2840–2845.

Park, R.Y., Sun, H.Y., Choi, M.H., Bai, Y.H., and Shin, S.H. (2005). *Staphylococcus aureus* siderophore-mediated iron-acquisition system plays a dominant and essential role in the utilization of transferrin-bound iron. J. Microbiol. *43*, 183–190.

Perkins-Balding, D., Rasmussen, A., and Stojiljkovic, I. (2004). Bacterial heme and hemoprotein receptors, In Iron Transport in Bacteria, J.H. Crosa, Mey, A.R., Payne, S.M., ed. (Washington, D.C.: ASM Press), pp. 66–85.

Pilpa, R.M., Fadeev, E.A., Villareal, V.A., Wong, M.L., Phillips, M., and Clubb, R.T. (2006). Solution structure of the NEAT (NEAr Transporter) domain from IsdH/HarA: the human hemoglobin receptor in *Staphylococcus aureus*. J. Mol. Biol. *360*, 435–447.

Pilpa, R.M., Robson, S.A., Villareal, V.A., Wong, M.L., Phillips, M., and Clubb, R.T. (2009). Functionally distinct NEAT (NEAr Transporter) domains within the *Staphylococcus aureus* IsdH/HarA protein extract heme from methemoglobin. J. Biol. Chem. *284*, 1166–1176.

Pishchany, G., Dickey, S.E., and Skaar, E.P. (2009). Subcellular localization of the *Staphylococcus aureus* heme-iron transport components IsdA and IsdB. Infect. Immun. *77*, 2624–2634.

Puri, S., and O'Brian, M.R. (2006). The *hmuQ* and *hmuD* genes from *Bradyrhizobium japonicum* encode heme-degrading enzymes. J. Bacteriol. *188*, 6476–6482.

Rahav, G., Volach, V., Shapiro, M., Rund, D., Rachmilewitz, E.A., and Goldfarb, A. (2006). Severe infections in thalassaemic patients: prevalence and predisposing factors. Br. J. Haematol. *133*, 667–674.

Reniere, M.L., and Skaar, E.P. (2008). *Staphylococcus aureus* haem oxygenases are differentially regulated by iron and haem. Mol. Microbiol. *69*, 1304–1315.

Reniere, M.L., Torres, V.J., and Skaar, E.P. (2007). Intracellular metalloporphyrin metabolism in *Staphylococcus aureus*. Biometals *20*, 333–345.

Schade, A.L. (1963). Significance of serum iron for the growth, biological characteristics, and metabolism of *Staphylococcus aureus*. Biochem. Z. *338*, 140–148.

Schmitt, M.P. (1997). Utilization of host iron sources by *Corynebacterium diphtheriae*: identification of a gene whose product is homologous to eukaryotic heme oxygenases and is required for acquisition of iron from heme and hemoglobin. J. Bacteriol. *179*, 838–845.

Schuller, D.J., Wilks, A., Ortiz de Montellano, P.R., and Poulos, T.L. (1999). Crystal structure of human heme oxygenase-1. Nat. Struct. Biol. *6*, 860–867.

Sebulsky, M.T., and Heinrichs, D.E. (2001). Identification and characterization of *fhuD1* and *fhuD2*, two genes involved in iron-hydroxamate uptake in *Staphylococcus aureus*. J. Bacteriol. *183*, 4994–5000.

Sebulsky, M.T., Hohnstein, D., Hunter, M.D., and Heinrichs, D.E. (2000). Identification and characterization of a membrane permease involved in iron-hydroxamate transport in *Staphylococcus aureus*. J. Bacteriol. *182*, 4394–4400.

Sebulsky, M.T., Shilton, B.H., Speziali, C.D., and Heinrichs, D.E. (2003). The role of FhuD2 in iron(III)-hydroxamate transport in *Staphylococcus aureus*. Demonstration that FhuD2 binds iron(III)-hydroxamates but with minimal conformational change and implication of mutations on transport. J. Biol. Chem. *278*, 49890–49900.

Sebulsky, M.T., Speziali, C.D., Shilton, B.H., Edgell, D.R., and Heinrichs, D.E. (2004). FhuD1, a ferric hydroxamate-binding lipoprotein in *Staphylococcus aureus*: a case of gene duplication and lateral transfer. J. Biol. Chem. *279*, 53152–53159.

Sharp, K.H., Schneider, S., Cockayne, A., and Paoli, M. (2007). Crystal structure of the heme–IsdC complex, the central conduit of the Isd iron/heme uptake system in *Staphylococcus aureus*. J. Biol. Chem. *282*, 10625–10631.

Singh, N., and Sun, H.Y. (2008). Iron overload and unique susceptibility of liver transplant recipients to disseminated disease due to opportunistic pathogens. Liver Transpl. *14*, 1249–1255.

Skaar, E.P., Gaspar, A.H., and Schneewind, O. (2004a). IsdG and IsdI, heme-degrading enzymes in the cytoplasm of *Staphylococcus aureus*. J. Biol. Chem. *279*, 436–443.

Skaar, E.P., Humayun, M., Bae, T., DeBord, K.L., and Schneewind, O. (2004b). Iron-source preference of *Staphylococcus aureus* infections. Science. *305*, 1626–1628.

Skaar, E.P., and Schneewind, O. (2004). Iron-regulated surface determinants (Isd) of *Staphylococcus aureus*: stealing iron from heme. Microbes Infect. *6*, 390–397.

Speziali, C.D., Dale, S.E., Henderson, J.A., Vines, E.D., and Heinrichs, D.E. (2006). Requirement of *Staphylococcus aureus* ATP-binding cassette-ATPase FhuC for iron-restricted growth and evidence that it functions with more than one iron transporter. J. Bacteriol. *188*, 2048–2055.

Stauff, D.L., Torres, V.J., and Skaar, E.P. (2007). Signaling and DNA-binding activities of the *Staphylococcus aureus* HssR-HssS two-component system required for heme sensing. J. Biol. Chem. *282*, 26111–26121.

Stocker, R., Yamamoto, Y., McDonagh, A.F., Glazer, A.N., and Ames, B.N. (1987). Bilirubin is an antioxidant of possible physiological importance. Science. *235*, 1043–1046.

Stranger-Jones, Y.K., Bae, T., and Schneewind, O. (2006). Vaccine assembly from surface proteins of *Staphylococcus aureus*. Proc. Natl. Acad. Sci. U.S.A. *103*, 16942–16947.

Taylor, J.M., and Heinrichs, D.E. (2002). Transferrin binding in *Staphylococcus aureus*: involvement of a cell wall-anchored protein. Mol. Microbiol. *43*, 1603–1614.

Thony-Meyer, L. (1997). Biogenesis of respiratory cytochromes in bacteria. Microbiol. Mol. Biol. Rev. *61*, 337–376.

Ton-That, H., Mazmanian, S.K., Faull, K.F., and Schneewind, O. (2000). Anchoring of surface proteins to the cell wall of *Staphylococcus aureus*. Sortase catalyzed *in vitro* transpeptidation reaction using LPXTG peptide and NH(2)-Gly(3) substrates. J. Biol. Chem. *275*, 9876–9881.

Torres, V.J., Pishchany, G., Humayun, M., Schneewind, O., and Skaar, E.P. (2006). *Staphylococcus aureus* IsdB is a hemoglobin receptor required for heme iron utilization. J. Bacteriol. *188*, 8421–8429.

Torres, V.J., Stauff, D.L., Pishchany, G., Bezbradica, J.S., Gordy, L.E., Iturregui, J., Anderson, K.L., Dunman, P., Joyce, S., and Skaar, E.P. (2007). A *Staphylococcus aureus* regulatory system that responds to host heme and modulates virulence. Cell Host & Microbe. *1*, 109–119.

Trivier, D., and Courcol, R.J. (1996). Iron depletion and virulence in *Staphylococcus aureus*. FEMS Microbiol. Lett. *141*, 117–127.

Unno, M., Matsui, T., and Ikeda-Saito, M. (2007). Structure and catalytic mechanism of heme oxygenase. Nat. Prod. Rep. *24*, 553–570.

Valenti, P., De Stasio, A., Seganti, L., Mastromarino, P., Sinibaldi, L., and Orsi, N. (1980). Capacity of staphylococci to grow in the presence of ovotransferrin or $CrCl_3$ as a character of potential pathogenicity. J. Clin. Microbiol. *11*, 445–447.

Vermeiren, C.L., Pluym, M., Mack, J., Heinrichs, D.E., and Stillman, M.J. (2006). Characterization of the heme binding properties of *Staphylococcus aureus* IsdA. Biochemistry. *45*, 12867–12875.

Watanabe, M., Tanaka, Y., Suenaga, A., Kuroda, M., Yao, M., Watanabe, N., Arisaka, F., Ohta, T., Tanaka, I., and Tsumoto, K. (2008). Structural basis for multimeric heme complexation through a specific protein–heme interaction: the case of the third neat domain of IsdH from *Staphylococcus aureus*. J. Biol. Chem. *283*, 28649–28659.

Wilks, A., and Schmitt, M.P. (1998). Expression and characterization of a heme oxygenase (Hmu O) from *Corynebacterium diphtheriae*. Iron acquisition requires oxidative cleavage of the heme macrocycle. J. Biol. Chem. *273*, 837–841.

Wu, R., Skaar, E.P., Zhang, R., Joachimiak, G., Gornicki, P., Schneewind, O., and Joachimiak, A. (2005). *Staphylococcus aureus* IsdG and IsdI, heme-degrading enzymes with structural similarity to monooxygenases. J. Biol. Chem. *280*, 2840–2846.

Zhu, H., Xie, G., Liu, M., Olson, J.S., Fabian, M., Dooley, D.M., and Lei, B. (2008). Pathway for heme uptake from human methemoglobin by the iron-regulated surface determinants system of *Staphylococcus aureus*. J. Biol. Chem. *283*, 18450–18460.

Iron Uptake in Yeasts

14

Pierre-Louis Blaiseau, Alexandra Seguin, Jean-Michel Camadro and Emmanuel Lesuisse

Abstract

Yeasts take up iron by three main mechanisms. In the reductive uptake mechanism, specialized flavohaemoproteins (Fre) dissociate extracellular ferric complexes by reduction involving trans-plasma membrane electron transfer. The resulting free iron is then imported by a high-affinity permease system (Ftr), coupled to a copper-dependent oxidase (Fet), which channels iron through the plasma membrane. As a consequence, iron uptake by this mechanism is dependent on the availability of copper. In the siderophore-mediated mechanism, siderophores excreted by the cells or produced by other bacterial or fungal species are taken up without prior dissociation, via specific, copper-independent high-affinity receptors. The iron is then dissociated from the siderophores intracellularly, probably by reduction. In the haem uptake mechanism, free haem or haem bound to haemoglobin is taken up as such, probably by endocytosis. Iron is released intracellularly after hydrolysis of the porphyrin ring catalysed by haem oxygenase. Within the cell, iron is stored in vacuoles or in siderophores. Iron can be mobilized from vacuoles by a reductive mechanism homologous to that found at the plasma membrane. Regulation of iron uptake and iron use are mediated by transcriptional regulators acting either as activators in iron-deficient conditions or as repressors in iron-rich conditions, according to the yeast species; these regulators thus adjust the iron uptake flux to the cell's requirements. In the baker's yeast, *Saccharomyces cerevisiae,* a post-transcriptional mechanism is active under low iron conditions, involving the degradation of RNAs encoding inessential iron-utilizing proteins. Other fungi have mechanisms serving a similar purpose at the transcriptional level. Studies in *S. cerevisiae* show that mitochondria are central to regulating cellular iron homeostasis, through the synthesis of iron–sulphur clusters.

Introduction

Various phyla of fungi include species with yeast and filamentous forms. Thus, the term 'yeast' does not refer to a homogeneous group of species, but rather to a morphological form that can be found in species of fungi that are distantly related evolutionarily. Moreover, many fungi have life cycles in which unicellular forms (yeasts) alternate with mycelial (or filamentous) forms, and environmental changes can also trigger a shift in the morphological form of many fungi. For example, the pathogenic fungus *Candida albicans* shifts from the yeast to the hyphal (filamentous) form in response to various stresses (temperature, contact with serum and others), and even the model baker's yeast *Saccharomyces cerevisiae* is able to adopt a filamentous form (pseudo-hyphal) when subjected to nitrogen starvation for example (Madhani and Fink, 1998). Here, we will use the term 'yeast' to refer to fungi that are unicellular for most of their life cycle or in defined growth conditions. These organisms live in a broad range of ecological niches, and the strategies they use to acquire iron vary according to the form and availability of iron in their environment. Despite this diversity, the mechanisms of iron uptake by yeasts conform to general rules that are well conserved between unrelated species. Pathogenic yeasts have developed strategies to obtain iron from their mammalian host, in which the most abundant sources of iron are transferrin and haemoglobin. Non-pathogenic yeasts take up iron from their environment (notably soil, water and fruit), where its concentration in solution may be very low, due to the strong tendency of iron to precipitate in aerobic conditions at neutral pH. All yeasts use essentially three main strategies to obtain iron from the host or from the natural environment: siderophore production and/or use (non-reductive iron uptake); reductive iron uptake; and haem uptake systems. The distinction between reductive and non-reductive uptake of iron is conventional: as a general rule, all iron uptake involves reduction to Fe(II), to facilitate ligand exchange. Consequently, the difference between siderophore uptake ('non-reductive') and reductive uptake systems is better described in terms of contrasted mechanisms with internal or external reduction, respectively (De Luca and Wood, 2000). Haem uptake itself does not involve reduction, because iron in either the ferrous or ferric state is tightly bound to the porphyrin ring. Another general rule is that the cellular systems involved in the uptake and utilization of iron by yeasts are precisely regulated according to the availability of iron and the cellular requirements for iron. In all yeasts, the regulation is mostly transcriptional (De Luca and Wood, 2000; Haas, 2003). However, post-transcriptional regulation has also been described in the yeast *S. cerevisiae* (Vergara and Thiele, 2008). We will describe these three main regulated strategies used by yeasts to obtain iron, and briefly describe the fate of intracellular iron, with constant reference to the model yeast *S. cerevisiae*, in which iron metabolism has been studied in greatest detail (Kosman, 2003; Philpott, 2006; Philpott and Protchenko, 2008). Iron metabolism in this yeast serves as a general model, because the corresponding studies both provide a molecular basis for a better understanding of iron metabolism in fungi, and also underpinned the development of the principle of 'reductive iron uptake', now recognized as a universal mechanism used by eukaryotic cells to import iron across the plasma membrane. Twenty years ago, when the studies on iron metabolism in *S. cerevisiae* started to progress, our collaborator A. Dancis wrote 'It is tempting to speculate that the iron uptake process in the simple eukaryote *S. cerevisiae*... provides a model for the missing link in mammalian cellular iron metabolism' (Dancis *et al.*, 1990). This speculation proved to be correct: work with baker's yeast not only resulted in a better understanding of the universal mechanism of reductive iron uptake, but also allowed the discovery of the molecular basis of the iron-copper connection in eukaryotes.

Reductive iron uptake

Principles of reductive iron uptake

Most – if not all – yeasts have a trans-plasma membrane electron transfer system that allows cells to reduce extracellular electron acceptors at the expense of intracellular NAD(P)H. This was first observed by Crane and coworkers (Crane *et al.*, 1982), who noted that *S. cerevisiae* cells were able to reduce the non-permeant reagent ferricyanide. The cell reductase activity is induced under conditions of iron deprivation and is required for iron uptake from various ferric chelates, indicating that ferric reduction at the cell surface is the first step of an inducible, high-affinity iron uptake pathway (Lesuisse and Labbe, 1989; Lesuisse *et al.*, 1987). In this reductive uptake pathway, iron is first reduced at the cell surface before being taken up by the cells.

The rationale for such a two-step mechanism of uptake is often presented as a means to solubilize ferric iron into the much more soluble ferrous species, thereby making it available to the cell. Confronted with the unavailability of insoluble ferric oxy-hydroxides in the extracellular medium, cells would have two possibilities for making this iron available: (1) excrete compounds binding Fe^{3+} with high affinity (siderophores) to displace the thermo-dynamical equilibrium towards the formation of soluble ferric chelates that could be taken up by the cell (see below, non-reductive uptake); (2) reduce insoluble ferric iron into the more soluble ferrous iron, which could then be taken up as such by the cell (reductive uptake). However, soluble ferric chelates, and not particles of insoluble ferric hydroxides, are the substrates of the cell ferrireductase and thus reduction of iron at the cell surface should not be considered as a physiological response to the problem of ferric iron insolubility. Actually, iron might be more readily available than expected to yeast cells in their ecological niche, because yeasts strongly acidify the surrounding medium due to the action of the plasma membrane H^+-ATPase (Perlin *et al.*, 1997), which greatly favours iron solubilization, and several species of yeast excrete large amounts of Krebs cycle intermediates, which are potential iron ligands. Moreover, the yeast cell wall itself can bind and concentrate relatively large amounts of ferric iron, making an extracellular iron pool directly available to the cells (Benchimol *et al.*, 1979; Lesuisse *et al.*, 1987). As pointed out by De Luca and Wood (2000), the need for reduction prior to uptake is related to the chemical concept of ligand exchange rates. Ligand exchange is much faster with Fe(II) than with Fe(III), and thus the reaction: $Fe(III)_{ext}L_6 + \text{permease} \rightarrow Fe(III)_{ext}\text{-permease} + 6\ L$ is not expected to occur because the turnover number (k_{cat}) would be unacceptably low (De Luca and Wood, 2000). Another possibility is to transport the ferric ligand $Fe(III)_{ext}L_6$ as such, without any ligand exchange prior to uptake. This indeed occurs as the non-reductive iron uptake mechanism (see below); however, this mechanism requires a specific permease for each different ferric chelate, because the ligands of Fe^{3+} can be structurally very different. Reductive iron uptake, by contrast, allows the cells to take up iron from virtually all ferric complexes with only one permease system: the cell ferrireductase is non-specific, and can reduce numerous different electron acceptors, and in particular –but not only – all ferric complexes, including siderophores (Lesuisse *et al.*, 1996) and even the ferric iron bound to transferrin (Knight *et al.*, 2005). Reduction of the ferric core allows iron ligands to be rapidly exchanged, and the ferrous iron can then bind to a specific high-affinity permease system. Thus, reductive iron uptake associates specific high-affinity iron uptake to non-specific reduction of iron.

Molecular features of reductive iron uptake

The molecular basis of reductive iron uptake in the model yeast *S. cerevisiae* is summarized in Fig. 14.1. The reduction step is mostly catalysed by members of the Fre family of metalloreductases. Excreted compounds, including anthranilate, 3-hydroxyanthranilate and 2,5-dimethylhydroquinone, may also contribute to extracellular iron reduction to some extent (De Luca and Wood, 2000; Lesuisse *et al.*, 1992). In *S. cerevisiae*, Fre1 (Dancis *et al.*, 1990, 1992) and Fre2 (Georgatsou and Alexandraki, 1994) account for most of the ferrireductase activity associated with the plasma membrane (Georgatsou and Alexandraki, 1994; Lesuisse *et al.*, 1996), although Fre3 and Fre4 may contribute to some extent to the reduction of extracellular siderophores (Yun *et al.*, 2001). Other Fre proteins are involved in intracellular iron homeostasis (Yun *et al.*, 2001). The plasma membrane Fre proteins are flavo-haemo-proteins that catalyse the transfer of one electron from intracellular NADPH to a wide variety of extracellular electron acceptors with low affinity (Lesuisse *et al.*, 1996; Shatwell *et al.*, 1996). Fre1 is homologous to the larger subunit (gp91phox) of the human neutrophil NADPH oxidase, and these two proteins share many structural and functional characteristics (Lesuisse *et al.*, 1996; DeLeo *et al.*, 1999). The reductase activity of cells mediated by the Fre proteins largely exceeds their iron uptake capacity and is less tightly regulated by iron availability than the uptake process itself (Eide *et al.*, 1992; Lesuisse and Labbe, 1992). This, and the lack of specificity of the Fre proteins, suggests that trans-plasma membrane electron transfers mediated by the Fre proteins may be involved in adapting the cell redox potential to the growth conditions more generally in addition to reductive iron uptake (Lesuisse and Labbe, 1992). When the electron acceptor is a ferric chelate, the ligands of iron can be rapidly exchanged, and the reduced iron can be taken up through a high-affinity transport complex that consists of a multicopper ferrooxidase and a permease.

In *S. cerevisiae*, the Fet3 multicopper oxidase (Askwith *et al.*, 1994) is a peripheral membrane protein anchored by a single transmembrane helix. Fet3 uses four copper atoms, organized in three sites within the extracellular domain, to couple the oxidation of four ferrous ions to the reduction of one dioxygen and two water molecules (Stoj and Kosman, 2003; Stoj *et al.*, 2006, 2007). Thus, high-affinity reductive iron uptake involves re-oxidation of iron for transport, and this process requires copper. This is a key point, and constitutes the molecular basis of the iron-copper connection in eukaryotes (Kaplan and O'Halloran, 1996). Because copper is required for high-affinity reductive iron uptake, mutations leading to defective copper uptake or intracellular copper mobility also result in defective iron uptake. In *S. cerevisiae*, extracellular Cu^{2+} is reduced to Cu^+ by the plasma membrane Fre reductases and is then

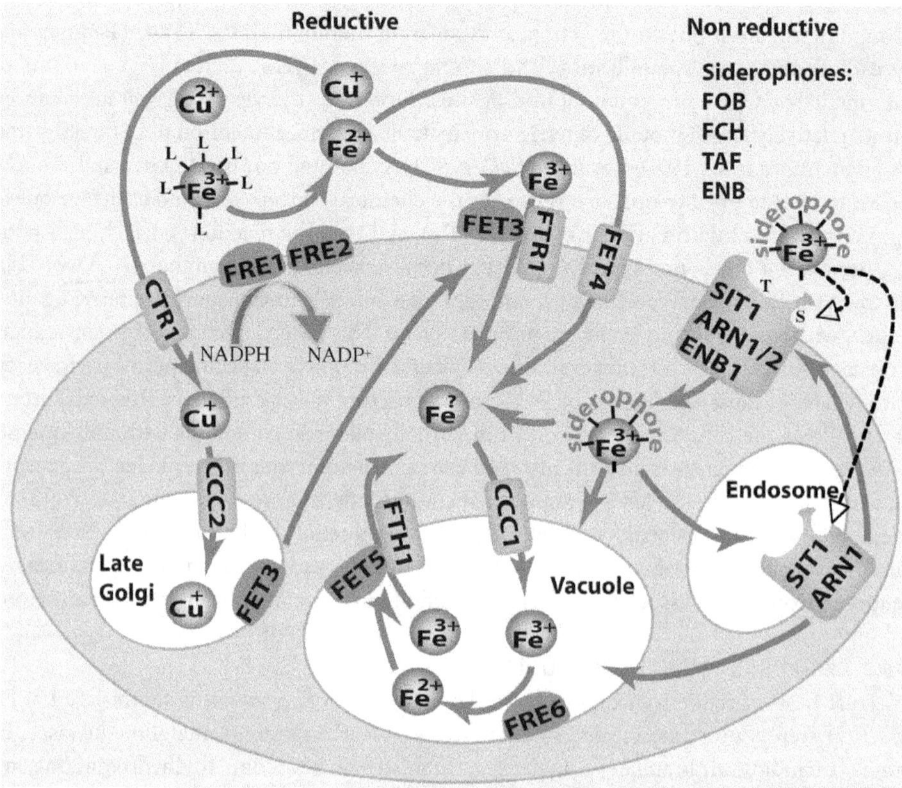

Figure 14.1 Iron uptake systems in the model yeast *Saccharomyces cerevisiae.* Reductive iron uptake (left part). During reductive uptake, extracellular ferric complexes are reduced by the plasma membrane reductases Fre1 and Fre2, and the more labile ferrous iron is taken up via the low-affinity divalent metal transporters Fet4 or Smf1 (not represented), or via the high-affinity oxidase-permease system (Fet3-Ftr1), which requires copper as a cofactor for Fet3. Extracellular cupric ions are reduced by the same reductases and the resulting cuprous ions enter the cell via Ctr1. Copper is loaded into Fet3 in a post-Golgi compartment via the Atx1 chaperone (not represented) and the Ccc2 copper transport system. Intracellular iron is stored in the vacuole (via Ccc1), from which it can be mobilized by a similar reductive mechanism involving homologous proteins (*i.e.* Fre6, Fet5 and Fth1). Non-reductive iron uptake (right part). Siderophore transporters (Sit1, Arn1, Taf1/Arn2, Enb1) mediate the uptake of various siderophores (FOB, ferrioxamine B; FCH, ferrichrome; TAF, triacetylfusarinine; ENB, enterobactin). In the absence of their substrates, some siderophore transporters (Sit1, Arn1) cycle from endosomes to the vacuole for degradation. In the presence of their substrates, which are sensed in a specialized domain of the protein (S), these transporters are massively translocated to the plasma membrane, and siderophores bind to the transport domain (T) of the transporters and are subsequently taken up. It is not known if siderophores are sensed at the plasma membrane or within the cell, in endosomes, after fluid-phase endocytosis (dotted arrows). Other siderophore transporters (Enb1) are constitutively targeted to the plasma membrane. Internalized siderophores can accumulate in the cytosol or in the vacuole.

taken up by the Ctr1 transporter (Dancis *et al.*, 1994; Georgatsou *et al.*, 1997). Copper is inserted post-translationally into Fet3 in a post-Golgi compartment of the secretory pathway. The copper chaperone Atx1 and the microsomal copper transporter Ccc2 are required for this copper insertion (reviewed in Kosman, 2003; Philpott, 2006). Cells defective in either of these pathways of copper metabolism are also defective in high-affinity iron uptake (Lin *et al.*, 1997; Weissman *et al.*, 2002).

The ferric iron produced by the ferrooxidase is transported into the cell by the iron permease (Ftr1 in *S. cerevisiae*; Stearman *et al.*, 1996). Ftr1 has seven transmembrane helices, and forms a stable complex with Fet3 at the yeast plasma membrane; this complex may be a heterodimer or higher order oligomer (Severance *et al.*, 2004). The oxidase step must be tightly coupled to iron permeation: if it were not, Fe^{3+}

would either precipitate by hydrolysis or be chelated by extracellular ligands, leading to a futile cycle of reduction-oxidation.

The question that immediately arises when examining the model of high-affinity reductive iron uptake in yeast (Fig. 14.1) is why should iron be re-oxidized for transport, after being reduced by cell surface ferrireductases? As pointed out by Kwok *et al.* (2006a), it is important to understand that the permease Ftr1 is not just a transporter of ferric ions: the oxidation step is genuinely coupled to the permeation step, and is essential for uptake (in the absence of the ferrooxidase Fet3, Fe^{3+} ions cannot be transported by the permease Ftr1; in contrast, in the absence of the Fre reductases, Fe^{2+} ions can be taken up via the Fet3–Ftr1 complex). Kwok *et al.* (2006a) compared this mechanism of uptake, where oxidation of iron is coupled to its permeation, to the binding of iron to transferrin or to its accumulation into ferritin, two processes where iron binding/storage is coupled to ferrooxidation. These authors provided evidence that the Fe^{3+} product of the Fet3 ferrooxidase reaction is channelled (and not simply transferred) to Ftr1 for uptake and the mechanism may be a conserved aspect of iron metabolism in aerobic organisms. This mechanism of channelling contrasts with the theoretical mechanism by which the ferric iron produced by the Fet3 oxidase equilibrates with the bulk phase (dissociative mechanism), from which it could be trapped by a chelator of Fe^{3+}. This last phenomenon has never been observed: reductive iron uptake can be inhibited by a ferrous chelator acting downstream from the reductase step (which indicates that the reduction step is not coupled to the oxidase step) but not by a ferric chelator acting downstream from the oxidase step (except in chimeric Fet3–Ftr1 constructs) (Kwok *et al.*, 2006b). In the channelling mechanism, the ferric iron produced by the oxidase is *directly* the substrate of the permease, and does not equilibrate with the bulk solvent, which makes uptake an inherently efficient process kinetically (Kwok *et al.*, 2006a,b; Stoj *et al.*, 2006). Even though not fully elucidated, this mechanism of oxidation of the more labile Fe^{2+} into Fe^{3+} coupled to transport seems to be required in a broad range of aerobic organisms for ionic iron transport through membranes, with both high affinity and high specificity.

Finally, yeast cells also have various systems for the transport of divalent metals, including Fe^{2+}. These systems (which are also reductive, because the substrate is ferrous iron) have lower affinity and lower specificity than the ferrooxidase-permease machinery. In *S. cerevisiae* (Fig. 14.1), Fet4 is one such transporter, with activity towards Fe^{2+}, Zn^{2+}, Cd^{2+}, Cu^{2+}, Co^{2+} and other cations (Dix *et al.*, 1997; Dix *et al.*, 1994). This iron uptake system might be important in this yeast, as *S. cerevisae* can grow anaerobically, in conditions in which iron is essentially in the ferrous form, and the Fet3-Ftr1 system cannot work (because of the oxygen requirement for ferrooxidase activity). Proteins homologous to Fet4 are not found in most other aerobic yeasts. Most yeasts also express members of the Nramp family of divalent metal transporters (which includes the mammalian ferrous iron transporter DMT1), which can play a role in the uptake of Fe^{2+} with low affinity; this is the case for Smf1 in *S. cerevisae*, (Agranoff *et al.*, 2005; Chen *et al.*, 1999).

Siderophore-mediated (non-reductive) iron uptake

Principles of non-reductive iron uptake

Non-reductive iron uptake involves ferric chelates entering the cells as such, without prior reductive dissociation. This does not mean that a reduction step is not required for iron assimilation from these chelates: in most cases, there is reduction after uptake to release iron from its ligands (De Luca and Wood, 2000). An exception to this rule is haem, for which the rate of ligand exchange is not greater when iron is in the ferrous than in the ferric form (see below). The ferric chelates taken up by yeasts are generally specialized low molecular weight compounds, called siderophores, with high affinity and high specificity for ferric iron. Some yeast species produce their own siderophores, mostly hydroxamates (Haas *et al.*, 2008). Most –if not all – yeast species are able to use siderophores (of fungal and bacterial origin) as iron sources. This means that many yeast species (including *S. cerevisiae*) have developed an opportunistic strategy of uptake allowing them to use siderophores produced by other organisms (xeno-siderophores). Siderophore users must have the machinery for siderophore recognition and uptake, and for intracellular iron release from the internalized siderophores. Siderophore producers must have in addition the machinery required for siderophore synthesis and secretion. As described above, the cells need to have a specific recognition/uptake system for each ferric chelate taken up non-reductively, because the ligands of Fe^{3+} can be structurally very different. This is true for the non-reductive uptake of siderophores, although some siderophores are structurally very similar, and some yeast siderophore transporters can mediate the uptake of several siderophores of the same chemical family.

Molecular features of siderophore-mediated iron uptake

Siderophore production

Biosynthesis of hydroxamate siderophores by fungi involves three main enzymatic steps (for recent reviews, see Miethke and Marahiel, 2007; Haas *et al.*, 2008). The first step is N^5-hydroxylation of ornithine catalysed by ornithine-N^5-monooxygenase, to give N^5-hydroxy-L-ornithine, a common intermediate for the synthesis of all hydroxamate-type siderophores. The second step, catalysed by N^5-transacylases, involves the transfer of an acyl group from acyl-coenzyme A to N^5-hydroxy-L-ornithine, to form the hydroxamate group. Finally, the hydroxamates are covalently linked via ester or peptide bonds, in a step catalysed by a group of important enzymes found only in bacteria and fungi: the non-ribosomal peptide synthetases (NRPS). From genomic data, it appears that the first enzyme required for hydroxamate synthesis, ornithine-N^5-monooxygenase, is not present in a large number of yeasts, particularly the *Saccharomycotina*, which includes the 'classical' yeast species *Saccharomyces, Candida, Pichia, Kluyveromyces, Yarrowia* and others. Therefore, these yeasts are presumably unable to produce their own hydroxamate siderophores (Haas *et al.*, 2008). Indeed, it has long been recognized that the baker's yeast *S. cerevisiae* does not secrete any siderophore (Atkin *et al.*, 1970). A few studies have reported the production of siderophores by *C. albicans*, (Ismail *et al.*, 1985; Sweet and Douglas, 1991), but we have been unable to reproduce these findings (E. Lesuisse, unpublished), and it is now generally accepted that this pathogenic yeast uses xenosiderophores but cannot produce siderophores itself (Haas, 2003). Indeed, with the exception of *Schizosaccharomyces pombe* (Labbe *et al.*, 2007), most siderophore producers are filamentous fungi, or fungi that alternate between yeast and filamentous forms in their life cycle (Winkelmann, 2007).

Siderophore uptake

The pathways involved in siderophore secretion (after their synthesis) are still largely unknown. In contrast, our knowledge of the molecular mechanisms involved in siderophore uptake by yeasts is expanding rapidly. Again (as for reductive iron uptake), the model yeast *S. cerevisiae* has played a decisive role, as it proved to be a very good tool for studying these mechanisms (illustrated in Fig. 14.1). We observed a long time ago that *S. cerevisiae* is able to take up some ferrisiderophores non-reductively: haem-deficient mutants, which are defective for reductive uptake (because haem is a cofactor of the surface ferrireductases), retain the ability to take up several ferrisiderophores with high affinity (Lesuisse and Labbe, 1989). The discovery of the main genes/proteins involved in reductive iron uptake (see above) provided the missing genetic tools to investigate the molecular mechanisms of siderophore uptake, because they facilitated the isolation of mutants defective for the various iron uptake pathways. This led to the cloning of the first yeast siderophore (ferrioxamine B) transporter (Sit1) by functional complementation (Lesuisse *et al.*, 1998).

The Sit1 protein belongs to the major facilitator superfamily, which is composed of transporters with 14 predicted transmembrane helices. The genome of *S. cerevisiae* contains six homologous genes of the *SIT* family, four of which encode different siderophore transporters: Sit1 (ferrioxamine B, ferrichrome, ferricrocin ...) (Lesuisse *et al.*, 1998), Arn1 (ferrichrome) (Heymann *et al.*, 2000b; Yun *et al.*, 2000), Taf1 (triacetylfusarinine) (Heymann *et al.*, 1999), and Enb1 (enterobactin) (Heymann *et al.*, 2000a). The Sit1 family of siderophore transporters is unique to fungi. Remarkably, these siderophore transporters do not show any structural similarity to any bacterial siderophore transporters, despite having identical substrates. Examination of fungal databases revealed about 140 sequences similar to those of the Sit family in Ascomycetes and 18 related sequences in Basidiomycetes (Philpott, 2006). Based on these sequence homologies, several groups have characterized siderophore transporters in different yeasts/fungi, including *Aspergillus nidulans* and *A. fumigatus* (Haas *et al.*, 2008; Haas *et al.*, 2003), *S. pombe* (Labbe *et al.*, 2007; Pelletier *et al.*, 2003), and *C. albicans* (Ardon *et al.*, 2001; Hu *et al.*, 2002).

There is still much to learn about the biochemistry of the mechanisms of siderophore uptake. Some transporters in *S. cerevisiae* are highly specific for a given siderophore (Tafl, Enb1), whereas others are less specific (Sit1); currently, no structure-function analysis allows the specificity of a given transporter for particular siderophores to be predicted. Siderophore transporters may function as proton symporters (Lesuisse *et al.*, 2001; Philpott, 2006), but this remains to be demonstrated. The mechanism of transport itself is not well understood, and it is not even clear whether siderophore transporters behave like classical permeases or need to be internalized together with their substrate for siderophore uptake (see below).

Intracellular trafficking of siderophore transporters

The siderophore transporters would obviously be expected to localize at the plasma membrane. However, studies on the Arn1 and Sit1 transporters in *S. cerevisiae* showed that they are found on intracellular vesicles rather than on the plasma membrane; massive plasma membrane sorting of these transporters only occurs when the cells are in contact with the corresponding siderophores (Froissard *et al.*, 2007; Kim *et al.*, 2002). Philpott and coworkers (Kim *et al.*, 2007; Kim *et al.*, 2005; Kim *et al.*, 2002) proposed a model for the ferrichrome transporter Arn1 in which, in the absence of siderophore in the medium, the transporter is directly sorted from the late Golgi to the vacuole for degradation with only transient accumulation in the late endosome/multivesicular bodies (MVB). This sorting to the vacuolar degradation pathway involves the recognition of Arn1 by a clathrin adaptor protein (Gga2), and its ubiquitination by Rsp5 (Erpapazoglou *et al.*, 2008; Kim *et al.*, 2007). Very low concentrations of ferrichrome in the medium (nanomolar) trigger the relocalization of Arn1 from the endosome to the plasma membrane via a receptor site, distinct from the uptake site present on the C-terminus of Arn1 itself (Kim *et al.*, 2005). Thus, Arn1 appears to contain two externally directed sites for ferrichrome binding: one corresponding to the uptake site (with a K_D of about 1 μM) and the other to the receptor –or sensing – site (with a K_D of about 10 nM). The binding of small amounts of ferrichrome to the receptor site may induce massive relocalization of Arn1 to the plasma membrane, allowing Arn1 to transport ferrichrome (if present at a concentration above the threshold for uptake) inside the cells. Uptake itself would occur via internalization of Arn1, followed by recycling of the transporter at the plasma membrane (Kim *et al.*, 2005). This model has two important features: (1) to trigger Arn1 relocalization at the plasma membrane, ferrichrome must first reach the receptor domain of Arn1 in the endosome, possibly by fluid-phase endocytosis; (2) Arn1 does not behave like a permease. In our view, this last point remains debatable for two reasons. First, we did not observe any dependence of ferrioxamine B uptake (via Sit1) on endocytosis (Lesuisse *et al.*, 2001) and, second, in our hands, there is always a small fraction of Arn1 and Sit1 associated with the plasma membrane under iron-deficient conditions, even in the absence of any siderophore in the medium (Froissard *et al.*, 2007). A massive relocalization of Sit1 to the plasma membrane occurs, as for Arn1, in the presence of its substrate, but also in the presence of a siderophore that is not a substrate of Sit1 (coprogen) (Froissard *et al.*, 2007). Thus, further work is required to resolve this question of siderophore sensing and trafficking. The issue is further complicated by the observation that different siderophore transporters use different regulated trafficking processes: for example, Enb1 and CaArn1 (the transporter of ferrichrome in *C. albicans*) are constitutively targeted to the plasma membrane, even in the absence of their substrates (Froissard *et al.*, 2007; Hu *et al.*, 2002). In these cases, there seems to be no mechanism of siderophore sensing, and the transporters are likely to behave like classical permeases. It would be interesting to know whether the complex mechanisms of siderophore sensing and trafficking are an adaptation of siderophore users that are not siderophore producers, or whether such mechanisms are also present in some siderophore producers. In these later species, both siderophore synthesis and the expression of siderophore transporters occur under iron-deficient conditions, so there is no evident physiological requirement for a mechanism of siderophore sensing.

Haem uptake

Haem is, together with transferrin, the most abundant iron source for pathogenic yeasts living in a mammalian host. Iron from transferrin can be taken up by *C. albicans* –and probably by other yeast species – via the reductive pathway (Knight *et al.*, 2005). But in haem, iron is tightly bound to the tetrapyrrole (porphyrin) ring in the ferric as well as in the ferrous state, and consequently, ligand displacement cannot be achieved by reduction. Iron assimilation from haem by yeast requires import of the haem molecule into the cell and subsequent enzymatic breaking (by haem oxygenase) of the porphyrin ring to release iron; note that no reverse ferrochelatase has yet been described in this microorganism (see also Chapter 2). Thus, in this case the uptake is truly non-reductive, because no reduction step is expected to occur before or after uptake. Some yeast species, for example *C. albicans*, can grow with haem or haemoglobin as the sole iron source. This requires a functional haem oxygenase, which is responsible for the physiological breakdown of haem into equimolar amounts of biliverdin, carbon monoxide, and iron (Santos *et al.*, 2003; Weissman *et al.*, 2002). However, the molecular mechanisms underlying haem uptake by yeasts remain largely unknown. In *C. albicans*, the first stage of haem and haemoglobin uptake probably requires the binding of haem to a family of glycosylphosphatidylinositol (GPI)-anchored plasma membrane proteins (CaRbt5, CaRbt51) (Weissman and Kornitzer, 2004), followed by endocytosis and vacuole degradation (Weissman *et al.*, 2008).

Intracellular fate of iron

This review addresses iron uptake mechanisms, but transport processes cannot be completely dissociated from the fate of the internalized iron, because the intracellular distribution of iron and its abundance in the various cell compartments is dependent on the uptake mechanisms, and retroactively controls them (see the regulation section below).

Iron storage in vacuoles and siderophores

As a general rule – and in contrast to bacteria, plants and animals – fungi and in particular yeasts do not produce ferritin. Instead, iron is stored either in vacuoles or in intracellular siderophores (Haas *et al.*, 2008; Howard, 1999). In *S. cerevisiae*, which is not a siderophore producer, iron is normally stored in the vacuole, probably associated with polyphosphates (Raguzzi *et al.*, 1988). The vacuole is the only compartment in which the iron concentration increases when yeast cells are subjected to an increased extracellular iron concentration (Raguzzi *et al.*, 1988). However, the vacuole is not simply a passive storage site for excess iron in the cell. Excess accumulated iron can be used by the cells when they are transferred to an iron-limiting environment (Eide *et al.*, 1992). Accordingly, vacuolar iron can be mobilized into the cytosol and mitochondria (Li and Kaplan, 2004). Iron uptake by the vacuole from the cytosol is mediated by the transporter Ccc1 (Li *et al.*, 2001). Nothing is known about the mechanism of uptake, or about the form of cytosolic iron that is taken up by the vacuole. Ccc1 also transports another divalent cation, Mn^{2+}, so presumably its substrate is Fe^{2+} (Li *et al.*, 2001). There is a remarkable lack of data concerning the form of iron in the cytosol of yeasts that are not siderophore producers, and this is a very promising field of investigation. For example, no iron chaperone has yet been discovered in yeast, whereas several copper chaperones, each responding to a particular function, have been extensively described (Kim *et al.*, 2008). Both metals are redox active, and are thus unlikely to stay in a 'free' form in the cytosol (Finney and O'Halloran, 2003). The iron taken up by cells can be detected in the vacuolar compartment very early (Raguzzi *et al.*, 1988). Iron crossing the plasma membrane via the Fet3–Ftr1 complex is expected to enter the cytosol as Fe^{3+}, and to subsequently enter the vacuole as Fe^{2+}, so there is presumably at least one redox step before vacuolar uptake.

The mechanisms of export of iron from the vacuole to the cytosol closely mimic the mechanisms of iron import through the plasma membrane. *In vitro*, iron can be mobilized from vacuoles by a reducing agent, for example ascorbate (E. Lesuisse, unpublished), which suggests that intra-vacuolar iron is in the oxidized state, and that iron export from the vacuole involves a reductive mechanism. Indeed, reductive export of the intra-vacuolar iron involves the vacuolar metalloreductase Fre6 (Singh *et al.*, 2007), generating ferrous iron which then crosses the vacuolar membrane either via the Fet5–Fth1 complex (Urbanowski and Piper, 1999), paralogous to the plasma membrane Fet3–Ftr1 complex, or through the divalent metal transporter Smf3 (Portnoy *et al.*, 2000), paralogous to the plasma membrane Smf1.

Siderophore-producing yeasts probably also use the vacuole as an iron-storage compartment, as orthologues of the *S. cerevisiae* Fet5, Fth1 and Ccc1 are found in several of them (Haas *et al.*, 2008). However, these yeasts have an additional form of intracellular iron storage, *i.e.* the siderophores themselves. In some species, specialized siderophores, that are not excreted, serve as a unique iron storage form (Johnson, 2008). This issue –more significant for filamentous fungi than for yeasts – is beyond the scope of the present article. Briefly, disrupting the siderophore biosynthesis pathway in siderophore-producing fungi often results not only in a decreased cell iron content, but also in alteration of vital functions related to the cell cycle, resistance to oxidative stress and other processes. This indicates that the role of siderophores is not only to bring extracellular iron to the cells, but also to optimize intracellular iron storage/detoxification, transfer and distribution. The role of intracellular siderophores is thus clearly established, but our knowledge of the mechanisms involved in iron release and redistribution to the various compartments from intracellular siderophores remains very poor; this is currently a developing field of investigation (Haas *et al.*, 2008; Johnson, 2008). The intracellular fate of xenosiderophores in yeasts that are not siderophore producers is also poorly understood. In *S. cerevisiae*, most of the intracellular ferrichrome (transported via Arn1) is found in the cytosol (Moore *et al.*, 2003), whereas ferrioxamine B (transported via Sit1) is stored in the vacuole (Froissard *et al.*, 2007). The intracellular fate of siderophores may therefore vary according to the siderophore and the transporter used for its import. This is consistent with the observation, described above, that different siderophore transporters use different trafficking processes. The (reductive?) mechanisms of iron release from internalized siderophores (either in the cytosol or in the vacuole) in yeast are still unknown. A better understanding of the mechanisms involved in intracellular processing of siderophores is a prerequisite for the development of new antifungal drugs based on the 'Trojan Horse' approach (siderophore-antifungal drug conjugates, which exploit the siderophore uptake system as a cellular entry gate; reviewed in Miethke and Marahiel, 2007).

Iron uptake by mitochondria

The mitochondrial compartment is the most important in terms of iron use, because it is the site of the biosynthesis of the vital cofactors haem and [Fe–S] clusters. The last enzyme of the haem synthesis pathway, ferrochelatase, is on the matrix side of the inner mitochondrial membrane, and the complex assembly machinery for [Fe–S] clusters is mainly in the mitochondrial matrix (reviewed in Lill and Muhlenhoff, 2006). Although iron is essential for mitochondria, this is also the compartment where it is potentially the most dangerous, because it can catalyse, via the Fenton reaction, the formation of the very reactive hydroxyl radical from normal byproducts of respiration (superoxide and hydrogen peroxide). Thus, mitochondria are expected to possess powerful mechanisms for fine regulation of their iron homeostasis. Paradoxically, the mechanisms of iron import, export and storage in this compartment are still very poorly understood in yeast (and more generally in all eukaryotes). Several proteins have been proposed as candidate iron transporters in *S. cerevisiae* mitochondria. They include the paralogous proteins Mmt1 and Mmt2 (Li and Kaplan, 1997), and the paralogous mitochondrial carrier proteins Mrs3 and Mrs4 (Froschauer *et al.*, 2009). The Mrs proteins (homologous to the mammalian mitoferrin) are the best candidates for the function of mitochondrial iron import (Foury and Roganti, 2002; Muhlenhoff *et al.*, 2003; Zhang *et al.*, 2006). However, the mechanism of uptake is still completely unknown, and will probably remain so as long as the iron substrate used by these proteins *in vivo* is not known – here again, we need further insight into the form of iron in the cytosol. Except in some pathological situations (see below, the regulation section), mitochondria normally do not accumulate iron, irrespective of the extracellular iron concentration (Raguzzi *et al.*, 1988), and this implies that mitochondrial iron import is tightly regulated. However, this does not fit with the hypothesis that mitochondrial iron uptake is exclusively controlled by the Mrs proteins, considered as 'iron permeases', because (i) the corresponding *MRS* genes are not tightly regulated by iron (Courel *et al.*, 2005; Foury and Talibi, 2001) and the abundance of the Mrs proteins does not vary substantially according to the iron concentration in the medium; (ii) overexpression of *MRS3*, *MRS4*, or both does not result in over-accumulation of iron in mitochondria (A. Dancis and E. Lesuisse, unpublished). One possibility is that the Mrs proteins couple mitochondrial iron import to intra-mitochondrial iron utilization (Zhang *et al.*, 2005, 2006), and thus, that mitochondrial iron import responds more to changes in cell physiology than to changes in iron concentration, by a mechanism that remains to be identified. Disruption of both *MRS* genes does not lead to massive iron depletion from the mitochondria (Muhlenhoff *et al.*, 2003), so it is plausible that mitochondria have other, still unknown, iron-uptake systems.

Regulation of high-affinity iron uptake

The mechanisms of iron uptake in the various yeast species are relatively homogeneous, with components of the reductive and of the siderophore-mediated iron uptake systems in different species being homologous proteins. By contrast, the way iron uptake is regulated is heterogeneous, with two contrasting transcriptional mechanisms: one positive (transcription of target genes is up-regulated under low iron conditions) and one negative (transcription of target genes is down-regulated under high iron conditions). Despite this major difference between species, a general rule is emerging: under low iron conditions, transcription of genes encoding components of the high-affinity iron uptake systems is induced, and expression of some iron-utilizing proteins is repressed, either by a transcriptional or by a post-transcriptional mechanisms; this ensures that iron is present within cells for essential biochemical reactions without wasting the iron resources available. Low-affinity iron uptake systems (Fet4 and Smf1 in *S. cerevisiae*) are less important for cell viability, and their regulation is often multifactorial (Liu and Culotta, 1999; Waters and Eide, 2002). Their regulation of will not be addressed in this chapter.

Positive transcriptional regulation of iron uptake in the model yeast *S. cerevisae*

In baker's yeast, a powerful approach to selecting mutants with specific abnormalities in iron uptake has been used to elucidate fundamental aspects of iron metabolism, including regulation (Dancis, 1998). The identification of the first iron-regulated gene in *S. cerevisiae FRE1*, encoding the main plasma membrane ferrireductase whose expression is induced in low iron conditions (Dancis *et al.*, 1990) provided the tool for this approach. The promoter of *FRE1* was fused to the coding sequence of *HIS3*, and inserted into the genome of a *his3* strain, auxotrophic for histidine. Histidine auxotrophy then became iron dependent (His^+ in low-iron media and His^- in high-iron media). Dancis and coworkers (Dancis, 1998) isolated spontaneous mutants able to grow in an iron-rich medium without histidine: these mutants, behaving as iron-deprived in an iron-rich environment, were expected to be deficient in iron uptake and/or in iron sensing. Three main categories of mutants were isolated in these experiments, constituting a mine for further studies: (1) mutants with low iron uptake or low copper uptake; (2) mutants with high iron uptake, and normal intracellular iron distribution; (3) mutants with high iron uptake, and abnormal intracellular iron distribution. The first category of mutants (deficient for iron uptake) allowed the characterization of components of the reductive iron uptake pathway and the iron–copper connection (see above). The second category of mutants (deficient for iron sensing) led to the characterization of Aft1, the main transcriptional activator of the iron-regulated genes (or 'iron regulon'). The third category of mutants, which were affected indirectly (via mitochondrial dysfunction) in transcription of iron-regulated genes, were instrumental in the discovery of the fundamental role played by mitochondria in regulation of iron uptake and metabolism in yeast.

Transcriptional regulation mediated by the activators Aft1 and Aft2

The primary response of *S. cerevisiae* to iron deprivation is the activation of the iron regulon by the transcriptional activator Aft1 (Yamaguchi-Iwai *et al.*, 1995). Aft1 is constitutively expressed, and shuttles from the cytosol to the nucleus in an iron-dependent manner (Ueta *et al.*, 2007). When Aft1 accumulates in the nucleus, it binds to consensus sequences in the upstream regions of target genes (PyPuCACCC) and subsequently activates their transcription (Yamaguchi-Iwai *et al.*, 1996; Yamaguchi-Iwai *et al.*, 2002). The target genes constitute the iron regulon, which essentially includes all genes encoding components of the high-affinity reductive and non-reductive iron uptake pathways and of the vacuolar reductive iron export pathway, and genes required for the copper loading of Fet3, for degradation of haem (*HMX1*) and for post-transcriptional regulation of iron metabolism (*CTH*) (Foury and Talibi, 2001). Aft2, a paralogue of Aft1, also participates – partially redundantly – in the transcriptional activation of target genes upon iron depletion (Blaiseau *et al.*, 2001; Rutherford *et al.*, 2001). In addition to the genes regulated by Aft1, Aft2 can activate two genes that are not the target of Aft1: *SMF3* (involved in vacuolar iron export) and *MRS4* (involved in mitochondrial iron import) (Courel *et al.*, 2005; Rutherford *et al.*, 2003). Aft2 appears to recognize a consensus sequence slightly different to that recognized by Aft1 (GGCACCC and CGCACCC, instead of TGCACCC), and may preferentially influence intracellular iron utilization over iron uptake (Courel *et al.*, 2005; Rutherford *et al.*, 2003).

In addition to Aft1 and Aft2, several pleiotropic co-regulators (for example Tup1 and Ssn6) are involved in transcriptional regulation of the iron regulon; the description of these factors is however beyond the scope of the present article (for a review, see Philpott and Protchenko, 2007). One important gene in iron homeostasis, *CCC1* (which encodes the vacuolar iron importer) is not transcriptionally activated by Aft1/2, but by Yap5 (belonging to the Yap family of regulators involved in the response to oxidative stress) in response to increased iron concentration, facilitating the storage of excess iron (Li *et al.*, 2008).

The cysteine residue at position 291 in the Aft1 protein seems to play an important role in iron sensing: mutation of this residue causes Aft1 to be retained in the nucleus and results in constitutive activation of Aft1-target genes (*AFT1*^{up} dominant mutant, which was the original mutant isolated with the genetic screening described above) (Yamaguchi-Iwai *et al.*, 1995). Moreover, iron induces intermolecular interactions between Aft1 molecules (leading to Aft1 oligomerization), and this interaction, for which Cys291 is essential, is required for export of Aft1 from the nucleus to the cytosol through the exportin Msn5 under iron-replete conditions (Ueta *et al.*, 2007). The complex features of Aft1 shuttling between the cytosol and the nucleus are also determined by the iron–independent interaction between Aft1 and two other proteins, Grx3 and Grx4, both nuclear monothiol glutaredoxins (Ojeda *et al.*, 2006). Grx3 and Grx4 themselves interact with Fra1 and Fra2, two cytosolic proteins of unknown function, and these complex interactions are essential for correct Aft1-dependent transcriptional regulation (Kumanovics *et al.*, 2008). The double mutant *grx3grx4* and the single mutants *fra1* and *fra2* behave like the constitutive *AFT1*^{up} mutant: the Aft1 activator remains in the nucleus and the iron regulon is activated even in the presence of an excess of iron (Kumanovics *et al.*, 2008; Pujol-Carrion *et al.*, 2006). In all of these mutants, iron over-accumulates in endosomal vesicles and in the vacuole; the cytosolic iron concentration is also higher than in the wild-type, but the mitochondrial iron concentration remains unchanged (E. Lesuisse and A. Dancis, unpublished).

Role of the mitochondria in regulation of iron homeostasis

Other mutants, corresponding to the third category described above, also show constitutive up-regulation of the iron regulon and over-accumulate iron; these mutants also show abnormal intracellular distribution of iron, most of it being accumulated in the mitochondria. They share the common feature of all being affected in some component of the complex biogenesis machinery of the mitochondrial [Fe–S] cluster, including cysteine desulphurase Nfs1, chaperone and scaffold proteins Ssq1, Isu1/2, Isa1/2, Jac1 and others, redox components Grx5, Yah1 and Arh1, the GTP transporter Ggc1, the yeast frataxin homologue Yfh1 and Atm1 which is the putative exporter of [Fe–S] clusters from the mitochondrial matrix to the cytosol (reviewed in Lill and Muhlenhoff, 2006; Rouault and Tong, 2005). Thus, a defect in mitochondrial [Fe–S] assembly or in [Fe–S] export from the mitochondria leads to constitutive, Aft1-dependent, activation of the iron regulon and to mitochondrial iron overload. A model has been proposed to explain the requirement of mitochondrial [Fe–S] biogenesis for correct Aft1-dependent regulation of the iron regulon (Philpott and Protchenko, 2007). In this model, iron-dependent export of Aft1 from the nucleus requires its binding to an unidentified [Fe–S] compound assembled in the mitochondrial matrix and exported from the mitochondria via Atm1. In wild-type cells, the concentration of this compound reflects the actual concentration of iron in the cell, and regulates Aft1 function accordingly (high iron would induce the synthesis of the unknown [Fe–S] compound, which, by binding to Aft1 and its associated proteins promote export of Aft1 from the nucleus). In pathological situations in which mitochondrial [Fe–S] assembly or export is impaired, Aft1 is unable to leave the nucleus, leading to constitutive activation of the iron regulon (Philpott and Protchenko, 2007). However, this model does not account for a fundamental observation: iron accumulates specifically in the mitochondria only in mutants affected in [Fe–S] assembly or export, not in *AFT1^{up}* and related mutants (*fra1*, *fra2*, *grx3grx4*). In addition, several yeast genes involved in mitochondrial [Fe–S] biogenesis/export have homologues in man (for example *YFH1* and *ATM1*); mutation of these genes also results in mitochondrial iron overload in man, although the mechanisms of the regulation of iron import in yeast and in human cells are completely different (reviewed in Rouault and Tong, 2005).

Another product of mitochondrial metabolism, haem, also plays a role in the transcriptional activation of Aft1 target genes, although to a much lesser extent than [Fe–S] clusters (Crisp *et al.*, 2003). The smaller influence of haem than [Fe–S] in the signalling pathway of iron sensing is unsurprising, because haem is generally much more stable than [Fe–S] clusters, and thus is less appropriate for triggering rapid physiological changes in response to environmental changes.

Thus, the mitochondrial compartment, which is the centre of iron use, participates in the regulation of iron homeostasis in yeast (and more generally in eukaryotes). The mechanisms by which this regulation occurs remain unclear. It is tempting to speculate that mitochondrial iron import could itself be regulated in a negative feedback loop by a labile byproduct (some [Fe–S] compound) of its iron metabolic activity, which may act as a signal to adjust mitochondrial iron import precisely to the need for iron in the mitochondrial matrix. In mutants unable to assemble or to export [Fe–S] clusters, this signal would be absent and the mitochondria would accumulate huge amounts of iron at the expense of the cytosolic iron pool sensed by the Aft1/2 activators, which would in turn activate the iron regulon. This model is speculative and requires experimental confirmation or refutation.

Post-transcriptional and post-translational regulation of iron homeostasis in the model yeast *S. cerevisae*

One member of the iron regulon, *CTH2*, encodes a member of the CCCH zinc finger mRNA-binding protein family, which has been demonstrated to bind specifically to consensus RNA sequences in untranslated regions (UTR) that are AU-rich (AREs) (reviewed in Vergara and Thiele, 2008). Cth2 indeed binds to AREs within the 3′-UTR of several yeast mRNAs, resulting in their targeted degradation (Puig *et al.*, 2005). About 80 mRNAs have been identified as the targets of Cth2. These mRNAs encode proteins involved in haem biosynthesis, [Fe–S] cluster biogenesis, oxidative phosphorylation, the Kreb's cycle, iron storage in the vacuole, and other enzymes of the intermediary metabolism (Puig *et al.*, 2005). *CTH2* transcription is induced (via Aft1 and Aft2) in low iron conditions, so iron restriction results in transcriptional activation of the genes involved in high-affinity iron uptake, and in post-transcriptional repression (via specific mRNA degradation) of genes involved in non-essential iron-requiring metabolic processes (Vergara and Thiele, 2008). This metabolic reprogramming of cells is a metabolic adaptation to iron deficiency, which probably allows the cells to save residual iron for essential processes. The product of *CTH1*, a paralogue of *CTH2*, also contributes to this iron-dependent reprogramming process, by targeting sub-categories of mRNAs (Puig *et al.*, 2008). A tentative scheme of transcriptional and post-transcriptional regulation of iron homeostasis in *S. cerevisiae* is presented in Fig. 14.2.

Finally, there is a third level of regulation of iron uptake in baker's yeast, which occurs post-translationally. As described above (intracellular trafficking of siderophore transporters), plasma membrane targeting of some siderophore transporters is positively regulated post-translationally by siderophores in

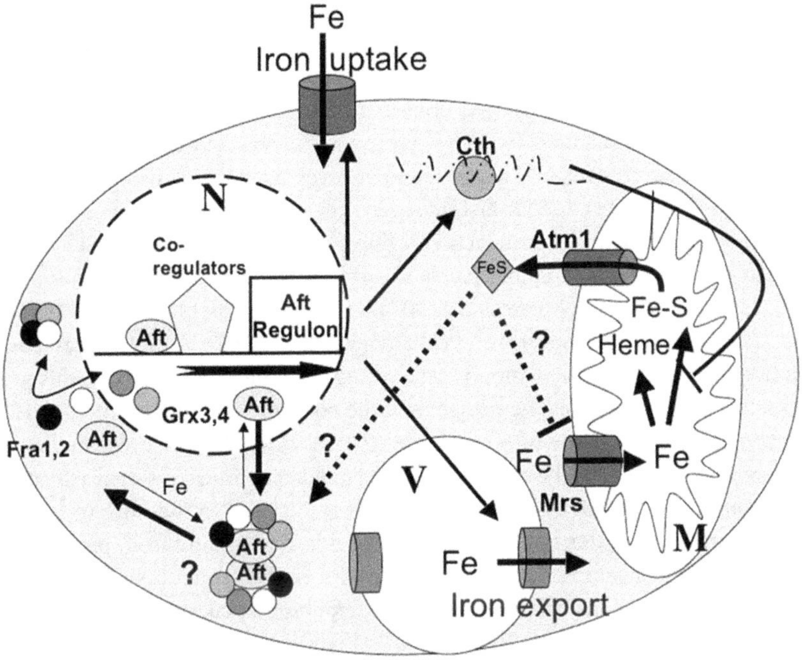

Figure 14.2 Regulation mechanisms of iron homeostasis in the model yeast *Saccharomyces cerevisiae.* The iron regulon is primarily regulated by the transcription activator Aft1. This tentative scheme represents a situation where the iron concentration is low, resulting in activation of the iron regulon. The activation function of Aft1 is modulated by shuttling of this protein between the nucleus and the cytosol, by a complex mechanism involving the nuclear proteins Grx3 and Grx4, the cytosolic proteins Fra1 and Fra2, and iron. Iron favors Aft1 oligomerization, which itself favors Aft1 inactivation via its export to the cytosol. Iron–independent association of Grx3 and Grx4 proteins to Aft1 is necessary to inactivate Aft1, by unknown mechanisms. The cytosolic proteins Fra1 and Fra2 can form a complex with the Grx proteins, and this association is also essential for appropriate Aft1-dependent regulation. The association of both Grx and Fra proteins to Aft1 oligomers is hypothetical. Activation of Aft1 function in iron-poor conditions leads to increased transcription of genes involved in iron uptake by the cells, and in iron export by the vacuole. It also leads to increased transcription of the *CTH* genes, resulting in targeted degradation of mRNAs encoding non-essential iron-utilizing proteins. An iron–sulphur cluster component produced by the mitochondria is involved in overall regulation, through unknown mechanisms. N, nucleus; V, vacuole; M, mitochondrion.

the extracellular medium. The amount of some siderophore transporters at the plasma membrane thus results from a precise adjustment between two counteracting regulatory mechanisms which operate at the transcriptional or post-translational levels: siderophores (as iron sources) block the Aft-dependent activation of siderophore transporters genes, but the same siderophores (as specific substrates) induce targeting of the corresponding siderophore transporters to the plasma membrane.

High-affinity reductive iron uptake is also regulated post-translationally: incubation of cells in high iron concentrations leads to the internalization and vacuolar degradation of both Fet3 and Ftr1 (Felice *et al.*, 2005). Internalization of the Fet3–Ftr1 complex by endocytosis is iron-specific and probably requires ubiquitination of Ftr1. A similar post-translational mechanism of degradation also controls the abundance of Fre1 and Ctr1 at the plasma membrane and acts in addition to transcriptional regulation of the corresponding genes (by Aft1 and Mac1, Mac1 being the main transcriptional activator of copper uptake genes; Jungmann *et al.*, 1993). Rapid degradation of cell surface transporters in response to an excess of their substrates may prevent the cells from accumulating toxic levels of metals before the adequate transcriptional and post-transcriptional responses become efficient.

Negative transcriptional regulation of iron uptake in other yeasts

In most fungi, transcriptional regulation of iron uptake (and of siderophore synthesis for the siderophore producer species) is mediated by an iron-dependent repressor of the GATA-factor family. Iron-responsive GATA-factors are evolutionarily well conserved transcription factors which bind to a 5′-(A/T) GATAA-3′ sequence; they contain two Cys_2-Cys_2 zinc fingers and a highly conserved cysteine-rich region (between the two zinc fingers) suggested to be involved in iron sensing (reviewed in Scazzocchio, 2000, and in Haas *et al.*, 2008). GATA-factors repress the transcription of genes involved in iron uptake and siderophore synthesis under iron-rich conditions. These factors have been described in several yeasts and filamentous fungi including *S. pombe* (Fep1) (Labbe *et al.*, 2007), *C. albicans* (Sfu1) (Pelletier *et al.*, 2007), *P. pastoris* (Fep1) (Miele *et al.*, 2007), *U. maydis* (Urbs1) (Banuett, 1995), *N. crassa* (Sre) (Zhou *et al.*, 1998), and *A. nidulans* (SreA) (Oberegger *et al.*, 2001). A model yeast for this mechanism of transcriptional regulation is the fission yeast *S. pombe* (Labbe *et al.*, 2007). In this yeast, the function of the GATA-factor Fep1 requires the presence of the co-repressors Tup11 and Tup12, orthologous to the Tup1 pleiotropic corepressor of *S. cerevisiae* (Znaidi *et al.*, 2004). Fep1-mediated transcriptional repression in *S. pombe* appears to be simpler than Aft1-mediated transcriptional activation in *S. cerevisiae.* Fep1 has an iron-dependent DNA-binding activity. It is localized to the nucleus in both iron-deficient and iron-rich conditions. Under iron-rich conditions, it is able to bind iron, probably in the conserved Cys-rich region, to form a complex able to associate with GATA sequences of target genes. This leads to transcriptional repression of the genes involved in siderophore uptake and in high-affinity reductive iron uptake (Jbel *et al.*, 2009; Labbe *et al.*, 2007).

In iron-poor conditions, Fep1-controlled genes are strongly expressed due to Fep1 inactivation, and a second transcriptional regulation mechanism operates. This mechanism is functionally similar to the post-transcriptional Cth-dependent regulation system of *S. cerevisiae*; it results in transcriptional down-regulation of genes encoding iron-dependent or iron-utilizing proteins and iron-consuming pathways. This second transcriptional mechanism of regulation, leading to cell metabolic reprogramming, is also well conserved in several fungi (but not in *S. cerevisiae*) (reviewed in Haas *et al.*, 2008). It is based on the recognition of CCAAT sequences by a conserved multimeric factor composed of orthologues of the Hap2, Hap3, Hap4 and Hap5 proteins of *S. cerevisae*, which are involved in this yeast in the metabolic adaptation of cells to non-fermentable carbon sources (Schuller, 2003). In *S. cerevisae*, Hap2, Hap3 and Hap5 form a CCAAT–binding complex interacting with DNA and with Hap4, which carries a functional activation domain. In other yeasts, interaction of the CCAAT–binding complex (composed of orthologues of Hap2, 3 and 5) with HapX (orthologous to Hap4) determines the down-regulation of genes encoding iron-utilizing proteins under conditions of iron starvation (reviewed in Haas *et al.*, 2008 and in Labbe *et al.*, 2007). Thus, in contrast to Hap4, which is an activator in *S. cerevisae*, HapX functions as a (co)repressor of gene transcription. In *S. pombe*, Php4 (orthologous to Hap4) is regulated by the GATA-factor Fep1, and is thus expressed in low iron conditions (Mercier *et al.*, 2008). Under such conditions, Php4 interacts with the CCAAT–binding complex bound to CCAAT boxes of genes encoding iron-using and iron-storage proteins to down-regulate their expression (Mercier *et al.*, 2008). In other yeast species, iron not only represses the expression of HapX via the GATA-factor, but also directly disrupts the interaction between HapX and the CCAAT–binding complex (Hortschansky *et al.*, 2007).

Conclusions and future trends

When studies on iron metabolism in *S. cerevisiae* started (in the middle of the eighties), there was a simple paradigm for iron uptake by microorganisms. As iron is largely insoluble in aerobic media, free-living unicellular organisms have to solubilize iron, which they mostly do by excreting specific iron-chelating compounds called siderophores, for which they have specific high-affinity receptors. Our knowledge of siderophores was already substantial, in particular due to work with bacteria and various fungi including *Neurospora, Aspergillus* and *Ustilago* species (Haas, 2003; Miethke and Marahiel, 2007). However, the molecular basis of siderophore uptake by yeasts remained unknown, partly because the eukaryotic siderophore transporters are structurally very different from those described in bacteria. No siderophore was found in various unicellular fungi tested, in particular in the model yeast *S. cerevisiae.* This probably substantially delayed major discoveries about siderophore uptake and use by fungi. Indeed, this yeast has proved to be a powerful tool for studies of molecular genetics, and molecular and cellular biology, due to several of its characteristics, including rapid growth and high growth yield, a budding pattern resulting in dispersed cells, easy replica plating and mutant isolation, a well-defined genetic system and a highly versatile DNA transformation system. Most importantly, the genome of this organism was the first to be fully sequenced, such that experimental data could rapidly be crossed with *in silico* studies.

S. cerevisae was later recognized to use a different system for high-affinity iron uptake, *i.e.* the reductive iron uptake system. The molecular characterization of this system led to major progress in the field of iron metabolism, not only concerning fungi, but also for plants and mammals (Hell and Stephan, 2003; Wessling-Resnick, 2006). The subsequent finding that this yeast is also able specifically to exploit xeno-siderophores by a non-reductive system, and the molecular characterization of this system, opened the path to molecular studies on siderophore uptake by other fungal species, and in particular pathogenic fungi. This field of investigation is now developing very rapidly and offers promising perspectives. The mechanisms of siderophore production, uptake and handling by pathogenic fungi are indeed of great significance for the analysis of pathogen–host interactions and the factors that determine virulence (Johnson, 2008). The molecular and structural characteristics of siderophore synthesis and uptake by fungi are unique. Consequently, siderophore transporters and the enzymes involved in siderophore synthesis are potentially good targets for new anti-fungal drugs. For example, fungal siderophore transporters offer the opportunity to develop new 'Trojan horse' antifungal drugs, exploiting the fungal siderophore transport systems as specific cellular entry gates (Miethke and Marahiel, 2007). There is still a lot to discover about siderophore secretion, uptake and use by yeasts. For example, little is known about the mechanisms of siderophore secretion, of intracellular iron release from internalized siderophores, and about the mechanisms of siderophore recycling (after iron release). The structural basis of siderophore transporter specificity and that of the structure–function relationships are also completely unknown, as are the thermodynamic and mechanistic properties of fungal siderophore transport.

During the last few years, the yeast *S. cerevisiae* has again proved to be a very useful tool for improving our understanding of iron metabolism in eukaryotes: the observation that numerous yeast mutants accumulate iron in mitochondria –reminiscent of several human diseases – led to detailed studies on the biogenesis of the [Fe–S] cluster and its role in normal and pathological iron metabolism (Lill and Muhlenhoff, 2008; Rouault and Tong, 2005). For example, studies in this field use the yeast model to decipher the role of mitochondrial frataxin (involved in [Fe–S] cluster biogenesis), the lack of which results in the rare human disease Friedrich ataxia (Bencze *et al.*, 2006). More generally, understanding mitochondrial iron metabolism/homeostasis (or homeodynamics) remains a challenge for the near future. Over the last few years, it has become clear (mostly due to the studies with yeast) that the mitochondrion is not only the main compartment of iron use, but is also a centre of regulation of iron homeostasis for the cell as a whole (Napier *et al.*, 2005; Schueck *et al.*, 2001). Among the questions arising, and that remain to be answered, are the following: how and in which form is iron taken up by the mitochondria? What are the molecular bases of the relationship between mitochondrial iron accumulation and defective [Fe–S] cluster assembly? There is no doubt that work with yeast will help to solve these questions within the next few years.

Finally, studies on iron uptake/metabolism in yeast – especially in species that are not siderophore producers – have only produced a black box so far: nothing is known about iron from the moment it crosses the plasma membrane to the moment it enters storage or iron-using organelles. The question of the form and of the amount of the cytosolic iron pool in the eukaryotic cell has always been a matter of debate and speculation (Kakhlon and Cabantchik, 2002). The cytosolic pool of iron (or 'labile iron pool') is in practice difficult to study, because what we call the 'cytosol' after cell fractionation is a mix of buffer, cytosol, and of the soluble content of a proportion of organelles that are necessarily broken in the course of cell fractionation. The identification of the form of cytosolic iron (associated with either organic compounds or with specialized chaperones, and that maybe different according to the uptake mechanism used by the cell) may well be key to understanding the mechanisms of iron sensing and the mechanisms of iron uptake by intracellular organelles.

References

Agranoff, D., Collins, L., Kehres, D., Harrison, T., Maguire, M., and Krishna, S. (2005). The Nramp orthologue of *Cryptococcus neoformans* is a pH-dependent transporter of manganese, iron, cobalt and nickel. Biochem. J. *385*, 225–232.

Ardon, O., Bussey, H., Philpott, C., Ward, D.M., Davis-Kaplan, S., Verroneau, S., Jiang, B., and Kaplan, J. (2001). Identification of a *Candida albicans* ferrichrome transporter and its characterization by expression in *Saccharomyces cerevisiae*. J. Biol. Chem. *276*, 43049–43055.

Askwith, C., Eide, D., Van Ho, A., Bernard, P.S., Li, L., Davis-Kaplan, S., Sipe, D.M., and Kaplan, J. (1994). The FET3 gene of *S. cerevisiae* encodes a multicopper oxidase required for ferrous iron uptake. Cell. *76*, 403–410.

Atkin, C.L., Neilands, J.B., and Phaff, H.J. (1970). Rhodotorulic acid from species of *Leucosporidium, Rhodosporidium, Rhodotorula, Sporidiobolus,* and *Sporobolomyces*, and a new alanine-containing ferrichrome from *Cryptococcus melibiosum*. J. Bacteriol. *103*, 722–733.

Banuett, F. (1995). Genetics of *Ustilago maydis*, a fungal pathogen that induces tumors in maize. Annu. Rev. Genet. *29*, 179–208.

Benchimol, M., de Souza, W., and Travassos, L.R. (1979). Distribution of anionic groups at the cell surface of different *Sporothrix schenckii* cell types. Infect. Immun. *24*, 912–919.

Bencze, K.Z., Kondapalli, K.C., Cook, J.D., McMahon, S., Millan-Pacheco, C., Pastor, N., and Stemmler, T.L. (2006). The structure and function of frataxin. Crit. Rev. Biochem. Mol. Biol. *41*, 269–291.

Blaiseau, P.L., Lesuisse, E., and Camadro, J.M. (2001). Aft2p, a novel iron-regulated transcription activator that modulates, with Aft1p, intracellular iron use and resistance to oxidative stress in yeast. J. Biol. Chem. *276*, 34221–34226.

Chen, X.Z., Peng, J.B., Cohen, A., Nelson, H., Nelson, N., and Hediger, M.A. (1999). Yeast SMF1 mediates H(+)-coupled iron uptake with concomitant uncoupled cation currents. J. Biol. Chem. *274*, 35089–35094.

Courel, M., Lallet, S., Camadro, J.M., and Blaiseau, P.L. (2005). Direct activation of genes involved in intracellular iron use by the yeast iron-responsive transcription factor Aft2 without its paralog Aft1. Mol. Cell. Biol. *25*, 6760–6771.

Crane, F.L., Roberts, H., Linnane, A.W., and Low, H. (1982). Transmembrane ferricyanide reduction by cells of the yeast *Saccharomyces cerevisiae*. J. Bioenerg. Biomembr. *14*, 191–205.

Crisp, R.J., Pollington, A., Galea, C., Jaron, S., Yamaguchi-Iwai, Y., and Kaplan, J. (2003). Inhibition of heme biosynthesis prevents transcription of iron uptake genes in yeast. J. Biol. Chem. *278*, 45499–45506.

Dancis, A. (1998). Genetic analysis of iron uptake in the yeast *Saccharomyces cerevisiae*. J. Pediatr. *132*, S24–29.

Dancis, A., Klausner, R.D., Hinnebusch, A.G., and Barriocanal, J.G. (1990). Genetic evidence that ferric reductase is required for iron uptake in *Saccharomyces cerevisiae*. Mol. Cell. Biol. *10*, 2294–2301.

Dancis, A., Roman, D.G., Anderson, G.J., Hinnebusch, A.G., and Klausner, R.D. (1992). Ferric reductase of *Saccharomyces cerevisiae*: molecular characterization, role in iron uptake, and transcriptional control by iron. Proc. Natl. Acad. Sci. U.S.A. *89*, 3869–3873.

Dancis, A., Yuan, D.S., Haile, D., Askwith, C., Eide, D., Moehle, C., Kaplan, J., and Klausner, R.D. (1994). Molecular characterization of a copper transport protein in *S. cerevisiae*: an unexpected role for copper in iron transport. Cell. *76*, 393–402.

De Luca, N.G., and Wood, P.M. (2000). Iron uptake by fungi: contrasted mechanisms with internal or external reduction. Adv. Microb. Physiol. *43*, 39–74.

DeLeo, F.R., Olakanmi, O., Rasmussen, G.T., Lewis, T.S., McCormick, S.J., Nauseef, W.M., and Britigan, B.E. (1999). Despite structural similarities between gp91phox and FRE1, flavocytochrome b558 does not mediate iron uptake by myeloid cells. J. Lab. Clin. Med. *134*, 275–282.

Dix, D., Bridgham, J., Broderius, M., and Eide, D. (1997). Characterization of the FET4 protein of yeast. Evidence for a direct role in the transport of iron. J. Biol. Chem. 272, 11770–11777.

Dix, D.R., Bridgham, J.T., Broderius, M.A., Byersdorfer, C.A., and Eide, D.J. (1994). The *FET4* gene encodes the low affinity Fe(II) transport protein of *Saccharomyces cerevisiae*. J. Biol. Chem. *269*, 26092–26099.

Eide, D., Davis-Kaplan, S., Jordan, I., Sipe, D., and Kaplan, J. (1992). Regulation of iron uptake in *Saccharomyces cerevisiae*. The ferrireductase and Fe(II) transporter are regulated independently. J. Biol. Chem. *267*, 20774–20781.

Erpapazoglou, Z., Froissard, M., Nondier, I., Lesuisse, E., Haguenauer-Tsapis, R., and Belgareh-Touze, N. (2008). Substrate- and ubiquitin-dependent trafficking of the yeast siderophore transporter Sit1. Traffic. *9*, 1372–1391.

Felice, M.R., De Domenico, I., Li, L., Ward, D.M., Bartok, B., Musci, G., and Kaplan, J. (2005). Post-transcriptional regulation of the yeast high affinity iron transport system. J. Biol. Chem. *280*, 22181–22190.

Finney, L.A., and O'Halloran, T.V. (2003). Transition metal speciation in the cell: insights from the chemistry of metal ion receptors. Science. *300*, 931–936.

Foury, F., and Roganti, T. (2002). Deletion of the mitochondrial carrier genes *MRS3* and *MRS4* suppresses mitochondrial iron accumulation in a yeast frataxin-deficient strain. J. Biol. Chem. *277*, 24475–24483.

Foury, F., and Talibi, D. (2001). Mitochondrial control of iron homeostasis. A genome wide analysis of gene expression in a yeast frataxin-deficient strain. J. Biol. Chem. *276*, 7762–7768.

Froissard, M., Belgareh-Touze, N., Dias, M., Buisson, N., Camadro, J.M., Haguenauer-Tsapis, R., and Lesuisse, E. (2007). Trafficking of siderophore transporters in *Saccharomyces cerevisiae* and intracellular fate of ferrioxamine B conjugates. Traffic *8*, 1601–1616.

Froschauer, E.M., Schweyen, R.J., and Wiesenberger, G. (2009). The yeast mitochondrial carrier proteins Mrs3p/Mrs4p mediate iron transport across the inner mitochondrial membrane. Biochim. Biophys. Acta *1788*, 1044–1050.

Georgatsou, E., and Alexandraki, D. (1994). Two distinctly regulated genes are required for ferric reduction, the first step of iron uptake in Saccharomyces cerevisiae. Mol. Cell. Biol. *14*, 3065–3073.

Georgatsou, E., Mavrogiannis, L.A., Fragiadakis, G.S., and Alexandraki, D. (1997). The yeast Fre1p/Fre2p cupric reductases facilitate copper uptake and are regulated by the copper-modulated Mac1p activator. J. Biol. Chem. *272*, 13786–13792.

Haas, H. (2003). Molecular genetics of fungal siderophore biosynthesis and uptake: the role of siderophores in iron uptake and storage. Appl. Microbiol. Biotechnol. *62*, 316–330.

Haas, H., Eisendle, M., and Turgeon, B.G. (2008). Siderophores in fungal physiology and virulence. Annu. Rev. Phytopathol. *46*, 149–187.

Haas, H., Schoeser, M., Lesuisse, E., Ernst, J.F., Parson, W., Abt, B., Winkelmann, G., and Oberegger, H. (2003). Characterization of the *Aspergillus nidulans* transporters for the siderophores enterobactin and triacetylfusarinine C. Biochem. J. *371*, 505–513.

Hell, R., and Stephan, U.W. (2003). Iron uptake, trafficking and homeostasis in plants. Planta *216*, 541–551.

Heymann, P., Ernst, J.F., and Winkelmann, G. (1999). Identification of a fungal triacetylfusarinine C siderophore transport gene (*TAF1*) in *Saccharomyces cerevisiae* as a member of the major facilitator superfamily. Biometals *12*, 301–306.

Heymann, P., Ernst, J.F., and Winkelmann, G. (2000a). A gene of the major facilitator superfamily encodes a transporter for enterobactin (Enb1p) in *Saccharomyces cerevisiae.* Biometals *13*, 65–72.

Heymann, P., Ernst, J.F., and Winkelmann, G. (2000b). Identification and substrate specificity of a ferrichrome-type siderophore transporter (Arn1p) in *Saccharomyces cerevisiae.* FEMS Microbiol. Lett. *186*, 221–227.

Hortschansky, P., Eisendle, M., Al-Abdallah, Q., Schmidt, A.D., Bergmann, S., Thon, M., Kniemeyer, O., Abt, B., Seeber, B., Werner, E.R., *et al.* (2007). Interaction of HapX with the CCAAT-binding complex-a novel mechanism of gene regulation by iron. EMBO J. *26*, 3157–3168.

Howard, D.H. (1999). Acquisition, transport, and storage of iron by pathogenic fungi. Clin. Microbiol. Rev. *12*, 394–404.

Hu, C., Bai, C., Zheng, X., Wang, Y., and Wang, Y. (2002). Characterization and functional analysis of the siderophore-iron transporter CaArn1p in *Candida albicans.* J. Biol. Chem. *277*, 30598–30605.

Ismail, A., Bedell, G.W., and Lupan, D.M. (1985). Siderophore production by the pathogenic yeast, *Candida albicans.* Biochem. Biophys. Res. Commun. *130*, 885–891.

Jbel, M., Mercier, A., Pelletier, B., Beaudoin, J., and Labbe, S. (2009). Iron activates *in vivo* DNA binding of *Schizosaccharomyces pombe* transcription factor Fep1 through its amino-terminal region. Eukaryot. Cell. *8*, 649–664.

Johnson, L. (2008). Iron and siderophores in fungal–host interactions. Mycol. Res. *112*, 170–183.

Jungmann, J., Reins, H.A., Lee, J., Romeo, A., Hassett, R., Kosman, D., and Jentsch, S. (1993). MAC1, a nuclear regulatory protein related to Cu-dependent transcription factors is involved in Cu/Fe utilization and stress resistance in yeast. EMBO J. *12*, 5051–5056.

Kakhlon, O., and Cabantchik, Z.I. (2002). The labile iron pool: characterization, measurement, and participation in cellular processes(1). Free Radic. Biol. Med. 33, 1037–1046.

Kaplan, J., and O'Halloran, T.V. (1996). Iron metabolism in eukaryotes: Mars and Venus at it again. Science *271*, 1510–1512.

Kim, B.E., Nevitt, T., and Thiele, D.J. (2008). Mechanisms for copper acquisition, distribution and regulation. Nat. Chem. Biol. *4*, 176–185.

Kim, Y., Deng, Y., and Philpott, C.C. (2007). GGA2- and ubiquitin-dependent trafficking of Arn1, the ferrichrome transporter of *Saccharomyces cerevisiae.* Mol. Biol. Cell. *18*, 1790–1802.

Kim, Y., Lampert, S.M., and Philpott, C.C. (2005). A receptor domain controls the intracellular sorting of the ferrichrome transporter, ARN1. EMBO J. *24*, 952–962.

Kim, Y., Yun, C.W., and Philpott, C.C. (2002). Ferrichrome induces endosome to plasma membrane cycling of the ferrichrome transporter, Arn1p, in *Saccharomyces cerevisiae.* EMBO J. *21*, 3632–3642.

Knight, S.A., Vilaire, G., Lesuisse, E., and Dancis, A. (2005). Iron acquisition from transferrin by *Candida albicans* depends on the reductive pathway. Infect. Immun. 73, 5482–5492.

Kosman, D.J. (2003). Molecular mechanisms of iron uptake in fungi. Mol. Microbiol. *47*, 1185–1197.

Kumanovics, A., Chen, O.S., Li, L., Bagley, D., Adkins, E.M., Lin, H., Dingra, N.N., Outten, C.E., Keller, G., Winge, D., *et al.* (2008). Identification of *FRA1* and *FRA2* as genes involved in regulating the yeast iron regulon in response to decreased mitochondrial iron-sulfur cluster synthesis. J. Biol. Chem. *283*, 10276–10286.

Kwok, E.Y., Severance, S., and Kosman, D.J. (2006a). Evidence for iron channeling in the Fet3p-Ftr1p high-affinity iron uptake complex in the yeast plasma membrane. Biochemistry *45*, 6317–6327.

Kwok, E.Y., Stoj, C.S., Severance, S., and Kosman, D.J. (2006b). An engineered bifunctional high affinity iron uptake protein in the yeast plasma membrane. J. Inorg. Biochem. *100*, 1053–1060.

Labbe, S., Pelletier, B., and Mercier, A. (2007). Iron homeostasis in the fission yeast *Schizosaccharomyces pombe.* Biometals *20*, 523–537.

Lesuisse, E., Blaiseau, P.L., Dancis, A., and Camadro, J.M. (2001). Siderophore uptake and use by the yeast *Saccharomyces cerevisiae*. Microbiology *147*, 289–298.

Lesuisse, E., Casteras-Simon, M., and Labbe, P. (1996). Evidence for the *Saccharomyces cerevisiae* ferrireductase system being a multicomponent electron transport chain. J. Biol. Chem. *271*, 13578–13583.

Lesuisse, E., and Labbe, P. (1989). Reductive and non-reductive mechanisms of iron assimilation by the yeast *Saccharomyces cerevisiae*. J. Gen. Microbiol. *135*, 257–263.

Lesuisse, E., and Labbe, P. (1992). Iron reduction and trans plasma membrane electron transfer in the yeast *Saccharomyces cerevisiae*. Plant Physiol. *100*, 769–777.

Lesuisse, E., Raguzzi, F., and Crichton, R.R. (1987). Iron uptake by the yeast *Saccharomyces cerevisiae*: involvement of a reduction step. J. Gen. Microbiol. *133*, 3229–3236.

Lesuisse, E., Simon, M., Klein, R., and Labbe, P. (1992). Excretion of anthranilate and 3-hydroxyanthranilate by *Saccharomyces cerevisiae*: relationship to iron metabolism. J. Gen. Microbiol. *138*, 85–89.

Lesuisse, E., Simon-Casteras, M., and Labbe, P. (1998). Siderophore-mediated iron uptake in *Saccharomyces cerevisiae*: the *SIT1* gene encodes a ferrioxamine B permease that belongs to the major facilitator superfamily. Microbiology *144*, 3455–3462.

Li, L., Bagley, D., Ward, D.M., and Kaplan, J. (2008). Yap5 is an iron-responsive transcriptional activator that regulates vacuolar iron storage in yeast. Mol. Cell. Biol. *28*, 1326–1337.

Li, L., Chen, O.S., McVey Ward, D., and Kaplan, J. (2001). CCC1 is a transporter that mediates vacuolar iron storage in yeast. J. Biol. Chem. *276*, 29515–29519.

Li, L., and Kaplan, J. (1997). Characterization of two homologous yeast genes that encode mitochondrial iron transporters. J. Biol. Chem. *272*, 28485–28493.

Li, L., and Kaplan, J. (2004). A mitochondrial-vacuolar signaling pathway in yeast that affects iron and copper metabolism. J. Biol. Chem. *279*, 33653–33661.

Lill, R., and Muhlenhoff, U. (2006). Iron-sulfur protein biogenesis in eukaryotes: components and mechanisms. Annu. Rev. Cell Dev. Biol. *22*, 457–486.

Lill, R., and Muhlenhoff, U. (2008). Maturation of iron-sulfur proteins in eukaryotes: mechanisms, connected processes, and diseases. Annu. Rev. Biochem. *77*, 669–700.

Lin, S.J., Pufahl, R.A., Dancis, A., O'Halloran, T.V., and Culotta, V.C. (1997). A role for the *Saccharomyces cerevisiae* ATX1 gene in copper trafficking and iron transport. J. Biol. Chem. *272*, 9215–9220.

Liu, X.F., and Culotta, V.C. (1999). Post-translation control of Nramp metal transport in yeast. Role of metal ions and the BSD2 gene. J. Biol. Chem. *274*, 4863–4868.

Madhani, H.D., and Fink, G.R. (1998). The control of filamentous differentiation and virulence in fungi. Trends Cell Biol. *8*, 348–353.

Mercier, A., Watt, S., Bahler, J., and Labbe, S. (2008). Key function for the CCAAT-binding factor Php4 to regulate gene expression in response to iron deficiency in fission yeast. Eukaryot. Cell. *7*, 493–508.

Miele, R., Barra, D., and Bonaccorsi di Patti, M.C. (2007). A GATA-type transcription factor regulates expression of the high-affinity iron uptake system in the methylotrophic yeast *Pichia pastoris*. Arch. Biochem. Biophys. *465*, 172–179.

Miethke, M., and Marahiel, M.A. (2007). Siderophore-based iron acquisition and pathogen control. Microbiol. Mol. Biol. Rev. *71*, 413–451.

Moore, R.E., Kim, Y., and Philpott, C.C. (2003). The mechanism of ferrichrome transport through Arn1p and its metabolism in *Saccharomyces cerevisiae*. Proc. Natl. Acad. Sci. U.S.A. *100*, 5664–5669.

Muhlenhoff, U., Stadler, J.A., Richhardt, N., Seubert, A., Eickhorst, T., Schweyen, R.J., Lill, R., and Wiesenberger, G. (2003). A specific role of the yeast mitochondrial carriers MRS3/4p in mitochondrial iron acquisition under iron-limiting conditions. J. Biol. Chem. *278*, 40612–40620.

Napier, I., Ponka, P., and Richardson, D.R. (2005). Iron trafficking in the mitochondrion: novel pathways revealed by disease. Blood *105*, 1867–1874.

Oberegger, H., Schoeser, M., Zadra, I., Abt, B., and Haas, H. (2001). SREA is involved in regulation of siderophore biosynthesis, utilization and uptake in *Aspergillus nidulans*. Mol. Microbiol. *41*, 1077–1089.

Ojeda, L., Keller, G., Muhlenhoff, U., Rutherford, J.C., Lill, R., and Winge, D.R. (2006). Role of glutaredoxin-3 and glutaredoxin-4 in the iron regulation of the Aft1 transcriptional activator in *Saccharomyces cerevisiae*. J. Biol. Chem. *281*, 17661–17669.

Pelletier, B., Beaudoin, J., Philpott, C.C., and Labbe, S. (2003). Fep1 represses expression of the fission yeast *Schizosaccharomyces pombe* siderophore-iron transport system. Nucleic Acids Res. *31*, 4332–4344.

Pelletier, B., Mercier, A., Durand, M., Peter, C., Jbel, M., Beaudoin, J., and Labbe, S. (2007). Expression of *Candida albicans Sfu1* in fission yeast complements the loss of the iron-regulatory transcription factor Fep1 and requires Tup co-repressors. Yeast *24*, 883–900.

Perlin, D.S., Seto-Young, D., and Monk, B.C. (1997). The plasma membrane H(+)-ATPase of fungi. A candidate drug target? Ann. NY. Acad. Sci. *834*, 609–617.

Philpott, C.C. (2006). Iron uptake in fungi: a system for every source. Biochim. Biophys. Acta. *1763*, 636–645.

Philpott, C.C., and Protchenko, O. (2007). The response to iron deprivation in *Saccharomyces cerevisiae*. Eukaryot. Cell *7*, 20–27.

Philpott, C.C., and Protchenko, O. (2008). Response to iron deprivation in *Saccharomyces cerevisiae*. Eukaryot. Cell. *7*, 20–27.

Portnoy, M.E., Liu, X.F., and Culotta, V.C. (2000). *Saccharomyces cerevisiae* expresses three functionally distinct homologues of the Nramp family of metal transporters. Mol. Cell. Biol. *20*, 7893–7902.

Puig, S., Askeland, E., and Thiele, D.J. (2005). Coordinated remodeling of cellular metabolism during iron deficiency through targeted mRNA degradation. Cell *120*, 99–110.

Puig, S., Vergara, S.V., and Thiele, D.J. (2008). Cooperation of two mRNA-binding proteins drives metabolic adaptation to iron deficiency. Cell. Metab. *7*, 555–564.

Pujol-Carrion, N., Belli, G., Herrero, E., Nogues, A., and de la Torre-Ruiz, M.A. (2006). Glutaredoxins Grx3 and Grx4 regulate nuclear localisation of Aft1 and the oxidative stress response in *Saccharomyces cerevisiae*. J. Cell. Sci. *119*, 4554–4564.

Raguzzi, F., Lesuisse, E., and Crichton, R.R. (1988). Iron storage in *Saccharomyces cerevisiae*. FEBS Lett. *231*, 253–258.

Rouault, T.A., and Tong, W.H. (2005). Iron–sulphur cluster biogenesis and mitochondrial iron homeostasis. Nat. Rev. Mol. Cell. Biol. *6*, 345–351.

Rutherford, J.C., Jaron, S., Ray, E., Brown, P.O., and Winge, D.R. (2001). A second iron-regulatory system in yeast independent of Aft1p. Proc. Natl. Acad. Sci. U.S.A. *98*, 14322–14327.

Rutherford, J.C., Jaron, S., and Winge, D.R. (2003). Aft1p and Aft2p mediate iron-responsive gene expression in yeast through related promoter elements. J. Biol. Chem. *278*, 27636–27643.

Santos, R., Buisson, N., Knight, S., Dancis, A., Camadro, J.M., and Lesuisse, E. (2003). Haemin uptake and use as an iron source by *Candida albicans*: role of CaHMX1-encoded haem oxygenase. Microbiology *149*, 579–588.

Scazzocchio, C. (2000). The fungal GATA factors. Curr. Opin. Microbiol. *3*, 126–131.

Schueck, N.D., Woontner, M., and Koeller, D.M. (2001). The role of the mitochondrion in cellular iron homeostasis. Mitochondrion. *1*, 51–60.

Schuller, H.J. (2003). Transcriptional control of nonfermentative metabolism in the yeast *Saccharomyces cerevisiae*. Curr. Genet. *43*, 139–160.

Severance, S., Chakraborty, S., and Kosman, D.J. (2004). The Ftr1p iron permease in the yeast plasma membrane: orientation, topology and structure–function relationships. Biochem. J. *380*, 487–496.

Shatwell, K.P., Dancis, A., Cross, A.R., Klausner, R.D., and Segal, A.W. (1996). The FRE1 ferric reductase of *Saccharomyces cerevisiae* is a cytochrome *b* similar to that of NADPH oxidase. J. Biol. Chem. *271*, 14240–14244.

Singh, A., Kaur, N., and Kosman, D.J. (2007). The metalloreductase Fre6p in Fe-efflux from the yeast vacuole. J. Biol. Chem. *282*, 28619–28626.

Stearman, R., Yuan, D.S., Yamaguchi-Iwai, Y., Klausner, R.D., and Dancis, A. (1996). A permease–oxidase complex involved in high-affinity iron uptake in yeast. Science *271*, 1552–1557.

Stoj, C., and Kosman, D.J. (2003). Cuprous oxidase activity of yeast Fet3p and human ceruloplasmin: implication for function. FEBS Lett. *554*, 422–426.

Stoj, C.S., Augustine, A.J., Solomon, E.I., and Kosman, D.J. (2007). Structure-function analysis of the cuprous oxidase activity in Fet3p from *Saccharomyces cerevisiae*. J. Biol. Chem. *282*, 7862–7868.

Stoj, C.S., Augustine, A.J., Zeigler, L., Solomon, E.I., and Kosman, D.J. (2006). Structural basis of the ferrous iron specificity of the yeast ferroxidase, Fet3p. Biochemistry *45*, 12741–12749.

Sweet, S.P., and Douglas, L.J. (1991). Effect of iron concentration on siderophore synthesis and pigment production by *Candida albicans*. FEMS Microbiol. Lett. *64*, 87–91.

Ueta, R., Fujiwara, N., Iwai, K., and Yamaguchi-Iwai, Y. (2007). Mechanism underlying the iron-dependent nuclear export of the iron-responsive transcription factor Aft1p in *Saccharomyces cerevisiae*. Mol. Biol. Cell. *18*, 2980–2990.

Urbanowski, J.L., and Piper, R.C. (1999). The iron transporter Fth1p forms a complex with the Fet5 iron oxidase and resides on the vacuolar membrane. J. Biol. Chem. *274*, 38061–38070.

Vergara, S.V., and Thiele, D.J. (2008). Post-transcriptional regulation of gene expression in response to iron deficiency: co-ordinated metabolic reprogramming by yeast mRNA-binding proteins. Biochem. Soc. Trans. *36*, 1088–1090.

Waters, B.M., and Eide, D.J. (2002). Combinatorial control of yeast *FET4* gene expression by iron, zinc, and oxygen. J. Biol. Chem. *277*, 33749–33757.

Weissman, Z., and Kornitzer, D. (2004). A family of *Candida* cell surface haem-binding proteins involved in haemin and haemoglobin-iron utilization. Mol. Microbiol. *53*, 1209–1220.

Weissman, Z., Shemer, R., Conibear, E., and Kornitzer, D. (2008). An endocytic mechanism for haemoglobin-iron acquisition in *Candida albicans*. Mol. Microbiol. *69*, 201–217.

Weissman, Z., Shemer, R., and Kornitzer, D. (2002). Deletion of the copper transporter *CaCCC2* reveals two distinct pathways for iron acquisition in *Candida albicans*. Mol. Microbiol. *44*, 1551–1560.

Wessling-Resnick, M. (2006). Iron imports. III. Transfer of iron from the mucosa into circulation. Am. J. Physiol. Gastrointest. Liver Physiol. *290*, G1–6.

Winkelmann, G. (2007). Ecology of siderophores with special reference to the fungi. Biometals *20*, 379–392.

Yamaguchi-Iwai, Y., Dancis, A., and Klausner, R.D. (1995). AFT1: a mediator of iron regulated transcriptional control in *Saccharomyces cerevisiae*. EMBO J. *14*, 1231–1239.

Yamaguchi-Iwai, Y., Stearman, R., Dancis, A., and Klausner, R.D. (1996). Iron-regulated DNA binding by the AFT1 protein controls the iron regulon in yeast. EMBO J. *15*, 3377–3384.

Yamaguchi-Iwai, Y., Ueta, R., Fukunaka, A., and Sasaki, R. (2002). Subcellular localization of Aft1 transcription factor responds to iron status in *Saccharomyces cerevisiae*. J. Biol. Chem. *277*, 18914–18918.

Yun, C.W., Bauler, M., Moore, R.E., Klebba, P.E., and Philpott, C.C. (2001). The role of the FRE family of plasma membrane reductases in the uptake of siderophore-iron in *Saccharomyces cerevisiae*. J. Biol. Chem. *276*, 10218–10223.

Yun, C.W., Tiedeman, J.S., Moore, R.E., and Philpott, C.C. (2000). Siderophore-iron uptake in *Saccharomyces cerevisiae*. Identification of ferrichrome and fusarinine transporters. J. Biol. Chem. *275*, 16354–16359.

Zhang, Y., Lyver, E.R., Knight, S.A., Lesuisse, E., and Dancis, A. (2005). Frataxin and mitochondrial carrier proteins, Mrs3p and Mrs4p, cooperate in providing iron for heme synthesis. J. Biol. Chem. *280*, 19794–19807.

Zhang, Y., Lyver, E.R., Knight, S.A., Pain, D., Lesuisse, E., and Dancis, A. (2006). Mrs3p, Mrs4p, and frataxin provide iron for Fe–S cluster synthesis in mitochondria. J. Biol. Chem. *281*, 22493–22502.

Zhou, L.W., Haas, H., and Marzluf, G.A. (1998). Isolation and characterization of a new gene, *sre*, which encodes a GATA-type regulatory protein that controls iron transport in *Neurospora crassa*. Mol. Gen. Genet. *259*, 532–540.

Znaidi, S., Pelletier, B., Mukai, Y., and Labbe, S. (2004). The *Schizosaccharomyces pombe* corepressor Tup11 interacts with the iron-responsive transcription factor Fep1. J. Biol. Chem. *279*, 9462–9474.

Index

I

K

L

M